D1250987

Signals, Sound, and Sensation

Springer
New York
Berlin
Heidelberg
Barcelona
Budapest
Hong Kong
London
Milan
Paris
Santa Clara
Singapore
Tokyo

AIP Series in
Modern Acoustics and Signal Processing

ROBERT T. BEYER, Series Editor-in-Chief
Physics Department, Brown University

EDITORIAL BOARD

YOICHI ANDO, Faculty of Engineering, Kobe University, Kobe, Japan
FLOYD DUNN, Bioacoustics Research Lab, University of Illinois, Urbana, Illinois
JOHN ERDREICH, Ostergaard Associates, West Orange, New Jersey
CHRIS FULLER, Department of Mechanical Engineering, Virginia Polytechnic Institute, Blacksburg, Virginia
WILLIAM HARTMANN, Department of Physics, Michigan State University, East Lansing, Michigan
IRA HIRSCH, Central Institute for the Deaf and the Department of Psychology, Washington University, St. Louis, Missouri
HERMAN MEDWIN, Naval Postgraduate School, Monterey, California
JOANNE L. MILLER, Department of Psychology, Northeastern University, Boston, Massachusetts
LARRY ROYSTER, Department of Mechanical and Aerospace Engineering, North Carolina State University, Raleigh, North Carolina
JULIA DOSWELL ROYSTER, Environmental Noise Consultants, Raleigh, North Carolina
WILLIAM A. VON WINKLE, New London, Connecticut

BOOKS IN SERIES

Producing Speech: Contemporary Issues for Katherine Safford Harris, edited by Fredericka Bell-Berti and Lawrence J. Raphael
Signals, Sound, and Sensation, by William M. Hartmann
Computational Ocean Acoustics, by Finn B. Jensen, William A. Kuperman, Michael B. Porter, and Henrik Schmidt
Pattern Recognition and Prediction with Applications to Signal Characterization, by David H. Kil and Frances B. Shin
Oceanography and Acoustics: Prediction and Propagation Models, edited by Alan R. Robinson and Ding Lee
Handbook of Condenser Microphones, edited by George S. K. Wong and Tony F. W. Embleton
Seismic Wave Propagation and Scattering in the Heterogeneous Earth, by Haruo Sato and Michael C. Fehler

BF 251 .H35 1997
Hartmann, William M.
Signals, sound, and
sensation

Signals, Sound, and Sensation

William M. Hartmann
Department of Physics and Astronomy
Michigan State University
East Lansing, Michigan

RITTER LIBRARY
BALDWIN-WALLACE COLLEGE

Springer

Library of Congress Cataloging-in-Publication Data
Hartmann, William M.
Signals, sound, and sensation / William M. Hartmann
p. cm.
Includes bibliographical references.
ISBN 1-56396-283-7
1. Auditory perception. 2. Psychoacoustics. 3. Signal theory (Telecommunication) 4. Fourier analysis. 5. Sound. 6. Signal processing—Mathematics. I. Title.
BF251.H35 1997 96-43808
152.1'5—dc21 CIP

Printed on acid free paper.

© 1998 Springer-Verlag New York, Inc.
All rights reserved. This work may not be translated or copied in whole or in part without the written permission of the publisher (Springer-Verlag New York, Inc., Fifth Avenue, New York, NY 10010, USA), except for brief excerpts in connection with reviews or scholarly analysis. Use in connection with any form of information storage and retrieval, electronic adaptation, computer software, or by similar or dissimilar methodology now known or hereafter developed is forbidden.
The use of general descriptive names, trade names, trademarks, etc., in this publication, even if the former are not especially identified, is not to be taken as a sign that such names, as understood by the Trade Marks and Merchandise Marks Act, may accordingly be used freely by anyone.

Printed and bound by United Book Press, Inc., Baltimore, MD.
Printed in the United States of America.

9 8 7 6 5 4 3 2 (Corrected second printing, 1998)

ISBN 1-56396-283-7 Springer-Verlag New York Berlin Heidelberg SPIN 10676706

Series Preface

". . . Soun is noght but air y-broke"
—GEOFFREY CHAUCER, *end of the 14th century*

Traditionally, acoustics has formed one of the fundamental branches of physics. In the twentieth century, the field has broadened considerably and become increasingly interdisciplinary. At the present time, specialists in modern acoustics can be encountered not only in Physics Departments, but also in Electrical and Mechanical Engineering Departments, as well as in Departments of Mathematics, Oceanography, and even Psychology. They work in areas spanning from musical instruments to architecture to problems related to speech perception. Today, six hundred years after Chaucer made his brilliant remark, we recognize that sound and acoustics is a discipline extremely broad in scope, literally covering waves and vibrations in all media at all frequencies and at all intensities.

This series of scientific literature, entitled *Modern Acoustics and Signal Processing (MASP)*, covers all areas of today's acoustics as an interdisciplinary field. It offers scientific monographs, graduate level textbooks, and reference materials in such areas as: architectural acoustics; structural sound and vibration; musical acoustics; noise; bioacoustics; physiological and psychological acoustics; speech; ocean acoustics; underwater sound; and acoustical signal processing.

Acoustics is primarily a matter of communication. Whether it be speech or music, listening spaces or hearing, signaling in sonar or in ultrasonography, we seek to maximize our ability to convey information and, at the same time, to minimize the effects of noise. Signaling has itself given birth to the field of signal processing, the analysis of all received acoustic information or, indeed, all information in any electronic form. With the extreme importance of acoustics for both modern science and industry in mind, AIP Press, now an imprint of Springer-Verlag, initiated this series as a new and promising publishing venture. We hope that this venture will be beneficial to the entire international acoustical community, as represented by the Acoustical Society of America, a founding member of the American Institute of Physics, and other related societies and professional interest groups.

It is our hope that scientists and graduate students will find the books in this series useful in their research, teaching, and studies.

James Russell Lowell once wrote: "In creating, the only hard thing's to begin." This is such a beginning.

Robert T. Beyer
Series Editor-in-Chief

To Marguerite Hartmann
and Christine Hartmann

Contents

---

*Asterisks throughout Contents refer to biographies.

Appendices

Preface

It is the very essence of our striving for understanding that, on the one hand, it attempts to encompass the great and complex variety of man's experience, and that on the other, it looks for simplicity and economy in the basic assumptions. The belief that these two objectives can exist side by side is, in view of the primitive state of our scientific knowledge, a matter of faith.

A. EINSTEIN

Signals, Sound, and Sensation is a book for people who are actively interested in the perception of sound. The subject of the book is psychoacoustics, which can be defined as the science that studies the transformation of a physical sound, as described mathematically, into the perception of the sound, as registered by a listener. Psychoacoustics is one of two disciplines in the foundation of hearing science; the other is auditory physiology.

The process by which a physical sound becomes a perception in the mind, with its capacity to inform and inspire, is highly complex. It is acoustical, electromechanical, neural, and psychological. The challenge of understanding how it all works has always attracted the attention of individuals with widely diverse backgrounds. Inevitably there are problems of communication between workers from different disciplines. There are also serious questions about how anyone who wants to work in the field ought to be educated in preparation for what is essentially a multidisciplinary career.

Signals, Sounds, and Sensation (*SS&S*) adopts the premise that a fundamental element in the science of perceived sound is the signal, described in physical terms. Scientists who understand the mathematics of the signal are better prepared to invent creative experiments and to develop realistic theoretical models of physiological and psychoacoustical effects.

There are at this time a number of excellent first texts on psychoacoustics and hearing science, and *SS&S* does not replace them. *SS&S* is a *second* text, and its primary intent is to empower. It helps the reader develop the quantitative skills necessary to solve signal problems that arise in his or her work. Accordingly, *SS&S* presents the topics studied by electrical engineers when they learn about signals and communications systems. Here are concepts of periodic signals, aperiodic signals, and noise, with their linear and nonlinear transformations. The difference is that the signals are all acoustical or audio, and what is most important about them is how they sound to a listener.

In matters of method too *SS&S* resembles an engineering text, though it is friendlier than most of that genre. It is concerned with ideas that are practical and yet fundamental enough to serve as the foundations of auditory models of enormous complexity. These ideas are given a mathematical formulation so that rigorous results can be derived and then applied to concrete quantitative examples.

The material in *SS&S* was chosen to cover the mathematics of signals in a systematic way. As a result, its approach to the facts of psychoacoustics is not as systematic or as balanced as that of other texts on hearing science. For example, there is very little on the important topic of binaural hearing because the subject itself is vast; there are already several second-level texts in the area, and the mathematics involved in binaural hearing can be found elsewhere in the book.

Because the goal of *SS&S* was to apply mathematics to psychoacoustics, psychoacoustical topics are everywhere mixed in with the applicable mathematical derivations. This has had the predictable consequence that it is hard to find any particular topic in the book. Therefore, the index was constructed with particular care. The index is the recommended place to start most searches.

The quantitative emphasis in *SS&S* extends to exercises at the end of each chapter. The reader who is not accustomed to exercises as an element of learning may need to recall from his or her earlier study of mathematics that doing exercises is the only sure road to successful understanding. To mitigate the pain, *SS&S* employs two techniques: First, most of the exercises are of the form, ''Show that X is true,'' so that exercises present new information along with the usual challenge. Second, some of the most entertaining applications to psychoacoustics are covered in the exercises.

The level of mathematical treatment in *SS&S* assumes that the reader is comfortable with the concepts and the practice of differential and integral calculus. An early chapter reviews the basic mathematics of complex numbers, and complex variables appear without further apology throughout the rest of the book. *SS&S* does not assume prior knowledge of differential equations, Fourier transforms, or linear algebra, though the reader will learn something about all of those topics, and a great deal about some of them, in going through the book. Some of the mathematics of linear-response systems depend logically upon the calculus of residues. Results of these calculations are given in the body of the text, and the calculations themselves appear in appendices.

Although *SS&S* is mathematical, the emphasis is on practical calculations, and no attempt has been made to provide full mathematical rigor. At the same time, *SS&S* is unique among current psychoacoustics books because it does not hide essential mathematical difficulties under the rug. Therefore, a reader who understands the signal theory in *SS&S* should be well prepared in matters of the signal for innovative work on sound and the perception of sound.

There is no necessary order for reading the chapters of this book, but the following guide may be helpful: Chapter 1 on the pure tone primarily introduces standard terminology for waves, though, like most other chapters in the book, it cannot resist implications for hearing. Chapter 2 is entirely devoted to the mathematics of complex numbers and phasors. Chapter 3 on the physics of signal power and the decibel scale is essential background for the psychophysics of loudness in Chapter 4. Chapter 5 begins the study of periodic signals in earnest with the mathematics of the Fourier series, and Chapter 6 is devoted to the perception of such periodic signals. Chapter 7 is on the Dirac delta function. It is short and entirely mathematical, and it is the essential bridge between the Fourier series (Chapter 5) and the Fourier integral in Chapter 8. Chapter 8 is central; the reader must have a good grasp of this material on the Fourier integral to understand chapters that follow: 9, 10, 13, 14, 15, 17, 18, 19, and 21. Chapter 9 is devoted to concepts of filters, including many concepts that hearing scientists frequently encounter but seldom have a chance to master. Chapter 10 continues with *auditory* filters. Chapters 11 and 12 are an interlude on pitch. Each may be read independently. Chapters 13, 14, and 15 continue with psychoacoustical applications of the Fourier integral transform. The most important of these is easily Chapter 14, which relates the power spectrum to the autocorrelation function. Chapter 16 is a pedagogical chapter on the probability density function, which plays an important role in Chapters 23 and 24 on noise and signal detection, but also applies to the modulation discussed in the chapters that immediately follow. Chapters 17, 18, 19, and 20 are on modulation, including beats, amplitude modulation, frequency modulation, and on the perception of these. Chapter 18 includes the envelope rules, and Chapter 20 describes mixed modulation. Chapter 21 is a thorough introduction to digital signal processing within the context of the Fourier integral. Chapter 22 is devoted to nonlinearity and distortion. Chapter 23 is on the mathematics of noise, and Chapter 24 is a practical guide to signal detection theory as it is currently applied in psychoacoustical experiments.

Many people have contributed to this book, some without actually knowing it at the time. I would particularly like to acknowledge Jont Allen, Ed Burns, Laurel Carney, Laurent Demany, Larry Feth, David Green, Steven Handel, Michael Harrison, Daniel Hartmann, Mitra Hartmann, Rhona Hellman, William Hellman, Jack Hetherington, Adrian Houtsma, Tom Kaplan, Larry Key, Ninamarie Levinsky, Jian-Yu Lin, Terry Lloyd, Roy Patterson, George Perkins, Jon Pumplin, Brad Rakerd, Wayne Repko, Daniel Sartor, Arnold Tubis, Gabriel Weinreich, Jamie Wing, and William Yost. Daniel Weber read the entire book and improved the presentation in numerous spots.

The National Science Foundation and The National Institute on Deafness and Other Communication Disorders of the National Institutes of Health have supported my research in psychoacoustics and made it possible for me to write this book. I am especially grateful to my colleagues in the Department of Physics and Astronomy at Michigan State University for creating a congenial working environment of intellectual excitement and friendship.

William M. Hartmann
East Lansing, Michigan, May 1996

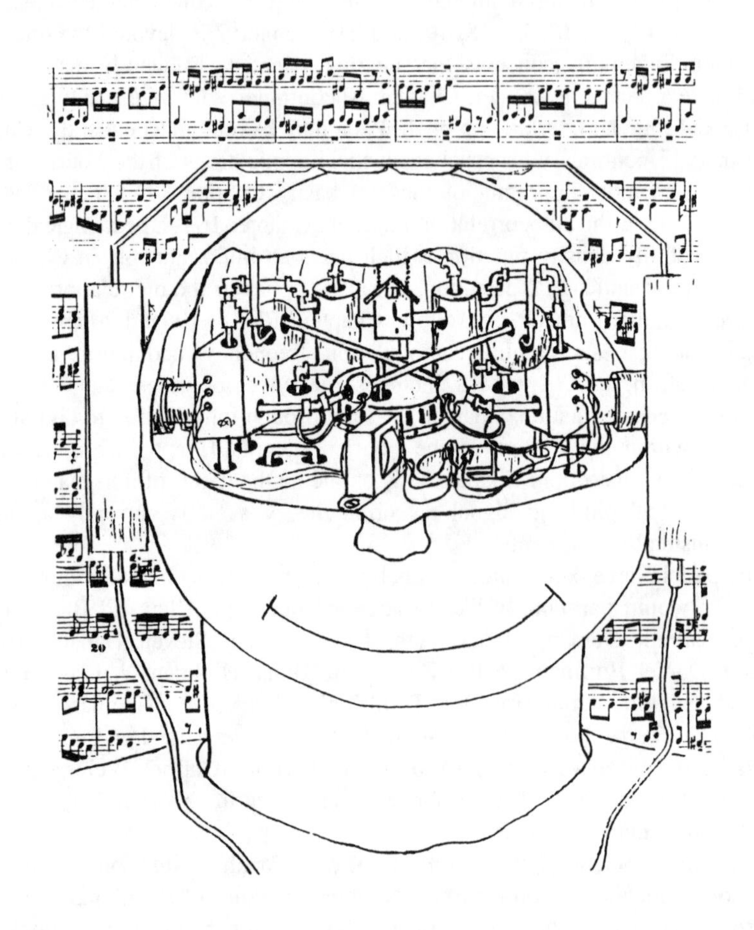

CHAPTER 1

Pure Tones

Take care of the Sense and the Sounds will take care of themselves.
LEWIS CARROLL

MATHEMATICS OF THE PURE TONE

The pure tone occupies a unique place in acoustics and other signal sciences. It is the most elementary of all signals. Mathematically the pure tone is known as a sine wave, which is a function of time t as shown in Fig. 1.1. It is described by the equation

$$x(t)=A\ \sin(2\pi t/T+\phi), \tag{1.1}$$

where A is the amplitude, T is the period in seconds, and ϕ is the phase in radians.

The units of x, whatever they might be (mechanical displacement, pressure, voltage), are the same as the units of the amplitude A. By convention, A is always a positive real number. Amplitude A can be zero, but in that case there is no wave at all, and there is nothing more to talk about. The amplitude multiplies the sine function—sin—which has a maximum value of 1 and a minimum value of -1. It follows that the sine wave $x(t)$ has a maximum value of A and a minimum value of $-A$.

The sine function—sin—is a trigonometric function of an angle, here given by $2\pi t/T+\phi$. This angle is the argument of the sine function; it is usually called the instantaneous phase. It is assumed that the angle is always expressed in units of radians. In a complete circle there are 2π radians, equivalent to 360 degrees.

The phase, ϕ, is the value of the instantaneous phase when $t=0$. In some cases we are interested in the function only for positive values of time t, it being presumed that the wave starts at time $t=0$. Then phase ϕ is often called the "starting phase."

The sine function is periodic; it repeats itself when the instantaneous phase of the function changes by 2π. Equation (1.1) shows that if the running time variable t starts at some value and then increases by T seconds the function comes back to its starting point. That is why we call T the period. We measure it in units of seconds per cycle.

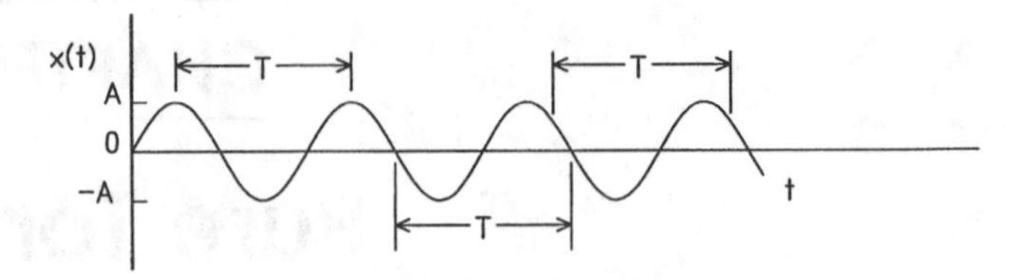

FIGURE 1.1. The sine wave has extreme values $\pm A$ and a characteristic waveform shape as shown here. The sine wave is periodic, so that any two equivalent points on the waveform are separated in time by the time interval T known as the "period." Theoretically, the sine wave extends indefinitely to infinite positive and infinite negative time, having no beginning or end.

The reciprocal of the period leads us to the definition of the frequency of the sine wave,

$$f = 1/T, \tag{1.2}$$

and its units are cycles per second, or Hertz (abbreviated Hz).

Because there are 2π radians in a cycle, the angular frequency, ω, measured in radians per second, is related to the frequency by

$$\omega = 2\pi f. \tag{1.3}$$

Therefore, one can write the sine wave in an alternative form

$$x(t) = A \sin(2\pi ft + \phi) \tag{1.4}$$

or

$$x(t) = A \sin(\omega t + \phi). \tag{1.5}$$

The latter form is simpler to write because it does not have the factor of 2π, and we shall use it often.

The units of cycles and the angular measures of degrees and radians are pure numbers. They do not have physical dimensions in the same sense that mass, length, and time have dimensions. It is often helpful to retain units of cycles or radians in defining period, frequency, or angular frequency, but one is under no obligation to do so. Hence, it is common to find period quoted in units of seconds and frequency in units of reciprocal seconds.

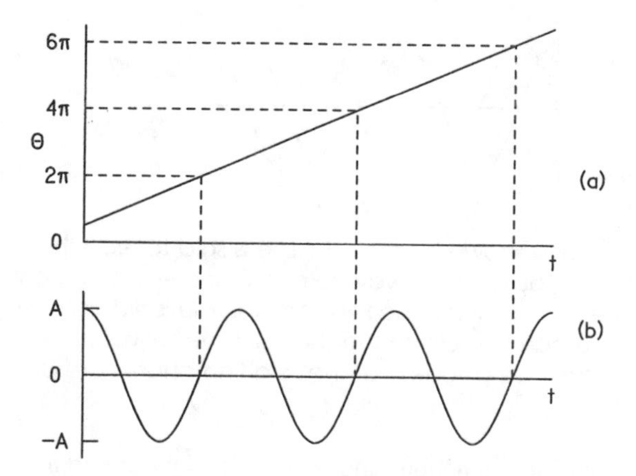

FIGURE 1.2. Panel (a) shows angle Θ, increasing from an initial value $\phi=\pi/2$ as time goes on. Over the duration shown, this angle advances through three multiples of the radian measure 2π. Panel (b) shows what happens when one takes the sine of angle Θ and multiplies by the amplitude A.

EXAMPLE: We consider the sine-wave vibration of an object whose position in millimeters is given by

$$x(t)=4\ \sin(2\pi 60t+\pi/2). \tag{1.6}$$

From Eq. (1.1), the amplitude of the vibration is 4 mm, which means that the maximum positive and negative excursions are ±4 mm. The frequency is 60 Hz so that the angular frequency is 120π or 377 radians/second. The phase is $\pi/2$ radians or 90 degrees.

The role of the sine function as a periodic function of a time-dependent instantaneous phase is emphasized by separating the aspects of periodicity and time dependence. One simply writes the sine wave in the form

$$x(t)=A\ \sin(\Theta), \tag{1.7}$$

where Θ is an angle, measured in radians. Angle Θ is a function of time

$$\Theta=\Theta(t)=\omega t+\phi. \tag{1.8}$$

As t increases, the angle Θ increases linearly, and the sine function goes through its periodic oscillations, as shown in Fig. 1.2. Figure 1.2 shows the special case where the phase is $\pi/2$ (90 degrees). As will be discussed in the exercises at the end of the chapter, this figure actually shows a cosine function, $x(t)=A\ \cos(\omega t)$. Confusing

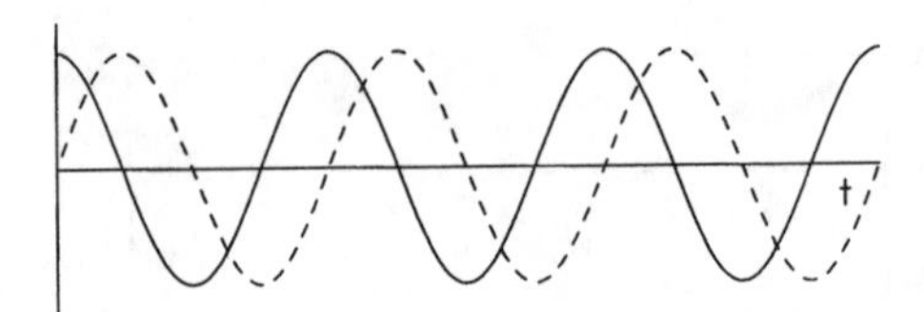

FIGURE 1.3. The wave shown by the solid line is said to lead the wave shown by the dashed line because every waveform feature—peak, positive-going zero crossing etc.—occurs at an earlier time for the solid line. Alternatively the dashed-line wave can be said to lag the solid-line wave. Both waves have the same frequency and amplitude, but their starting phases are different.

though it may seem, sine functions and cosine functions are both instances of what is often generically called a "sine wave."

Phase Lead—Phase Lag

When two sine waves have the same frequency but different starting phases one of them is said to lead or to lag the other. Figure 1.3 is an example. The lagging wave is given by $x(t)=A\ \sin(\omega t)$ and the leading wave is given by $x(t)=A\ \sin(\omega t + \pi/2)$. According to Eqs. (1.7) and (1.8) the starting phase angle for the leading wave is $\phi=\pi/2$. This angle is positive and less than π (180 degrees), which corresponds to a condition for leading. It should be evident that if ϕ had been equal to 180 degrees then neither wave would lead or lag the other.

H. R. HERTZ (1857–1894)

Heinrich Rudolf Hertz was born in Hamburg, Germany on February 22, 1857. He studied physics with von Helmholtz in Berlin. Between 1885 and 1889, while living in Karlsruhe, he found experimental proof that oscillating electrical charges produced propagating electromagnetic waves, as had been suggested by Maxwell's field equations. He had actually been alerted to this possibility by his former mentor, von Helmholtz. He showed that his electromagnetic waves were subject to the same laws of propagation, reflection, refraction, and polarization as light waves, thereby establishing their equivalence. It was only a short step from Hertz's experiments to the invention of radio communication, but Hertz did not live to see that. He died in Bonn on January 1, 1894.

Unique Nature of a Pure Tone

The pure tone (sine or cosine) waveform is special because it consists of only a single frequency. The significance of this idea becomes clear when one considers Fourier analysis, to be covered in later chapters, by which a waveform can be

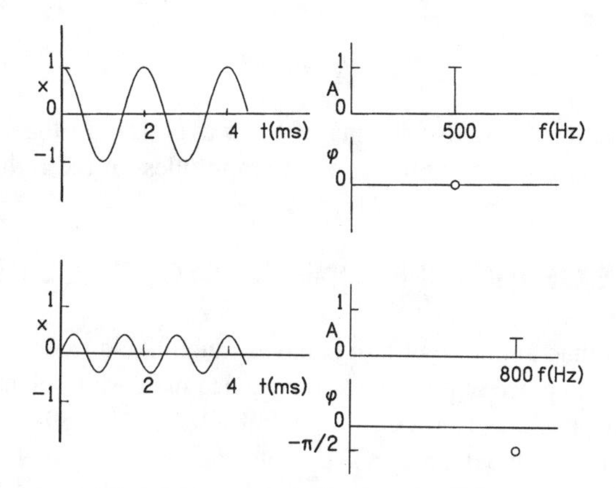

FIGURE 1.4. Pure tone waveforms, with frequencies of 500 Hz and 800 Hz, are shown on the left as functions of time. The corresponding amplitude and phase spectra are shown on the right. The phase reference is chosen so that a cosine function has zero starting phase.

analyzed into its constituent frequencies. All waveforms have at least two constituent frequencies (or components), except for the pure tone which has only one. That is what makes it pure.

Because the pure tone waveform has only a single frequency, any linear operation (i.e. any filtering) that is performed on this waveform leaves the wave shape unchanged. The amplitude may be increased or reduced; the phase may be shifted, but the wave retains its sine-wave shape. The pure tone is the only waveform for which this statement can be made. To put this statement into perspective, we might imagine that we have an electronic filter and an oscilloscope that can display both the input and the output of the filter. If the input is a sine wave then we cannot tell which trace on the oscilloscope is showing the input and which is showing the output. They both have the same frequency and shape.

THE SPECTRUM

A pure tone has an infinite duration. In fact, as will be shown in the chapters on Fourier analysis, a tone with a beginning or an end is not, strictly speaking, a pure tone. As a consequence, a figure showing a sine or cosine function cannot represent a pure tone unless the figure is infinitely long. An alternative representation makes it possible to describe a pure tone exactly. This is the spectral representation. The amplitude spectrum shows the amplitude of the wave plotted against frequency. The phase spectrum shows the starting phase, also plotted against frequency. Several examples are given in Fig. 1.4, where the phase reference has been chosen to be zero for a cosine wave, that is, the wave is given by the function

$$x(t)=A\cos(\omega t+\phi). \tag{1.9}$$

It should be noticed how the heights of the waveforms in the left-hand figures directly translate into the representations of amplitudes on the right.

IMPORTANCE OF THE PURE TONE IN PSYCHOACOUSTICS

Because the human auditory system behaves much like a frequency analyzer, the pure tone occupies a place of enormous importance in hearing science. The German physiologist and physicist Herman von Helmholtz (1821–1894) conjectured that the inner ear (the cochlea) analyzes a tone according to its frequencies, with different neurons excited by different frequencies. The experimental work of Georg von Békésy (1899–1972) confirmed this conjecture and showed how the mechanism works. The basilar membrane vibrates as it is forced by the fluids within the canals of the cochlea. The vibration pattern appears to be a traveling wave, moving from the region of the oval window to the apex of the cochlea (helicotrema). For a high-frequency pure tone the maximum vibration occurs near the oval window and there is little vibration elsewhere on the basilar membrane. For a low-frequency pure tone the entire basilar membrane vibrates to some extent, with a maximum near the apex.

Along the length of the basilar membrane are hair cells which convert mechanical vibrations to neural impulses that can be used by higher centers of the nervous system. Those hair cells near the oval window are excited by high-frequency pure tones, those near the apex by low-frequency pure tones. The neurons that contact the hair cells and transmit the excitation to higher auditory centers maintain the frequency specificity of the system. There are about 3,500 hair cells in the inner row, and about ten neurons that enervate each of them. This one-to-many mapping of hair cells to the fibers of the auditory nerve suggests that there is something very important about preserving the frequency analysis originally established by the electromechanical properties of the basilar membrane.

The idea that place of excitation serves to encode frequency in a way that is preserved is known as the "place theory of hearing." The organization of the auditory system whereby different frequencies preferentially excite different neurons is known as "tonotopic organization." It appears that the entire auditory system, from the auditory nerve up to the auditory cortex of the brain is tonotopically organized (Romani, Williamson, Kaufman, 1982).

The point of particular interest here is that the pure tone is the signal that leads to the simplest excitation of the auditory nervous system because the pure tone is "tonotopically local." At any stage of the auditory system it plays the role of the elemental excitation. The appeal of the pure tone for experimental work is obvious. The psychoacoustician or physiologist uses pure tones in order to retain control, in other words, for the same reasons that the chemist uses pure chemicals.

SOUND OF A PURE TONE

Like any other periodic tone, the pure tone can be characterized by the psychological dimensions of pitch, loudness, and tone color. The pure tone actually serves as the reference for pitch. The reason is that pitch is closely related to (but not identical to!) the physical quantity of frequency. Because the pure tone is "pure frequency" its role as the standard reference for pitch was a sensible choice.

The standard textbook range of audible frequencies is from 20 Hz to 20,000 Hz. The upper limit depends greatly on age and otological history. For the majority of young adults the upper limit is likely to be closer to 17,000 Hz. The lower limit of 20 Hz is also problematical. In order to make a 20-Hz tone audible one must use a strong signal. But a strong 20-Hz signal runs the risk of creating harmonic distortion, particularly the third harmonic distortion component at 60 Hz. Although this distortion product may be small, the hearing organ is so much more sensitive at 60 Hz than at 20 Hz that one must worry about the third harmonic when trying to establish a threshold for 20 Hz.

The term "tone color" refers to that part of timbre that is attributable to the steady state part of a tone, i.e., the tone without transients associated with onset or offset or ongoing aperiodic fluctuations. The tone color of a pure tone depends on frequency. For low frequencies the pure tone is described as "dull." For high frequencies the pure tone is said to be "bright" or "piercing." What is meant by low and high in this connection appears to be established by the frequency ranges of the human voice and of melody-carrying musical instruments. A pure tone with a frequency below 200 Hz (the fundamental frequency of the adult female human spoken voice) is dull, a judgement that likely results from the fact that we almost never hear tones with this fundamental frequency in the absence of significant harmonic content that adds brightness to the tone.

A pure tone with a frequency greater than or equal to 2000 Hz is bright. If it is intense it may be called "shrill." There are several facts that bear on this judgement: First, 2000 Hz approaches the frequency range where the hearing organ is most sensitive, namely 3000 to 5000 Hz. The greater sensitivity makes a tone in this frequency range sound louder. Second, sounds for which the strongest components have frequencies near or greater than 2000 Hz are encountered comparatively infrequently in the everyday environment. Most of the energy in speech signals is below 2000 Hz, and although numerous medieval woodwind instruments had spectra that extended to high frequencies, most modern musical instruments act as low-pass filters that attenuate harmonics with frequencies greater than 1000 Hz. This is true of strings, brass, and woodwind instruments. There are exceptions, the piccolo for example, and there are organ pipes with fundamental frequencies as high as 8372 Hz ($=C_9$). The fact that the tone color of the pure tone (one particular waveform) runs the gamut from dull to piercing shows that it is the frequency content and not the shape of the waveform that determines tone color.

EXERCISES

Exercise 1: 90-degree phase shifts

Phase shifts can best be handled mathematically with trigonometric identities.

(a) Use the trigonometric identity

$$\sin(C+B)=\sin(C)\cos(B)+\sin(B)\cos(C)$$

to show that

$$A\ \sin(\omega t+\pi/2)=A\ \cos(\omega t),$$

and that

$$A\ \sin(\omega t-\pi/2)=-A\ \cos(\omega t).$$

This explains why Fig. 1.2 might be called a "cosine wave."
(b) Use the trigonometric identity

$$\cos(C+B)=\cos(C)\cos(B)-\sin(B)\sin(C)$$

to show that

$$A\ \cos(\omega t+\pi/2)=-A\ \sin(\omega t)=A\ \sin(\omega t+\pi)$$

and that

$$A\ \cos(\omega t-\pi/2)=A\ \sin(\omega t).$$

This equation is what led to the starting phase of $-\pi/2$ in the phase spectrum of the 800-Hz tone in Fig. 1.4. Many more trigonometric identities can be found in Appendix B at the back of this book.

Exercise 2: Choice of reference phase

What are the amplitudes and phases of (a) $x=\sin(\omega t)$, (b) $x=\cos(\omega t)$, (c) $x=-\cos(\omega t)$, if you use the sine wave as the reference for zero phase? What if you use the cosine wave as the reference?

Exercise 3: What goes around comes back

Use a calculator to show that $\sin(\pi/6)=0.5$. Use this result, and the fact that there are 2π radians in a complete cycle, to find $\sin(37\pi/6)$.

Exercise 4: Amplitudes are always positive numbers

Careful mathematical treatments of signals are relentless in insisting that amplitudes are always positive numbers. This leads to useful results when the discussion turns to modulation. For the present: (a) show that any wave that is claimed to have a negative amplitude can be represented as a wave with a positive amplitude and a phase shift of 180 degrees (π radians). (b) Explain why a representation scheme wherein amplitudes can be both positive and negative and phases can range from zero to 360 degrees is redundant.

Exercise 5: Experiment

In June 1993, a French radio station announced plans to make summer radio listening more pleasant by broadcasting an ''ultrasonic'' tone to chase away female mosquitoes, the only kind that bite humans. Radio Fugue FM, based in Compiegne, decided to broadcast a 16-kHz tone, said to be inaudible to humans, but the same frequency as emitted by male mosquitoes, in order to keep the bloodsucking females away.

Use a function generator and headphones to discover whether you could hear the 16-kHz tone. Is it possible that Radio Fugue would drive away as many listeners as mosquitoes?

Exercise 6: More of the same

Consider the sum of two tones having the same frequency, but different amplitudes (positive, of course) and phases,

$$x(t) = A_1 \cos(\omega t) + A_2 \cos(\omega t + \phi_2).$$

The angular frequency of the sum is again ω.

(a) Use the trigonometric identity from Eq. (B7) in Appendix B to find amplitude A and phase ϕ so that $x(t)$ can be written as a simple pure tone,

$$x(t) = A \cos(\omega t + \phi).$$

(b) Show that $0 \le A \le A_1 + A_2$

(c) Show that $\phi \le \phi_2$.

CHAPTER 2

Complex Representation

Don't let us make imaginary evils, when you know we have so many real ones to encounter.

OLIVER GOLDSMITH

This chapter is an introduction to complex numbers, namely to numbers that have both an imaginary part and a real part. This complex representation is used extensively in the treatment of signals in chapters which follow. In itself, this chapter is purely mathematical, without immediate application. It introduces rectangular and polar forms of complex numbers, operations of addition and multiplication, and the concepts of complex conjugate, magnitude (absolute value), and reciprocal. These concepts are frequently needed in signal processing, particularly in the study of the Fourier transform and of filters.

COMPLEX NUMBERS—RECTANGULAR FORM

A complex number z has a real part x and an imaginary part y:

$$z = x + iy, \tag{2.1}$$

where $i = \sqrt{-1}$, and both x and y are real numbers, possibly negative.

The engineering literature frequently uses the symbol j to indicate $\sqrt{-1}$; mathematicians and physicists tend to use i. The symbol i is used in this book to be consistent with the *Journal of the Acoustical Society of America*, historically the most important journal for psychoacoustics and physiological acoustics.

The (+) sign in Eq. (2.1) has an unusual interpretation, but it is universally understood, and it works out mathematically in a way that never causes problems. The (+) symbol cannot indicate an operation like the addition of real numbers because the real and imaginary parts of z are along different dimensions. Rather, the (+) symbol means an addition similar to the addition of vectors (arrows). Therefore, the number z can be represented on a cartesian coordinate system, as in Fig. 2.1(a). We understand $x + iy$ as instructions to go x units along the horizontal axis, to turn 90 degrees upward (symbol i), and then to go y units along the vertical axis.

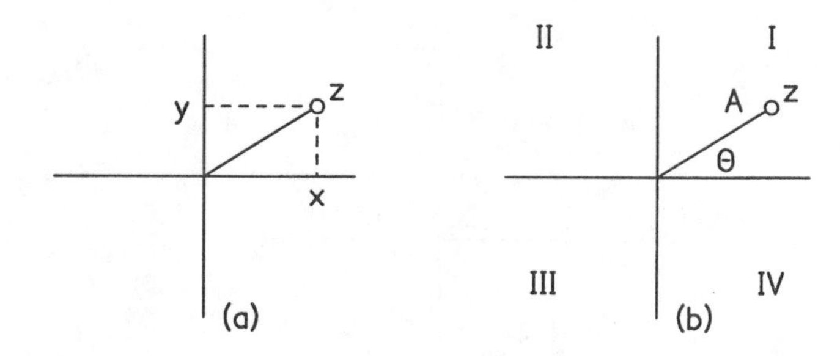

FIGURE 2.1. A complex number z is represented in the complex plane where there is a real axis (horizontal) and an imaginary axis (vertical). (a) The cartesian representation of z is in terms of real part x and imaginary part y. (b) The polar representation of z consists of amplitude A and phase θ. The two representations are related by some fundamental trigonometry to be described below. The quadrants of the complex plane are labelled with roman numerals.

Complex Numbers—the Meaning of Equality

The vector representation for complex numbers emphasizes that two complex numbers are equal if their real parts are equal and their imaginary parts are equal. In mathematical terms, if z_1 and z_2 are two complex numbers, where $z_1=x_1+iy_1$ and $z_2=x_2+iy_2$, then $z_1=z_2$ if and only if $x_1=x_2$ and $y_1=y_2$.

COMPLEX NUMBERS—POLAR FORM

A complex number is uniquely specified by two real numbers, namely x and y rectangular components. It can also be specified by two different real numbers, namely by the length of the line in Fig. 2.1(a) and the angle that the line makes with the x-axis, as shown in Fig. 2.1(b).

The length of the line is called the "absolute value" of z, and is given the symbol $|z|$. The length is also called the "magnitude" of z because it is only in terms of such a real number as $|z|$ that one complex number can be said to be greater than or less than another. By convention, $|z|$ is always positive (or zero). Frequently $|z|$ is given the symbol r, to signify "radius," the point being that all numbers with the same magnitude lie on a circle with a radius of $|z|$, as shown in Fig. 2.2. We are interested primarily in signals and therefore we will refer to the magnitude as the "amplitude" and give it the symbol A.

The angle θ is called the "phase" or the "argument" of the complex number. By convention, θ takes on values between 0 and 2π radians (0 and 360 degrees). It is just because θ can take on all possible angle values that we never have to consider the possibility of a negative magnitude A. The vector direction can be completely specified by θ. In practice, we have to be prepared to see any real number for θ.

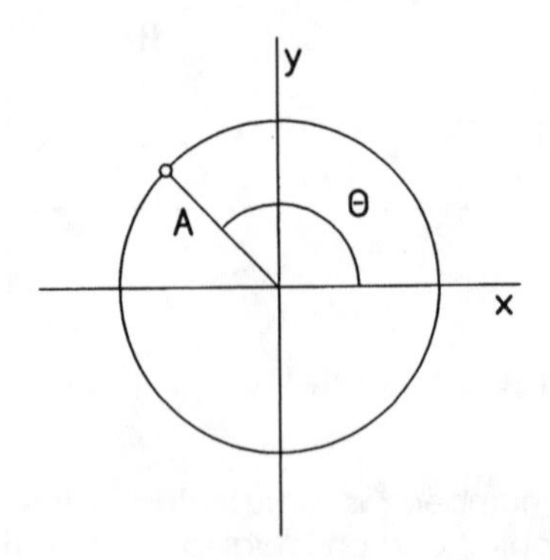

FIGURE 2.2. All complex numbers with magnitude A lie on the circle with radius A. The vector (with a tiny circle at its end) shows a particular complex number with magnitude A and phase $3\pi/4$=135 degrees.

Numbers outside the conventional range ($0-2\pi$ radians) can be brought into the conventional range by subtraction or addition. For example, a complex number with an argument of 450 degrees is the same as a complex number with an argument of 90 degrees.

Conversion

By the simple trigonometry of Fig. 2.1 one can find the rectangular coordinates $[x,y]$ of complex number z if one knows the polar coordinates $[A,\theta]$. The conversion is:

$$x=A\cos\theta, \tag{2.2}$$

and

$$y=A\sin\theta.$$

Therefore, the complex number z can be written as

$$z=A[\cos\theta+i\sin\theta]. \tag{2.3}$$

The reverse conversion, finding A and θ given the rectangular coordinates x and y, is slightly more complicated. There is no problem in finding the magnitude, A. By the Pythagorean theorem we have

$$A=\sqrt{x^2+y^2}. \tag{2.4}$$

To find the phase, or argument, one can write

$$y/x = \sin\theta/\cos\theta = \tan\theta. \tag{2.5}$$

Equation (2.5) is unambiguous, but the next step, actually finding angle θ poses a problem. One can take the inverse tangent (arctan or $\tan^{-1}$) of y/x to find θ, but this calculation does not necessarily put θ into the correct quadrant. The arctangent function has a principal value range from $-\pi/2$ to $+\pi/2$. Therefore, it will return values of θ in quadrants I and IV, but never in II or III (see Fig. 2.1b). Therefore, the arctangent operation gives the right answer only if x is positive. If x is negative one needs to add π to the answer given by the inverse tangent.

In general then, we can write

$$\theta = \mathrm{Arg}(y,x), \tag{2.6}$$

where the "Arg" function is defined by

$$\mathrm{Arg}(y,x) = \arctan(y/x) \quad \text{for } x>0 \tag{2.7}$$

and

$$\mathrm{Arg}(y,x) = \arctan(y/x) + \pi \quad \text{for } x<0.$$

EXAMPLE: The polar form of complex numbers can be illustrated by finding the polar form of the number $-3+i5$. The magnitude is given by $\sqrt{9+25} \approx 5.83$, and the phase is $\tan^{-1}(-5/3)+\pi \approx -1.03+3.14=2.11$ radians. An equivalent description of the phase in degrees is: $\tan^{-1}(-5/3)+180 \approx -59+180=121$ degrees.

The polar form of complex numbers is made more powerful by an important identity, known as Euler's (pronounced "Oilers") formula:

$$e^{i\theta} = \cos\theta + i\sin\theta. \tag{2.8}$$

The way to persuade oneself that Eq. (2.8) is true is to look at the series expansion (C2) in Appendix C for $\exp(i\theta)$ and compare with the series expansions (C4) and (C5) for sine and cosine.

From Eq. (2.3) it follows that any complex number can be written as

$$z = Ae^{i\theta}. \tag{2.9}$$

Equation (2.9) is a wonderfully compact way to represent a complex number with amplitude A and phase θ. We shall see it often in the chapters that follow.

ADDITION OF COMPLEX NUMBERS

Two complex numbers add like vectors. The horizontal components add algebraically to give the horizontal component of the sum, and the vertical components add algebraically to give the vertical component of the sum. This is easily expressed in rectangular coordinates. If $z_3 = z_1 + z_2$, then

$$x_3 = x_1 + x_2, \tag{2.10}$$

and

$$y_3 = y_1 + y_2. \tag{2.11}$$

Addition is expressed only awkwardly if the complex numbers are in polar coordinates, as in Eq. (2.9).

EXAMPLE: If $z_1 = 4 + i3$ and $z_2 = -1 + i2$, then, applying the above rules, we find that the sum z_3 is given by $z_3 = 3 + i5$. The addition of these two numbers is represented graphically in Fig. 2.3.

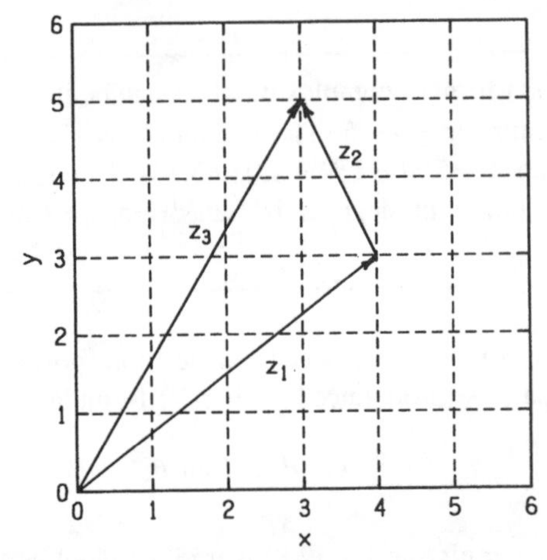

FIGURE 2.3. This figure shows how two vectors, representing complex numbers z_1 and z_2, are added tail to tip to create the sum z_3.

MULTIPLICATION OF COMPLEX NUMBERS

Two complex numbers are most conveniently multiplied if they are in polar form. If complex number z_1 equals $A_1 e^{i\theta_1}$ and complex number z_2 equals $A_2 e^{i\theta_2}$, then their product, $z_4 = z_1 z_2$, is given by

$$z_4 = A_4 e^{i\theta_4}, \tag{2.12}$$

where

$$A_4 = A_1 A_2, \tag{2.13}$$

and

$$\theta_4 = \theta_1 + \theta_2. \tag{2.14}$$

The product, therefore, has a simple form. The magnitudes of the two numbers are multiplied together, Eq. (2.13). The phases of the two are added together, Eq. (2.14), just as one would expect for the multiplication of two exponentials having a common base.

The product z_4 can also be expressed in rectangular coordinates,

$$z_4 = x_4 + iy_4 = x_1x_2 - y_1y_2 + i(x_1y_2 + x_2y_1). \tag{2.15}$$

The product in this form does not have a simple interpretation like the product in polar form. From the rules for equality of complex numbers we interpret Eq. (2.15) as saying that the real part of z_4 is $x_4 = x_1x_2 - y_1y_2$ and the imaginary part of z_4 is $y_4 = x_1y_2 + x_2y_1$. [Note: It should be noted that although two complex numbers add like vectors, they do not multiply like vectors. There are two kinds of vector products, dot products and cross products. Neither of these corresponds to the multiplication operation for complex numbers.]

EXAMPLE: To multiply the complex numbers $z_1 = 4 + i3$ and $z_2 = -1 + i2$, we first put the numbers in polar form, $z_1 = 5e^{i0.644}$ and $z_2 = \sqrt{5}\, e^{i2.034}$, and then apply the above rules, to find that the product z_4 is given by $z_4 = 5\sqrt{5}\, e^{i2.678}$.

COMPLEX CONJUGATE AND ABSOLUTE VALUE

The complex conjugate of a complex number z is given the symbol z^*. It is the same as z itself except that the sign of the imaginary part is reversed. The rule for finding the complex conjugate of any complex number or expression is straightforward: One just reverses the sign of the symbol i each time that it occurs. Therefore,

$$\text{If } z = x + iy \text{ then } z^* = x - iy. \tag{2.16}$$

$$\text{If } z = A(\cos\theta + i\sin\theta) \text{ then } z^* = A(\cos\theta - i\sin\theta). \tag{2.17}$$

$$\text{If } z = Ae^{i\theta} \text{ then } z^* = Ae^{-i\theta}. \tag{2.18}$$

The last equation shows that if z is represented by a vector above the horizontal axis by θ degrees, then z^* is a vector of the same length below the horizontal axis by an angle of the same size. The reverse is also true, of course.

The complex conjugate provides a way to find the absolute value of a complex number. The product of a complex number and its complex conjugate is equal to the square of the magnitude of the complex number:

$$zz^* = |z|^2. \tag{2.19}$$

This fact is most easily proved from Eqs. (2.12–2.14) where

$$zz^* = Ae^{i\theta}Ae^{-i\theta} = A^2, \tag{2.20}$$

and, by definition, $A^2 = |z|^2$. The proof has made use of several rules about exponents. The first is that to multiply two numbers with the same exponent base (here a base of e) one adds the exponents. The second is that $e^0 = 1$.

The fact that $zz^* = |z|^2$ can also be proved from Eq. (2.15) where

$$zz^* = (x+iy)(x-iy) = x^2 - i^2y^2 + ixy - ixy = x^2 + y^2, \tag{2.21}$$

and, by the Pythagorean theorem, $x^2 + y^2$ is $|z|^2$.

It should be noted that when one multiplies a complex number by its complex conjugate, the result is always a positive real number. This is true no matter how complicated the original complex number might be.

RECIPROCAL OF A COMPLEX NUMBER

One frequently encounters complex numbers in the form

$$\frac{1}{x+iy}. \tag{2.22}$$

This number is not in the standard form for complex numbers given in Eq. (2.1), and this prevents us from representing the number in the complex plane. The problem is that the denominator is complex.

We can put this number into the standard form by making the denominator real by multiplying by its complex conjugate. We must similarly multiply the numerator, of course. Therefore,

$$\frac{1}{x+iy} = \frac{1}{x+iy}\frac{x-iy}{x-iy}, \tag{2.23}$$

or

$$\frac{1}{x+iy} = \frac{x-iy}{x^2+y^2}. \tag{2.24}$$

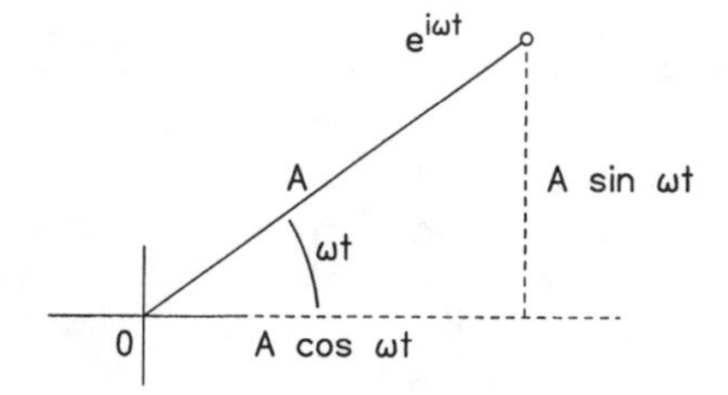

FIGURE 2.4. Real and imaginary parts of $\exp(i\omega t)$, plotted for a particular value of time t.

The expression on the right is now in the form of Eq. (2.1) and we can deal with it.

PHASORS

A phasor is a vector in the complex plane that rotates in time. It is described by the complex number

$$z = Ae^{i\omega t}, \tag{2.25}$$

and a snapshot of it is given in Fig. 2.4. Complex number z has a magnitude A and a phase angle that increases linearly with time, just like the instantaneous phase in Fig. 1.2(a). The real part and the imaginary part change together to keep the length constant and equal to A.

The phasor representation is useful because phasors can be added like vectors. As an example, we consider the sum

$$z = 6e^{i\omega_1 t} + 3e^{i\omega_2 t},$$

where ω_2 is somewhat greater than ω_1. The result for z is shown in Fig. 2.5 for several values of time.

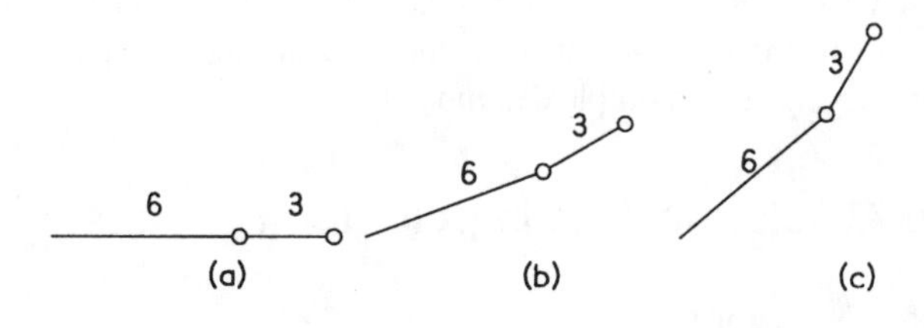

FIGURE 2.5. The sum of two phasors at three instants of time: (a) time t=0, (b) later, (c) still later.

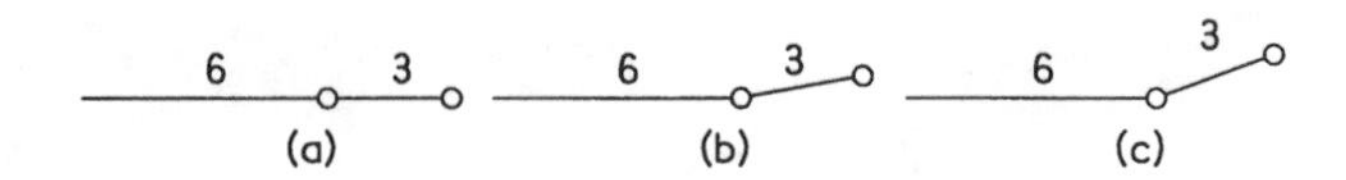

FIGURE 2.6. A replot of Fig. 2.5 viewed from the coordinate frame established by the phasor with amplitude 6. The smaller phasor rotates with an angular frequency of $\omega_2-\omega_1$.

Sometimes it is helpful to view the sum of phasors with respect to one of the phasors itself, i.e. from a coordinate frame that rotates with a phasor. Figure 2.6 is an example.

EXERCISES

Exercise 1: Exponential forms for trigonometric functions

(a) Use Euler's formula, Eq. (2.8), to show that

$$\cos\theta=\frac{e^{i\theta}+e^{-i\theta}}{2}$$

and

$$\sin\theta=\frac{e^{i\theta}-e^{-i\theta}}{2i}.$$

[Hint: Write an equation for $e^{-i\theta}$ and consider the evenness and oddness of cosine and sine functions.]

(b) Show that it is easy to derive the trigonometric identities

$$\cos(A+B)=\cos(A)\cos(B)-\sin(A)\sin(B)$$

$$\sin(A+B)=\sin(A)\cos(B)+\cos(A)\sin(B)$$

by using Euler's formula, Eq. (2.8), and the fact that $e^{i(A+B)}=e^{iA}e^{iB}$. [Hint: Use the fact that $\cos(A+B)=\mathrm{Re}\, e^{i(A+B)}$ and that $\sin(A+B)=\mathrm{Im}\, e^{i(A+B)}$. Here, the symbols Re and Im stand for operations which extract the real part and the imaginary part, respectively, of a complex number.]

Exercise 2: Absolute values in polar form

(a) Show that $|e^{i\theta}|=1$ for all θ.

(b) Use Euler's formula to show that

$$\cos^2\theta + \sin^2\theta = 1.$$

Exercise 3: De Moivre's Theorem

Use Euler's formula to prove De Moivre's Theorem:

$$(\cos\theta + i\sin\theta)^n = \cos(n\theta) + i\sin(n\theta).$$

[Abraham de Moivre (1667–1754) was born in France but lived most of his life in England. He was a friend of Isaac Newton and a fellow of the Royal Society.]

Exercise 4: Generalized complex conjugate

(a) Show that the complex conjugate of the sum is equal to the sum of the complex conjugates. In other words, if $z_3 = z_1 + z_2$ and $z_4 = z_1^* + z_2^*$, then $z_4 = z_3^*$.

(b) Show that the complex conjugate of the product is equal to the product of the complex conjugates. In other words, if $z_3 = z_1 z_2$ and $z_4 = z_1^* z_2^*$, then $z_4 = z_3^*$.

(c) For a simple complex number, such as $x + iy$, one finds the complex conjugate by changing the sign of the imaginary part. Equivalently, one can replace i by $-i$ everywhere that i appears. It is a happy fact that this rule generalizes to complex expressions of arbitrary complexity: To find the complex conjugate one replaces each occurrence of i by $-i$. For example, if

$$z = e^{i\cos[i\tanh(\sqrt{2i+p})]}$$

then

$$z^* = e^{-i\cos[-i\tanh(\sqrt{-2i+p^*})]},$$

where p is some complex variable. [Note: The above equations have no physical significance. Their only purpose is to illustrate the mathematical rule for finding the complex conjugate of a complicated function.]

Show that the generalized rule (c) for complex conjugates follows from the addition and multiplication rules in parts (a) and (b) above.

Exercise 5: Complex conjugates

(a) Show that the sum $z+z^*$ is a real number.

(b) Show that the difference $z-z^*$ is an imaginary number.

Exercise 6: Powers of i

(a) Given that $i=\sqrt{-1}$, show that $i^2=-1$, and $i^3=-i$, and $i^4=1$, and $i^5=i$, and $i^6=-1$, and so, cyclically.

(b) Prove that the product of two imaginary numbers is a real number.

Exercise 7: Multiplication of complex numbers

The multiplication example in the text above found the product of two complex numbers z_1 and z_2 by first putting the numbers in polar form. The numbers were $z_1=4+i3$ and $z_2=-1+i2$. In this exercise you should multiply the numbers z_1 and z_2 in their rectangular coordinate form by using Eq. (2.15) to show that the product is $z_4=-10+i5$. Compare this with the result $z_4=5\sqrt{5}e^{i2.678}$ from the example.

Exercise 8: Conversion between rectangular and polar form

(a) Show that $3+i3=3\sqrt{2}\,e^{i\pi/4}$.

(b) Show that $-3+i3=3\sqrt{2}\,e^{i3\pi/4}$.

(c) Show that $5-i7=8.60\,e^{-i0.95}$.

(d) Show that $10\,e^{i\pi/2}=10i$.

(e) Show that $10\,e^{-i\pi/2}=-10i$.

(f) Show that $10\,e^{-i\pi}=-10$.

(g) Show that $10\,e^{i\pi/4}\approx 7.07+i7.07$.

(h) Show that $10\,e^{i3\pi/4}\approx -7.07+i7.07$.

(i) Show that $10\,e^{i419\pi/4}\approx -7.07+i7.07$.

Exercise 9: Reduction to the range 0 to 2π radians

Put the following angles in the range $0\leqslant \text{angle}<2\pi$:

(a) $-\pi$ rad.

(b) -18π rad.

(c) -2 rad.

(d) 50 degrees.

(e) -90 degrees.

(f) 1000 degrees.

Exercise 10: Find the absolute values of the following complex numbers

(a) $3+i4$

(b) $10+i8$

(c) $10-i8$

(d) $10\ e^{i\pi/3}$

Exercise 11: Where's the error?

Bozo wants to use Eq. (2.7) to find the phase of the complex number $-3+i4$. Because x is negative he uses the equation

$$\mathrm{Arg}(y,x)=\tan^{-1}(y/x)+\pi \quad \text{for } x<0.$$

He calculates: $\tan^{-1}(4/3)=0.927$ rad, and then adds π or 3.142 rad to get 4.069 rad. But this answer cannot be right. It is evident that the number $-3+i4$ lies in the second quadrant, and so the phase must be less than π, or 3.142. (In fact, the right answer is 2.214 rad.) Where's the error in Bozo's calculation?

Exercise 12: Reciprocals

Consider the number $z_1=1+i$: (a) Show that the reciprocal $1/z_1$ is equal to $(1-i)/2$. (b) Show that the phase of z_1 is 45 degrees, and the phase of the reciprocal of z_1 is -45 degrees. (c) Show that in general the phase of the reciprocal of a number is equal to the negative of the phase of the number. (d) When is the reciprocal of a complex number z equal to the complex conjugate z^*?

Exercise 13: Powers

Use the rules of multiplication for complex numbers in polar form to show that if $z=Ae^{i\theta}$, and if z is raised to the power p (doesn't have to be an integer), then

$$z^p = A^p e^{ip\theta}.$$

Note that this formula says that if the number z makes an angle of 25 degrees with the horizontal axis, then the number z^2 makes an angle of 50 degrees, and z^3 makes an angle of 75 degrees, and so on. It also says that the magnitudes of these powers multiply just like ordinary positive numbers.

Exercise 14: Roots

Roots, like the square root, are just fractional powers, so most of what one can say about roots has already been said in Exercise 13.

(a) Show that if complex number z makes an angle of 40 degrees with the horizontal then the square root of z makes an angle of 20 degrees.

(b) Show that the square root of -9 is $3i$.
But there is one tricky thing about roots, and that is that there is more than one of them. You are already familiar with the fact that a number like 16 has two square roots, $+4$ and -4, in the sense that $4 \times 4 = 16$ and $(-4) \times (-4) = 16$. The multiplicity of roots extends to complex numbers too. Suppose that complex number z has a phase of 40 degrees, as in part (a). We don't change z at all if we add 360 degrees and consider the phase to be 400 degrees. Now when we take the square root, we get one result with a phase of 20 degrees as before, but another possibility has a phase of 200 degrees. These are the two square roots of a number with phase of 40 degrees.

(c) Show that in general the number of n-th roots of a complex number is n, if n is an integer.

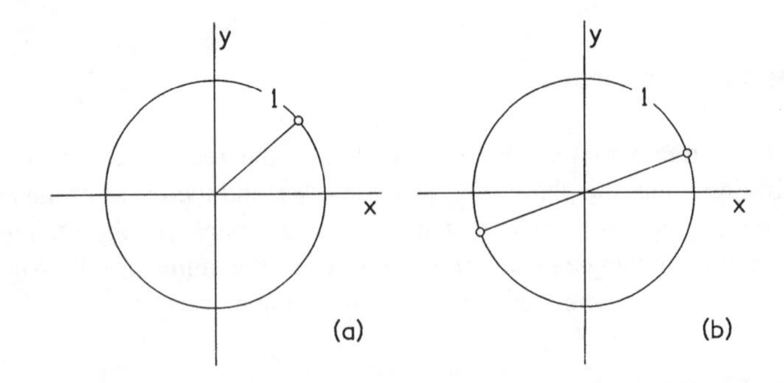

FIGURE E.14. (a) A complex number with a magnitude of 1 unit and a phase of 40 degrees. (b) The two square roots of the complex number in (a). The two square roots of a complex number are always 180 degrees apart.

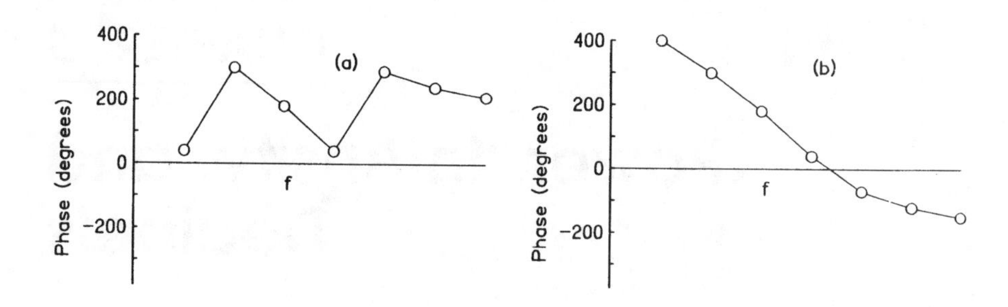

FIGURE E.15. (a) Measured phases as a function of frequency in a hypothetical experiment. (b) The same phases, unwrapped to make a smoother function.

(d) Show that there are two square roots of -9, namely $3i$ and $-3i$.

Exercise 15: Phase unwrapping

Suppose that you are doing an experiment where phase angles change continuously as a function of frequency f. You measure these phase angles for equally spaced values of f and find the following phases in degrees: 40, 300, 180, 40, 290, 240, and 210. Plotted on graph paper, your data look like Fig. E.15(a). You suspect (though you don't really know) that the jumps in Fig. E.15a are artifacts that arise because phase data cannot be known outside the range from 0 to 360 degrees. Show that you can get a smoother looking graph if you add or subtract 360 degrees from some of the phases, as shown in Fig. E.15(b). The process of guessing additive multiples of 360 degrees with the goal of smoothing data, or using other information not available in the original phase measurements, is known as "phase unwrapping." It is a way of coping with the ignorance that one always has about measured phase data.

CHAPTER 3

Power, Intensity, and Decibels

Ring out ye crystal spheres
Once bless our human ears
(If ye have power to touch our senses so.)
JOHN MILTON

Signals transport power, energy, and ultimately information. As J. J. Gibson noted, information and energy are not the same, but there is no information without energy. Power and energy are closely related because power is the rate of producing or using energy. An equivalent statement of the relationship is that energy is the integral of power over time.

Energy and power are useful concepts in physics, in part because they suppress details. For example, to gain a kilowatthour of energy one can burn a certain amount of coal, or a certain amount of gasoline, or a certain number of oil paintings by Rembrandt. It's all the same so far as matters of energy are concerned.

Previous chapters in this book have regarded the signal $x(t)$ as a disturbance of some kind. To deal with energy and power we must give the signal a somewhat more physical representation. Here we consider two cases. The first is electrical, where the signal is a voltage in an electrical circuit. The second case is acoustical, where the signal is a pressure variation in a fluid medium like air.

Voltage and Power

If signal $x(t)$ represents a voltage in a circuit, then the power that is provided can be easily described only if the voltage drives a load that is purely resistive, i.e. if there is no reactive (capacitive or inductive) part to the load. If the resistance of the load is R (a real positive number measured in units of ohms) then, by Ohm's law, the current in amperes is $i(t)=x(t)/R$, and the instantaneous power at time t is the product of the instantaneous voltage and the instantaneous current:

$$P(t)=x(t)i(t)=x^2(t)/R. \tag{3.1}$$

AVERAGE POWER AND RMS VALUE

The concept of instantaneous power, given by Eq. (3.1), is not as useful as the idea of average power. The average power, averaged over a duration of time T_D, is given by

$$\bar{P} = \frac{1}{T_D} \int_0^{T_D} dt\, P(t), \tag{3.2}$$

$$= \frac{1}{T_D} \int_0^{T_D} dt\, x^2(t)/R. \tag{3.3}$$

In order to represent the average fairly, the duration T_D must be longer than the durations of characteristic time variations of the signal. For instance, if the signal happens to be noise passed through a filter with a bandwidth (range of frequencies) of only 1 Hz, then the signal will fluctuate on the time scale of 1 second. The averaging time T_D must be longer than this in order to represent the average power correctly.

If the signal is periodic, then T_D must be long compared to the period or else it must be equal to an integral number of periods. For a periodic signal the best choice of averaging duration is a single period T. All other periods are identical to any single period.

The root-mean-square value of the signal is defined as the numerical value of a constant which would lead to the same average power $\bar{P}$ as the signal. In other words, by definition

$$x_{\mathrm{RMS}}^2/R = \bar{P} = \frac{1}{T_D} \int_0^{T_D} dt\, P(t). \tag{3.4}$$

Therefore,

$$x_{\mathrm{RMS}} = \sqrt{\frac{1}{T_D} \int_0^{T_D} dt\, x^2(t)}. \tag{3.5}$$

Equation (3.5) gives the definition of the RMS value of x in detail. The operations of R (root), M (mean), and S (square) are done in reverse order. One first squares the signal, $x(t)$, then takes the mean by averaging over time, and finally takes the square root.

EXAMPLE 1: *Find the RMS value of a sine wave*

For a sine wave signal, $x(t)=A\sin(\omega t)$. We find the RMS value by first finding the average power. Assuming a resistance of $R=1$ ohm, as usual, and using Eq. (3.3) we find an average power of

$$\bar{P}=\frac{1}{T}\int_0^T dt\,A^2\sin^2(\omega t), \tag{3.6}$$

where T is the period. The sine function and its square are shown in Fig. 3.1 for unit amplitude ($A=1$). (The square of the sine function, which is the integrand, can be visualized by thinking about the trigonometric identity $\sin^2\omega t=1/2-1/2\cos 2\omega t$.)

It is easy to average the $\sin^2$ function over a period. We first recall that for any angle θ, $\sin^2\theta+\cos^2\theta=1$. Next we realize that the average of $\cos^2$ must be the same as the average of $\sin^2$, because one is just a phase shifted version of the other. Therefore, the average of either one must be 1/2, and, for a sine of amplitude A,

$$\bar{P}=A^2/2. \tag{3.7}$$

The root-mean-square (RMS) value is the square root of the average power, per Eq. (3.4), and therefore,

$$x_{\mathrm{RMS}}=\sqrt{A^2/2}=A/\sqrt{2}=0.707A. \tag{3.8}$$

This equation says that the RMS value of a sine is 0.707 times the amplitude of the sine. In general we expect that the RMS value of a periodic signal is less than the maximum value, but the ratio is not always 0.707. The ratio depends on the waveform.

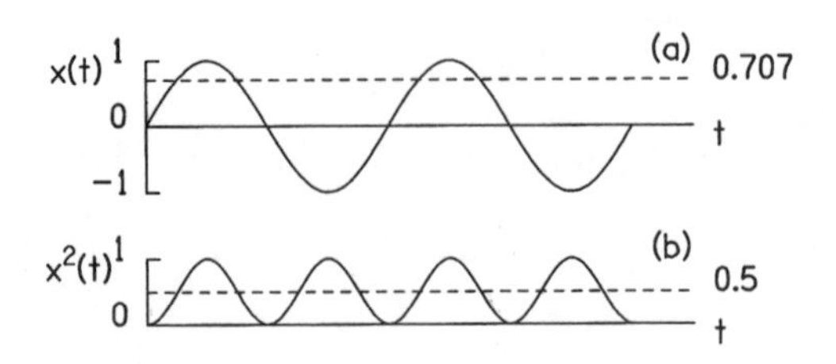

FIGURE 3.1. Figure (a) shows a sine wave with unit amplitude. The RMS value is shown by a dashed line. Figure (b) shows the instantaneous power which is the square of Fig. (a). The average power is shown by a dashed line. It is the mean square of the sine.

CREST FACTOR

The crest factor, defined by symbol C, is the ratio of the extreme value that the signal has, to the RMS value,

$$C = \frac{\mathrm{Max}[|x(t)|]}{x_{\mathrm{RMS}}}. \tag{3.9}$$

The absolute value signs mean that either the positive extreme or the negative extreme is used, depending on which has the larger absolute value. In either case the crest factor is always a positive number by convention. If the signal is random, one may be unsure of how large the signal might become, and so the crest factor is uncertain. For a periodic signal, $\mathrm{Max}[|x(t)|]$ is the largest value that the absolute value of the signal has during a period.

The crest factor is of interest when a signal with tall spikes or deep valleys is something that one wants to avoid, mainly because of limits in physical systems that receive or process the signal. A mechanical system, such as the diaphragm of a loudspeaker, a headphone, or a microphone, has a maximum limit of displacement beyond which the system distorts badly. An electrical system, such as a transistor amplifier, distorts by clipping when the signal approaches the power supply voltage. If the crest factor of a signal is very large then one can avoid distortion by these physical systems only by reducing the overall level of the signal. But then one cannot transfer much power. On the other hand, if the crest factor is small then one can transfer a lot of power (the RMS value of the signal can be large), but there is never a spike that causes the system to distort. To put it in the form of a problem, we suppose that there is a certain amount of energy that we need to transmit. We can do it all in a single short burst (large crest factor) or we can do it with a smoother signal that never becomes very large, but never becomes very small either (small crest factor). With the burst signal we are more likely to distort the signal or break something. Put still another way, signals with small crest factors are useful for transmitting information, large crest factors are used as weapons.

The concept of crest factor is also used to rate voltmeters that are called "true RMS" voltmeters. Here the goal is to indicate an accurate value of the RMS voltage. It is easy for the meter to do that if the signal is smooth like a sine wave, but hard if the signal is impulsive. Therefore, specifications for RMS voltmeters include a crest factor rating, which is the crest factor of the most impulsive signal that the meter can read without making a large error. In the case of meters then, the larger the crest factor rating the better.

EXAMPLE 2: *Crest factor of a rectangular pulse*

A periodic rectangular pulse wave, or pulse train, can be defined by its duty factor, the fraction of the period for which the signal is in a high state. There are two natural ways to describe a rectangular pulse with a duty factor p. The first is in Fig. 3.2(a). The signal is in a high state (value 1) for duration pT and a low state (value 0) for the rest of the time, duration $(1-p)T$. Acoustical or audio signals, however, are likely to have zero average value. The pulse train shown in (b) has been displaced vertically so that it has zero average value. It also has been scaled so that it has an average square (average power) of one. It is left as an exercise to show that the average power is, in fact, one.

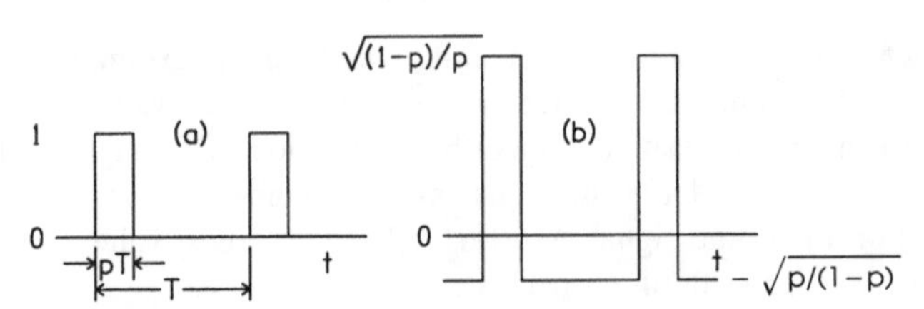

FIGURE 3.2. (a) A rectangular pulse train with duty factor p. (b) The same, displaced vertically so that the average value of the waveform is zero.

The crest factor is $C=\sqrt{1/p}$ for the pulse of Fig. 3.2(a), because the maximum absolute value is 1 and the RMS value is $\sqrt{p}$. For the pulse of Fig. 3.2(b) the crest factor is given by

$$C=\sqrt{(1-p)/p} \quad \text{for } p<\tfrac{1}{2} \tag{3.10}$$

and

$$C=\sqrt{p/(1-p)} \quad \text{for } p>\tfrac{1}{2}. \tag{3.11}$$

The crest factor is given by the maximum absolute value itself because the RMS value has been made equal to 1, by scaling the waveform. For the square wave, $p=1/2$, and the crest factor has the smallest possible value, namely $C=1$.

DECIBELS

It is common to refer to the power of a signal or the intensity of a sound as a "level." Levels are measured in decibels, which is a logarithmic measure. Therefore, when we quote levels, we are using the exponent of the measure of a physical quantity (i.e. we are using a power of ten) instead of using the quantity itself. In practice, physical quantities might be electrical (power or voltage amplitude) or they might be acoustical (intensity or pressure amplitude). For definiteness, we use power and voltage amplitude, P and A, in the development below.

The basic log relation for signal levels (measured in Bels) is given by the equation

$$L_2-L_1(\text{Bels})=\log(\bar{P}_2/\bar{P}_1), \tag{3.12}$$

where L_2 is the level of signal number 2 with average power $\overline{P}_2$, and L_1 is the level of signal number 1 with average power $\overline{P}_1$. Because the unit of Bels is inconveniently large, one normally uses a unit that is 10 times smaller, the decibel (dB), for which the level equation becomes

$$L_2 - L_1(\text{decibels}) = 10 \log(\overline{P}_2/\overline{P}_1). \tag{3.13}$$

Equation (3.12), which measures levels in Bels, is no longer used by anyone. Henceforth we shall only use Eq. (3.13) which measures levels in dB, and we will drop the parenthetical notation (decibels) from the equations that follow.

Equation (3.13) emphasizes that the dB scale is a relative scale by which one signal can be compared with another. There is nothing in the equation that gives an absolute value to signal levels, only relative levels. The equation shows that a physical ratio, $\overline{P}_2/\overline{P}_1$, is converted by the log operation to a difference, $L_2 - L_1$.

Because the power P is proportional to the square of the voltage x, it is possible to derive a formula for relative levels in terms of voltages, so long as the two signals that are compared have the same waveform. For example, if signal 1 and signal 2 are both sine waves, with voltage amplitudes A_1 and A_2 respectively, then the level difference is

$$L_2 - L_1 = 20 \log(A_2/A_1). \tag{3.14}$$

To prove this relationship, one only has to remember that the log of the square of a quantity is equal to twice the log of the quantity. [For any positive real A, $\log A^2 = 2 \log A$. The reader will notice that the exponent has simply fallen off the log.] That is why 20 replaces 10 in going from Eq. (3.13) to Eq. (3.14).

It is important to realize that the equations above do not mean that there are "power decibels" and "voltage decibels." Signal number 1 and signal number 2 differ in level by a certain number of dB. One can obtain that number either by measuring the powers and using Eq. (3.13) or else by measuring the voltage amplitudes and using Eq. (3.14). Either technique must give the same answer for the level difference.

EXAMPLE 3: *A factor of two in voltage amplitude*

If the voltage amplitude of sine signal 2 is twice that of sine signal 1 then the level difference in dB is

$$L_2 - L_1 = 20 \log(A_2/A_1) = 20 \log 2 = 6 \text{ dB}. \tag{3.15}$$

This level calculation can be done in terms of power too. If the voltage amplitude of signal 2 is twice that of signal 1 then the average power of signal 2 is four times that of signal 1. It follows that

$$L_2 - L_1 = 10 \log(\overline{P}_2/\overline{P}_1) = 10 \log 4 = 6 \text{ dB}. \tag{3.16}$$

This example illustrates the equivalence of voltage and power calculations of level differences.

Absolute Electrical dB

The decibel scale is foremost a relative scale whereby any physical magnitude must be compared with a reference. The notion of absolute dB simply means that there is a reference that has been agreed on in advance. The level is therefore written as

$$L = 10 \log(\overline{P}/\overline{P}_{\text{Ref}}). \tag{3.17}$$

In audio technology there are several such common references. The measure called dBv is a logarithmic measure of voltage with respect to an RMS reference voltage of 1 volt. Therefore,

$$L(\text{dBv}) = 20 \log(x_{\text{RMS}}/1), \tag{3.18}$$

where the denominator in the argument of the log is 1 volt.

It is also common to measure signals in units called dBm, where the reference $\overline{P}_{\text{Ref}}$ is a power of 1 milliwatt. Therefore,

$$L(\text{dBm}) = 10 \log(\overline{P}), \tag{3.19}$$

where $\overline{P}$ is a power expressed in milliwatts. It is, however, unusual to measure electrical power. It is usual to measure voltage instead, and to infer a power from it. By convention, the conversion is done by assuming a resistive load of 600 ohms. Therefore, the reference voltage, corresponding to the reference power, obeys the equation

$$\overline{P}_{\text{Ref}} = x_{\text{Ref}}^2/600 = 1 \text{ mW}. \tag{3.20}$$

The reference voltage is therefore $x_{\text{Ref}}=\sqrt{0.6}=0.7746$ volts, and therefore

$$L(\text{dBm})=20\log(x_{\text{RMS}}/0.7746), \tag{3.21}$$

where x_{RMS} is the RMS voltage of the signal.

ACOUSTICS: PRESSURE AND INTENSITY

The above introduction of the concepts of RMS value, crest factor, and levels was done in terms of electrical signals: voltage x and power P. It is possible to run an entirely parallel development for acoustical signals, in terms of pressure x and intensity I. We follow that idea below.

The instantaneous intensity of a sound $I(t)$ is defined as acoustical power per unit area. The instantaneous intensity is given by the product of the instantaneous values of the pressure, $x(t)$, and the fluid velocity, $u(t)$,

$$I(t)=x(t)u(t). \tag{3.22}$$

In the MKS system of units the intensity is measured in watts/m^2, and the fluid velocity in m/s. The pressure is measured in Newtons/m^2, or Pascals,

$$1\ \text{Pascal}=1\ \text{Pa}=1\ \text{Newton/meter}^2. \tag{3.23}$$

Further information on metric units of pressure can be found in the note at the end of this chapter just before the exercises.

For a plane wave, fluid velocity is related to the pressure by the equation

$$u(t)=x(t)/(\rho c), \tag{3.24}$$

where ρc is the specific impedance of the medium in which the plane wave propagates. It is the product of the density ρ (units of kg/m^3) and the speed of sound c (units of m/s). Substituting, we find an expression for the intensity as a function of the pressure

$$I(t)=x^2(t)/(\rho c). \tag{3.25}$$

The parallel between acoustical and electrical cases should be clear: In both cases there is a variable that represents the fundamental disturbance, a voltage $x(t)$ that can be either positive or negative, or a pressure $x(t)$ that can be either a compression (positive) or a rarefaction (negative). By squaring $x(t)$, we obtain an energy-like quantity, a power or an intensity [as given in Eq. (3.1) or (3.25)], that can only be positive or zero.

Sound Levels

Sound levels come from acoustical measurements of intensities or pressure amplitudes. The levels are ratios, comparing one sound with another, and are expressed in decibels. By analogy with the electrical case, the difference in levels between sound 2 and sound 1 is:

$$L_2 - L_1(\text{decibels}) = 10 \log(I_2/I_1), \tag{3.26}$$

where I_2 and I_1 (without explicit dependence on time t) represent the average intensities of the two sounds. Also,

$$L_2 - L_1 = 20 \log(x_{\text{RMS},2}/x_{\text{RMS},1}), \tag{3.27}$$

where $x_{\text{RMS},2}$ and $x_{\text{RMS},1}$ represent the RMS pressures of the two sounds.

EXAMPLE 4: ***Ordinary ranges***

The range of sound intensities in our everyday acoustical world is about six orders of magnitude (six powers of ten). On an unusual day, when we do psychoacoustical experiments and go to a rock concert, the range may be as large as 12 orders of magnitude. The corresponding level differences are

$$L_2 - L_1 = 10 \log(10^6) = 60 \text{ dB} \tag{3.28}$$

and, for the unusual day,

$$L_2 - L_1 = 10 \log(10^{12}) = 120 \text{ dB}. \tag{3.29}$$

The example makes it clear that the human auditory system is capable of functioning over a wide range of intensities.

Absolute Acoustical dB

An absolute acoustical decibel scale is referenced to a nominal threshold of hearing. There are two such references, one for pressure and the other for intensity.

The nominal threshold of hearing, approximating the average threshold for normal listeners for a 1000-Hz sine tone, corresponds to an RMS pressure which we define either as x_{Ref} or as x_0,

$$x_{\text{Ref}} = x_0 = 2 \times 10^{-5} \text{ Pa} = 20 \ \mu\text{Pa}. \tag{3.30}$$

Therefore, a signal with RMS pressure x_{RMS} has a sound pressure level of

$$L = 20 \log(x_{RMS}/x_0) = 20 \log[x_{RMS}/(2\times 10^{-5})]. \qquad (3.31)$$

One says that this level L is in dB SPL, or dB *re* 20 microPascals. There is a reference for sound intensity as well. For a plane wave, the threshold intensity I_0 is obtained from threshold RMS pressure by an RMS version of Eq. (3.25)

$$I_0 = x_0^2/(\rho c). \qquad (3.32)$$

In MKS units the impedance of air ρc is 415 Rayls [pronounced ''rails,'' and named after Lord Rayleigh (John William Strutt)]. Therefore, the value of I_0 is

$$I_0 = (2\times 10^{-5})^2/415 = 0.964\times 10^{-12} \text{ watts/m}^2. \qquad (3.33)$$

This value is always approximated by the value 10^{-12} watts/m^2. Therefore, the level of a signal with an acoustical intensity of I is

$$L = 10 \log(I/I_0) = 10 \log(I/10^{-12}). \qquad (3.34)$$

In principle, one can measure either intensity or pressure to find the sound level. Both lead to the same result in decibels. In practice one normally uses a microphone that measures pressure and converts to level or infers the intensity. In fact, the acoustical situation is not so simple. The relationship between pressure and intensity in terms of specific acoustical impedance applies for plane waves; it does not apply at places that are near a reflecting or absorbing boundary. Even for plane-wave conditions there is the fact that the impedance of air is not absolutely constant. With extreme conditions of atmospheric pressure and temperature, the impedance ρc can differ by as much as $\pm 20\%$ from the nominal value of 415 MKS Rayls. Thus, levels computed from actual intensity measurements or actual pressure measurements might differ by as much as 1.5 dB. Fortunately, there are few things in psychoacoustics that require one to know the absolute intensity to within ± 2 dB.

WAVES IN TIME AND SPACE

The treatment of sound intensity would not be complete if it did not deal with the common observation that distant sources of sound tend to be less loud than sources that are close. That is because the intensity of distant sources tends to be less. To deal with sources in space requires a brief introduction to room acoustics, which describes how sound propagates from a source to a receiver.

A signal in electrical form is considered to be a function of time. A periodic electrical signal repeats when time advances by period T. A signal in acoustical form is a function of both time and space. A periodic acoustical signal repeats in time (period T), and it also repeats as a function of a spatial coordinate. The repeat

RITTER LIBRARY
BALDWIN-WALLACE COLLEGE

distance in space is called the wavelength and is given the symbol λ. It is the distance that a wave travels before starting over again.

Because a wave repeats itself after a time interval equal to a period, the wavelength is the product of the wave speed c and the period. Therefore,

$$\lambda = cT. \tag{3.35}$$

Normally, the wave speed c is the speed of sound in air. Equation (3.35) is nothing more than an application of the well-known fact that the distance that something goes is equal to the speed of travel multiplied by the duration of the travel. It is more common to write this equation using the frequency f, which is the reciprocal of T,

$$c = f\lambda, \tag{3.36}$$

which emphasizes the inverse relationship between frequency and wavelength. When the frequency is large the wavelength is small and vice versa.

EXAMPLE 5: *Wavelength of a 1000-Hz tone*

Calculate the wavelength of a 1000-Hz tone in meters and feet.

Solution: The speed of sound in dry air at 20 degrees Celsius is 343.2 meters per second. The wavelength is

$$\lambda = c/f = 343.2/1000 = 0.3432 \text{ meters} \tag{3.37}$$

The conversion from meters to feet is exactly 0.3048 meters per foot.* Therefore, the wavelength of a 1000-Hz tone is 1.126 feet.

LORD RAYLEIGH (1842–1919)

John William Strutt was born on November 12, 1842 in Essex, England, the eldest son of the second Baron Rayleigh. He attended Cambridge University where he developed great skills in applied mathematics. Shortly after his marriage in 1871 he contracted rheumatic fever. As part of his recovery he took a trip down the Nile River in a houseboat, and it was there that he began to write the *Theory of Sound*, his best-known acoustics work. His interest in sound had been sparked by reading Helmholtz, and his interest in acoustics continued throughout his career, during which he also made contributions to every other

*This is now the official definition of the "foot." If the standard meter changes at some time then the standard foot will change with it.

aspect of classical physics. In 1879 he returned to Cambridge for a five-year period where he instituted a course of study in experimental physics that had widespread influence. His own experiments were mostly done on his estate in Essex after he became the third Baron Rayleigh. As an experimentalist, Rayleigh was adept at obtaining precise results from modest equipment. From precise measurements of the density of air he discovered the element argon as a constituent of the atmosphere, for which he won the Nobel Prize in 1904. He died on June 30, 1919, an unfinished paper on acoustics still on his desk.

Equation (3.36) is the link between two pictures of a wave, a space picture and a time picture. This concept is shown in Fig. 3.3, where four snapshots show the wave in space at four closely spaced instants, separated by one-sixth of a period.

The concept of wavelength becomes important whenever a sound wave interacts with massive objects. A sound wave will be diffracted, i.e. it will bend around an object, if the wavelength is large compared to the dimensions of the object. Otherwise, the sound wave will be significantly obstructed by the object. For example, a

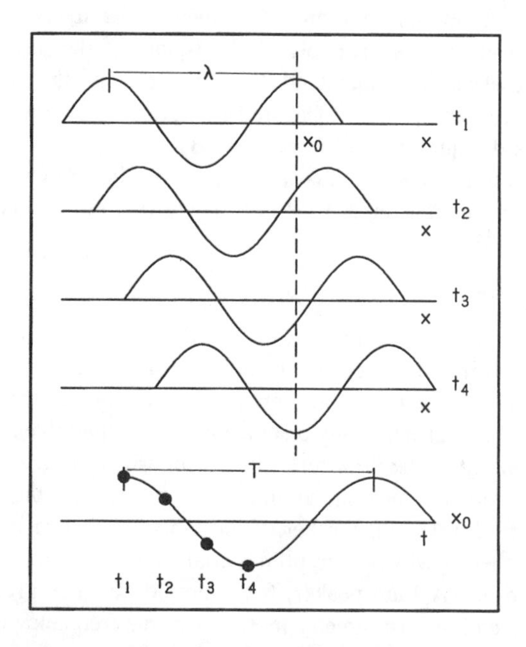

FIGURE 3.3. The wave, as a function of spatial coordinate x, travels to the right with speed c. It is shown at four successive instants in time. As the wave passes an observer, located at point x_0, the observer sees the succession of waveform features as a function of time t, shown in the bottom graph. If the length of the wave is λ then it takes a time $T=\lambda/c$ for the entire wave to pass point x_0.

refrigerator acts as a significant barrier to a 1000-Hz tone, but a bottle does not. That's because the refrigerator is large compared to the wavelength of one foot but the bottle isn't. For a 100-Hz tone, the wavelength is ten times longer and neither the bottle nor the refrigerator is large compared to that wavelength. Therefore, a 100-Hz tone easily diffracts around both of them.

Radiation

The topic of sound radiation, as it applies to the human listening experience, could easily fill several books on its own. In this section, we consider the simple case of free-field propagation, where the inverse square law applies, together with a few deviations from that law.

The first step in the study of sound radiation is to start with a completely idealized situation, the spherically symmetrical radiator in free field. This radiator produces a sound field that is the same in all directions; it is a "monopole." A uniformly pulsating spherical balloon is an example of a monopole radiator. The concept of a free field refers to a motionless homogeneous medium (air) which has no boundaries. Therefore, there are no reflections. An anechoic room, with absorbing wedges on all surfaces, approximates a free field.

The sound intensity in this situation is described by the inverse square law, which says that the intensity varies inversely as the square of the distance between the source and the observer. To understand this law, we recall that the intensity is the power per unit area. The power (measured in watts) originates at the source and propagates outward, equally in all directions. By the time the radiated power has reached the observer at distance r it is spread out over the surface of a sphere whose radius is r. This surface has area $A = 4\pi r^2$. Therefore, the observer finds a sound field with intensity P/A, or

$$I = P/(4\pi r^2). \tag{3.38}$$

This is an inverse square law because r^2 is in the denominator.

If the sound source is on the floor, then it radiates only into the upper half of the space. Therefore, the intensity at any distance r is twice that given by Eq. (3.38), the extra power coming from the floor reflection. Similarly, a source on the floor in the corner of a room radiates into one-eighth of the space, and the intensity is eight times that given by Eq. (3.38). Further, if the vibrating source is itself not spherically symmetric, then it will radiate preferentially in some directions, even if it is suspended in mid air. A loudspeaker, for example, radiates preferentially in the forward direction, and this asymmetry increases as the frequency increases because there is less diffraction from the loudspeaker cone.

Although the directionality due to nearby surface reflections or limited diffraction may mean that Eq. (3.38) does not predict the intensity correctly, it does not necessarily mean that the inverse square law is invalid. For any particular direction in space the intensity still depends inversely on the square of the distance from the source to the observer, but different directions are different. The loudspeaker is a common example; the human mouth is another.

Sources like loudspeakers are characterized by their directionality in the forward direction, otherwise known as "on axis." The directivity factor, $Q(f)$, gives the ratio of the intensity on axis to the intensity that would be produced at that point by a spherical radiator emitting the same power (Beranek, 1988). Normally $Q(f)$ is an increasing function of frequency f. Therefore, for a point in the forward direction,

$$I_d = Q(f)P/(4\pi r^2), \tag{3.39}$$

where the subscript d refers to the direct sound in the forward direction. We observe that directionality does not violate the inverse square law.

Serious deviations from the inverse square law occur when the free-field conditions are upset by reflections from the surfaces in a room. In the limit of highly reverberant conditions, the intensity of sound does not depend at all on the distance to the source! In such conditions, the intensity in the reverberant field is given by

$$I_r = 4P/R_c, \tag{3.40}$$

where R_c is a number known as the "room constant."

The room constant is approximately equal to the total absorbing area of the room surfaces, which is the total surface area S multiplied by the average absorption coefficient α; i.e. $R_c = \alpha S$. The absorption coefficient is a measure of the fraction of power impinging on the room surfaces that is absorbed (sound energy is turned into heat energy). Because the room constant is approximately proportional to α, the intensity in the reverberant field is approximately inversely proportional to α (Eq. (3.40)). However, when the absorption coefficient is large, one needs a better approximation to the room constant, namely

$$R_c = \alpha S/(1-\alpha). \tag{3.41}$$

This has the logical advantage that if α is as large as 1 (100% of the impinging power is absorbed) then the room constant is infinite and the intensity in the reverberant field becomes zero. The case of 100% absorption characterizes an ideal anechoic room, a room with fiberglass or foam wedges to absorb all the incident sound so that there are no echoes (an-echoic).

The total intensity in a sound field is the sum of the intensity in the direct sound and the intensity in the reverberant field, from Eq. (3.39) through (3.41),

$$I = P[Q(f)/(4\pi r^2) + 4(1-\alpha)/\alpha S]. \tag{3.42}$$

Figure 3.4 shows how the level of a one-milliwatt source decreases as a function of distance to the observer. The straight line for infinite room constant shows the inverse square law (−6 dB per doubling of distance, or −20 dB per decade) that applies in free field. Other curves show how the reverberant field establishes the long-distance limit.

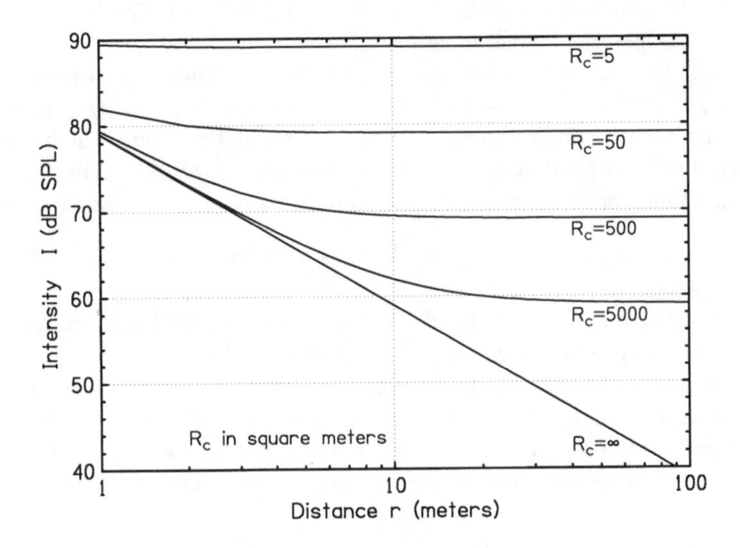

FIGURE 3.4. A spherically symmetric source, namely $Q(f)=1$, is at the origin ($r=0$) emitting 1 milliwatt. The graph shows the intensity as a function of distance from the source to the observer for five different values of the room constant. The plot for $R_C=\infty$ is free field.

According to Eq. (3.42) one would expect the inverse square law to hold for small values of distance r. However, the theory behind Eq. (3.42) is incomplete because it fails to consider special complications that arise when r is small. When r is smaller than a characteristic size of the source, or when r is smaller than a wavelength, then the observer is in the "near field" of the source. It is difficult to generalize about the r-dependence of intensity in the near field. We consider a few examples:

Suppose that a listener stands with his ear 10 cm away from a wall (r small compared to source size) and that the entire wall vibrates uniformly with a very low frequency, so that the wavelength (on the wall and in the air) is long compared to the size of the wall. All parts of the vibrating wall contribute in a constructive way to the sound field at the ear. If the listener now moves to a distance of 20 cm from the wall, nothing is changed. The intensity does not depend at all on distance r.

Suppose, now, that the frequency of the sound increases so that the wavelength in air is not large compared to the size of the wall. Then the sound field at the ear includes some contributions from remote parts of the wall that arrive out of phase. At any point, the sound field will show the effects of cancellation and reinforcement of waves that originate at different parts of the wall. The spatial dependence of the field is complicated, and if the listener moves from 10 cm to 20 cm, the intensity at the ear might decrease or it might even increase.

Finally, suppose that the size of the source is small compared to distance r, but that the wavelength is not small compared to r. So long as the source is a spherical radiator we can say that the listener is in the far field; in other words, the listener is not in the near field. However, if the vibration pattern of the source is complicated, having different phases at different points on the surface, then the source creates a multipolar radiation pattern. Monopole, dipole, and quadrupole radiation patterns decrease with distance r at different rates, and only the monopole radiation obeys the inverse square law. The monopole radiation dominates the others depending on the ratio of r to the wavelength. Thus, even if the source is small, the listener might be in the near field, where the inverse square law does not apply, if distance r is smaller than the wavelength.

This section began with the concept of the monopole radiator in free field, where a simple inverse square law describes the intensity of the sound field as a function of the distance to the source. However, this condition is not normally realized when a sound field propagates indoors. In large rooms the reverberant field normally dominates, so that the intensity is better described as distance-independent than as inverse square law. The reverberant sound is relatively less important compared to the direct sound when one is very close to the source, but then one may well be in the near field of the source where the monopole approximation is not valid. The ideal free-field conditions are approximated outdoors and in an anechoic room so long as the distance to the source is large compared to the size of the source and to the wavelength.

MEASURING SOUND PRESSURE LEVELS

Sound pressure levels are measured by a sound level meter, consisting of a pressure-sensitive microphone, amplifier, weighting filter, temporal averaging mechanism, and display, all in a convenient package. The display is an analog or digital meter that reads out in decibel units *re* 20 μPa. Because of the enormous physical range of sound pressures encountered in ordinary life, the sound level meter has a range switch to change the sensitivity.

The weighting filters attenuate high- and low-frequency signals before they get to the display. The filter responses for A and C weighting scales are shown in Fig. 3.5. By convention, the two scales are the same at 1000 Hz. The C-scale frequency response is almost flat. That is, it weights all frequencies almost the same. The A scale, by contrast, attenuates low frequencies dramatically. For example, if a 1000-Hz sine tone is measured on an A-weighted sound level meter, and then the frequency is decreased to 100 Hz while the intensity remains constant, the sound-level meter will be found to read almost 20 dB lower. What the weighting filter does to low-frequency pure tones it also does to the low-frequency components of complex tones or noise.

The A-weighting filter response function resembles the response of the human ear to sine signals of low to intermediate power, as will be described in Chapter 4. Therefore, it is A-weighted readings that are most often used in studies of sound

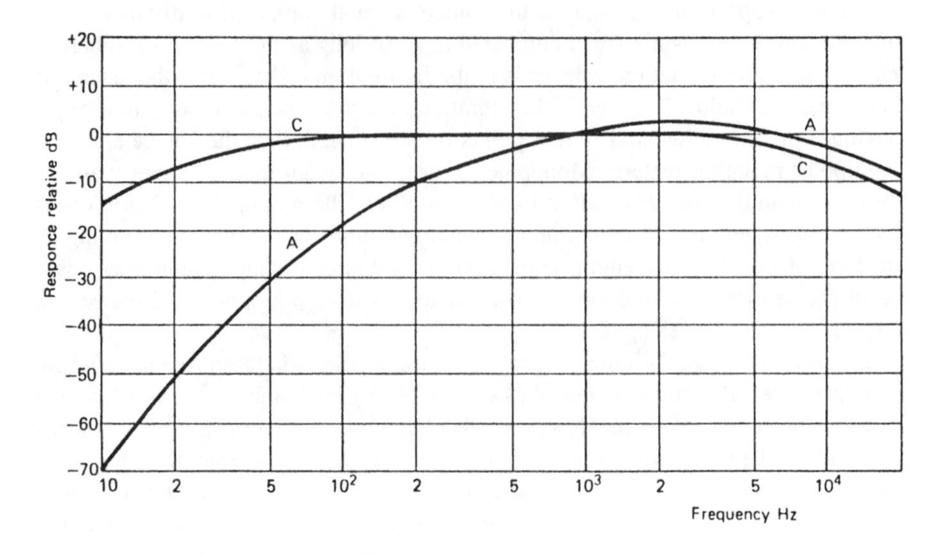

FIGURE 3.5. Amplitude response of the A and C filters used in sound-level meters.

annoyance; for example, around airports or in neighborhoods where rowdy students have most excellent parties.

The A-weighted scale also makes it possible to measure sound levels in environments with appreciable added ultra-low-frequency sound, for example in a building with an ambitious heating, ventilating, and air conditioning system. Acousticians who work underground in central Paris must use A-weighted scales to avoid contamination of their acoustical measurements by the subsonic rumble of the *Metro*, which permeates the nether regions of the city.

ADDING SIGNALS

Previous sections of this chapter have considered individual electrical or acoustical signals $x(t)$, their instantaneous power or intensity $x^2(t)$, the average power $\overline{P}$, or intensity I, and the RMS values of x. The average power, the intensity, and the RMS values are commonly expressed as levels L, measured in dB.

The parallel treatment of electrical and acoustical quantities is not a mere curiosity; it is an essential feature of electroacoustics as it is applied to audio technology or psychoacoustical experiments. The connections between electrical and acoustical worlds are transducers, notably microphones and loudspeakers (or headphones). These devices are approximately linear, and this approximation is becoming increasingly good as the technology improves. Therefore, if one wants to reduce the RMS acoustical pressure created by a loudspeaker by a factor of two, one has only to reduce the electrical voltage that drives the speaker by a factor of two. In the language of sound levels, a 6-dB decrease in sound pressure is accomplished by a 6-dB attenuation of the electrical signal. In fact, one becomes so accustomed to the idea that acoustical signals can be modified (attenuated, gated, enveloped, mixed, filtered, or modulated) by the corresponding operations on their electrical predecessors, that only if one is being quite careful does one pay much attention to the technical difference.

Having learned to deal with the power and level of individual signals, one naturally wonders about signals in combination. The basic rule is that signal voltages or pressures add linearly. For instance, if a radio announcer's microphone generates an electrical signal $x_1(t)$, and this is combined with a recorded musical background signal $x_2(t)$, then the result is the linear sum

$$x_3(t)=g_1x_1(t)+g_2x_2(t), \tag{3.43}$$

where g_1 and g_2 are the gain constants for each of the two input signals. The sum is done in an electronic device known as an "adder" or "mixer" which establishes each gain. Any combination of signals x_1 and x_2 other than a linear combination would constitute distortion. For simplicity in what follows, we set the gains equal to one, so

$$x_3(t)=x_1(t)+x_2(t). \tag{3.44}$$

Similarly, if a violin creates an acoustical pressure wave $x_1(t)$ and a flute creates a wave $x_2(t)$ then when both instruments play in a duet, the air acts as a linear mixer and the resulting waveform is

$$x_3(t)=x_1(t)+x_2(t). \tag{3.45}$$

When two signals are added like this, the instantaneous power is the square of the sum, namely

$$P_3(t)=x_1^2(t)+x_2^2(t)+2x_1(t)x_2(t). \tag{3.46}$$

A similar equation describes the intensity I of an acoustical signal which is the sum of $x_1(t)$ and $x_2(t)$. All of the steps that follow apply equally in the electronic and acoustical domains.

The instantaneous power can be written in terms of the power in individual signals, plus a cross-product term,

$$P_3(t)=P_1(t)+P_2(t)+2x_1(t)x_2(t), \tag{3.47}$$

where $P_1=x_1^2(t)$ and $P_2=x_2^2(t)$.

Things start to get interesting when we calculate the average power by averaging Eq. (3.47),

$$\overline{P_3}=\overline{P_1}+\overline{P_2}+\frac{2}{T_D}\int_0^{T_D} dt\ x_1(t)x_2(t). \tag{3.48}$$

There is no simple way to deal with the integral of the cross product in Eq. (3.48) that will work in all cases. If signals $x_1(t)$ and $x_2(t)$ are completely uncorrelated, as in the audio and musical examples given above, then the integral is zero. Then the output power is just the sum of the two input powers,

$$\overline{P_3}=\overline{P_1}+\overline{P_2}. \tag{3.49}$$

On the other hand, if the two input signals $x_1(t)$ and $x_2(t)$ are identical, then Eq. (3.48) degenerates to

$$\overline{P_3}=\overline{P_1}+\overline{P_1}+2\overline{P_1}=4\overline{P_1}, \tag{3.50}$$

which is another statement of Example 2, where we found that if a signal amplitude is doubled then the power is multiplied by four. Alternatively, signal $x_2(t)$ might be the negative of $x_1(t)$ (correlation is -1) so that the two signals cancel perfectly, and $\overline{P_3}=0$.

It follows that if there are two signals $x_1(t)$ and $x_2(t)$ of equal power $\overline{P_1}$ then the power of the sum could be as small as zero or as large as $4\overline{P_1}$. It could be anything in between, depending on the correlation of the two signals.

More discussion of correlation between signals will be found in later chapters of this book. For the present, we observe that it is very common for two signals to have no long-term correlation. This is the case, for example, when a signal is added to noise. Indeed, unless something special causes signals x_1 and x_2 to be correlated, one normally expects that the cross-product integral is zero and the powers simply add, $\overline{P_3}=\overline{P_1}+\overline{P_2}$.

Adding Signals and Levels

The next step is to consider the addition of powers in terms of sound levels. To calculate levels, we take the logarithm of the power, which is a highly nonlinear transformation. If $\overline{P_3}=\overline{P_1}+\overline{P_2}$, then it is definitely *not* the case that $L_3=L_1+L_2$. Levels don't add like that. The following examples show how to find the final level when uncorrelated signals are added.

EXAMPLE 6: *Mixing two signals*

Tone 1 has an RMS voltage of 2 volts, and tone 2 has an RMS voltage of 3 volts. Calculate the level of the sum of these two tones, expressed in dBm.

Solution:

For the power of the sum,

$$P_3=2^2+3^2=13.$$

Therefore, $x_{RMS}=\sqrt{13}=3.61$.

We recall that for dBm, the reference is 0.7746 v. Therefore,

$$L_3=20 \log \frac{3.61}{0.7746}=13.4 \text{ dBm}.$$

EXAMPLE 7: *An unequal duet*

A trombone causes a sound-level meter to read $L_1=80$ dB SPL, and a flute causes it to read $L_2=68$ dB SPL. What is the reading if the trombone and flute play together?

Solution:

We begin by writing expressions for the individual intensities, I_1 for the trombone and I_2 for the flute.

$$I_1=10^{-12}\times 10^{80/10}=10^{-4}$$

$$I_2=10^{-12}\times 10^{68/10}=10^{-5.2}=10^{-4}\times 10^{-1.2}=0.063\times 10^{-4}.$$

Therefore,

$$I_3=I_1+I_2=1.063\times 10^{-4},$$

and the level is

$$L_3=10 \log(I_3/10^{-12})=10 \log 1.063+80=80.3 \text{ dB}.$$

This example illustrates the point that the level of a sound is only slightly increased by adding another sound having an appreciably smaller level.

EXAMPLE 8: *The violin section*

A single violin produces a level of 70 dB SPL. What level is produced by 30 violins if all play with equal intensity? Assume that the signals are uncorrelated.

Solution:

The power of one violin is P_1. The power of 30 violins is $30P_1$. Therefore, the level of 30 violins relative to the level of one is

$$\Delta L=10 \log(30P_1/P_1)=10 \log 30=14.8 \text{ dB},$$

and the sound pressure level of 30 violins is $70+14.8=84.8$ dB SPL.

The Ensemble Average

The violins of Example 8 serve to introduce a further concept in signal averaging, the ensemble average. Some of the ideas in this discussion depend on facts from later chapters, notably Fourier Series and beats. The reader may wish to defer the present section until later.

To continue with the violins, suppose there are only two of them, playing in unison. Further, let us consider only the fundamental components of the tones, which are cosine waves. If the two unison violin tones differ by only 1 Hz, then the fundamentals will beat once per second. Over the course of 1 second the intensity will be maximum and minimum, but the average over that 1 second will be correctly given by the sum of the intensities of the two violins, just like the assumption in Example 8.

Suppose, however, that the violins, still playing 1 Hz apart, play a tone that is considerably shorter than 1 second. There is then no opportunity for a complete beat cycle to occur and the actual intensity of the fundamental component will depend on the difference in starting phase for the fundamentals from the two instruments. If the phase difference is small, the average intensity will be greater than the sum of the individual intensities. If the phase difference is close to 180 degrees, the average intensity will be less than that sum. Unless one wants to deal with the starting phases of the instruments there is no way to know what will happen to that short tone.

Our ignorance concerning the actual intensity of the short tone can be made to appear less serious by the fact that the two violinists will play the passage with the short tone more than once. Over the course of numerous rehearsals and concerts they will play it often, and across those occasions the phase difference will be a random variable, uniformly spread over the entire range from 0 to 360 degrees. Therefore, although it has not proved possible to calculate an average over time for any single instance of the short tone, it is possible to calculate an average over different performances of that short tone. This is the concept of the ensemble average. The collection of different performances of the tone is the ensemble. Because the random phase difference causes the cross term in the intensity to vanish, the ensemble average of the intensity of the short tone is equal to the sum of the intensities of the contributions from the individual violins, just as calculated in Example 8.

The remainder of this section describes mathematically what has been said above about the ensemble average: The two tones that are summed are $x_1(t)=A_1\cos(\omega_1 t+\phi_1)$ and $x_2(t)=A_2\cos(\omega_2 t+\phi_2)$. Their sum is $x_3=x_1+x_2$. The instantaneous power in the sum is

$$P_3(t)=A_1^2\cos^2(\omega_1 t+\phi_1)+A_2^2\cos^2(\omega_2 t+\phi_2)+2A_1A_2\cos(\omega_1 t+\phi_1)\cos(\omega_2 t+\phi_2). \quad (3.51)$$

Even in a short tone there are lots of cycles of signals x_1 and x_2. Therefore, it is possible to do a time average of the first two terms in the equation for P_3. The averages are, of course, simply $\overline{P_1}$ and $\overline{P_2}$. The cross term can be rewritten in terms of sum and difference frequencies by using a trigonometric identity from Appendix B, whereupon the time average power becomes

$$\overline{P_3}=\overline{P_1}+\overline{P_2}+A_1A_2\,\overline{\cos[(\omega_1+\omega_2)t+\phi_1+\phi_2]}$$
$$+A_1A_2\,\overline{\cos[(\omega_1-\omega_2)t+\phi_1-\phi_2]}. \tag{3.52}$$

There are many cycles of the component at frequency $\omega_1+\omega_2$, and this component averages to zero because the average value of the cosine function is zero. The problem arises with the component at frequency $\omega_1-\omega_2$, which varies only slowly if the two frequencies are similar. The time average then depends on the phase difference $\phi_1-\phi_2$. It is at this point that we have recourse to the ensemble average, typically notated by the symbol $\langle ... \rangle$. The ensemble average power is given by

$$\langle\overline{P_3}\rangle=\overline{P_1}+\overline{P_2}+\langle A_1A_2\,\overline{\cos[(\omega_1-\omega_2)t+\phi_1-\phi_2]}\rangle. \tag{3.53}$$

The ensemble average is taken over all possible values of the phase difference $\phi_1-\phi_2$, and the cosine function averages to zero. As a result, we finally have

$$\langle\overline{P_3}\rangle=\overline{P_1}+\overline{P_2}. \tag{3.54}$$

Thus, the ensemble average power is found to be the sum of the powers of the two violins.

Psychoacoustics Experiments and Frozen Noise

The concept of ensemble average has obvious application to psychoacoustical experiments, where the results are based on many separate trials. For example, in an experiment to study the detection of a signal in noise, there are trial-to-trial variations in the relative phase between the signal and noise components with frequencies near the frequency of the signal. These variations lead to intensity variations like those in Eq. (3.52). If the signal and noise are uncorrelated, one expects that the average over the ensemble of all the trials will result in the cancellation of these variations and that the ensemble average intensity will be the sum of the intensity of the signal and the intensity of the noise, just as in Eq. (3.54). Such a lack of correlation is guaranteed if the noise and signal sources are functionally independent, for example if one uses an analog noise source. If, however, the signal and noise are generated by a computer, it may happen, either by choice or by careless programming, that the signal and noise are not phase randomized.

Deliberately choosing a noise that is phase correlated with the signal is known as a "frozen noise" experiment. Hanna and Robinson (1985) performed a series of experiments with a specific 150-ms sample of noise. A 100-ms sine signal was added to the noise, temporally centered in it. The threshold for detecting the signal was found to depend on the starting phase of the signal, as shown in Fig. 3.6 below.

Hanna and Robinson actually used ten different samples of 150-ms frozen noise, leading to nine more plots like Fig. 3.6. A systematic variation in threshold versus signal phase was evident in all the plots. The overall 10-dB variation shown here was typical, though a few plots showed overall variations less than 5 dB. For all noise samples, the variation was approximately a sinusoidal function of the starting phase, but the patterns were naturally different for different particular frozen noise samples.

The explanation for these masking data is that the signal adds linearly to components of the noise with frequencies in a narrow band around the signal frequency. This narrow band can be thought of as a noisy sine tone, and such a tone will have a particular phase. Adding a signal in phase or out of phase with the ''noise-tone'' can make a big difference in the detectability of the signal. The phase of the noisy sine depends on the haphazard amplitudes and phases of the noise components in the narrow band of the frozen noise. It is different across different exemplars of frozen noise. The strength of the noisy sine vector is also random across different samples of frozen noise. For some noise samples it is large, leading to a large threshold variation; for others it is small.

SPECTRAL DENSITY AND SPECTRUM LEVEL

Previous sections have considered the power'and the intensity of signals with discrete spectral components. Each component could be identified by its frequency

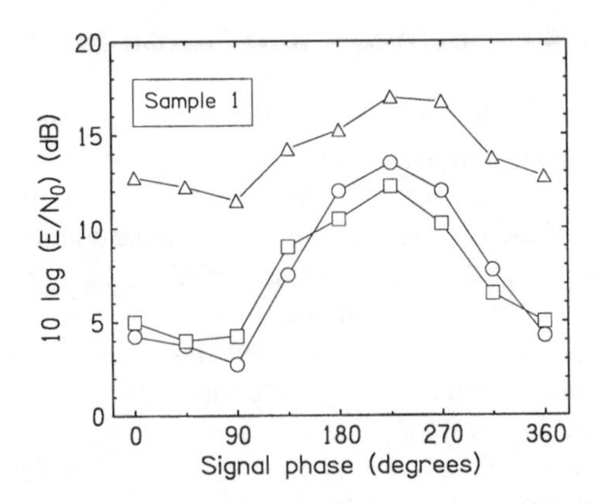

FIGURE 3.6. Signal thresholds versus signal starting phase measured by Hanna and Robinson. Different symbols show the results for three different listeners. The thresholds are given in terms of (E/N_0), which is the ratio of the energy in the signal pulse to the noise spectral density. These thresholds are plotted in dB on the vertical axis. The starting phases are given in degrees along the horizontal axis.

and amplitude. When dealing with thermal noise we need to modify our perspective slightly because thermal noise is not made up of discrete frequency components. Instead it is made from a continuum, i.e. from components that are continuously distributed in frequency. The continuum leads to the concept of the power spectrum or spectral density.

The power spectrum, symbolized by N_0, is a density. It is measured by the amount of power, expressed in watts, in a band that is 1-Hz wide. Therefore, the units of N_0 are watts/Hz. This concept is analogous to the density of matter. Within a homogeneous block of material the matter is continuously distributed. The density of matter is measured by the amount of mass, expressed in grams, in a sample that is 1 cubic centimeter (cc) in volume. Therefore, the units of matter density are grams/cc.

In general, the power spectrum is a function of frequency. The total power in a band between a lower frequency f_l and an upper frequency f_u is therefore,

$$\bar{P}=\int_{f_l}^{f_u} df\, N_0(f). \tag{3.55}$$

Figure 3.7 illustrates this point.

There are two other ways in which the power spectrum is frequently quoted. The first is in terms of an RMS voltage-like quantity with units of volts-per-root-Hertz. For instance, if the RMS noise voltage is $v_0=10^{-6}$ volts/$\sqrt{\text{Hz}}$ then the noise power in a one-ohm resistor is

$$N_0=v_0^2/R=10^{-12} \text{ watts/Hz}. \tag{3.56}$$

The second commonly used quantity is called the *spectrum level*. It provides a way to put the power spectrum onto a dB scale. For reasons of computational

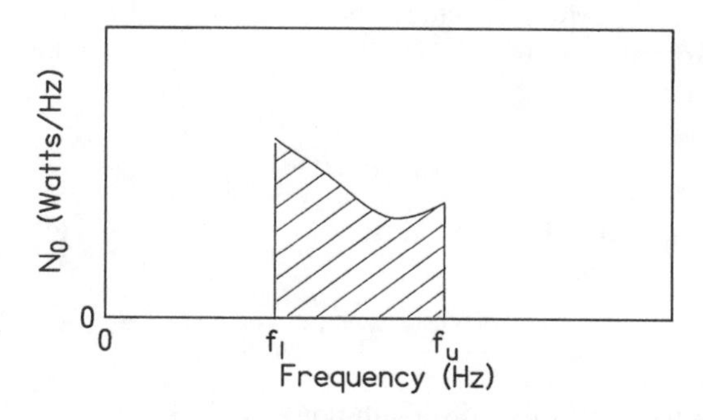

FIGURE 3.7. The plot shows a power spectrum, namely power per unit bandwidth, as a function of frequency. The shaded area is the total power between a lower frequency f_l and an upper frequency f_u.

convenience the reference density is taken to be 10^{-12} watts/Hz. Therefore, the spectrum level in dB is given by

$$L_0 = 10 \log(N_0/10^{-12}), \tag{3.57}$$

where N_0 is the density of the noise in units of watts per Hz.

White Noise

If the power spectrum is constant, independent of frequency, then the noise is called ''white,'' in analogy to white light which contains photons of all visible frequencies. In fact, the analogy to white light is something of an understatement. The frequencies of photons from the visible spectrum span a range that is less than an octave (from red $\approx 4\times10^{14}$ Hz to violet $\approx 7\times10^{14}$ Hz). Therefore, it does not take a large relative bandwidth to get light that is white. By contrast, the range of audible frequencies (20 Hz to 20,000 Hz) is about 10 octaves ($2^{10} = 1024 \approx 1000$).

It is evident that no noise can be truly white because, according to Eq. (3.55), noise with unlimited bandwidth would have infinite power. It is possible, however, for noise to be white over a given frequency band, from a lower frequency to an upper frequency. In that case, the integral in Eq. (3.55) for the total power is easy to do, and the result is

$$\overline{P} = N_0(f_u - f_l) = N_0(BW), \tag{3.58}$$

where BW is the bandwidth over which the power spectrum is constant, $(BW = f_u - f_l)$.

EXAMPLE 9: *Calculation of spectrum level for white noise*

This example shows how to calculate the spectrum level for the special case of white noise. According to Eq. (3.58), the spectral density of white noise is given by

$$N_0 = \overline{P}/(BW). \tag{3.59}$$

Therefore, the spectrum level in dB is

$$L_0 = 10 \log\left[\frac{\overline{P}/(BW)}{10^{-12} \text{ watts/Hz}}\right]. \tag{3.60}$$

We recall that the level of a noise with power $\overline{P}$ is

$$L = 10 \log(\overline{P}/10^{-12} \text{ watts}). \tag{3.61}$$

Therefore, the numerical value of the spectrum level is given by

$$L_0 = L - 10 \log(BW). \tag{3.62}$$

This formula means that the spectrum level can be calculated from the sound pressure level, as would be measured on a sound level meter, or on a voltmeter that reads out on a decibel scale, by subtracting ten times the log of the bandwidth in Hertz.

The reader will find additional information on spectral density and the color of noise in Chapter 23 on noise.

OBSERVATIONS ON THE DECIBEL SCALE

It is sometimes said that we measure sound levels on a log scale because of the wide range of intensities over which the auditory system functions. This explanation is not entirely persuasive. There are many quantities that vary over a wide range: The electrical resistivity of different materials at room temperature varies over 25 orders of magnitude, and astronomical distances have a well-deserved reputation for requiring astronomical numbers. But no one uses a log scale for these quantities.

Instead, log scales (or ratio scales) seem to be particularly applied to quantities with perceptual significance. Log scales are used in both acoustics and optics to describe quantities that relate directly to sensations of magnitude. A counter example is the Ph scale for acidity and basicity of chemical solutions. This is a log scale that does not seem to be related to perception.

The American psychophysicist S. S. Stevens pointed out that monetary income is perceptually a ratio scale. Additional annual income of $100,000 would change the lifestyle of most individuals but would hardly be noticed by a large corporation. On the other hand, increasing income by a factor of two (3 dB) would be an important event for both the individual and the corporation. The ratio scale makes the point that it does not necessarily matter how much more income one acquires. What matters is how much more compared to how much one already has. And so it seems to be for sound levels, though, as will be seen in the next chapter, the ratio scale is not really an accurate scale for the percept known as "loudness."

The ubiquitous use of the dB scale in acoustics is a perennial stumbling block for students. The confusion is not confined to students; professionals too have occasional difficulties with the scale. It is worth noting though, that things could be worse. Astronomers classify stars according to their apparent magnitudes, defined such that a factor of 100 in brightness corresponds to a difference of five magnitude units. Further, the scale is backward so that brighter stars have smaller magnitudes. Therefore, the apparent magnitude of an astronomical object is given by

$$M = -2.5 \log(B/B_0), \tag{3.63}$$

where B is the brightness of the object and B_0 is the brightness of an object with magnitude zero. (The star Sirius has a magnitude that is approximately zero and serves as a reference.) Compared to the astronomer's awful magnitude scale, the acoustician's dB scale does not seem so bad.

A NOTE ON METRIC UNITS OF PRESSURE

The term ''MKS'' stands for meter, kilogram, second, which are units for the fundamental physical quantities of length, mass, and time. All units in the science of mechanics (e.g. energy, momentum, torque, pressure, etc.) can be expressed in terms of just these three fundamental units.

An alternative metric system is called ''CGS,'' standing for centimeter, gram, second. These units also represent length, mass, and time. As an exercise in converting between CGS and MKS units we show that the nominal threshold of hearing in MKS units, namely 20 μPa, is equivalent to the CGS measure, 0.0002 dynes per square centimeter. Older papers in the acoustical literature use this CGS terminology.

First, we note that a Pascal is a unit of pressure, namely force per unit area. The unit of force is the Newton (N), and the unit of area is m^2. One Pascal is equal to one Newton/m^2. Therefore 20 $\mu\text{Pa}=2\times 10^{-5}\ \text{N/m}^2$.

The conversion among the basic units is that 1 kg=1000 g, and 1 m=100 cm. Force has dimensions of mass times acceleration, or mass times length divided by time squared. It follows that the CGS unit of force (the dyne) is related to the MKS unit of force (the Newton) by a factor of 10^5, which comes from a factor of 10^3 for mass and a factor of 10^2 for length, i.e.

$$1\ \text{N}=10^5\ \text{dyne}.$$

Because pressure is force per unit area, we need to convert between MKS and CGS units of area to get the denominator right. Area has dimensions of length squared, so this leads to a factor of 10^4,

$$1\ \text{m}^2=10^4\ \text{cm}^2.$$

Finally, starting with $2\times 10^{-5}\ \text{N/m}^2$, we find

$$2\times 10^{-5}\ \frac{\text{N}}{\text{m}^2}\times 10^5\ \frac{\text{dynes}}{\text{N}}\times 10^{-4}\ \frac{\text{m}^2}{\text{cm}^2}=2\times 10^{-4}\ \text{dynes/cm}^2.$$

This is equal to 0.0002 dyne/cm^2, as advertised.

EXERCISES

Exercise 1: The square

A square wave has limits of plus and minus 1. Show that the RMS value is 1.

Exercise 2: The triangle

A triangle wave has a maximum value of 1 and a minimum value of -1. Show that the RMS value is $\sqrt{1/3}$.

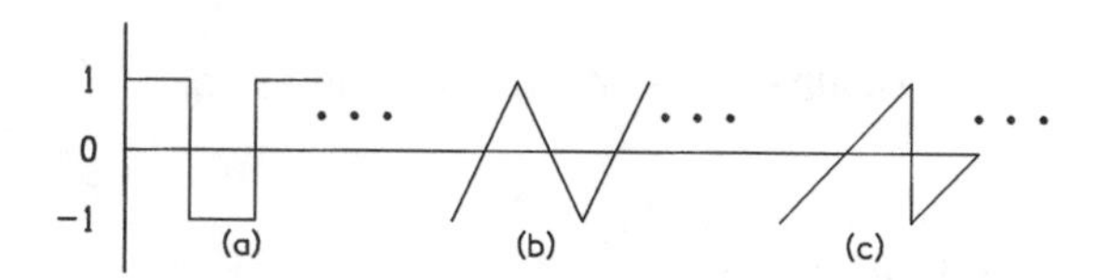

FIGURE E.2. (a) A square wave for Exercise 1. (b) A triangle wave for Exercise 2. (c) A sawtooth for Exercise 3.

Exercise 3: Sawtooth surprise

A sawtooth wave has a maximum value of 1 and a minimum value of -1. Show why its RMS value is the same as for the triangle from Exercise 2.

Exercise 4: Crest factors

(a) Given the answers to Exercises 1, 2, and 3, find the crest factors for a square wave, and triangle and sawtooth waves. Also find the crest factor for a sine wave.

(b) Part (a) shows that crest factor depends on the shape of the waveform. Prove that the crest factor does not depend on a scale factor that multiplies the waveform. To do this, start by proving that if a waveform is multiplied by a constant number then the RMS value is also multiplied by the same constant number.

Exercise 5: The half-wave

The half-wave rectified sine is a sine signal with the bottom cut off, as shown in Fig. E.5. Therefore, the power in a half-wave rectified sine is just half the power in the sine. Show, therefore, that the RMS value of a half-wave rectified sine is 1/2 because the RMS value of a sine is $\sqrt{1/2}$.

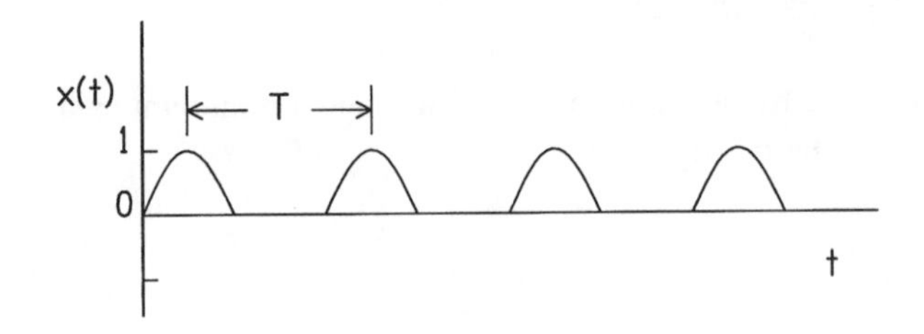

FIGURE E.5. Half-wave rectified sine.

Exercise 6: The pulse

(a) Show that the pulse of Fig. 3.2(b) has zero average value, i.e. the average of x is zero. Show that its average square is unity.

(b) Compare a pulse with a duty factor $p=0.2$ with a pulse that has a duty factor $p=0.8$. Show that these two pulses have the same crest factor, because $0.8=1-0.2$.

Exercise 7: Essence of elegance—proving a lot starting with very little

Many important facts about logarithms can be proved starting with the single fact that the log of a product is equal to the sum of the logs. To be specific, we say

$$\log(AB)=\log(A)+\log(B).$$

Here, and below, it is assumed that A and B are positive real numbers. Use this equation to prove the following:

(a) Prove that $\log(1)=0$, because any number A is unchanged if multiplied by one.

(b) Prove that $\log(1/A)=-\log(A)$, because $A\times(1/A)=1$.

(c) Prove that $\log(A/B)=\log(A)-\log(B)$, using statement (b).

(d) Prove that $\log(ABC)=\log(A)+\log(B)+\log(C)$.

(e) Prove that $\log(A^q)=q\log(A)$, where q is any number.

(f) Prove that $\log(\sqrt{A})=(1/2)\log(A)$, because $\sqrt{A}\times\sqrt{A}=A$.

Exercise 8: Logarithm conversions

Logarithms with base 10 are used to create the dB scale. These are the so-called "common logs." There are also natural logs, where the base is the number e, where

e is approximately 2.71828. To convert from logs in one base to logs in the other simply requires multiplying by a constant. It is not hard to find that constant. Suppose that the common log of number A is given by x_c, and that the natural log of number A is given by x_n. Because of the definition of the logarithms we have two expressions for A,

$$A = 10^{x_c} = e^{x_n}.$$

This equation gives us what we need, a relation between x_c and x_n, but not in a very practical form. We get a better form by taking the logs (common logs) of both sides.

$$\log(10^{x_c}) = \log(e^{x_n}),$$

or

$$x_c = x_n \log(e).$$

(a) Use a calculator to show that this gives the formula,

$$x_c \approx 0.4343 x_n.$$

Thus the scale factor for converting natural logs to common logs is just the common log of the number e. The common log of any number is equal to the natural log of that number multiplied by 0.4343.

(b) The natural log is the basis of a logarithmic scale called the ‘‘neper’’ scale, sometimes used in electrical engineering. If r represents a voltage ratio, then the relative level in nepers is given by

$$L(\text{nepers}) = \ln(r),$$

where ln is the natural log. This can be contrasted with the decibel scale,

$$L(\text{dB}) = 20 \log(r).$$

Use the two equations above to show that 1 neper is approximately 8.7 dB.

Exercise 9: Be a logarithm wizard

The log of 2 is approximately 0.30103. Normally we further approximate this by saying $\log(2)=0.3$. Knowing the log of 2, explain why you then know that $\log(4)=0.6$, and $\log(8)=0.9$, and $\log(5)=0.7$.

Exercise 10: Measure of attenuation

Units of decibels are frequently used when describing attenuation. For instance, a low-pass filter with a cut-off frequency of 1000 Hz may attenuate a 2000-Hz sine tone by 12 dB. Or the loss of light intensity in an optical fiber may be quoted as 3 dB per kilometer. Why doesn't the strength of the input signal matter in these descriptions of loss? What is the basic assumption about a transmission system when one uses dB units to measure loss? Is this assumption identical to the assumption that the transmission system is linear?

Exercise 11: Ocean absorption for low-frequency tones

The attenuation of sound in the ocean due to absorption is small at low frequency. Baggeroer and Munk (1992) report an attenuation of 0.5×10^{-3} dB/km at 57 Hz. Their experiment used a huge path length, 18,000 km. Show that the absorption loss over this path was only 9 dB.

This experiment, the Heard Island test, was part of a continuing study of sound propagation over long distances in the ocean. Transmitters and receivers, separated by half the globe, are placed in the SOFAR channel, a region of ocean depth where the combination of salinity and temperature leads to a minimum in the speed of sound. Sound is refracted back into the channel both from above and below. A long-term goal of this study is to measure global warming by precise measurements of the speed of sound in the ocean. The speed changes with temperature by 4.6 m/sec per centigrade degree. The appeal of the idea is that one expects local fluctuations to be averaged out if one uses the entire ocean as a thermometer. Any temperature change that is detected should be truly global.

Exercise 12: dBm versus dBv

Show that for a given voltage, the difference between dBm and dBv values is simply 2.22 dB, i.e.

$$L(\mathrm{dBm}) - L(\mathrm{dBv}) = 2.22.$$

Exercise 13: Impedance of air

Given that the density of air is $\rho=1.21$ kg/m^3 and that the speed of sound is 343 m/s, show that the specific acoustical impedance of air is 415 MKS units, i.e. 415 Rayls.

Exercise 14: The Voyager

In 1986, two pilots, Richard Ruttan and Jeanna Yeager (no relation to Chuck) flew nonstop around the world without refueling. The trip took nine days. Their specially constructed airplane, the Voyager, had a minimal cabin for the pilots with little isolation from engine noise. Sound pressure levels *re* 20 μPa were measured in the cockpit area. The levels in octave bands are given in the table from Stephenson *et al.* (1990),

31.5 Hz	100 dB
63.0 Hz	108 dB
125 Hz	107 dB
250 Hz	100 dB
500 Hz	95 dB
1000 Hz	90 dB
2000 Hz	85 dB
4000 Hz	82 dB
8000 Hz	73 dB

(a) Add up the power contributions from all the octave bands to show that the unweighted sound pressure level (*re* 20 μPa) was 111 dB. Because so much of the sound energy was low frequency, an A-weighted reading was considerably less.

(b) Use the A attenuation curve of Fig. 3.5 to show that the A-weighted level was 97 dBA.

Although both pilots experienced considerable temporary threshold shift (TTS of 30 dB for some frequencies), neither experienced any permanent loss of hearing. The Voyager is now in the Smithsonian Air and Space Museum in Washington, DC.

Exercise 15: Atmospheric pressure

Atmospheric pressure at the surface of the earth varies from day to day and from place to place. An approximate average is the standard atmospheric pressure, taken to be exactly equal to the pressure of a column of mercury 76 cm high (76 cm=29.92 inches). This pressure is equal to 1.01325×10^5 Pa.

(a) The pressure of 10^5 Pa is defined as a pressure of *one bar*. It follows that standard atmospheric pressure is 1.01325 bar or approximately 1013 millibars. Show that the threshold of hearing (20 μPa) corresponds to an RMS pressure change of 2×10^{-7} millibar or 2×10^{-10} bar. This means that the auditory system can respond to a pressure variation that is about two ten-billionths of the ambient pressure.

(b) The Lufft company makes an aneroid barometer (measures air pressure without the use of mercury or any other fluid) that hangs on a wall, as a convenient short-term predictor of the weather. The dial is calibrated from a minimum of 989 millibars to a maximum of 1043 millibars. Calculate the sound pressure level (dB SPL) of a sine wave with its peak and valley values equal to the maximum and minimum values on the barometer. [Ans: 159.6 dB]

(c) Hurricane Andrew caused enormous devastation in south Florida in August of 1992. The air pressure at the eye of that hurricane was measured at 926 millibars, which means that at the eye there was a partial vacuum. Calculate the peak sound pressure level (dB SPL) of a pulse where the air pressure changes from the standard value of 1013 millibars to 926 millibars. [Ans: 172.8 dB]

(d) Due to a fault in the helmet of a space suit an astronaut's ears are briefly exposed to the vacuum of outer space. Calculate the peak sound pressure level of a pulse where the air pressure changes from the standard value of 1013 millibars to vacuum. [Ans: 194.1 dB]

Exercise 16: Edison's engine

In 1878 Thomas Edison received a patent on a ''vocal engine'' designed to convert sound energy into useful work. Calculate the level of a tone necessary to light a 60-watt light bulb, assuming that the intensity is collected over an area of one square meter and that power is converted from acoustical to electrical form with an efficiency of 100%. [Ans: 138 dB]
Note: The power in the normal human voice is between 10 and 100 microwatts. Therefore, to light a 60-watt bulb would require collecting the vocal output of about one million people, all talking at once. By contrast, collecting the *heat* output from a single person with normal metabolism would be enough to light the bulb.

Exercise 17: Large loudspeakers

The Mirage Resort in Las Vegas has built a simulated volcano in its front yard (De Togne, 1992). Sound effects for the volcano are mainly generated by broad-band horn-loaded loudspeakers with a maximum electrical power rating of 500 watts. The maximum acoustical output is 127 dB SPL, measured at a distance of 1 meter. These two specifications allow an estimate of the efficiency of the speaker, defined as the ratio of the output acoustical power over the input electrical power. Assume a radiation pattern in which the speaker radiates uniformly into the front half space and not at all into the back half space, and show that the estimated efficiency is 6.3%.

Exercise 18: Big voice in the sky

In 1956 the Baldwin Company installed a 9000-watt (equal to 12 horsepower) audio system in a B26 bomber to test the concept of air-to-ground voice transmission from an altitude of 6000 feet (1.83 km) (Martin *et al.*, 1956a). The system used 240 drivers loaded by three twin-horns. The horns were highly directional so that the acoustical footprint appeared to cover an area of about half a square mile (1.3 square km). (My estimate, using a frequency of 1 kHz, not the authors'.)

(a) Show that the directivity factor of the system was about 32.

(b) Assume that the speaker system was 10 percent efficient in converting electrical power to acoustical, and calculate the peak sound intensity level on the ground, neglecting the absorption and turbulence of the air. (Ans: 88 dB)

(c) In ten words or less, explain the appeal of speaking from the sky using 9000 watts.

Despite the abundance of power and high directionality, experimental tests found only marginal intelligibility on the ground. The degradation was attributed to a combination of absorption by the air (especially at high frequencies), turbulence, multipath interference, and Doppler distortion (Martin *et al.*, 1956b).

Exercise 19: A matter of units

In an otherwise informative article on compression drivers, Michael Klasco (1993) writes, "...at the conductor's position, at full tilt, an orchestra may reach about 1 acoustic watt of power." Use concepts of sound power and intensity to criticize this statement.

Exercise 20: Huh?

One occasionally hears mathematical slang which expresses the spectrum level in units of "dB per Hz." It is a curious expression, which is completely understandable, but at the same time completely illogical.
(a) Using the definition of spectrum level, show that the spectrum level has units of dB. A verbose description is "dB *re* 10^{-12} watts per Hertz," or "dB *re* 1 picowatt per Hertz."
(b) Show that a spectrum level of 30 "dB per Hz" would be equivalent to a spectrum level of 30,000 "dB per kHz." What do you think of that?

Exercise 21: Spectrum level

A common hyperbole used to describe an audio system with a wide bandwidth is that the response extends "from DC to daylight."
Consider a white noise with a spectrum level of 30 dB. Suppose that the bandwidth of this noise really does extend from DC to daylight, i.e. from 0 Hz to 10^{15} Hz. Show that the total power in the noise is 1 Megawatt.

Exercise 22: Parallel DACs

A digital-to-analog converter (DAC) includes some noise in its analog output. One can try to decrease the importance of this noise by using several DACs to convert the same digital signal and adding their analog outputs. The Accuphase company (Gardena, CA) makes a compact disc player that uses 16 DACs to convert the signal for each channel. To compare with a typical CD player that uses a single DAC, one can figure that if the noise is independent in each DAC, then the noise power from the 16 DACs is 16 times that from a single DAC. On the good side, however, the signal from the 16 DACs should have 16 times the amplitude of the signal from a single DAC. Compare 16 times the noise power with 16 times the signal amplitude to show that the 16 DACs give a theoretical improvement in signal-to-noise ratio of 12 dB. Whether an improvement of this magnitude is really achieved in practice depends on the independence of the noise in each DAC, a questionable assumption. [In 1994, the list price of the Accuphase CD player was in excess of $20,000 (King, 1994).]

Exercise 23: Wavelength

The vampire bat and the common dolphin emit sounds with frequencies as high as 200 kHz (Guinness, 1990). Calculate the wavelength in each case. Remember that the dolphin lives in sea water where the speed of sound is about 1500 m/s.

Exercise 24: Increment detection

A common psychoacoustical experiment is the increment detection task. One begins with a sine tone, called the pedestal, and increments its amplitude to find the smallest detectable increment. This experiment has sometimes been represented as adding a sine tone to another sine having the same frequency. It is important to realize, however, that the result of the addition is phase sensitive. It is quite possible to begin with a sine tone with amplitude A and to add a sine tone with amplitude B and to end up with a sine tone having amplitude A again. Show that the signal

$$x(t)=A\ \sin(\omega t)+B\ \sin(\omega t+\phi)$$

has amplitude A if

$$\cos\ \phi=-B/(2A).$$

It is evident that adding the sine increment with amplitude B does not change the power in the tone, and the increment cannot be detected.

CHAPTER 4

Intensity and Loudness

The problem with the human observer is that he or she is human.
S. S. STEVENS

Given all the interesting things that one could study about perception of sound, the topic of loudness seems banal. A talker can speak loudly or softly; the radio volume control can be up or down. What's interesting about that? The message or the music are the same.

In actuality, what makes loudness interesting is that it is a window that enables scientists to peer into the encoding of magnitude by the auditory system. Facts learned from loudness studies lead to insights into probable physiological mechanisms of sensory coding. Further, the variation of this magnitude along the tonotopic coordinate is responsible for our ability to recognize different vowels and to perceive different tone colors. As a result, the study of loudness occupies a prominent position in psychoacoustics.

LOUDNESS LEVEL

Sound intensity is a physical quantity; it can be measured with acoustical instrumentation. Loudness, by contrast, is a psychological quantity, and it can only be measured by a human listener, or by something that purports to emulate a listener. On physical grounds alone, one expects that loudness should be different from intensity because the ear does not transmit all frequencies equally, i.e. it does not have a flat frequency response. The outer ear, with its complicated geometry, has a transfer function with half a dozen overlapping resonances ranging from 3 kHz to 12 kHz. These can be measured with a probe microphone in a human ear canal or in a manikin with molded ears (Shaw and Teranishi, 1968, 1979).

The transfer function of the human *middle* ear can also be calculated and measured (Zwislocki, 1962; Wada *et al.*, 1992). It has a maximum near 1 kHz. This result agrees with the transfer function measured between the ear drum and the cochlea by Nedzelnitsky (1980). Together, the outer and middle ear lead to a peaked transfer function with a broad maximum near 3 kHz. Therefore, the magnitude of the input to the nervous system does not depend on intensity alone; the frequency also matters.

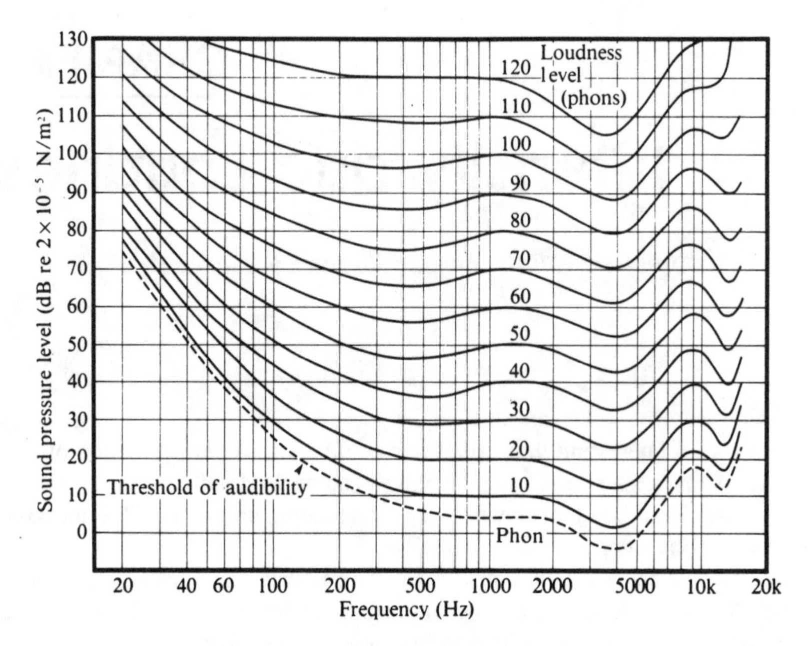

FIGURE 4.1. Equal loudness contours.

Phons

The first step on the road to a scale for loudness is a measure of intensity that compensates for the effect of frequency on the loudness of sine tones. Such a measure is the loudness level, and it is expressed in units of phons [rhymes with Johns, though Fletcher (1953/1995, p. 177) recommended that it be pronounced as "phones" because the unit was invented by Barkhausen, and that's how he would have pronounced it]. By definition, any two sine tones that have equal phons are equally loud. The curves of equal phons are called equal-loudness contours.

To construct an equal-loudness contour, one begins with a standard sine tone having a frequency of 1000 Hz and a level that can be freely chosen. One then changes the frequency, say to 200 Hz, and asks what the level of the 200-Hz tone has to be to make it as loud as the 1000-Hz tone. If, for example, the level of the 1000-Hz sine is 30 dB SPL, then it is found that the level of the 200-Hz sine must be 35 dB SPL for equal loudness. These two points (1000 Hz, 30 dB SPL and 200 Hz, 35 dB SPL) are on the same equal-loudness contour. By definition, this contour is called the 30-phon loudness level because 1000 Hz is always the reference frequency for the phon scale. Equal-loudness contours, recommended by the International Standards Organization, are shown in Fig. 4.1.

The loudness level, expressed in phons, is not a measure of loudness. It is only a frequency-compensated decibel scale. A true measure of loudness would scale with the sensation of magnitude, but there is no reason to suppose that the numbers 10 phons, 20 phons, etc. will do that. In fact, as we will see below, they don't. To find a measure of loudness requires psychophysical techniques that lead to numerical values of perceived magnitudes.

Threshold of Audibility

The lowest equal-loudness contour is the threshold of hearing, measured in free field with a loudspeaker in front of the listener. The minimum near 4000 Hz is caused by the resonances of the outer ear. Thresholds measured with headphones are different in this frequency region because they modify the outer ear geometry. Also, when headphones are used, thresholds are higher at low frequencies because the occlusion effect leads to an increase in low-frequency noise, perhaps caused by blood flow (Yost, 1994). The threshold of hearing involves extremely minute physical displacements. The threshold for a 1-kHz tone corresponds to an eardrum displacement less than the diameter of a hydrogen atom ($\approx 0.5\times 10^{-10}$ m).

If, as is generally supposed, the strong frequency dependence of the threshold curve is caused by filtering by the outer and middle ears, then the flattening of the equal-loudness contours with increasing intensity requires some form of compensating neural response. This is because the peripheral filtering can be expected to apply equally at all levels (until the acoustic reflex is activated). One possibility is that there is an upward spread of excitation on the basilar membrane that occurs for frequencies below 1 kHz. As the pattern spreads, more neurons are excited, and the total count rate for the entire system increases. This might cause the sensation of loudness to grow in a way that happens to balance much of the low-frequency peripheral attenuation. An alternative possibility is that the distribution of high- and low-threshold fibers may be different at low frequencies compared to high.

It is also not certain that the frequency dependence of threshold can be attributed entirely to acoustical and mechanical filtering. Even if we knew these physical properties precisely, we would not know the connection between them and the detection of a signal. Between a hypothetical just-adequate velocity and a hypothetical just-adequate displacement there is a difference of 6 dB per octave, because the velocity is the derivative of the displacement (as will be described in Chapter 8). Possibly, mechanical filtering is responsible for only part of threshold elevation at low frequency (Zwislocki, 1965).

DIFFERENCE LIMENS

A difference limen (DL), or just-noticeable difference (JND), is the smallest detectable change in a quantity. The difference limen measures the resolving power of the reception process and thereby determines the amount of information that can be transferred by modulating the particular quantity. If the DL for intensity were so large that a listener could only distinguish between on and off, then auditory

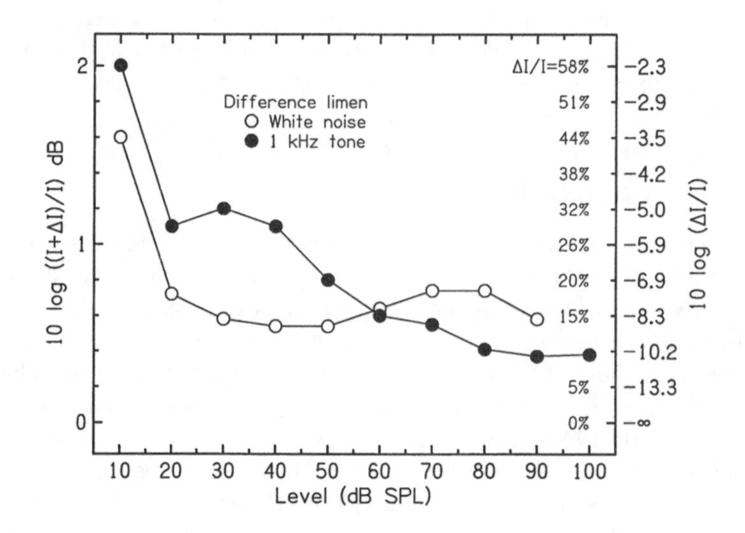

FIGURE 4.2. Difference limens measured by Houtsma *et al.* (1980) for white noise and for a 1-kHz tone. The scale on the left vertical axis shows the DL in decibels (1 tic mark=0.2 dB). There are two scales shown on the right (also with 0.2 dB tic marks). The inner scale gives the percentage intensity change, e.g. 0.6 dB corresponds to a change of about 15%. The outer scale, based on the formula 10 log($\Delta I/I$), is used by some authors. It is logically confusing but has the advantage of spreading the relevant numerical scale. It may be considered a harmless measure, so long as one does *not* append the unit dB to the numerical values.

communication would need to be in a binary format like Morse code. Fortunately, however, the auditory system can distinguish many different intensities over an enormous range.

Measured intensity DLs are generally compared with an idealization known as Weber's law. Weber's law says that the difference limen in intensity ΔI is proportional to the intensity I itself. This is equivalent to saying that $\Delta I/I$ is a constant, or that the just-noticeable increment in intensity is a fixed percentage. Equivalently, the law says that the just-noticeable change in level is a constant number of decibels, no matter what level is used as a starting point.

Values of the intensity difference limen as measured by Houtsma *et al.* (1980) are shown in Fig. 4.2 for white noise and for a 1-kHz tone. The data show that for white noise, Weber's law is rather well obeyed. For levels greater than 20 dB SPL the intensity DL is approximately constant at 0.6 dB. For the tone, however, the intensity DL decreases slowly, but steadily, as the intensity increases. The latter behavior is known as the "near miss to Weber's law" (McGill and Goldberg, 1968). Dozens of different measurements of intensity difference limens reveal that the pattern shown in Fig. 4.2 can be generalized. When the stimulus is broad band, so that all the tuned elements of the auditory system are engaged, Weber's law holds rather well. When the stimulus is narrow band, one sees the near miss.

Actual values of the DL depend on the experimental technique. The DLs in Fig. 4.2 were obtained by finding the error made by listeners in a two-interval (each 500 ms) loudness matching task. Difference limens measured in a two-interval forced-choice discrimination task turn out to be somewhat higher; DLs from an increment-detection task come out to be somewhat lower (Viemeister, 1988). But although the precise values depend on method, the conclusions remain the same: Weber's law holds for broad-band stimuli; the near miss holds for narrow-band stimuli.

The psychoacoustical study of difference limens has had particular impact on physiological modeling. The current paradigm, as described by Viemeister (1988), is that each tuned region of the auditory system obeys Weber's law. For a broad-band stimulus, such as white noise or a click, Weber's law is observed for the auditory system overall because each tuned region of the system individually obeys Weber's law. But as the level is decreased below 20 dB SPL, the excitation approaches the noise floor in the less-sensitive frequency regions of the auditory system. This causes the difference limen to increase (Fig. 4.2) because fewer neurons contribute information.

For a narrow-band stimulus, such as a sine tone or a one-third-octave band of noise, discrimination improves with increasing level, contrary to Weber's law, because the excitation pattern spreads with increasing intensity. The additional neurons that begin to fire improve the counting statistics and decrease the DL. This explanation for the near miss to Weber's law makes further predictions. It predicts that Weber's law should hold better for high-frequency tones than for low because the opportunity for upward excitation pattern spread is less. It also predicts that if high-frequency masking noise is added above the frequency of a test tone then Weber's law should hold. These predictions have been borne out by experiment (e.g. Florentine *et al.*, 1987; Zwicker and Fastl, 1990, p. 175).

Although the above model seems to be consistent with DL data, it is not at all clear why Weber's law should hold in narrow tuned regions. In fact, until recently it was thought that Weber's law could not possibly hold in a physiologically tuned region over the huge dynamic range (100 dB) for which it is observed to hold psychophysically. The problem was that individual auditory nerve fibers have a dynamic range (from threshold to saturation) that is less than 50 dB. This physiological dynamic range problem was solved, in principle, by the discovery of fibers with high thresholds. Liberman and Oliver (1984) have found small diameter fibers that have thresholds as much as 90 dB above the thresholds of larger diameter fibers. Just how neural counts from high-threshold fibers are combined with counts from low-threshold fibers to encode signal intensity is unknown, but the idea that Weber's law might hold over each little patch of the tonotopic coordinate no longer seems inconceivable.

PSYCHOPHYSICS

When you go to the emergency room with your heart attack, the staff will ask you to describe your pain by giving it a numerical value. This is an example of magnitude estimation, a psychophysical method in which observers report the magnitude of a sensation by assigning a number. Psychophysics itself may be said to be the study of the relationship between the magnitude of sensation and the magnitude of a stimulus as measured in conventional physical units. The assumption of the magnitude estimation procedure is that observers are able to make reliable reports of their sensations on a numerical scale. Many different stimuli, from the brightness of lights to the smell of coffee, and from the roughness of sandpaper to the sweetness of sucrose solution, have been scaled in this way (Stevens, 1971).

A magnitude estimation experiment for loudness presents listeners with a series of stimuli having the same relative spectrum but different intensities. The listener assigns numbers on any scale he chooses. After many trials it is found that the numerical estimates settle into a regular monotonic relationship with stimulus intensity. Such experiments show that equal ratios of intensities lead to equal ratios of numerical estimates. In other words, the estimates are proportional to a power of the intensity. This result agrees with previous numerical estimates of loudness based on a task in which a listener is asked to adjust a stimulus to be twice as loud, or half as loud, as a standard (Richardson and Ross, 1930).

If the estimates are regarded as a measure of psychological magnitude Ψ, then the power law is

$$\Psi = kI^p, \tag{4.1}$$

where I is the intensity in watts per square meter and k and p are constants. The experimental analysis takes the logarithms of the magnitude estimates, from Eq. (4.1)

$$\log \Psi = \log k + p \log I. \tag{4.2}$$

Because the signal level in dB is $L = 10 \log(I/I_0)$,

$$\log \Psi = \log(kI_0^p) + (p/10)L. \tag{4.3}$$

This equation says that the log of the loudness estimates should be a linear function of the level in dB. The slope of the line gives the exponent in the power law.

Loudness of Broad-band Stimuli

Loudness grows somewhat differently for broad-band stimuli and for narrow-band stimuli. This section begins with broad-band noise. Figure 4.3 shows magnitude estimates from 64 subjects for 9 different noise levels over a range of 40 dB (Hartmann, 1993a). Different parts of the figure test different possible psychophysical laws. Part (a) tests the linear law, loudness proportional to sound pressure. The pronounced curvature of this plot shows that the law fails. A plot of loudness versus intensity is curved in the same way but more so. Part (b) tests the logarithmic

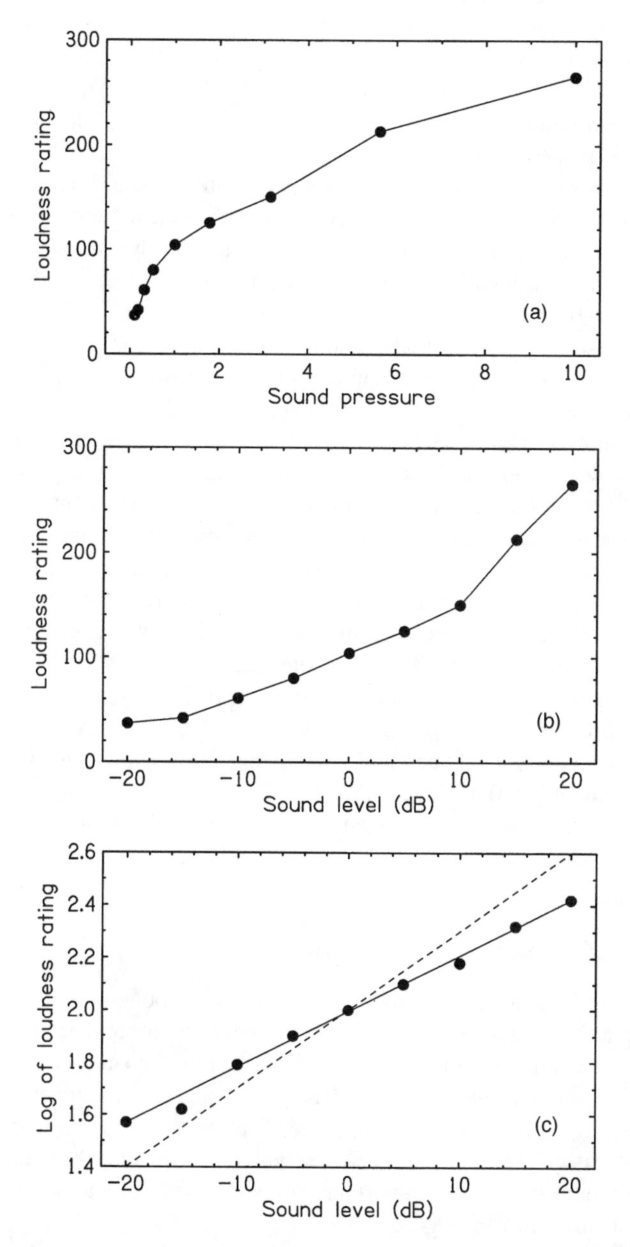

FIGURE 4.3. Loudness ratings for broad-band noise averaged over 64 listeners plotted in three ways: (a) Loudness versus sound pressure, obtained by inverting $L=20 \log(p)$. (b) Loudness versus level in dB. (c) The logarithm of loudness versus level in dB. The slope of the line in (c) establishes the exponent in the power law, 0.22. For comparison, the dashed line has a slope of 0.3.

law, which says that loudness should be proportional to level as measured on the decibel scale. This plot is also curved, but in the opposite way. Finally, part (c) tests the power law. It is evident that the power law gives the best straight line. The slope of this line corresponds to an exponent $p=0.22$. Data from Houtsma and from Bilsen (personal communications, 1993) using the same stimuli led to exponents of 0.22 and 0.25, respectively.

The ultimate broad-band signal is ''uniform exciting noise.'' This is a noise that has been filtered so that equal power occurs in each critical band (see Chapter 23). Experiments using uniform exciting noise over a wide dynamic range show that loudness grows as a negatively accelerated function of level with an asymptotic exponent of $p=0.23$ (Zwicker and Fastl, 1990, p. 185).

There is reason to believe that the magnitude estimation procedure underestimates exponent p. A general psychophysical rule, known as the regression effect, says that observers compress the scale over which they have control in cross-modality matchings (Marks, 1974). In estimating loudness, listeners have control of the numerical scale (vertical axis), and compressing it leads to a value of p that is too small. To compensate, the magnitude estimation experiment can be supplemented with a magnitude production task, where the listener has control over the level of the sound. In a production experiment, the listener's task is to adjust the level to correspond to numbers given by the experimenter. Normally, these numbers are chosen to be in the range that the listener previously gave in a magnitude estimation task. This time, the regression effect causes the listener to compress the horizontal scale and to overestimate the exponent p. Finally, the exponent is taken to be the average from magnitude estimation and production experiments. The loudness experiments of Stevens and Greenbaum (1966) found $p=0.27$ in magnitude estimation and $p=0.33$ in magnitude production for an average of $p=0.3$. It is possible that this value is somewhat too large because their noise band only extended to 2 kHz.

Loudness of Tones

Whereas broad-band signals exhibit a loudness exponent of about 0.25, one would expect the loudness to grow more rapidly for tones because the upward spread of excitation with increasing level gives tones an additional process for increasing neural count rate. In the 1920s Fletcher measured loudness by balancing two-ear listening with one-ear listening (see Fletcher, 1953). He found an exponent $p=1/3$ for levels greater than 40 dB SPL. Zwicker and Fastl (1990), using loudness doubling and halving methods, report an exponent of 0.3 for a tone, to be compared with 0.23 for uniform exciting noise.

Most important is the growth of loudness for a 1-kHz tone which sets the standard for the conventional scale of loudness. Taking the average of many experiments led Stevens (1955) to the value that is most often quoted in the literature, $p=0.3$. Whether or not this value is exactly right, it is certainly convenient. It predicts that in order to double the loudness of a 1-kHz tone the intensity should be increased by 10 dB. By the definition of loudness level, increasing the loudness level of a tone with any frequency by 10 phons should double its loudness.

EXAMPLE: ***Find the relationship between the level change required for loudness doubling and the exponent p in the psychophysical law***

If loudness Ψ_2 is twice Ψ_1, then

$$\Psi_2/\Psi_1 = 2 = (I_2/I_1)^p. \tag{4.4}$$

Taking logarithms, we find

$$\log 2 = p \log(I_2/I_1) \tag{4.5}$$

or

$$0.3 = (p/10)\Delta L \tag{4.6}$$

so that

$$\Delta L(\text{for loudness doubling}) = 3/p. \tag{4.7}$$

For example, if $p = 1/3$ (Fletcher and Steinberg, 1924) then ΔL for loudness doubling is 9 dB.

Actual measured values of loudness for a tone are shown in Fig. 4.4. They are typical of all such measurements in that the loudness follows a compressive power law above 40 dB, but grows more rapidly below 40 dB. The rapid growth of loudness at low levels seen for normal listeners, occurs at higher levels for listeners with hearing impairment. This kind of impairment can be simulated by adding masking noise for normal listeners. Thus, rapid loudness growth is characteristic of low sensation levels, i.e. levels that are barely above threshold.

Other experimental approaches to loudness have given different values of the exponent. By 1972 Stevens actually preferred Fletcher's original value of 1/3. Warren (1977) asked 390 listeners to make one loudness comparison each. He obtained a loudness-doubling level change of 6 dB, corresponding to $p = 0.5$. Although Warren's range of test intensities was far smaller than normally used, the exponent of 0.5 is intriguing because it is reminiscent of the inverse square law for intensity as a function of distance. If loudness judgements actually reflect a listener's estimate of the distance to the source, then such an exponent might be expected. This line of reasoning is complicated by the fact that sound levels in ordinary rooms do not obey the inverse square law.

Support for a large exponent was also found by Schneider (1981), using a method of paired comparisons. For two pairs of tones with four different levels and an overall range from 50 to 100 dB SPL, Schneider asked listeners to decide which pair had the greater loudness change. The strength of this approach is that it

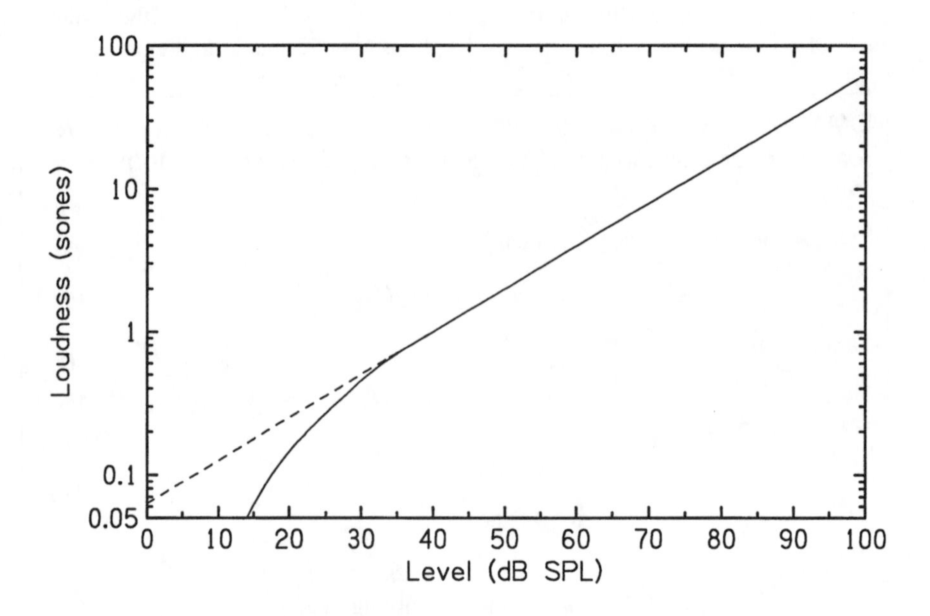

FIGURE 4.4. The measured loudness of a 1000-Hz tone, idealized by Stevens and Davis (1938/1983) and Zwicker and Fastl (1990), is shown by the solid line. It can be compared with a power law having an exponent p=0.3, shown dashed.

involved no cross-modality comparison. Schneider's five listeners gave exponents of 0.22, 0.40, 0.49, 0.52, and 0.56. The average is 0.44, and the s.d. is 0.14.

The Sone Scale

The exponent $p=0.3$ is the basis for the current national and international loudness scale known as the *sone* scale. The reference for this absolute scale is that a 1000-Hz sine tone with a level of 40 dB SPL shall have a loudness of one sone. This reference sets the value of k in Eq. (4.1). Therefore,

$$\Psi(\text{sones}) = \frac{1}{15.849}\left(\frac{I}{I_0}\right)^{0.3}, \tag{4.8}$$

or

$$\log \Psi(\text{sones}) = -\log(15.849) + 0.03 \times 10 \log \frac{I}{I_0}. \tag{4.9}$$

It follows by definition that any tone with a loudness level of 40 phons has a loudness of one sone, and sones can be calculated from phons by the equation

$$\log \Psi(\text{sones}) = -1.2 + 0.03 L_\phi, \tag{4.10}$$

where L_ϕ is the loudness level in phons.

INTENSITY DIFFERENCE LIMENS AND LOUDNESS

A recurring concept in psychophysics is that one ought to be able to estimate sensations of magnitude by counting difference limens. For instance, one would expect that the loudness of a sound could be calculated if it were known that it is a certain number of difference limens above threshold. Thoughts like this led Gustav Fechner to propose that perceived magnitudes should be logarithmic functions of physical magnitudes. In particular, this means that the decibel scale should be a correct measure of loudness.

Fechner's argument went as follows: If loudness can be measured in terms of number of difference limens (DL) then an increase in intensity, dI, should lead to an increase in loudness $d\Psi$ according to the equation

$$d\Psi = dI/(\Delta I), \tag{4.11}$$

where ΔI is the DL for intensity. If the DL is small then changing the intensity should lead to a large increase in loudness, and vice versa. The loudness can then be found by integrating, with primes representing integrated variables,

$$\int_0^\Psi d\Psi' = \int_{I_0}^I \frac{dI'}{\Delta I'}. \tag{4.12}$$

Fechner assumed that Weber's law holds, namely $\Delta I = k_1 I$, where k_1 is constant, so that

$$\int_0^\Psi d\Psi' = \int_{I_0}^I \frac{dI'}{k_1 I'}, \tag{4.13}$$

or

$$\Psi = \frac{1}{k_1} \ln(I/I_0), \tag{4.14}$$

where I_0 is the threshold of hearing, and we have assumed that the loudness at threshold is zero. Because logarithms with different bases are related by simple constants, changing from a natural log to a common log above only changes the constant k_1. Therefore, Fechner's calculation predicted that loudness corresponds to a decibel scale. Unfortunately, this prediction disagrees with every known experiment on the subject.

What was wrong with Fechner's calculation was the original assumption that loudness could be measured by counting the number of difference limens. That assumption required all difference limens to correspond to equivalent changes in perceptual magnitude, which is untrue. An intensity change of one difference limen leads to a greater change in loudness the higher the initial intensity of the signal. Recently, Hellman and Hellman (1990) proposed a variation on Fechner's idea that appears to be highly promising. A condensed version of their argument is described below:

First, it is assumed that loudness is linearly related to a neural count rate, R,

$$\Psi = k_2 R - k_3, \tag{4.15}$$

where k_2 is a scale factor, and k_3 is a reduction that accounts for spontaneous firing rate and threshold. The count rate is taken to be a monotonic function of signal intensity, $R = R(I)$, that is smooth enough that the difference limen in count rate ΔR is related to the difference limen in intensity ΔI by the slope of $R(I)$, i.e.,

$$\Delta R = \frac{dR}{dI} \Delta I. \tag{4.16}$$

The next step is key to the argument. It is assumed that the difference limen ΔR is proportional to the square root of the count rate. This proportionality holds if the neural firing process is Poisson (Teich and Khanna, 1985). Thus,

$$\Delta R = k_4 \sqrt{R}, \tag{4.17}$$

where k_4 is a constant with dimensions of square-root of reciprocal time.

Rearranging Eq. (4.16) leads to

$$\frac{dR}{\Delta R} = \frac{dI}{\Delta I}, \tag{4.18}$$

or, by substituting from Eq. (4.17),

$$\frac{dR}{\sqrt{R}} = k_4 \frac{dI}{\Delta I}. \tag{4.19}$$

If R_0 corresponds to the firing rate at the threshold of hearing, then integrating Eq. (4.19) leads to

$$\int_{R_0}^{R} \frac{dR'}{\sqrt{R'}} = k_4 \int_{I_0}^{I} \frac{dI'}{\Delta I'} \tag{4.20}$$

or

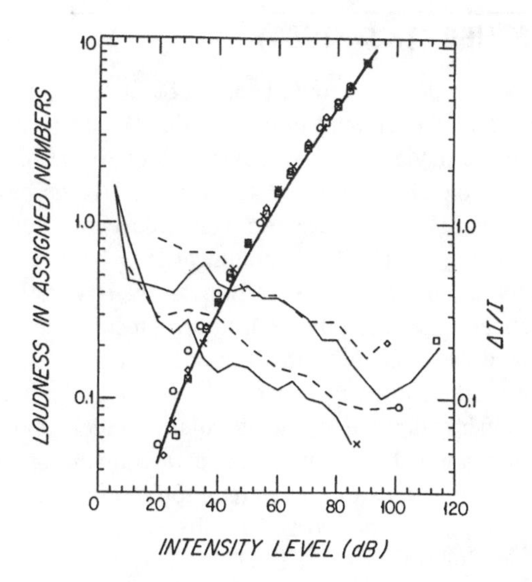

FIGURE 4.5. The measured loudness function (Hellman and Zwislocki, 1963) is shown by the heavy solid line. Loudness functions calculated from difference limen data are shown by symbols that match the symbols for the DL data. The DL data, referenced to the right-hand axis, are: circles for Houtsma *et al.* (1980); diamonds for Florentine *et al.* (1987); crosses and squares for incremented tones and paired comparisons respectively, from Viemeister and Bacon (1988).

$$\sqrt{R} = \sqrt{R_0} + k_4 \int_{I_0}^{I} \frac{dI'}{\Delta I'}. \tag{4.21}$$

Therefore, from Eq. (4.15) loudness is given by

$$\Psi(I) = k_2 \left[\sqrt{R_0} + k_4 \int_{I_0}^{I} \frac{dI'}{\Delta I'} \right]^2 - k_3. \tag{4.22}$$

Hellman and Hellman numerically integrated experimental DLs for a 1000-Hz tone according to Eq. (4.22) to predict the loudness. The results of their calculations are shown in Fig. 4.5. The smoothing caused by the integration is evident. Although the difference limen data are ragged and variable, the calculated loudness plots are smooth and similar. The agreement with the measured loudness data, with asymptotic slope of about 0.3, is very good.

HARVEY FLETCHER (1884–1981)

Harvey Fletcher was born in Provo, Utah on September 11, 1884. From 1908 to 1911, he was a graduate student in the Physics Department at the University of Chicago, where he made important contributions to the oil-drop experiment that first determined the charge on the electron. In 1916 he began work for the Research Department at Western Electric, later to become Bell Labs. Assuming the task of optimizing telephone communications, he took a first-principles approach to speech and hearing, especially hearing. His work featured the first systematic acoustical application of electronics, a technology that was uniquely available at Bell and which changed the face of hearing research forever. Among the concepts that he developed there were the scaling of loudness, equal-loudness contours, loudness summation, the articulation index, and the critical band. His research group applied its experience in the invention of the electronic audiometer and hearing aid, as well as stereophonic sound reproduction. In 1952 he returned to Provo, where he worked on the electronic synthesis of musical sounds. He died in Utah on July 23, 1981.

LOUDNESS SUMMATION ACROSS FREQUENCY

The way that loudness sums across frequency can be illustrated with an example in which we add together two tones that individually have equal loudness. If the tones are well separated in frequency then the loudness of the sum is double the loudness of either one. If the tones are close together in frequency, so that they fall into the same critical band, then the loudness of the sum can be calculated by applying the power law formula to the intensity of the sum. Because the power law is compressive (exponent 0.3), the loudness calculated from the physical sum of the tones (closely spaced tones) is less than the sum of the two loudnesses (well-separated tones).

Figure 4.6 shows the total loudness as a function of the frequency separation between two tones, each of which has a loudness of 4 sones. When the separation is small the tones beat, and listeners judge the loudness based on the peak intensity. This peak is four times the intensity of each individual tone, so that the loudness is greater by a factor of $4^{0.3}=1.5$. The loudness of the beating tone is then the original 4 sones times 1.5, or 6 sones. As the frequency difference exceeds 10 or 20 Hz the beats disappear, and listeners perceive a complex with the intensity that is the sum of the intensities of the two individual tones, giving a loudness of 5 sones ($2^{0.3}=1.23$). When the frequency difference grows beyond a critical bandwidth, the loudness increases to 8 sones, twice the original 4.

The loudness of a signal with a continuous, or quasi-continuous, spectrum behaves according to similar principles: pressures add within critical bands, and loudness values add when the signal occurs in several bands separated by wide gaps. When signal bands are wide and continuous, something between these two extremes occurs. A convincing demonstration of this effect is made by increasing

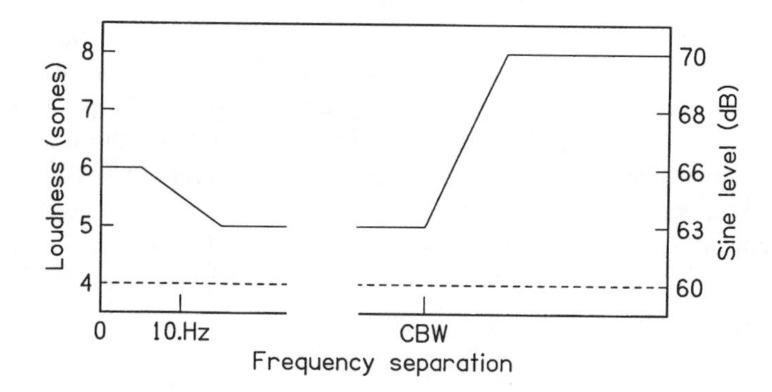

FIGURE 4.6. Loudness of a two-tone signal as a function of the frequency separation between the tones. Each tone has a loudness of four sones. The coordinate on the right gives the level of a single sine tone that matches the loudness of the two-tone signal.

the bandwidth of a noise while reducing the spectrum level to keep the total power constant. So long as the noise bandwidth remains smaller than a critical bandwidth, the loudness depends only on the power and remains constant. As the noise bandwidth grows beyond the critical bandwidth, the loudness starts to increase with increasing bandwidth (Zwicker *et al.* 1957; Scharf, 1970).

The calculation of loudness for continuous bands is made complicated because of interaction effects. Two methods of calculating loudness for broad-band sounds have become international standards. A method by S. S. Stevens (1961) is ISO R532A, and a method by E. Zwicker (Zwicher and Fastl, 1990) is ISO R532B. The former is also an American national standard, ANSI 3.4-1980.

According to the method by Stevens, loudness is calculated by summing loudness indices (S_m) for 26 one-third-octave bands from 40 to 12,500 Hz (see Appendix J for the center frequencies of these bands).

The total loudness is given by the formula

$$\Psi = S_{\max} + F \sum_{m=1, m \neq \max}^{26} S_m , \tag{4.23}$$

where $S_{\max}$ is the loudness index for the loudest band. Factor F is 0.15, which heavily discounts the bands with lesser loudness indices because of mutual interaction effects such as masking. The loudness indices themselves are found by measuring sound pressure levels in each band and then applying an equal-loudness contour, given in the standard, that converts to sone values appropriate for noise bands. An alternative calculation uses octave bands, for which $F=0.3$.

The method by Zwicker is rather similar in practice, but it is based on the idea of summing neural excitation in different bands instead of summing acoustical power.

In Zwicker's method the bands are critical bands on the bark scale. As described in Chapter 10, bands on the bark scale are similar to one-third-octave bands except at low frequency. The method begins with the physical power spectrum and calculates the excitation in each critical band. This depends not only on the power spectrum but also on the slopes of the auditory filters responsible for the critical bands. Excitation spreads upward to higher frequency bands. Therefore, it is quite possible that the excitation in band m due to acoustical energy in band m is overshadowed by the excitation in band m due to acoustical energy in band $m-1$. The method treats this interaction by taking the larger excitation to represent the contribution of band m to the loudness sum.

TEMPORAL INTEGRATION

If we want to know how hard it's raining, we put an empty can outdoors and let it fill with rain. If the level of rain water is 3 mm after 15 minutes, we say that the average rate is 12 mm per hour. This measurement corresponds to a perfect temporal integration of rain drops. The rate of rain R and the height of water in the can h_r are related by the equation

$$h_r = R t_I, \tag{4.24}$$

where t_I is the integration time, in this case 1/4 hour.

If we want to monitor rainfall on a continuous basis we must do something about the fact that our can will soon fill to the top (it's a small can) and no longer measure correctly. Therefore, we drill small holes in the can so that it leaks at a controlled rate. The rate at which the can fills is then given by

$$\frac{dh_r}{dt} = R - l, \tag{4.25}$$

where R is the rainfall rate and l is the leak rate.

To improve the chances of not overflowing we drill holes up the sides of the can so that more holes are leaking as the level grows higher. We arrange the holes so that the leak rate is proportional to the height of water,

$$l = h_r / \tau. \tag{4.26}$$

The constant of proportionality is τ, which has the dimensions of time to get the physical units right. Finally, from Eqs. (4.25) and (4.26), we have a differential equation for the height of water in the can,

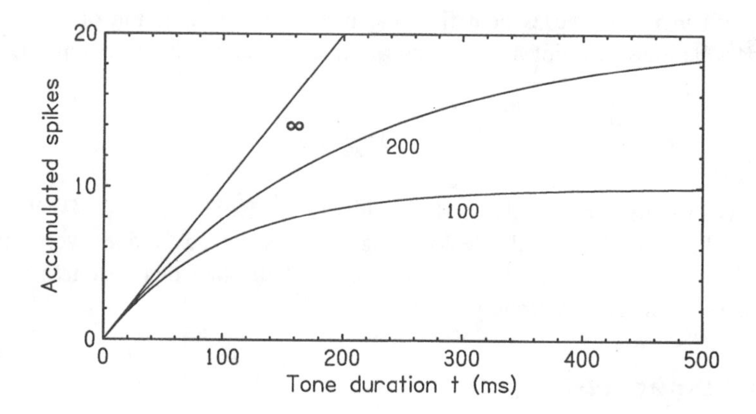

FIGURE 4.7. The number of accumulated spikes in a tone of duration t, given a firing rate $R = 100/s$. Three integration time constants are shown: τ equal to 100 ms, 200 ms, and infinity. A loudness integration model suggests that the percept of loudness grows with tone duration according to curves like these with a time constant possibly depending on signal frequency.

$$\frac{dh_r}{dt} = R - h_r/\tau. \tag{4.27}$$

This has the solution

$$h_r(t) = R\tau(1 - e^{-t/\tau}) + h_{r0}e^{-t/\tau}, \tag{4.28}$$

where h_{r0} is the height of water in the can before we started. If the can is initially empty then

$$h_r(t) = R\tau(1 - e^{-t/\tau}). \tag{4.29}$$

Equation (4.29) describes the output of a system known as a leaky integrator.

Loudness Integration

In 1947, Munson suggested that the auditory system integrates loudness according to a leaky-integrator model. Plomp and Bouman (1959) also employed this model and made the further assumption that the rate to be integrated was constant. This idea is shown in Fig. 4.7, using the assumption that an accumulation of neural spikes represents loudness. Therefore, the response to a sound with duration t is given by a cumulative number of spikes $h_r(t)$.

There are two ways in which Eq. (4.29) can be applied. First, assuming that loudness grows monotonically with neural spike count, as in Eq. (4.15), we can use

Eq. (4.29) to model the evident fact that increasing the duration of a short tone increases its loudness. For a tone duration much less than the time constant ($t \ll \tau$) Eq. (4.29) becomes

$$h_r(t) = Rt, \tag{4.30}$$

which is just the same as the perfect integration given in Eq. (4.24). When the duration becomes longer, the leaky integrator has the advantage over a perfect integration model because it correctly represents the fact that loudness does not continue to increase indefinitely.

Signal Detection

A second application of the leaky-integrator model is to signal detection. The model describes how the level of a barely detectable tone varies with the duration of the signal. Detection threshold is assumed to correspond to a fixed number of accumulated spikes, h_{r0}, so Eq. (4.29) gives the threshold firing rate R_0 as

$$h_{r0} = R_0(t)\tau(1 - e^{-t/\tau}), \tag{4.31}$$

where t is the duration of the signal. Then

$$10 \log R_0(t) = \text{constant} - 10 \log(1 - e^{-t/\tau}). \tag{4.32}$$

If it is assumed that firing rate R is proportional to the signal power in watts then the threshold level in decibels is given by

$$L_0(t) = L_0(\infty) - 10 \log(1 - e^{-t/\tau}), \tag{4.33}$$

where $L_0(\infty)$ is the threshold for asymptotically long duration.

Watson and Gengel (1969) measured the level increase for short duration, specifically $L_0(t) - L_0(\infty)$ for several different signal frequencies. As shown in Fig. 4.8 they were able to fit their data, for $t > 16$ ms, with a leaky-integrator model by assuming a frequency-dependent integration time. As found in numerous other studies, reviewed by Gerken *et al.* (1990), the integration time increases for decreasing signal frequency.

The above description of temporal integration is not without controversy. There is obviously a large discrepancy between model and experiment for low frequency at a duration of 16 ms. Possibly this can be attributed to the fact that when tone durations are short some of the energy is splattered outside the range of a single auditory filter. Calculations using Cambridge filter bands from Chapter 10 find that this explanation works for 125 Hz, but does not give a large enough effect at 1000 Hz. Further, not everyone agrees that integration time should be such a strong

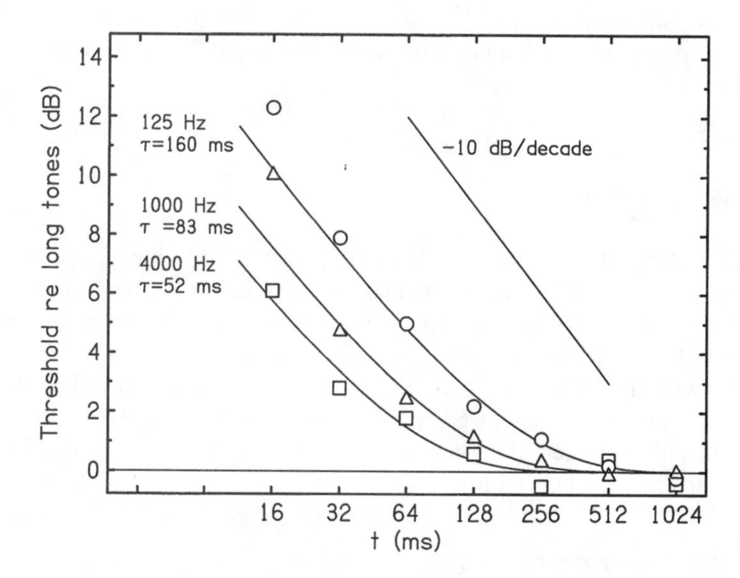

FIGURE 4.8. Symbols show threshold levels $L_0(t) - L_0(\infty)$ measured by Watson and Gengel (1969) for tones of three different frequencies: circles for 125 Hz, triangles for 1000 Hz, and squares for 4000 Hz, as a function of tone duration. The zero line $L_0(\infty)$ was determined by the data at the two longest durations. The straight line labelled "−10 dB/decade" is the prediction of a perfect temporal integration model. Other curves correspond to the leaky-integrator model, with time constants as labelled. Time constants were chosen to fit the data at intermediate durations.

function of frequency (e.g. Florentine *et al.*, 1988). Finally, Viemeister and Wakefield (1991) have challenged the entire concept of integration times that are as long as hundreds of milliseconds. Their experiments on the detection of tone pulse pairs suggest instead a multiple-looks model in which signal power is integrated for only a few milliseconds in a single observation, and the data from successive observations are combined. The multiple-looks model is, however, less appealing as an explanation of the duration dependence of loudness for tones above threshold. Temporal integration occurs over long durations there too (Munson, 1947; McFadden, 1975). The possible role of compression in changing the details of temporal integration is explored in Exercise 10 and in a section of Chapter 20 on modulation transfer functions.

It is possible that work with cochlear impaired listeners will ultimately prove to be useful in finding the right answer. Impaired listeners show anomalously small temporal integration (Carlyon *et al.*, 1990; Zwicker and Fastl, 1990, p. 303). Finally, we note that loudness does not always increase with duration. Sine tones that are weak, 20 dB SPL or less, and constant, decay in loudness over the course of tens or hundreds of seconds. The effect is particularly noticeable at high frequency; it has been thoroughly reviewed by Ward (1973) and Scharf (1983).

EXERCISES

Exercise 1: Hum

(a) Show that a 60-Hz hum with a level of 50 dB SPL has a loudness level of only 20 phons. (b) What is the corresponding loudness in sones according to the 0.3 power law? (Ans. 0.25 sones). (c) Explain why this answer overestimates the loudness. (d) What is the loudness level of a 60-Hz hum with a level of 90 dB SPL? Compare with the answer for part (a). (In the United States electrical power is distributed with a frequency of 60 Hz. Therefore, 60-Hz hum is ubiquitous. Also, the whistles of the Queen Mary have a fundamental frequency of 60 Hz. There are three of them; each weighs a ton.)

Exercise 2: *A*-weighted scale

Figure 3.5 shows the *A*-weighted scale for measuring sound levels, intended to compensate for the external and middle ear. Compare with Fig. 4.1 to check the statement that the *A* scale corresponds to the 40-phon equal-loudness contour.

A-weighting is thought to be appropriate in damage risk assessment, for example in setting noise standards for the factory floor. The low-frequency sounds that the *A*-weighted meter discounts are not only less audible, they are less likely to cause damage to hearing because they are physically attenuated by the external and middle ear before they reach the delicate inner ear (Hellström, 1993). However, there is not complete agreement on this point. Moore, Peters, and Glasberg (1990) have suggested that the high thresholds at low frequency are due to an especially large internal noise at low frequency. To the extent that they are right, the threshold curve underestimates the risk of low-frequency sounds.

Exercise 3: Equal ratios law

Prove that if sensation is proportional to stimulus magnitude raised to a power, then equal ratios of stimulus magnitude lead to equal ratios of sensation, independent of the magnitude of the stimulus. Prove that if sensation were proportional to the logarithm of the stimulus magnitude, according to Fechner's incorrect law, then equal ratios of stimulus magnitude would lead to equal *differences* in sensation, independent of the magnitude of the stimulus.

Exercise 4: Weber's Law

The text says that if $\Delta I/I$ is constant (Weber's Law) then the intensity difference limen is a constant number of decibels. Prove that this is so.

Exercise 5: Definition of difference limens

Psychoacousticians say that an experimentally measured difference limen represents the smallest change in a physical quantity such as the intensity of a tone or noise. They do not say that the DL measures the discrimination of a psychological quantity such as loudness. (a) Explain the distinction and why it matters. (b) Is this distinction compromised in attempts, by Fechner and others, to relate loudness to difference limens?

Exercise 6: Plus ça change, plus c'est la même chose

ANSI standard S3.4-1980 relates sones (Ψ) to loudness level in phons by the equation

$$\Psi = 2^{0.1(L_0 - 40)}.$$

Show that this corresponds to

$$\Psi(\text{sones}) = \frac{1}{16}\left(\frac{I}{I_0}\right)^{0.301}$$

or

$$\log \Psi(\text{sones}) = -1.203 + 0.0301 L_\phi,$$

where L_ϕ is the loudness level in phons. Compare with Eq. (4.10).

Exercise 7: Hellman-Hellman formula

Suppose that Weber's law holds so that $\Delta I = k_1 I$. Show that the Hellman-Hellman formula for loudness gives

$$\Psi(I) = k_2[\sqrt{R_0} + k_5 L]^2 - k_3,$$

where $k_5 = 0.04343 k_4 / k_1$, and L is the level in dB. Therefore, loudness is a parabolic function of level. Could such a function agree with Fig. 4.3(b)?

Exercise 8: Warren's exponent and the perception of distance

Exactly why is a loudness exponent $p = 0.5$ reminiscent of the inverse square law for intensity?

Exercise 9: Plomp-Bouman temporal integration

Suppose that the integration time is τ=200 ms. What duration of tone produces a total neural count that is half the count for a tone of infinite duration? Assume a constant firing rate. (Ans. 139 ms)

Exercise 10: Temporal integration for a power-law rate

Equation (4.33) gives the threshold for a temporal integration model in which firing rate is proportional to signal intensity. Show that if the rate is taken to be a power of the intensity, $R \propto I^p$, then

$$L_0(t) = L_0(\infty) - (10/p)\log(1 - e^{-t/T}).$$

Therefore, if $p<1$, as expected for a compressive system, the slope of the function is steeper. Can this improve the fit to the Watson and Gengel data in Fig. 4.8? [Note: A classic paper by Penner (1978) explored the effect of compressive signal encoding on temporal integration. A difference from the suggestion made in this exercise is that Penner also changed the form of the integration law.]

Exercise 11: Prothetic and metathetic continua

Stevens (1966) drew the distinction between prothetic and metathetic quantities. Prothetic continua are associated with intensity, magnitude, or degree. Metathetic continua are described by a point on a scale. Loudness is prothetic; we say, ''more loudness.'' Pitch is metathetic; we do not say ''more pitch,'' instead we say ''higher pitch.'' It is found that category rating (e.g. scale of 1 to 10) leads to a bias in the description of prothetic quantities whereas it does not for metathetic quantities.

(a) Which kind of quantity is better represented by a bar graph?

(b) Classify the psychological correlates of light intensity, air temperature, humidity, speed, weight, and several color parameters.

(c) Find other examples of prothetic and metathetic quantities.

Exercise 12: Intensity discrimination variable

Buus, Florentine, and Zwicker (1995) find that for both normal and hearing-impaired listeners, the decision variable d' is proportional to ΔL, the change in level in dB, i.e. $10\log[(I+\Delta I)/I]$. Explain how this observation offers support for Weber's Law. (See Chapter 24 for a review of d'.)

CHAPTER 5

Fourier Series

Henceforth a series of new time began,
The mighty year in long procession ran.
JOHN DRYDEN

The Fourier transformation is an idea of enormous theoretical importance and practical significance. It says that a waveform, such as an acoustical signal, can be represented as a sum of pure tones, each with its own frequency, amplitude, and phase. If the waveform is periodic then the constituent pure tones are harmonics, which means that the frequencies are integer multiples of a fundamental frequency. In this chapter we are particularly concerned with periodic functions like that. Such functions can be represented by a special case of the Fourier transformation, the Fourier series.

The operation known as ''Fourier analysis'' is a way to make evident the individual frequency components in a tone. Fourier analysis can be a highly theoretical pursuit. For instance, the exercises at the end of this chapter illustrate how a waveform can be Fourier-analyzed mathematically using pencil and paper. Fourier analysis can also be highly physical. Many objects in the physical world behave as resonators, which means that these objects respond with special enthusiasm to particular frequencies. Such an object performs a Fourier analysis in the sense that it causes particular frequencies to stand out whenever they occur in a waveform that the object encounters. A hollow tube is an acoustical example; a tuned filter made from capacitors and inductors is an electrical example. Even more generally, some form of Fourier analysis is done by any information transmission system because any system acts as a filter, passing some frequencies and attenuating others.

The auditory systems of all mammalian animals, and of many nonmammals, perform a Fourier analysis. The fundamental process by which acoustical vibrations are transduced into neural impulses is tuned like a bank of filters. A single sound, such as a vowel sound, is analyzed into different frequency bands because different neural channels are sensitive to different frequencies.

J. B. J. FOURIER (1768–1830)

Jean Baptiste Joseph Fourier was born in France on March 21, 1768. He learned mathematics in a military school where he subsequently taught. He went with Napoleon to Egypt in 1798, and served briefly as governor of lower Egypt, returning to France in 1803. The mathematical series which bears his name was first announced in 1807. The series expresses a function as a sum of sines and cosines and was invented specifically to solve certain boundary value problems, notably on the conduction of heat. Fourier died in Paris on May 16, 1830.

The converse of analysis is synthesis. The operation known as Fourier synthesis creates complex waveforms by adding together pure tones of particular frequencies, amplitudes, and phases. Contemporary digital signal generation techniques allow an experimenter to create a waveform in just that way.

DEFINITION OF FOURIER SERIES

Consider a function of time $x(t)$ that is periodic with period T. An example is shown in Figure 5.1.

Periodicity means that we can add any integral multiple m of T (i.e. mT) to the running time variable t and the function will have the same value as at time t, i.e.

$$x(t+mT)=x(t), \quad \text{for all integral } m. \tag{5.1}$$

Because m can be either positive or negative and as large as we like, it is clear that x is periodic into the infinite future and past. Then Fourier's theorem says that x can be represented as a Fourier series like this:

$$x(t)=A_0+\sum_{n=1}^{\infty} [A_n \cos(\omega_n t)+B_n \sin(\omega_n t)]. \tag{5.2}$$

The series in Eq. (5.2) expresses function $x(t)$ as a sum of cosines and sines. The reader will note that, except for the constant term A_0, each term in Eq. (5.2) is either a sine or a cosine function. It is important to remember that the series itself is a function of time, not a function of frequency. Time is the running variable in the series and all the other parameters are regarded as constants. All the cosines and sines have frequencies ω_n that are harmonics, i.e. they are integral multiples of a fundamental angular frequency ω_0,

$$\omega_n=n\omega_0=2\pi n/T, \tag{5.3}$$

where n is integer.

The fundamental frequency f_0, is given by $f_0=\omega_0/(2\pi)$. A peculiarity of the notation in Eq. (5.3) is that there are actually two correct ways to write the funda-

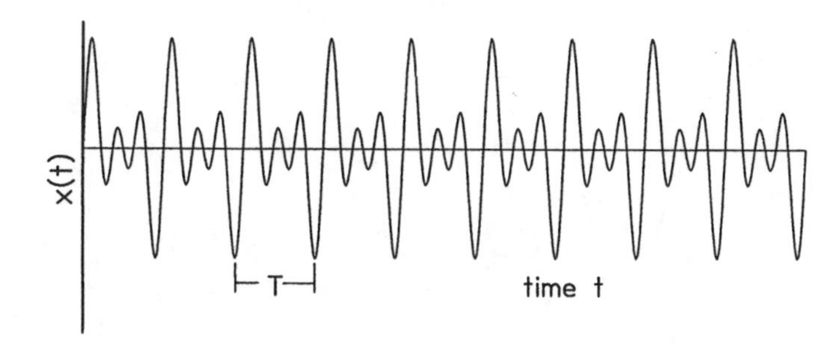

FIGURE 5.1. A periodic waveform with period T repeats itself indefinitely through all past and future time. A part of such a waveform (nine cycles) is shown here.

mental angular frequency, either ω_1 or ω_0. This mathematical oddity results from the fact that for $n=1$, $\omega_1=(1)\omega_0$. Correspondingly, there are two ways to write the fundamental frequency in Hz, f_1 or f_0. The notation f_0 or $F0$ normally denotes the fundamental frequency in published research on speech.

The fundamental frequency is the lowest frequency wave that a sine or cosine wave can have and still fit exactly into one period of the function $x(t)$ because $f_0=1/T$, as shown in Fig. 5.2 below. The fact that only harmonic frequencies can contribute to the series will be proved when we study the Fourier integral. The essential idea is that in order to make a function $x(t)$ which has period T, the only sines and cosines that are allowed to enter the sum are those that fit exactly into the same period T. A sine or cosine with the fundamental frequency fits, of course. So do those sines and cosines with frequencies that are integral multiples of the fundamental. Figure 5.2 shows this concept. After a period of time T has elapsed, the sines in parts (b) and (c) are ready to start over again, as they must, in order to fit with the complex wave in (a). But the sine in part (d) is not ready to begin again after time T. It cannot be a part of the complex wave in (a).

The factors A_n and B_n in Eq. (5.2) are the Fourier coefficients. They give the strength of the contribution of the n-th cosine and sine terms to the series. In principle, an infinite number of terms are required to synthesize the general wave-

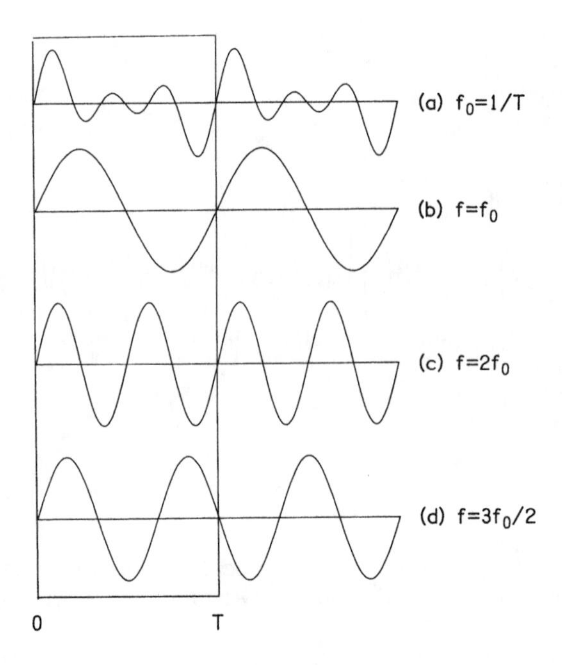

FIGURE 5.2. A complex wave together with sine waves that might possibly be harmonics. (a) Two cycles of the complex wave from Fig. 5.1. (b) The fundamental sine wave has the same period as the complex wave. It fits in the box, which is exactly one period long. (c) A sine that has half the period of the complex wave. It too fits and will be a harmonic (the second) if B_2 is not zero. (d) A sine that has a period that is 2/3 of the period of the complex wave. It does not fit and cannot possibly be a harmonic in the complex wave.

form, but many important waveforms can be exactly characterized by a series with a finite number of terms, and others can be adequately approximated by only a few terms.

Suppose that we have a periodic function $x(t)$ and we wish to find a Fourier series to represent it. Once we know the period of $x(t)$, the only remaining problem is to find the coefficients A_n and B_n. These coefficients can be found by projecting the function $x(t)$ onto trigonometric functions that have the period of T/n.

PROJECTING OF FUNCTIONS

The concept of projecting a function has a formal resemblance to the concept of projecting a vector onto a reference coordinate system. For example, if the vector $v=4\hat{j}+3\hat{k}$ is projected on to the $[\hat{j},\hat{k}]$ coordinate system, we find that the projection of v onto $\hat{j}$ is 4 and the projection of v onto $\hat{k}$ is 3.

To project a function onto another function means to multiply the two functions together and then integrate over the space of interest. For example, in Fourier analysis we consider cos $\omega_n t$ to be a reference function. Then the projection of function x on that reference function is the integral

$$\text{Projection}=\int dt\ x(t)\cos\ \omega_n t, \tag{5.4}$$

where we have left the limits of the integral undefined for the moment, because we are not sure what the space (or range) of interest along the time axis might be.

Orthogonality

Two functions are said to be "orthogonal" if the projection of one on the other is zero. Function $x(t)$ is orthogonal to function $y(t)$ if

$$\text{Projection}=\int dt\ x(t)y(t)=0. \tag{5.5}$$

This is a sensible definition. A vector that is entirely along the $\hat{j}$ axis is orthogonal (perpendicular) to the $\hat{k}$ axis because it has zero projection along the $\hat{k}$ axis.

Two functions are orthogonal if either one of them is zero, but that is not particularly interesting. More important is the idea that two functions are orthogonal if their product is positive as much as it is negative, so that when one does the integral over the space of interest, the positive and negative portions cancel and the integral reduces to zero.

EXAMPLES

The reader might like to verify a few things at this point: First, a function can never be orthogonal to itself because the product is a square and a square can never be negative. Therefore, there can be no cancellation of positive and negative contributions to the integral. Second, a constant, like the number 5, is orthogonal to a cosine function if the space of interest is a part of the time axis with length equal to an integral number of periods of the cosine.

In Fourier analysis we are only interested in functions that all have a particular period T. Therefore, the space of interest is a region of the time axis with duration T, such as the range from $-T/2$ to $T/2$. The reason is that for periodic functions, all the information about the function is contained in a range with duration T and it is superfluous to extend the space of interest outside that range.

There are three important orthogonality integrals in Fourier analysis. These are integrals of sine and cosine functions having the frequencies of harmonics, as given by Eq. (5.3).

The first integral says that the sine is orthogonal to the cosine,

$$\frac{2}{T}\int_{-T/2}^{T/2} dt\ \sin(\omega_n t)\cos(\omega_m t) = 0, \tag{5.6}$$

for all m and n. This is shown in Fig. 5.3(a) for the special case of $n=m=1$, namely for the fundamental.

The next orthogonality integral says that two cosines are orthogonal, so long as they have different frequencies.

$$\frac{2}{T}\int_{-T/2}^{T/2} dt\ \cos(\omega_n t)\cos(\omega_m t) = 0 \quad (n \neq m). \tag{5.7}$$

This is shown in Fig. 5.3(b) for the special case that $m=1$ and $n=2$, first and second harmonics.

The third orthogonality integral says that two sines are orthogonal, so long as they have different frequencies,

$$\frac{2}{T}\int_{-T/2}^{T/2} dt\ \sin(\omega_n t)\sin(\omega_m t) = 0 \quad (n \neq m). \tag{5.8}$$

This is shown in Fig. 5.3(c) for the case that $m=1$ and $n=3$, first and third harmonics.

Normalization

The above orthogonality integrals take care of all the interesting integrals for sine and cosine functions except for the case where the two functions being integrated are actually the same. Then we have the projection of a function on itself. There are two such projections,

$$\frac{2}{T}\int_{-T/2}^{T/2} dt\ \cos(\omega_n t)\cos(\omega_m t) = 1 \quad (n = m \neq 0), \tag{5.9}$$

and

$$\frac{2}{T}\int_{-T/2}^{T/2} dt\ \sin(\omega_n t)\sin(\omega_m t) = 1 \quad (n = m \neq 0). \tag{5.10}$$

It is common to define projection integrals so that the projection of a function on itself is normalized to unity, as in Eqs. (5.9) and (5.10). That is the reason for the prefactors $2/T$ in all the integrals above.

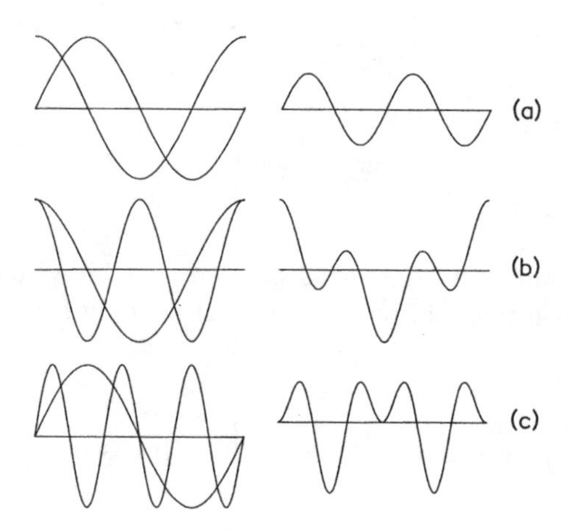

FIGURE 5.3. Three cases of orthogonal functions. (a) The fundamental sine is orthogonal to the fundamental cosine. (b) The fundamental cosine is orthogonal to the second harmonic cosine. (c) The fundamental sine is orthogonal to the third harmonic sine. In each case the panel on the left shows the two functions, the panel on the right shows their product. The area under each product is zero.

These facts about the projection of functions are an adequate basis for deriving the formalism of Fourier analysis of periodic functions.

To find the coefficients A_n and B_n in the Fourier series for a function $x(t)$ we project $x(t)$ onto individual cosine and sine functions, respectively. For example, suppose that we have some signal defined by function $x(t)$ and we want to find the contribution of the third harmonic cosine term, i.e. we want to find A_3. We project $x(t)$ onto the function cos $\omega_3 t$, where $\omega_3 = 2\pi 3/T$, i.e. we calculate

$$\frac{2}{T}\int_{-T/2}^{T/2} dt\; x(t)\cos(\omega_3 t). \tag{5.11}$$

We now show that this projection gives us the coefficient A_3, as advertised. We start by making the assumption that Eq. (5.2) correctly represents function $x(t)$. We then use this assumption by substituting Eq. (5.2) for $x(t)$ in Eq. (5.11) to get Eq. (5.12),

$$\frac{2}{T}\int_{-T/2}^{T/2} dt\ x(t)\cos(\omega_3 t)=\frac{2}{T}\int_{-T/2}^{T/2} dt\left\{A_0\cos(\omega_3 t)+\sum_{n=1}^{\infty}[A_n\cos(\omega_n t)\cos(\omega_3 t)\right.$$

$$\left.+B_n\sin(\omega_n t)\cos(\omega_3 t)]\right\}. \tag{5.12}$$

The important structure of the right-hand side is revealed if we use the fact that the integral of a sum is the sum of integrals, so that the right-hand side of the above equation becomes

$$A_0\frac{2}{T}\int_{-T/2}^{T/2} dt\ \cos(\omega_3 t)+\sum_{n=1}^{\infty}A_n\frac{2}{T}\int_{-T/2}^{T/2} dt\ \cos(\omega_n t)\cos(\omega_3 t)$$

$$+\sum_{n=1}^{\infty}B_n\frac{2}{T}\int_{-T/2}^{T/2} dt\ \sin(\omega_n t)\cos(\omega_3 t). \tag{5.13}$$

Because of the infinite sums, there appear to be an infinite number of integrals on the right-hand side of Eq. (5.13), but actually, all of them are zero except for one, because of the orthogonality rules described in the box above. The only non-zero integral is the one that multiplies A_3 ($n=3$), and that integral has the value 1. Therefore, the entire expression in Eq. (5.13) becomes equal to A_3, which is what we set out to prove. Similarly, projecting $x(t)$ onto cosine functions of other harmonic frequencies immediately gives values for all other coefficients, the set $\{A_n\}$.

An entirely parallel argument shows that projecting $x(t)$ onto sine functions gives the values of the set $\{B_n\}$. The analog to Eq. (5.12), for the general harmonic m, is

$$B_m=\frac{2}{T}\int_{-T/2}^{T/2} dt\ x(t)\sin(\omega_m t)=\frac{2}{T}\int_{-T/2}^{T/2} dt\left\{A_0\sin(\omega_m t)\right.$$

$$\left.+\sum_{n=1}^{\infty}[A_n\cos(\omega_n t)\sin(\omega_m t)+B_n\sin(\omega_n t)\sin(\omega_m t)]\right\}. \tag{5.14}$$

The first term (in A_0) is clearly zero, because the integral of a sine function (odd) over even limits is zero. (If this is not clear, the reader should look at Appendix D on even and odd functions.) Otherwise, the orthogonality conditions for the projection onto a sine function insure that Eq. (5.14) gives us B_m.

To summarize, the equations for A_n and B_n are:

$$A_n = \frac{2}{T} \int_{-T/2}^{T/2} dt\ x(t) \cos(\omega_n t) \quad \text{for } n > 0. \tag{5.15a}$$

$$B_n = \frac{2}{T} \int_{-T/2}^{T/2} dt\ x(t) \sin \omega_n t \quad \text{for } n > 0. \tag{5.15b}$$

It is these equations that constitute the Fourier transform, so far as the Fourier series is concerned.

Equation (5.15) gives the coefficients for $n>0$. The only other possibility is $n=0$. The coefficient A_0 is simply a constant that shifts function $x(t)$ up or down. Constant A_0 is the only term in the Fourier series (Eq. (5.2)) that could possibly have a non-zero value when averaged over a period. All the other terms are sines and cosines; they are negative as much as they are positive, and average to zero. Therefore, A_0 is the average value of $x(t)$. If $x(t)$ is an electrical signal, we would call A_0 the DC (direct current) component of x. To find A_0 we project function $x(t)$ onto a cosine of zero frequency, i.e. onto the number 1, which is an obscure way to say that we simply find the average value of x,

$$A_0 = \frac{1}{T} \int_{-T/2}^{T/2} dt\ x(t). \tag{5.16}$$

The limits of the integrals in Eqs. (5.15) and (5.16) are from $-T/2$ to $T/2$. These symmetrical limits are often convenient, but they are not necessary. Because all the functions involved, $x(t)$ and the trigonometric functions, repeat themselves after a duration T, the values of A_n and B_n are independent of the limits of integration, so long as the upper and lower limits differ by one period, T. For example, it may be convenient to use the limits 0 to T.

THE SPECTRUM

The Fourier series is a function of time, where A_n and B_n are coefficients that weight the cosine and sine contributions to the series. Coefficients A_n and B_n are real numbers that may be positive or negative.

An alternative approach to function x deemphasizes the time dependence and considers mainly the coefficients themselves. This is the spectral approach. The

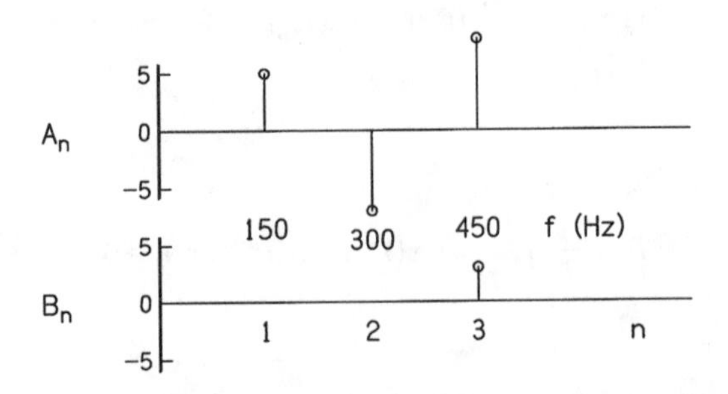

FIGURE 5.4. Coefficients A_n and B_n as a function of harmonic number n (or of frequency nf_0) constitute the Fourier spectrum of $x(t)$ from Eq. (5.17).

spectrum simply consists of the values of A_n and B_n, plotted against frequency, or, what is the same thing, plotted against harmonic number n. For example, if we have a signal given by

$$x(t)=5\cos(2\pi 150t)-7\cos(2\pi 300t)+8\cos(2\pi 450t)+3\sin(2\pi 450t), \tag{5.17}$$

then the spectrum consists of only a few terms. The period of the signal is 1/150 s, the fundamental frequency is 150 Hz, and there are two additional harmonics, a second harmonic at 300 Hz and a third at 450 Hz. The spectrum is shown in Fig. 5.4.

The individual terms from Eq. (5.17), corresponding to the individual spectral lines in Fig. 5.4, are plotted in the top four lines of Fig. 5.5. The bottom line shows the sum of all the four terms, which is the complex wave $x(t)$ given by Eq. (5.17).

Amplitude and Phase Parameters

The harmonics that are the ingredients in a function $x(t)$ are characterized by their particular angular frequencies ω_n. It is evident that if there are N harmonics with non-zero strengths then there may be as many as $2N$ non-zero coefficients, one value of A (for cos) and one value of B (for sin) for each harmonic. In most cases, this is not the most useful form for the spectrum. More useful is the amplitude-and-phase form, where we define a real non-negative variable C_n to be the amplitude and an angle ϕ_n to be the phase. In these terms, the Fourier series becomes

$$x(t)=A_0+\sum_{n=1}^{N} C_n\cos(\omega_n t-\phi_n). \tag{5.18}$$

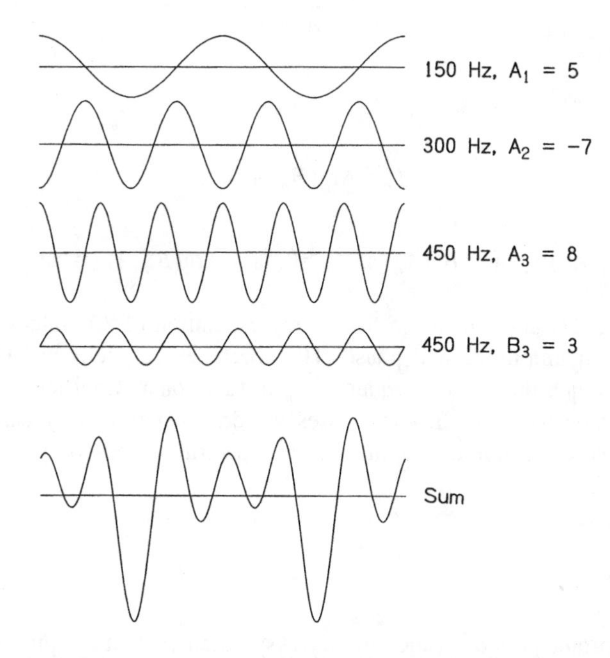

FIGURE 5.5. The top four functions show Fourier component waves with frequencies of 150, 300, and 450 Hz, and with amplitudes given by Eq. (5.17) and Fig. 5.4. Both cosine and sine components of 450 Hz are required. The bottom function shows the sum of the components, as given by Eq. (5.17).

It is easy to relate this form to what has gone before. By expanding the cosine above $[\cos(\omega_n t - \phi_n) = \cos(\omega_n t)\cos(\phi_n) + \sin(\omega_n t)\sin(\phi_n)]$ we can turn Eq. (5.18) into the form

$$x(t) = A_0 + \sum_{n=1}^{N} [C_n \cos(\omega_n t)\cos(\phi_n) + C_n \sin(\omega_n t)\sin(\phi_n)]. \tag{5.19}$$

Comparing, term by term, with Eq. (5.2) we can relate the two forms of spectral coefficients:

$$A_n = C_n \cos(\phi_n) \tag{5.20a}$$

and

$$B_n = C_n \sin(\phi_n). \tag{5.20b}$$

These equations can be inverted:

$$C_n = \sqrt{A_n^2 + B_n^2} \tag{5.21a}$$

and

$$\phi_n = \text{Arg}(B_n, A_n), \tag{5.21b}$$

where $\text{Arg}(B_n, A_n) = \tan^{-1}(B_n/A_n)$ if $A_n > 0$, and $\text{Arg}(B_n, A_n) = \tan^{-1}(B_n/A_n) + \pi$ if $A_n < 0$.

Equations (5.21) allow us to go from a representation of $2N$ values of A and B to $2N$ values of amplitude and phase. The coefficient C_n tells us, with a single number, how much there is of frequency ω_n in function x. Coefficient C_n must be a positive number or zero. In some cases we do not particularly care about the phases, and therefore our description of the spectrum needs only N values, the values of the C_n.

SYMMETRY

Many important periodic functions have symmetries that simplify the Fourier series. If the function $x(t)$ is an even function $[x(-t) = x(t)]$ then the Fourier series for x contains only cosine terms. All coefficients of the sine terms, B_n, are zero. If $x(t)$ is odd $[x(-t) = -x(t)]$, the Fourier series contains only sine terms, and all coefficients A_n are zero. Sometimes it is possible to shift the origin of time to obtain a symmetrical function. For instance, the square wave of Fig. 5.6(a) can be turned into an odd function by shifting the origin of time as in Fig. 5.6(b). Such a time shift is allowed if the physical situation at hand does not require that $x(t)$ be synchronized with some other function of time or with some other time-referenced process.

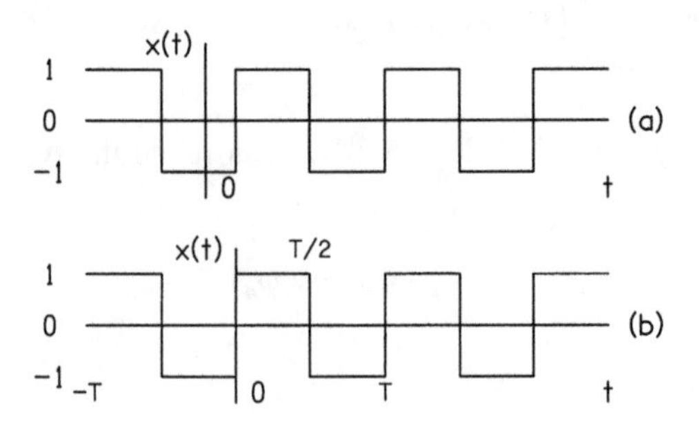

FIGURE 5.6. (a) A square wave. It is in a high state for half of the period and in a low state for the other half. It is neither an even nor an odd function of time t. (b) The origin of time is shifted to make the square wave an odd function of time.

EXAMPLE 1: *Odd-symmetry square wave*

Because the function of Fig. 5.6(b) is odd in time, all coefficients $\{A_n\}$ are zero, and only sine terms contribute to the Fourier series. Therefore, we can find the Fourier series for $x(t)$ by finding only the coefficients B_n. Substituting the square wave values ± 1 into Eq. (5.15b) we find

$$B_n = \frac{2}{T}\int_0^{T/2} dt\ \sin(\omega_n t)(+1) + \frac{2}{T}\int_{-T/2}^{0} dt\ \sin(\omega_n t)(-1). \qquad (5.22)$$

Because the square wave is odd and the sine function is odd, their product is even and the second integral (over a negative half-time period) is the same as the first (over a positive half-time period). We only need to evaluate the first integral and multiply by two. Remembering that $\omega_n = 2\pi n/T$, we find

$$B_n = \frac{2}{\pi}\left[\frac{1-\cos(n\pi)}{n}\right] \qquad (5.23)$$

or

$$B_n = \frac{4}{\pi n} \quad \text{for } n \text{ odd} \qquad (5.24a)$$

and

$$B_n = 0 \quad \text{for } n \text{ even}. \qquad (5.24b)$$

Now that we have the coefficients B_n, we can plug them into Eq. (5.2) to find the Fourier series for the square wave in Fig. 5.6(b),

$$x(t) = \frac{4}{\pi}\sum_{n=1,3,5\ldots}^{\infty} \frac{1}{n}\sin(2\pi n t/T). \qquad (5.25a)$$

The series can also be written as a sum on all integers,

$$x(t) = \frac{4}{\pi}\sum_{n=1}^{\infty} \frac{1}{2n-1}\sin[2\pi(2n-1)t/T]. \qquad (5.25b)$$

The reader will note that the odd harmonics in Eq. (5.25b) are given by the quantity $2n-1$, where n is the parameter in the sum. The Fourier series for a square wave with even symmetry is the subject of Exercise 2.

EXAMPLE 2: ***Even-symmetry triangle wave***

The triangle wave is shown in Fig. 5.7. It has even symmetry. Therefore, to find the Fourier series we only need to find the coefficients A_n.

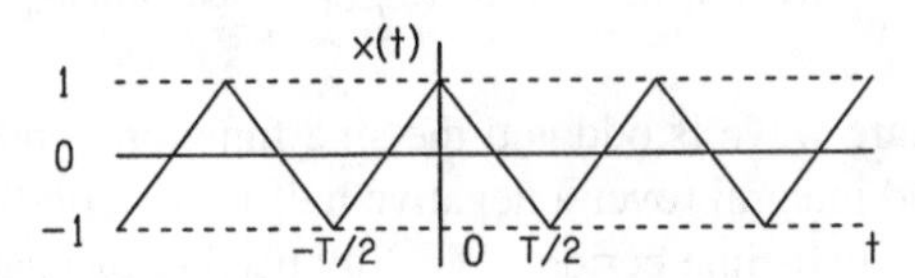

FIGURE 5.7. The triangle wave with even symmetry about t=0.

The first thing to note is that the function is negative as much as it is positive. Therefore it has zero average value and coefficient A_0 is zero. The other values of A_n can be found from the fact that $x(t)=1-4t/T$ on the interval of time from $t=0$ to $t=T/2$. As in the case of the square wave above, the part of the integral for negative time is just equal to the integral for positive time. Therefore, from Eq. (5.15)

$$A_n=\frac{4}{T}\int_0^{T/2} dt\ \cos(\omega_n t)[1-4t/T]. \tag{5.26}$$

The integral turns out to be zero unless harmonic number n is odd. Then

$$A_n=8/(\pi^2 n^2) \quad \text{for } n \text{ odd}. \tag{5.27}$$

From Eq. (5.2) we find the Fourier series for the even-symmetry triangle of Fig. 5.7,

$$x(t)=\frac{8}{\pi^2}\sum_{n=1,3,5\ldots}^{\infty}\frac{1}{n^2}\cos(2\pi nt/T). \tag{5.28}$$

The Fourier series for a triangle wave with odd symmetry is the topic of Exercise 3.

Figure 5.8 shows the evolution of a triangle wave as components are added into the Fourier sum. The smoothest waveform there is simply the first component in the series, a sine with the fundamental frequency and amplitude $8/\pi^2$. The next trace shows the sum of harmonics 1 and 3. Finally, the trace with the sharpest peak and the straightest sides is the sum of harmonics 1, 3, 5, 7, 9, 11, 13, and 15. If an infinite number of harmonics were added, the function would be a perfect triangle wave that would touch the horizontal lines in the figure.

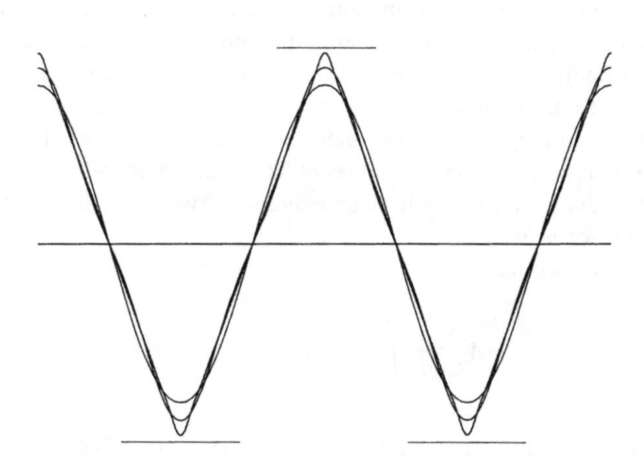

FIGURE 5.8. Successive approximations to the triangle wave based on one, two, or eight Fourier components. The greater the number of components the sharper the peak of the waveform.

Half-Wave Symmetry

The function shown in Fig. 5.9 has what is known as "half-wave symmetry." It is not an even or odd function of time. Instead, its symmetry is of the form,

$$x(t) = -x(t + T/2), \tag{5.29}$$

where T is the period. This symmetry means that if one moves along the time axis by half a period, the function is reversed in sign. Each half cycle is an inverted version of the previous half cycle. The circled points on the function in Fig. 5.9 illustrate that symmetry. Because the function has period T, Eq. (5.29) also implies that

$$x(t) = -x(t - T/2). \tag{5.30}$$

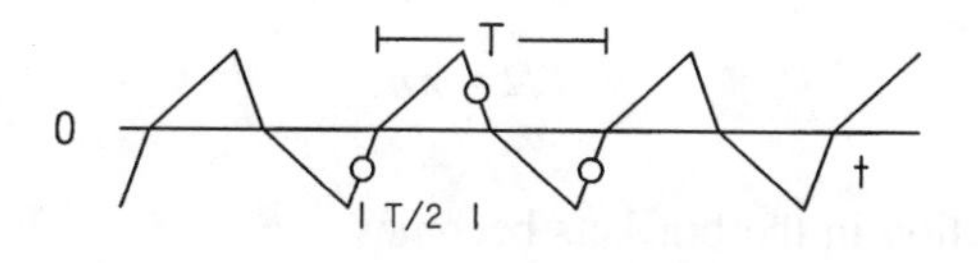

FIGURE 5.9. A function with half-wave symmetry.

The special nature of half-wave symmetry is that a periodic function $x(t)$ that has half-wave symmetry has no even-numbered harmonics in its spectrum. The spectrum has zero amplitude for all the even harmonics, and the Fourier series for $x(t)$ becomes a sum of odd harmonics only.

To prove that the amplitudes of even harmonics are zero, we need to prove that A_n is zero and B_n is zero for all even values of harmonic number n. Below we prove that $A_n=0$ for n even, and it is left as an exercise to work out the (quite similar) proof that $B_n=0$ for n even.

From Eq. (5.15a) we have

$$A_n=\frac{2}{T}\int_0^T dt\ x(t)\cos\ \omega_n t. \tag{5.31}$$

We prepare to take advantage of half-wave symmetry by dividing the integral into two parts, split at $T/2$.

$$A_n=\frac{2}{T}\int_0^{T/2} dt\ x(t)\cos\ \omega_n t+\frac{2}{T}\int_{T/2}^{T} dt'x(t')\cos\ \omega_n t'. \tag{5.32}$$

(Note that the change to t' in the second integral is only a notational device. It has no significance because t' is a variable of integration that is consumed by the integral.)

We change variables in the second integral,

$$t'=t+T/2, \tag{5.33}$$

whereupon

$$A_n=\frac{2}{T}\int_0^{T/2} dt\ x(t)\cos\ \omega_n t+\frac{2}{T}\int_0^{T/2} dt\ x(t+T/2)\cos\ \omega_n(t+T/2). \tag{5.34}$$

By half-wave symmetry $x(t+T/2)=-x(t)$, and so

$$A_n=\frac{2}{T}\int_0^{T/2} dt\ x(t)[\cos\ \omega_n t-\cos(\omega_n t+\omega_n T/2)]. \tag{5.35}$$

Because ω_n is a harmonic,

$$\omega_n T/2=\pi n. \tag{5.36}$$

Therefore, the function in the brackets becomes

$$[\cos\ \omega_n t-\cos(\omega_n t+n\pi)]. \tag{5.37}$$

When the harmonic number n is even, the phase difference between the two cosines functions in the bracket is 0, or 2π or 4π or 6π, etc., equivalent to no phase difference at all. We end up subtracting identical functions and the bracket equals zero, for even n. Therefore A_n is zero for even values of n. This is what we set out to prove.

If n is odd then the phase difference is π or 3π or 5π, etc., all equivalent to 180 degrees. As a result,

$$A_n = \frac{4}{T} \int_0^{T/2} dt \, x(t) \cos \omega_n t \quad (n \text{ odd}). \tag{5.38}$$

The reader should show that

$$B_n = \frac{4}{T} \int_0^{T/2} dt \, x(t) \sin \omega_n t \quad (n \text{ odd}). \tag{5.39}$$

Double Symmetry—Odd

The square wave of Fig. 5.6 has both odd symmetry and half-wave symmetry. Therefore, all values of A_n are zero and all the values of B_n are zero for n even. It is evident that such a lot of symmetry greatly simplifies the Fourier series. Only the B_n for odd n are non-zero, and they can be calculated from a reduced integral, as we now show.

The values of B_n are given by Eq. (5.39), which integrates from 0 to $T/2$. We can reduce it to an integral from 0 to $T/4$. First, from half-wave symmetry $x(t) = -x(t-T/2)$. Next, from odd symmetry $x(t-T/2) = x(T/2-t)$. Therefore, $x(t) = -x(T/2-t)$, which is a new symmetry. Because sine functions of odd harmonics also have this symmetry, the integral from 0 to $T/2$ consists of two identical integrals, one from 0 to $T/4$, the other from $T/4$ to $T/2$. We only need to evaluate one of them and multiply by two. In summary,

For functions with odd-double symmetry:

$$A_n = 0 \quad \text{for all } n$$

$$B_n = 0 \quad \text{for } n \text{ even} \tag{5.40}$$

$$B_n = \frac{8}{T} \int_0^{T/4} dt \, x(t) \sin \omega_n t \quad \text{for } n \text{ odd.}$$

Double Symmetry—Even

A function like the triangle wave of Fig. 5.7 has both even symmetry and half-wave symmetry. Using methods similar to the above, the reader can show that $x(t) = -x(T/2 - t)$. This symmetry also simplifies the calculation of the Fourier coefficients.

For functions with even-double symmetry:

$$B_n = 0 \quad \text{for all } n$$

$$A_n = 0 \quad \text{for } n \text{ even} \tag{5.41}$$

$$A_n = \frac{8}{T}\int_0^{T/4} dt\, x(t)\cos\omega_n t \quad \text{for } n \text{ odd.}$$

In cases where double symmetry exists, the equations above can lead to considerable simplication in calculating the series. Because the time interval of integration has been divided by four, one can end up doing one-fourth as much work to calculate the Fourier coefficients. This economy is illustrated in several exercises below, for the square wave and triangle wave, as well as the parabola wave and the circle wave.

COMPUTER EVALUATION OF FOURIER SERIES COEFFICIENTS

For most periodic functions that one can imagine, it is not possible to evaluate the Fourier series coefficients in closed form. But one can always do the integral numerically. The schematic program in this box calculates the first five coefficients and the average value (DC term) for the periodic function (period T),

$$x(t) = \sqrt{1 + r\cos(\omega t)},$$

where $\omega = 2\pi/T$, and r is a constant.

Function $x(t)$ is shown in Fig. 5.10(a) for the particular case where $r = 1$. Program EVEN takes advantage of the fact that $x(t)$ is even. It calculates only coefficients A_n because all values of B_n are zero. The spectrum for harmonics 0 through 5 is shown in Fig. 5.10(b).

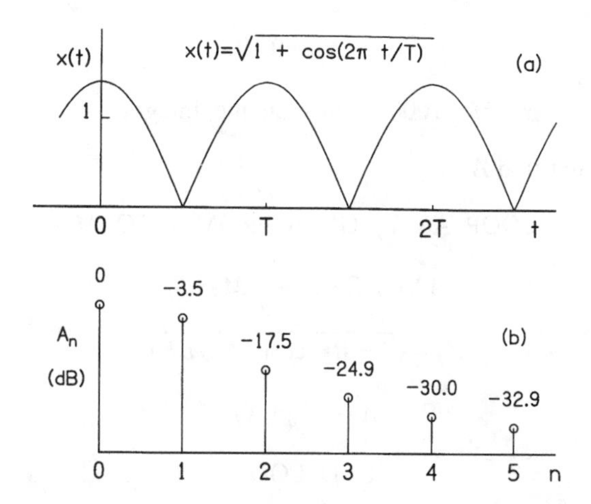

FIGURE 5.10. An even periodic function $x(t)$ and its spectrum, calculated by program EVEN. Levels of harmonics are relative to the level of the DC component.

[Note: Exercise 7 will show that because the function has a discontinuity in its derivative, we can predict that the spectrum is asymptotically -12 dB per octave. In fact, the -12 dB/octave rule is successful after the fifth or sixth harmonic. Therefore, by knowing the first five harmonic amplitudes one essentially knows them all to within a few tenths of a dB. For example, based on a 5th harmonic amplitude of -39.9 dB one would predict a 20th harmonic amplitude of -63.9 dB ($-39.9-24$), very close to the correct answer of -64.1 dB. The strength of high harmonics is enormously sensitive to parameter r when r is close to 1. For $r \geqslant 1$, function $x(t)$ has a discontinuous derivative; for $r<1$ it does not. For example, if r is changed from 1 to 0.95, the level of the 20th harmonic falls from -64.1 dB to -110 dB.]

Program EVEN can be used to calculate the first five Fourier components for any even function of time. One only needs to substitute the desired function for X in the loop. The loop should cover precisely one cycle of function X.

The program works by accumulating 1000 increments to the integral as the angle ωt in $x(t)$ goes from 0 to 2π, equivalent to one period. The increments for the n-th harmonic are accumulated in the amplitude variable A_n. The value of A_n is not normalized correctly because, in the end, the values of A_n are expressed in decibels *re* A_0.

PROGRAM EVEN

INPUT PARAMETER R

Comment: Set A_N equal to zero as N goes from 0 to 5.

$$A_N=0$$

$M=1000$ Points on the integral

Comment: First find A_0.

LOOP AS T_I GOES FROM 1 TO M

$$ANGLE=(2*\pi/M)*T_I$$

$$X_I=\sqrt{1+R*\cos(ANGLE)}$$

$$A_0\rightarrow A_0+X_I$$

END LOOP

Comment: Next find A_N as N goes from 1 to 5.

LOOP AS T_I GOES FROM 1 TO M

$$ANGLE=(2*\pi/M)*T_I$$

$$X_I=\sqrt{1+R*\cos(ANGLE)}$$

$$A_N\rightarrow A_N+2.*X_I*\cos(N*ANGLE)$$

END LOOP

Comment: Finally, find the level of A_N in dB relative to A_0.

$$LEVEL_N=20*\log_{10}(A_N/A_0)$$

END

EXERCISES

In these exercises the diligent reader will calculate the Fourier transforms of periodic signals by doing Fourier integrals analytically. Fortunately, the integrals are all easy ones. Most of the exercises give a figure for a waveform in time and also give the Fourier series. It is the reader's task to show that the Fourier series is correct. Because of this style of presentation, this exercise section becomes one of the larger menageries of periodic functions and their Fourier transforms. Here will be found such exotic beasts as the saber-sawtooth wave and the parabola wave, as well as a number of variations on familiar species. Note that although the time-dependent figures may show only a few cycles, the waveforms actually continue into the infinite past and the infinite future.

Exercise 1: Even and odd symmetries

(a) Prove that if $x(t)$ is an odd function, then the Fourier series for x does not include any cosine terms, i.e. coefficients A_n are zero for all n.

(b) Prove that if $x(t)$ is an even function then the Fourier series for x does not include any sine terms. Your proofs will make use of the cancellation of equivalent positive and negative areas in the integral.

Exercise 2a: Even square wave by time shifting

Figure E.2a has been made from Fig. 5.6(b) for the square wave by shifting the time origin by $T/4$ so that the square wave is an even function of time instead of an odd function of time.

By starting with Eq. (5.25b) and substituting for the time argument, $t' = t + T/4$, show that the Fourier series of Fig. E.2a becomes

$$x(t) = \frac{4}{\pi}\left\{\cos\left(\frac{2\pi t}{T}\right) - \frac{1}{3}\cos\left(\frac{2\pi 3t}{T}\right) + \frac{1}{5}\cos\left(\frac{2\pi 5T}{T}\right) - \frac{1}{7}\cdots\right\} \qquad (5.42a)$$

or

$$x(t) = \frac{4}{\pi}\sum_{n=1}^{\infty}\frac{(-1)^{(n-1)}}{2n-1}\cos[2\pi(2n-1)t/T].$$

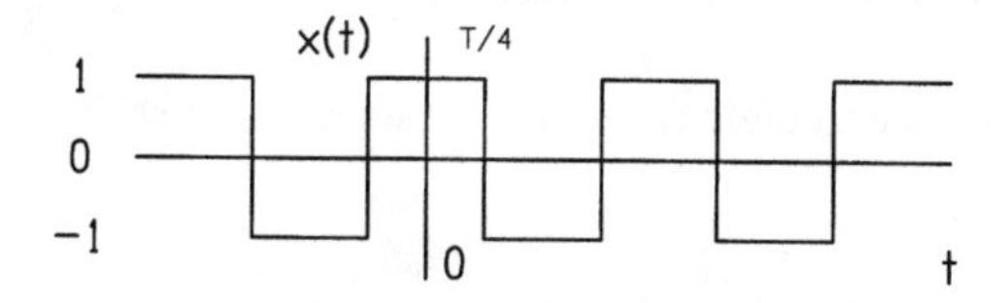

FIGURE E.2a. The square wave with even symmetry.

Exercise 2b: Even square wave by double symmetry

The square wave in Fig. E.2a is a case of even double symmetry. Calculate its Fourier coefficients from Eq. (5.41), using the fact that on the interval from 0 to $T/4$ the function $x(t)$ is simply the number 1. Compare your answer with the series in Eq. (5.42a).

Exercise 2c: Odd square wave by subtracting a dc offset (A_0)

The square wave of Fig. E.2c does not have even or odd symmetry. However, Fig. E.2c differs from Fig. 5.6(b) by only a factor of 1/2 and a vertical shift of 1/2. Use Eq. (5.25a) to show how the Fourier series for the wave of Fig. E.2c is

$$x(t)=\frac{1}{2}+\frac{2}{\pi}\sum_{n=1,3,5\ldots}^{\infty}\frac{1}{n}\sin(2\pi nt/T). \tag{5.42c}$$

In doing this part of the exercise you have used the fact that multiplying function $x(t)$ by a constant scale factor simply has the effect of multiplying all the Fourier coefficients, A_n or B_n or C_n by that scale factor. This is a characteristic of all transformations of the kind that are called "linear."

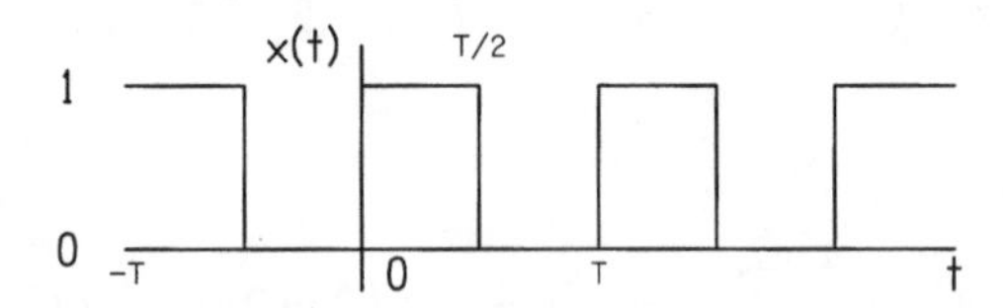

FIGURE E.2c. A square pulse train that is always positive.

Exercise 3: Odd double symmetry

Figure E.3 shows the triangle from Fig. 5.7 with a time shift to make it odd about $t=0$.

(a) Show that this function has half-wave symmetry.

(b) Show that the triangle wave of Fig. E.3 is represented by the series,

$$x(t)=\frac{8}{\pi^2}\sum_{n=1}^{\infty}\frac{(-1)^{(n-1)}}{(2n-1)^2}\sin[2\pi(2n-1)t/T]. \tag{5.43}$$

(c) Explain why parameter n in the sum above takes on all positive values when the only existing harmonics are the odd ones.

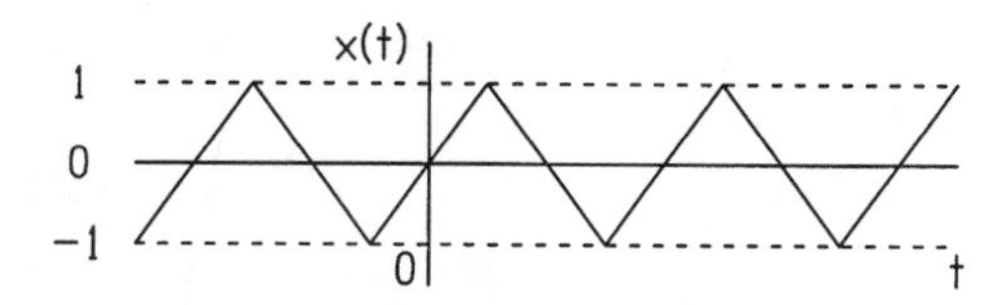

FIGURE E.3. The triangle wave with odd symmetry.

Exercise 4: Triangle wave as the integral of the square wave

There is an alternative way to calculate the Fourier series for the triangle wave of Fig. E.3 because the square wave in Fig. E.2a is simply the derivative of the triangle of Fig. E.3, apart from a simple scale factor. Therefore, the triangle wave is essentially the integral of the square wave. Prove that the series in Eq. (5.43) is correct by starting with the Fourier series in Eq. (5.42a) and integrating it term by term. [Note: The hard part in this exercise is getting the constant factor. To find the factor, first note that the derivative of the triangle wave is $\pm 4/T$. Next, note that it is particularly easy to evaluate the Fourier series in Eq. (5.42) at the point $t=0$. You can find the answer for the infinite sum in Appendix C.]

The converse to the above is that the square wave is the derivative of the triangle wave. Sellick and Russell (1980) showed the importance of tympanic membrane velocity by applying a triangular wave displacement to it. The resulting cochlear microphonic then resembled a square wave. The reason is that the velocity is the derivative of the displacement.

Exercise 5: Spectral shapes

The tone color of a complex tone does not depend on all the details of the spectrum. Often the spectral shape, defined by the spectral envelope, is enough to characterize the tone. The spectral envelope is a smooth curve that passes through the peaks of the spectrum, tracing out the shape of the spectral amplitudes.

(a) Show that the spectral envelope of the square wave (Exercise 2) decreases at a rate of −6 dB/octave and that the spectral envelope of a triangle (Exercise 3) wave decreases at a rate of −12 dB/octave.

(b) Figure E.5 shows the first 16 non-zero harmonics of a triangle wave, numbers 1 through 31. The plot at the left shows the harmonic levels in decibels (a log scale) on the vertical axis and the harmonic frequencies on the horizontal axis. Sketch a line connecting the peaks to plot the spectral envelope. The plot on the right is a log-log plot, levels in decibels vertically and log frequency horizontally. Explain why the spectral envelope is a perfectly straight line.

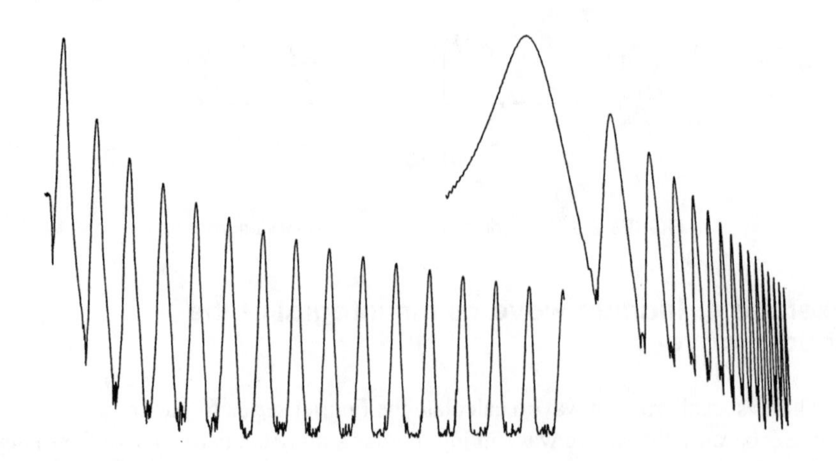

FIGURE E.5. Two measurements of the spectrum of a triangle wave, showing harmonics 1–31. On the left the frequency scale is linear, on the right it is logarithmic. The levels of the harmonics are identical on both plots. The widths of the peaks are an artifact of the spectrum analyzer.

Exercise 6: Parabola wave

The parabola wave is a series of alternating upward and downward parabolas, linked at zero crossings to make a periodic wave with no average value, as shown in Fig. E.6.

(a) Use double symmetry, and the fact that the function on the interval from 0 to $T/4$ is $x(t)=1-(4t/T)^2$, to show that the Fourier series for the parabola wave is

$$x(t)=\frac{32}{\pi^3}\sum_{n=1,3,5\ldots}^{\infty}(-1)^{(n-1)/2}\frac{1}{n^3}\cos(2\pi nt/T). \tag{5.44}$$

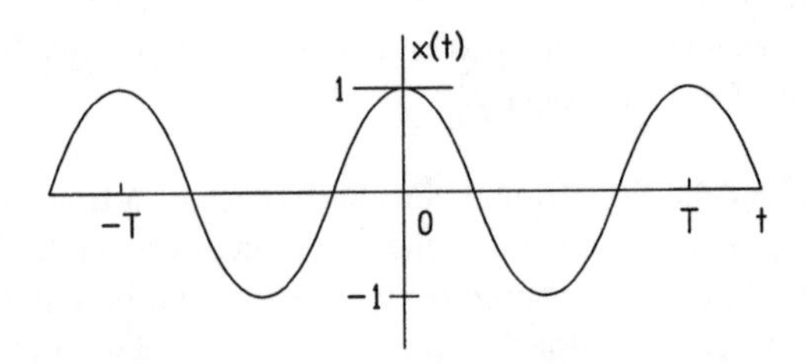

FIGURE E.6. The parabola wave.

(b) Criticize the following reasoning: "To compute the Fourier coefficients for the parabola wave we do the integral

$$A_n = (8/T)\int_0^{T/4} dt\, \cos(2\pi nt/T)[1-(4t/T)^2]. \quad (5.45a)$$

We discover that this integral is not zero for even values of n. But by the double symmetry argument we know that there are no finite Fourier coefficients for even values of n. Clearly there is some contradiction."

(c) Criticize the following reasoning: "We could also consider the integral

$$B_n = (8/T)\int_0^{T/4} dt\, \sin(2\pi nt/T)[1-(4t/T)^2] \quad (5.45b)$$

and we discover that it is not necessarily zero. But by the fact that the parabola wave is an even function, we know that there are no sine terms (no finite B_n) in its Fourier series. Clearly there is some contradiction."

Exercise 7: Discontinuities and asymptotic dependence on *n*

The square wave is a function with a discontinuity. Its Fourier series is proportional to $1/n$. These two facts go together. A function $x(t)$ has a discontinuity if and only if its Fourier series decreases as $1/n$ for asymptotically large n. This idea can be extended to other functions. The triangle wave is a function with a discontinuity in its first derivative. Its Fourier series is proportional to $1/n^2$. These facts go together too because if we do a term-by-term differentiation of a Fourier series with coefficients proportional to $1/n^2$, we must get a series with coefficients proportional to $1/n$. (Recall Exercise 4 in this connection.) The next stage in this development is, of course, the parabola wave. Exercise 6 shows that the Fourier coefficients decrease as $1/n^3$. Therefore, you have reason to suspect that the parabola wave has a discontinuity in its second derivative. Consider the time derivatives of the parabola wave and show that the discontinuity is there as expected.

The above argument is rather powerful. Just by looking at a function you can immediately predict the spectral envelope at high frequency, -6 dB/octave, or -12 dB/octave, or, as for the parabola wave, -18 dB/octave.

Exercise 8: Circle wave

The circle wave is a series of equivalent upward and downward semicircles, linked at zero crossings, as shown in Fig. E.8.

Use double symmetry to determine which coefficients of the Fourier series are not zero. With a computer, determine the first five non-zero Fourier coefficients using the fact that the function on the interval from 0 to $T/4$ is given by

$$x(t) = \sqrt{1-16(t/T)^2}. \quad (5.46)$$

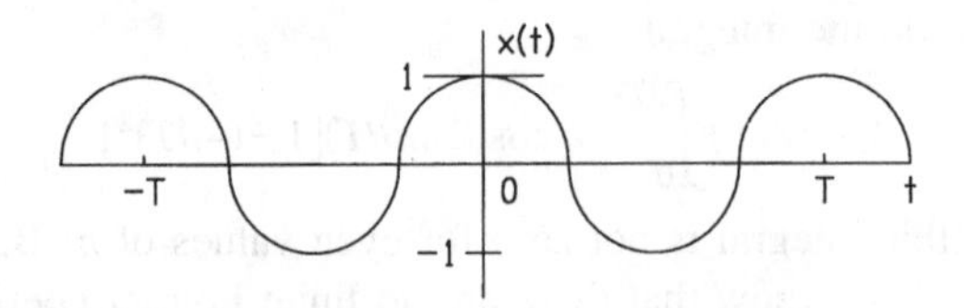

FIGURE E.8. The circle wave.

Exercise 9: Half-wave-rectified sine

The half-wave-rectified sine can be drawn as an even function of time, as in Fig. E.9a, by choosing a starting phase that actually makes it a half-wave-rectified cosine. It cannot be drawn as an odd function.

Show that if the minimum and maximum are 0 and 1, as shown, the Fourier series is

$$x(t)=\frac{1}{\pi}+\frac{1}{2}\cos(\omega_0 t)-\frac{2}{\pi}\sum_{n=2,4,6\ldots}^{\infty}(-1)^{n/2}\frac{1}{n^2-1}\cos(n\omega_0 t) \qquad (5.47)$$

where $\omega_0=2\pi/T$.

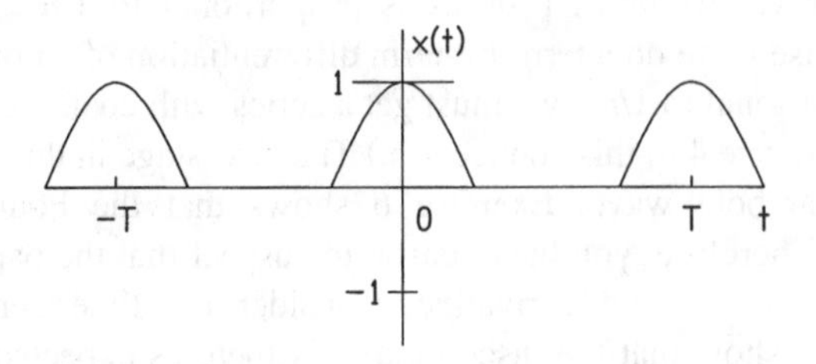

FIGURE E.9a. The half-wave rectified cosine.

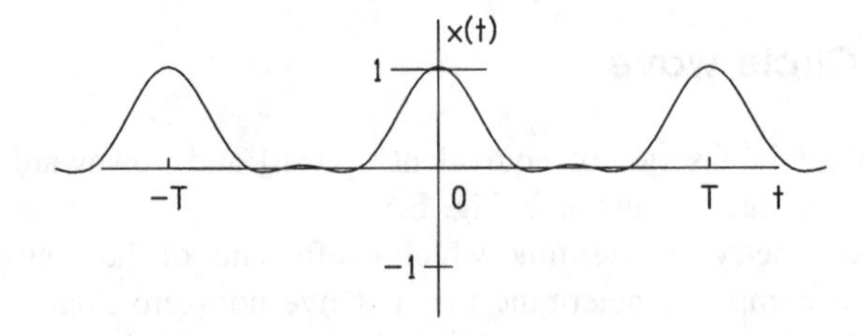

FIGURE E.9b. The sum of the constant term and the first two harmonics (n=1 and n=2) in the Fourier series for a half-wave rectified sine.

Exercise 10: Full-wave-rectified cosine

The full-wave-rectified cosine is given by the equation $x(t)=|\cos(\omega_0 t)|$. The function is shown by the solid line in Fig. E.10. The dashed line shows the original cosine function before the absolute value operation.

The figure makes it evident that by taking the absolute value we have decreased the period of the waveform by a factor of 2, i.e., we have doubled the fundamental frequency. Therefore, all of the harmonics in the Fourier series for the full-wave-rectified cosine have frequencies that are multiples of $2\omega_0$.

(a) Show that the Fourier series is given by

$$x(t)=\frac{2}{\pi}-\frac{4}{\pi}\sum_{n=1}^{\infty}(-1)^n\frac{1}{(2n)^2-1}\cos(2n\omega_0 t) \tag{5.48}$$

where $\omega_0=2\pi/T$, and T is the period of the original cosine (twice the period of the full-wave signal).

(b) Another way to calculate the Fourier series for the full-wave is to begin with the half-wave. If we call the full-wave function $FW(t)$ and call the half-wave function $HW(t)$ then, Figs. E.9a and E.10 make it clear that $FW(t)=2HW(t)-\cos(\omega_0 t)$. Show that the Fourier series for the full-wave in Eq. (5.48) is just twice the series for the half-wave, given in Eq. (5.47), minus its first harmonic.

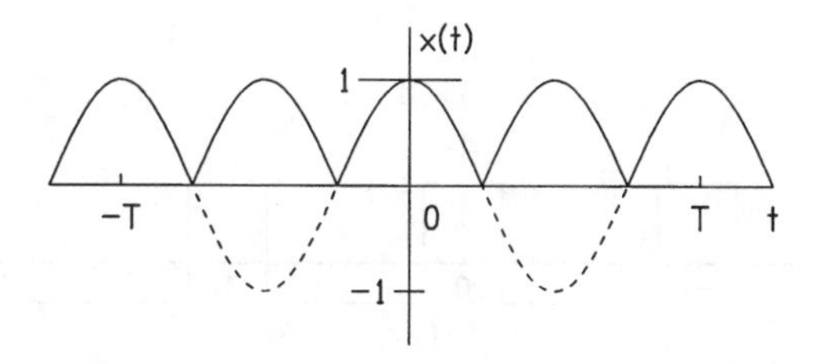

FIGURE E.10. The full-wave-rectified cosine.

Exercise 11: Rectangular pulse wave

The periodic rectangular pulse wave is defined by the duty factor, p. The waveform is in a high state ($x=1$) for a fraction p of the period and in a low state ($x=0$) for a fraction $(1-p)$, as shown in Fig. E.11a.

(a) Show that the Fourier series is given by

$$x(t)=p+\frac{2}{\pi}\sum_{n=1}^{\infty}\frac{1}{n}\sin(n\pi p)\cos(2\pi nt/T). \tag{5.49}$$

The rectangular pulse wave shown above has a finite average value. The average value can be calculated by averaging the function, and finding that the answer is p. Alternatively, one can evaluate A_0

$$A_0=\lim_{n\to 0}\frac{\sin(n\pi p)}{(n\pi)}=p, \tag{5.50}$$

which also shows that the average value is p. We therefore find a function with zero average value by subtracting p from the function in Fig. E.11a to obtain the function in Fig. E.11b.

Its Fourier series is then

$$x(t)=\frac{2}{\pi}\sum_{n=1}^{\infty}\frac{1}{n}\sin(n\pi p)\cos(2\pi nt/T), \tag{5.51}$$

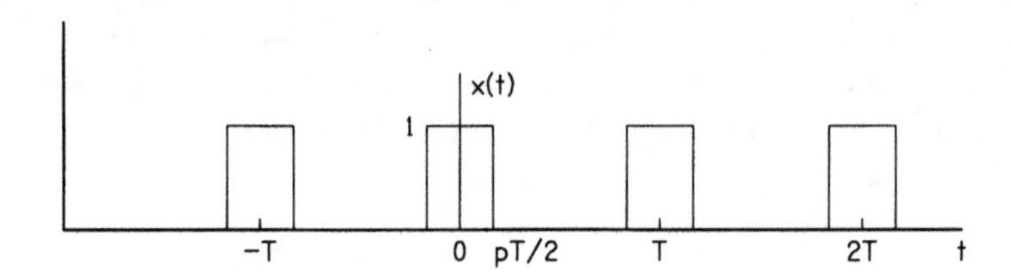

FIGURE E.11a. The rectangular pulse wave, with p equal to 1/3.

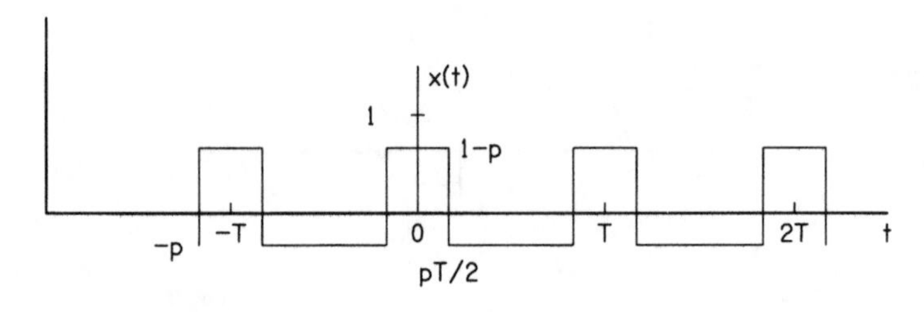

FIGURE E.11b. The function of E.11a, displaced to have zero average value.

identical to Eq. (5.49) except for the constant term.

(b) *Square-wave limit:* Show that in the limit that $p=1/2$, where the pulse train becomes a square wave, Eq. (5.51) agrees with the Fourier series for a square wave from Exercise 2.

(c) *Missing harmonics:* Part (b) above showed that for $p=1/2$, the even harmonics (integer multiples of 2) are missing from the spectrum. This fact can be generalized: if the duty factor p is of the form 1/(integer) then the harmonic number equal to the (integer) is missing from the spectrum and so are integral multiples of it. For example, if the duty factor is 1/3, then there is no third harmonic; there is no sixth or ninth or twelfth either. If the duty factor is 1/5 then there is no fifth harmonic or tenth or fifteenth, etc. Show that this result follows from the expression for the harmonic amplitudes $[\sin(n\pi p)]/n$.

(d) *Experiment:* Many electronic function generators are able to generate a rectangular pulse wave, with a variable duty factor as controlled by a knob on the front panel. Duty factors of the precise form $1/n$, where n is an integer, can be achieved by nulling the n-th harmonic. Given a favorable fundamental frequency, one can even do an impressive job of nulling harmonics with low harmonic number by ear. Use the method of bracketing to adjust the duty factor of a rectangular pulse to 1/3 by nulling the third harmonic of a 200-Hz pulse train. By checking your adjustment with a spectrum analyzer you will probably find that you can make the third harmonic as small as -40 dB with respect to the level of the fundamental. That corresponds to setting $p=1/3$ to a precision of 1 part in 100, far better than could be done by adjusting the duty factor while watching the waveform on an oscilloscope. Next try nulling the second harmonic to create a precise square wave ($p=1/2$). You may find that you can adjust the level to be -60 dB with respect to the fundamental, a precision of one part in 1000 in the duty factor.

Exercise 12: The double ramp

The double ramp is shown in Fig. E.12. It is made from two straight lines, each of constant slope. It is therefore evident that if we differentiate the double ramp we get the rectangular pulse wave shown in Fig. E.11b, apart from a constant of proportionality. Except for this prefactor, then, the double ramp is the integral of the rectangular pulse wave in Eq. (5.51). [Note: It is important that the rectangular pulse wave has no average value; otherwise its integral would grow linearly with time, i.e. the double ramp function would be superimposed on a straight line, with slope equal to the average value of the derivative.]

(a) Integrate the Fourier series in Eq. (5.51), term by term, to find the Fourier series for the double ramp of Fig. E.12. The answer is

$$x(t)=\frac{T}{\pi^2}\sum_{n=1}^{\infty}\frac{1}{n^2}\sin(n\pi p)\sin(2\pi nt/T). \tag{5.52}$$

(b) To get the prefactor right we need to scale the function, eliminating the prefactor T. According to Fig. E.11b the positive slope is $1-p$. According to Fig. E.12 the positive slope must be $2/(pT)$. Set these equal to show that the Fourier series for the function in Fig. E.12 is

$$x(t)=\frac{2}{p(1-p)\pi^2}\sum_{n=1}^{\infty}\frac{1}{n^2}\sin(n\pi p)\sin(2\pi nt/T). \tag{5.53}$$

Note: The limit $p=0$ is the sawtooth wave, discussed in Exercise 14, except for an overall sign change.

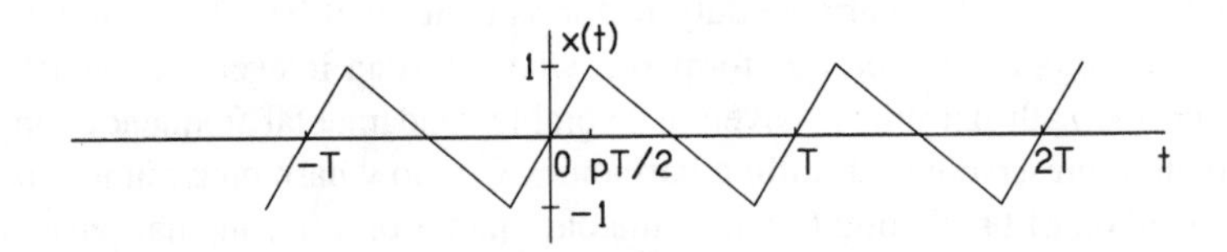

FIGURE E.12. The double ramp, with p equal to 1/3.

Exercise 13: The symmetrized pulse

The odd-symmetrized pulse is shown in Fig. E.13. Show that it can be represented by the Fourier series,

$$x(t)=\frac{2}{\pi}\sum_{n=1}^{\infty}\frac{\cos(n\pi p)-1}{n}\sin(2\pi nt/T). \tag{5.54}$$

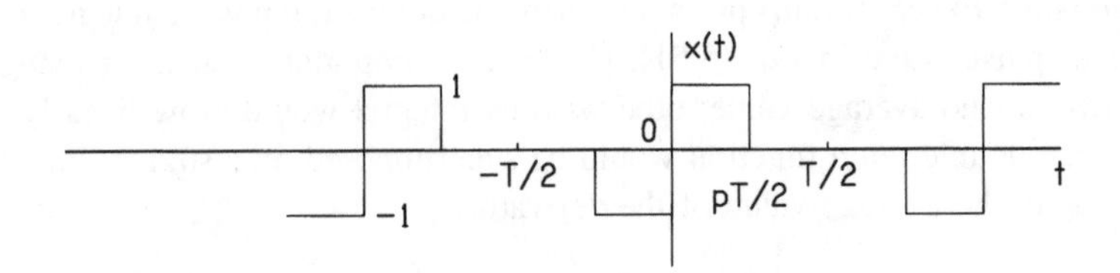

FIGURE E.13. The symmetrized pulse.

Exercise 14: The sawtooth

The sawtooth is a wonderful function. Everywhere that its slope is defined, the slope is positive and constant. The function is directly proportional to time. That is why the sawtooth is the waveform used for the time base in an oscilloscope. (a) Use the odd symmetry, and the fact that on the interval from 0 to $T/2$ the sawtooth waveform is $x(t)=2t/T$, to show that the Fourier series for the sawtooth in Fig. E.14a is

$$x(t)=\frac{2}{\pi}\sum_{n=1}^{\infty}\frac{(-1)^{(n+1)}}{n}\sin(2\pi nt/T). \tag{5.55}$$

The alternating sign in the series above can be eliminated by changing the time origin as in Fig. E.14b.

(b) Use the transformation (Fig. E.14a to Fig. E.14b), namely $t\rightarrow t+T/2$, in the Fourier series of Eq. (5.55) to show that the Fourier series for Fig. E. 14b is

$$x(t)=\frac{-2}{\pi}\sum_{n=1}^{\infty}\frac{1}{n}\sin(2\pi nt/T). \tag{5.56}$$

(c) The double ramp in Fig. E.12 becomes equal to the sawtooth of Fig. E.14b, except for an overall change in sign, in the limit that the positive-going ramp is infinitely steep. Show that the series in Eq. (5.56) can be obtained from the series for the double ramp function [Eq. (5.53)] in the limit that $p=0$.

(d) For reasons known only to himself, Bozo says that the negative sign at the front

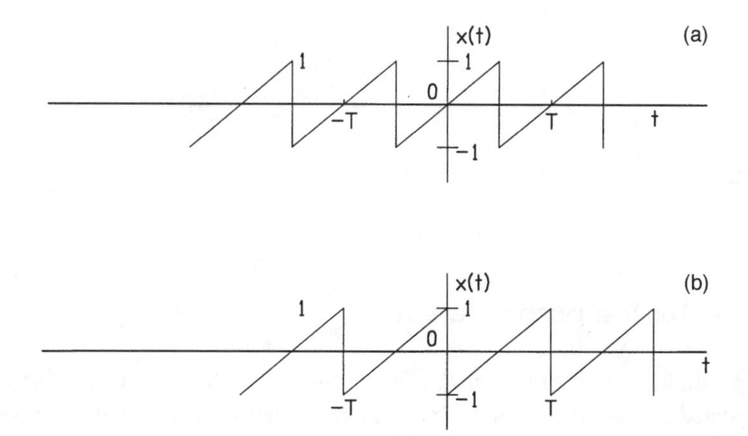

FIGURE E.14. (a) The sawtooth. (b) The time-shifted sawtooth.

of Eq. (5.56) looks wrong. Show him how the first few terms of the series, including the negative sign, fit with the waveform in Fig. E.14b.

Exercise 15: The half-saw

The half-saw (or ramp) in Fig. E.15 was made famous by a classic series of Wavetek function generators. It is a half-wave-rectified sawtooth and it has no particular symmetry at all.

(a) Do the integrals to find the Fourier coefficients to show that the Fourier series for the half-saw in Fig. E.15 is given by

$$x(t)=\frac{1}{4}-\sum_{n=1,3,5\ldots}^{\infty}\frac{2}{\pi^2 n^2}\cos(2\pi nt/T)+\sum_{n=1}^{\infty}\frac{(-1)^{(n+1)}}{n\pi}\sin(2\pi nt/T). \tag{5.57}$$

(b) The Fourier series for the half-saw is revealing. It is the sum of two parts. The second part is an odd function (sum of sine terms) that appears to have a discontinuity (see Exercise 7). The first part is an even function (sum of cosine terms) that appears to have a discontinuity only in its derivative. From such reasoning we realize that it is possible to construct a half-saw by adding together a sawtooth waveform and a triangle waveform that is offset so that its minimum value is zero. Make a sketch showing how the half-saw can be constructed in this way.

(c) Fourier series for the sawtooth and triangle waveforms appear in Eqs. (5.55) and (5.28). Show that these series can be summed to give Eq. (5.57) as expected.

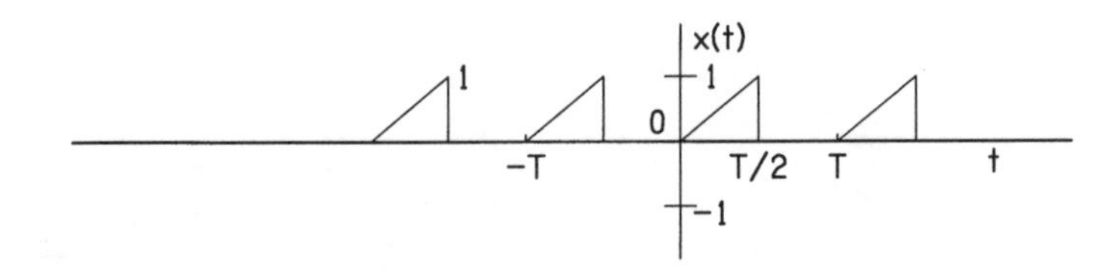

FIGURE E.15. The half-saw.

Exercise 16: The parabolic saw

The parabolic saw, with period T, is given by $x(t)=(2t/T)^2$ for $|t|\leq T/2$, and is then repeated for all other periods. It is an even function of time and its maximum value is 1.

(a) Show that the Fourier series for the parabolic saw is

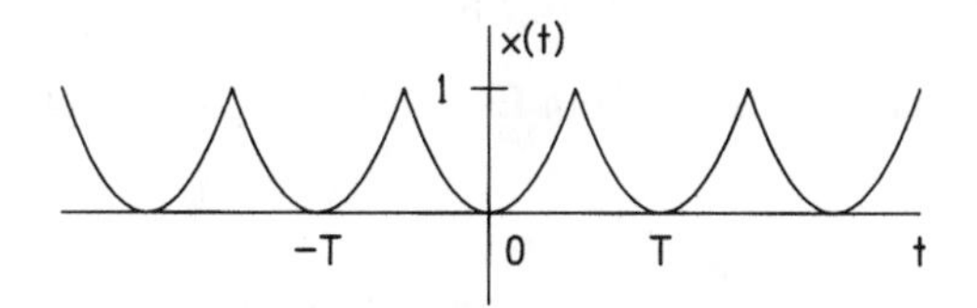

FIGURE E.16. The parabolic saw.

$$x(t)=\frac{1}{3}+\frac{4}{\pi^2}\sum_{n=1}^{\infty}\frac{(-1)^n}{n^2}\cos\left(\frac{2\pi nt}{T}\right). \qquad (5.58)$$

(b) Notice that the derivative of the parabolic saw looks like the standard sawtooth wave because the derivative of a parabola is a straight line. Compare Fig. E.16 above with Fig. E.14 to show that the derivative of the parabolic saw is $4/T$ times the sawtooth wave of Fig. E.14a.

(c) Do a term-by-term differentiation of the Fourier series for the parabolic saw obtained in part (a) and show that the result agrees with Eq. (5.55) for the sawtooth, multiplied by $4/T$ from part (b).

Exercise 17: Exponential calculation of the Fourier coefficient

Equations (5.15) give the coefficients A_n and B_n by projecting the function $x(t)$ onto cosine and sine functions, using an integral from $-T/2$ to $T/2$ or, equivalently, from 0 to T. Sometimes it is more convenient to project onto an exponential function.

Use Eqs. (5.15) to prove that

$$A_n+iB_n=\frac{2}{T}\int_0^T dt\ e^{i\omega_n t}\ x(t). \qquad (n\neq 0) \qquad (5.59)$$

The proof makes use of the fact that $x(t)$ is real.

Equation (5.59) means that by evaluating a single integral, one gets both the coefficients, A_n and B_n. The integral on the right-hand side is generally a complex number. Its real part is A_n and its imaginary part is B_n.

Exercise 18: The sabersaw

The sabersaw function is similar to the sawtooth except that the teeth are exponential. The period of the function is T. On the interval 0 to T the function is given by

$$x(t)=\exp(\alpha t/T) \quad (0\leq t<T). \tag{5.60}$$

Use the exponential calculation from Exercise 17 to show that the Fourier series for the sabersaw is

$$x(t)=\frac{(e^{\alpha}-1)}{\alpha}+\sum_{n=1}^{\infty}\frac{2(e^{\alpha}-1)}{\alpha^2+(2\pi n)^2}[\alpha\cos(2\pi nt/T)-(2\pi n)\sin(2\pi nt/T)]. \tag{5.61}$$

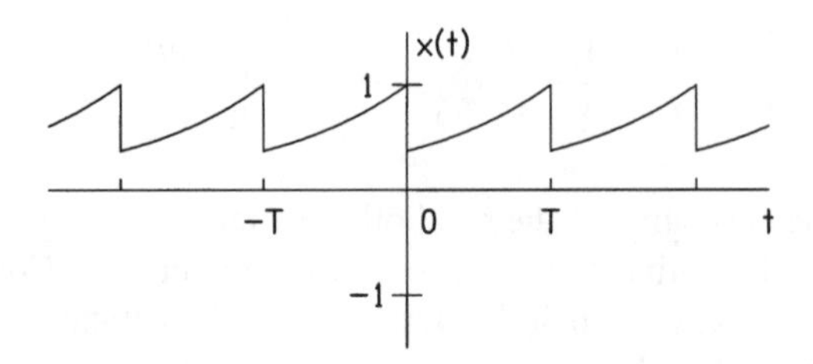

FIGURE E.18. The sabersaw for $\alpha > 1$.

Exercise 19: Fourier coefficients as optimum curve fitting parameters

Suppose that the infinite Fourier series for a periodic function $x(t)$ exists, but we want to approximate $x(t)$ by a sine and cosine series with a finite number of terms, N. A measure of the error that we make by truncating the series is the squared difference averaged over a period,

$$E=\frac{1}{T}\int_0^T dt\left|x(t)-A_0-\sum_{n=1}^{N}[A_n\cos(\omega_n t)+B_n\sin(\omega_n t)]\right|^2. \tag{5.62}$$

It is reasonable to choose coefficients A_n and B_n to minimize this error, that is

$$dE/dA_n=0 \tag{5.63a}$$

and

$$dE/dB_n=0. \tag{5.63b}$$

A priori, there is no reason to think that the coefficients A_n and B_n determined in this way should be related to the coefficients A_n and B_n that appear in the infinite Fourier series, which computes $x(t)$ exactly. However, it turns out that they are exactly the same coefficients! Prove that minimizing the error, via Eqs. (5.63) simply leads back to the definitions of the Fourier coefficients A_n and B_n in Eq. (5.15).

This result proves that the coefficients from Eq. (5.15) are not only the correct coefficients for the infinite series, they are also the optimum coefficients for a *finite* series. This means that if one wants to approximate a periodic function by a single sine wave then the best choice is to use the wave that has the largest Fourier amplitude C_n. Its amplitude and phase are given by the Fourier transform. If one wants an approximation that is the sum of two sine waves, then the best choice is to add up the components with the two largest amplitudes as given by the Fourier transform.

Exercise 20: The complement trick—the trapezoid and the triangular pulse

When one is looking for the Fourier transform of a periodic function $x(t)$, it may be useful to consider the complementary function $1-x(t)$. Sometimes it is easier to find the Fourier transform of the complementary function. Finally, of course, the two Fourier transforms differ only by a sign and by a DC term (frequency $f=0$) of unity. For example, consider the periodic trapezoid function shown in Fig. E.20(a). The function $1-x(t)$ is the triangular pulse wave shown in (b).

Show that the Fourier series for the triangular pulse is given by

$$tp(t)=\frac{\tau}{T}+\frac{T}{\pi^2\tau}\sum_{n=1}^{\infty}\frac{1}{n^2}\left[1-\cos\left(\frac{2\pi n\tau}{T}\right)\right]\cos\left(\frac{2\pi nt}{T}\right).$$

Therefore, you can show that the Fourier series for the periodic trapezoid is given by

$$x(t)=1-\frac{\tau}{T}-\frac{T}{\pi^2\tau}\sum_{n=1}^{\infty}\frac{1}{n^2}\left[1-\cos\left(\frac{2\pi n\tau}{T}\right)\right]\cos\left(\frac{2\pi nt}{T}\right).$$

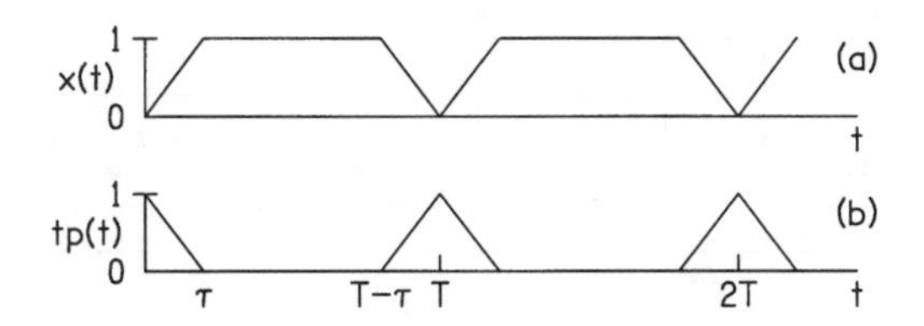

FIGURE E.20. (a) The periodic trapezoid function. (b) The complementary function, which is the periodic triangular pulse.

Exercise 21: The almost-square wave

The square wave is a standard option on electronic function generators. Ideally, the square function is in a high state for precisely 50% of the period and in a low state for the other 50%. As we have seen, such a function has half-wave symmetry and has no even-numbered harmonics. But real-world function generators are not ideal and the duty factor is not exactly 50%. Therefore, there is some even harmonic distortion that can be seen on a precision spectrum analyzer, as shown in Fig. E.21. Interestingly, the spectrum of the distortion is almost flat... that is, the 20th harmonic is essentially just as large as the second harmonic.

Explain why the even harmonic distortion spectra should be flat, especially if the square-wave quality is rather good.

[Hint: Consider that the difference between the actual, imperfect, square wave, which is high for slightly more than 50% of the cycle (a) and the ideal square wave (b) is the function shown in (c). Use the linearity of the Fourier transform to add the spectrum of (c) to the spectrum of (b).]

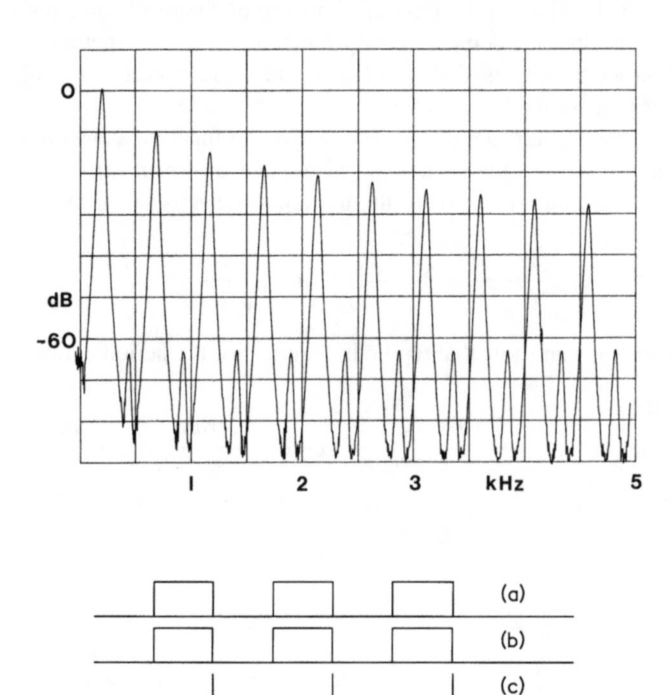

FIGURE E.21. (top) Spectrum of a 242-Hz ''square'' wave from a Wavetek function generator. Odd harmonics 1–19 are strong. Even harmonics 2–20 all have the same amplitude, 63 dB down from the fundamental. (a) The slightly imperfect square wave, (b) the perfect square wave, and (c) the difference between them.

CHAPTER 6

Perception of Periodic Complex Tones

Mr. Holmes, there have come to my ears several incidents which are hard to reconcile with the settled order of Nature.

A. CONAN DOYLE

Chapter 5 developed the concept of the periodic complex tone. Such a tone has the special property that if one knows the waveform for a single period, one knows the waveform for all time because a periodic complex tone repeats a single cycle indefinitely. According to Fourier's theorem, this tone is completely specified by the amplitudes and phases of its harmonics. The harmonics constitute the spectrum of the tone; they occur at discrete frequencies that are integer multiples of a fundamental frequency. Because of its rigid periodicity, the periodic complex tone is not similar to running speech or music, and yet it is prototypical of both of these.

The present chapter presents some facts about the perception of periodic complex tones. It is organized around the central dichotomy that the individual harmonics of the tone are accessible to the auditory system, and yet listeners normally do not hear them. Instead, listeners perceive a periodic complex tone as a single entity, characterized by a pitch and a tone color.

Once we leave the ideal world of mathematics and enter the real world of physical measurement and human perception, we must confront the fact that there can never be a periodic complex tone according to the strict mathematical definition, because all tones in the real world have a finite duration. This practical problem has an equally practical solution in the concept of asymptotic duration. A tone is asymptotically long if we find that making it longer doesn't change anything about the way that the tone is measured or perceived *while the tone is sounding*. For most practical purposes, a one-second steady tone from an electronic function generator can be called periodic. Although it doesn't last forever, each cycle is identical to the previous cycle.

Periodic complex tones, like those from the function generator, stand in contrast to aperiodic tones. Both kinds of tone may be constructed by adding up *partials*, defined as discrete sine wave components. Whereas periodic tones have harmonic partials, aperiodic tones have inharmonic partials. Harmonic partials tend to fuse together to make an integrated perceptual entity. Inharmonic partials tend to segregate and to be heard out individually. Periodic tones are the characteristic stuff for making melodies and harmony in music.

MUSICAL TONES AND THEIR SOURCES

Sustained-tone instruments, the bowed strings, brass, and woodwind instruments, as well as the human singing voice, make tones that carry melodies and can be used in close harmony. These tones do not have the cycle-to-cycle redundancy of the periodic complex tones from a function generator. They include intentional expressive variations such as glides and vibrato (frequency modulation) and tremolo (amplitude modulation), as well as unintended noise and glitches. These tones might be called ''adequately periodic,'' in the sense that the perceived attributes of fusion and pitch that apply to the ideal tones from precision electronic function generators also apply to them.

Sustained musical instrument tones are close enough to periodic that the standard physical description of them begins with a perfectly periodic prototype and then deals with variations as deviations from that prototype. The prototype is not just an abstract idealization. With special care it is possible to produce a sustained tone with a brass or woodwind instrument that is stable on an oscilloscope screen for many seconds. That means that the relative phases among the harmonics do not change over the course of many seconds. That, in turn, means that the harmonic frequencies deviate from integer multiples of a fundamental frequency by much less than 1 Hz.

Thus, if wind instrument tones (woodwinds and brass) differ at all from perfect periodicity, it is because the performer makes it so. The basic mechanism of sound production is capable of making ideally periodic tones. The situation is slightly different for the bowed strings, where jitter (random frequency modulation) appears to be an intrinsic character of the bowing process (McIntyre, Schumacher, and Woodhouse, 1981).

The harmonics of a wind instrument tone, essentially perfect, should not be confused with the resonances of the instrument itself. Wind instruments are hollow tubes that have been fashioned to have resonance frequencies that are approximately in a harmonic relationship. These resonances are responsible for generating the harmonics of tones, but they may easily deviate from harmonicity by 5 or 10 percent. How a musical instrument generates a tone with partials that are in a perfect harmonic relationship from resonances that are only approximately harmonic, is an essential part of the nonlinear tone generation process in wind instruments (Benade, 1976). Similarly, although the natural modes of a stretched string have frequencies that are progressively too large to be harmonic, the process of bowing the string creates a tone with harmonic partials.

Percussion instruments are best described in contrast to sustained-tone instruments. Whereas energy from the performer is continually fed into the production process in a sustained tone, a percussive tone gets all its energy at the onset. After the onset it is simply a free vibrator. Percussive tones are all more or less aperiodic. Nevertheless, many of them are capable of playing melodies. Percussive tones with the least inharmonicity, particularly the plucked and struck strings, can play close harmony. The partials of these instruments are much closer to a harmonic relationship than the partials of highly aperiodic instruments such as drums.

There is a long and wonderful history of techniques and devices that have been used to get the partials of percussion instruments to be closer to a harmonic relationship (Rossing, 1990). We list some of them here: The kettle of air that loads the membrane of timpani, the wax blob that loads the membrane of tabla drums, the plug that loads the end of a chime bar, the undercutting and mounting of glockenspiel bars, the resonators under the bars of marimba or xylophone, the tuning of church bells, and the winding of piano strings or guitar strings. When the science and the art are working right, the tones of these percussion instruments are adequately periodic, while the residual inharmonicities impart an evolving quality that makes the tones interesting (Fletcher, 1964).

HARMONIC ANALYSIS BY THE AUDITORY PERIPHERY

The study of the perception of periodic complex tones begins, as it should, at the auditory periphery. Physiological measurements on the basilar membrane and on the auditory nerve show that the peripheral auditory system analyzes tones according to frequency. This analysis is an integral part of the initial transduction of a tone from mechanical vibrations to neural impulses. Therefore, different harmonics of a complex tone end up in different neural channels.

The effects can be seen psychophysically in masking experiments using a sine tone probe and a complex tone masker. The idea of the masking experiment is that the sine tone cannot be heard if the excitation it causes in the auditory system is swamped by the masker excitation. The amount of masking, measured in dB, is defined as the threshold level of the sine tone probe (at which the sine tone is barely audible) in the presence of the complex tone, minus the threshold level when the sine tone is presented alone. By measuring the amount of masking as a function of the probe frequency, the experiment traces out the excitation pattern of the masker.

Excitation patterns for complex tones can be measured with nonsimultaneous masking techniques, where the probe and the masker occur at slightly different times. Plomp (1964) reported the results of forward masking experiments using a two-alternative forced-choice (2AFC) method. The masker was a 200-ms, 60-dB-SPL complex tone with a 500-Hz fundamental frequency and 12 harmonics of equal sensation level. Each experimental trial consisted of two presentations of the complex tone, separated by 1s. Either the first or second presentation, chosen at random, was followed by a 20-ms burst of the sine probe. The frequency of the probe was fixed at one of 75 values, from 300 Hz to 4000 Hz. The listener's task

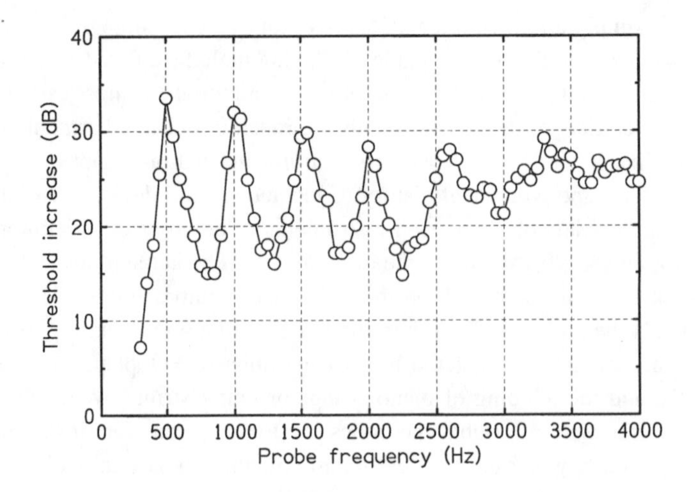

FIGURE 6.1. The increase in threshold for detecting a 20-ms sine tone probe in the presence of a 500-Hz complex tone forward masker compared to detection in quiet. The masker had twelve harmonics. Data from Plomp (1964).

was to indicate, by means of pushbuttons, whether the probe was heard on the first or second presentation. Trials were separated by 5s. The threshold for the probe was taken to be the level at which 75% of the listener's responses were correct.

The results, averaged over four listeners, are given in Fig. 6.1. The peaks and valleys in the threshold show the tonotopic organization of the auditory system. There are regions of strong excitation caused by spectral peaks of the complex tone, where the threshold for the sine probe could increase by almost 20 dB, and between these there are regions of reduced excitation where the threshold for the probe is minimum. Because neural channels are ordered tonotopically, the data indicate that the auditory system analyzes components of the complex tone masker into separate frequency channels, in this case up to about the fifth harmonic.

A similar peak and valley structure was found using another nonsimultaneous masking technique known as pulsation threshold (see the first Methods box) introduced by Houtgast (1974). For complex tones with a fundamental frequency of 250 Hz, and harmonic levels of 40 or 70 dB SPL, the pulsation threshold revealed six peaks at frequencies of harmonics of the complex tone.

The most straightforward masking technique is direct masking, where masker and probe signal are simultaneous. The complex tone studied by Plomp in forward masking was also studied by direct masking using a Békésy tracking method (see the second Methods box). Each harmonic had a level of 40 dB SPL. The tracking experiment covered Plomp's range, 300 to 4000 Hz, in 15 minutes, which can be compared with the ten hours needed for one listener in Plomp's 2AFC experiment. The midline from the tracking experiment is plotted in Fig. 6.2.

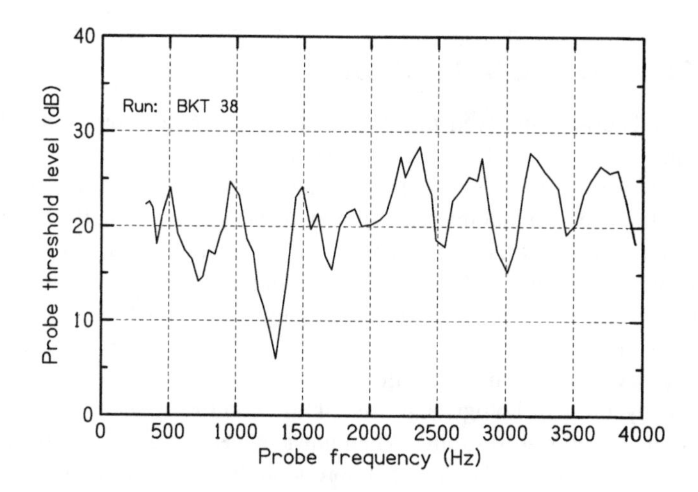

FIGURE 6.2. Data from Békésy trace of the threshold of a sine tone given a continuous 50-dB-SPL 500-Hz complex tone masker with 10 harmonics. The midline shown here has been reduced from 192 points to 70 by a spline fit.

The direct masking data in Fig. 6.2 show that the first three harmonics (500, 1000, and 1500 Hz) stand out as expected, but the fourth harmonic cannot be seen. Harmonics 5, 6, and 7 actually produce dips in the threshold and not peaks. These dips are reproducible for this listener, but they do not occur for all listeners. The dips can be attributed to interaction between the complex tone and the probe. A linear interaction leads to beats, which call attention to a harmonic, leading to a reduced threshold. A nonlinear interaction produces difference tones which may also reduce threshold.

In conclusion, frequency analysis by the auditory system can be seen in direct masking, but the data are not as impressive as the data for nonsimultaneous masking methods. Auditory bandwidths determined by direct masking are wider than those determined by forward masking or pulsation threshold (Houtgast, 1972). Fewer harmonics of a complex tone appear to be resolved. Second, the beats and difference tones that occur with direct tone-on-tone masking lead to some peculiar effects. Because of these problems, tone-on-tone masking is not normally used to determine auditory filter bandwidths.

Both nonsimultaneous masking experiments and direct masking experiments give evidence that low-numbered harmonics of a periodic complex tone are distinctly separated by the auditory system. One might predict that these harmonics are individually audible when one listens to the tone. However, this is not what happens when one listens to the complex tones of speech or music, or even to the tones of electronic function generators. Normally such tones are perceived as single entities. Some process, higher than the periphery, evidently puts the harmonics back together again.

METHOD: PULSATION THRESHOLD

In 1957, Thurlow observed that if a weak tone and a strong tone alternate several times per second, the weak tone may appear to be continuous, extending through the strong tone (see also Thurlow and Elfner, 1959). Thurlow drew an analogy with the perceptual interpretation of visual scenes where the background is presumed to be continuous even when it is obscured by a foreground. For example, when a book is on the table, one supposes that the table continues without interruption under the book, even though one cannot see the part of the table that is under the book.

The analogy with vision is appealing, but it is incomplete because it does not really convey the psychological power of the auditory effect. The continuity effect can be unbelievably strong. When music alternates with masking noise in a 50-50 division, one has the distinct impression that the rhythm of the music is maintained; the melodic line and embellishments, such as pitch glides, become continuous. Essentially half the music comes entirely from within the listener's head, and the listener hardly notices which half it is. With speech the effect can also be impressive. If speech is electronically interrupted several times per second by silent intervals, there is a considerable loss of intelligibility. But much of the intelligibility can be restored by filling the silent intervals with intense random noise (see Warren, 1984, for a review). For a comparably dramatic visual analogy one needs the picture of fragmented letters with and without blobs of ink from Bregman (1981) shown in Fig. 6.3. Even if one knows that part (a) is letters of the alphabet, one cannot recognize the letters because of the gaps. In part (b) the gaps are filled with ''noise'' which would mask the letters if they were actually continuous. Recognition there is easy.

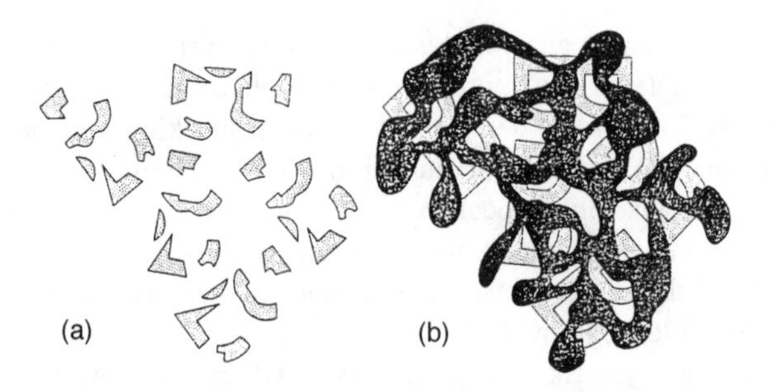

FIGURE 6.3. (a) A pattern of letters, fragmented by blank gaps. (b) The same, with ink blobs covering the gaps.

The continuity effect was termed ''auditory induction'' by Warren *et al.* (1972), who suggested that the effect occurs when the level difference between the alternating sounds is such that the weak sound would have been completely masked by the strong sound if it actually were continuous. A similar observation was made by Houtgast (1972), who turned the effect into an experimental method as an alternative to masking. The experiment arranges a weak signal to alternate with a strong masker, for example noise. The experiment looks for the most intense signal that can just barely be heard as pulsing, because, according to Houtgast's interpretation, there then exists a neural channel in which the excitation caused by the signal is just barely greater than the excitation caused by the masker. Hence, the method is named ''pulsation threshold.''

To find the pulsation threshold the listener adjusts either the level of the signal or the level of the masker. By adjusting the level of the signal, the listener probes the masker. By adjusting the level of the masker, the listener probes the excitation caused by the signal. A typical sequence of events for a pulsation threshold experiment is shown in Fig. 6.4.

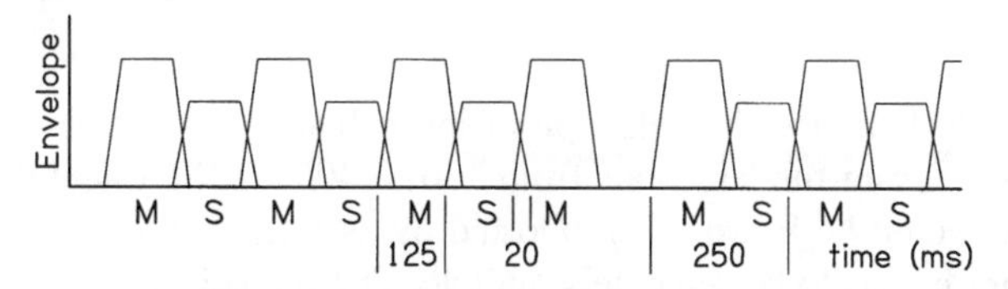

FIGURE 6.4. Amplitude envelopes of alternating signal (S) and masker (M) in a pulsation threshold experiment. There are four cycles per second. Signal and noise onset and offset ramps are 20 ms in duration and coincide, leading to some overlap. Note that every fourth signal pulse is missing. Houtgast found that this modification makes it easier to adjust the pulsation threshold.

HEARING OUT HARMONICS

Although a periodic complex tone is normally perceived as a single entity, the frequency analysis begun by the peripheral auditory system is not entirely lost. There are special signal conditions in which listeners do hear the harmonics of a periodic tone individually. This section describes such conditions. By studying them, psychoacousticians hope eventually to come to some understanding of the process by which harmonics are integrated in the normal case.

Spectrally Prominent Harmonics

The spectra of simple waveforms tend to have amplitudes that decrease smoothly with increasing frequency, and they tend not to have large spectral gaps. Using computer synthesis, it is possible to generate periodic tones that violate these rules, and to make a particular harmonic be unusually prominent in the spectrum. That harmonic is then heard out individually. For example, the two spectra shown in Fig. 6.5 both have a strong pitch at 1000 Hz in addition to the expected low pitch near the fundamental frequency of 200 Hz.

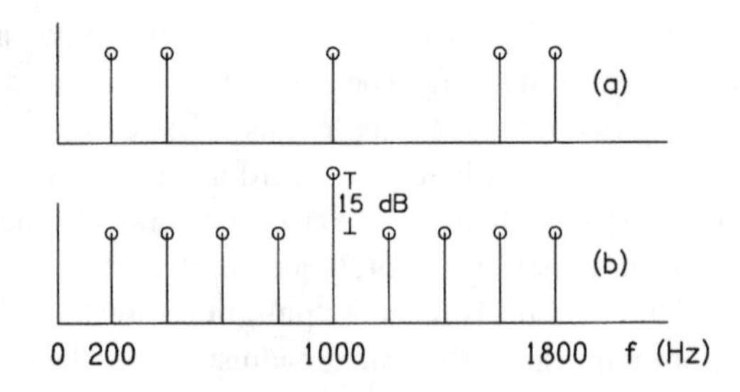

FIGURE 6.5. Two periodic complex tones with fundamental frequency of 200 Hz that produce a strong pitch near 1000 Hz because the fifth harmonic is spectrally prominent. In (a) the fifth harmonic is isolated in a gap. In (b) the fifth harmonic is boosted in level.

Martens (1981) studied the detection of individual boosted harmonics as in part (b) of Fig. 6.5. Using fundamentals from 100 to 400 Hz, he found that harmonics near 1500 Hz could be heard when boosted by as little as 5 dB. Other harmonics required larger boosts to be separately audible. These results are roughly consistent with the measurements made by Deng, Geisler, and Greenberg (1987) on the increment that is required for a harmonic to begin to dominate the synchrony of an auditory nerve fiber that is not tuned to that harmonic.

Popping Out Vocal Harmonics

The Gyuto and Gyume monks of Tibet perform a "harmonic chant" that requires them to tune their vocal tracts to produce a sharp resonance that emphasizes a single harmonic of a sung vowel sound. The listener hears the tonal vowel, with a low pitch corresponding to the fundamental frequency of the voice, accompanied by the whistling of the tuned harmonic. The whistle tracks the fundamental tone through vocal fluctuations, leading to an artistic ambiguity about whether there is one source or two. Westerners, on first exposure to this harmonic chanting, literally cannot believe their ears. It is so different from anything that we do in singing, or thought of doing, or even thought possible to do.

The technique has been brought to the west by David Hykes, whose performances reveal a remarkable control of the frequency and sharpness of the resonance. On a single sustained vowel sound he is able to slide the resonance up and down the harmonic series, emphasizing individual harmonics and causing them to pop out in succession. Alternatively he can move the fundamental and the emphasized upper harmonic in parallel, or hold a steady emphasized frequency while moving the fundamental.

An example of a sound with an emphasized harmonic is given in Fig. 6.6(a). This figure shows the spectrum of a one-second sample of a sustained tone sung by David Hykes in the song "Raised by a Power." The fundamental frequency is 208

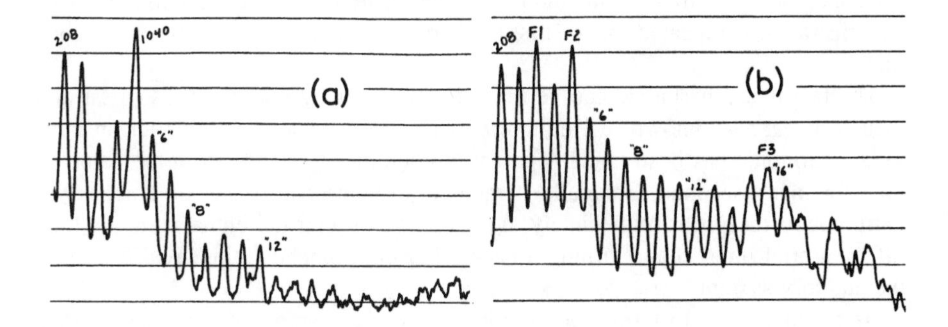

FIGURE 6.6. Spectra of sung vowels. The horizontal axis is linear with frequency, from 0 to about 4500 Hz. Each division on the vertical scale is 10 dB (overall dynamic range of 80 dB) Fig. (a) shows a harmonic chant. Fig. (b) shows the western vowel "AH." Labels indicate the fundamental component at 208 Hz, the positions of three formants, and a few harmonic numbers. The isolated strong harmonic at 1040 Hz in (a) is heard as a high whistle accompanying the low-pitch tone.

Hz. Twelve harmonics can clearly be seen in the figure, dominated by an emphasized fifth harmonic. The fifth harmonic at 1040 Hz is about 8 dB higher than the fundamental, and it has more power than all the other harmonics combined. This tone evokes two pitches, one at the frequency of the fundamental, the other at the frequency of the fifth harmonic. The particular tone chosen for analysis here is typical.

Figure 6.6(b) is the spectrum of the vowel "AH" sung by a western singer in a natural singing voice. It also has a fundamental frequency of 208 Hz. The third and fifth harmonics are emphasized by the first and second formants of this vowel sound. There is a broad third formant centered on the 16th harmonic at 3328 Hz.

A comparison between the harmonic chant and the western vowel shows that both strongly emphasize upper harmonics. In both cases the oral filter resonances are sharp. The harmonic chant, however, uniquely isolates a harmonic in the spectrum, with spectral valleys on either side of the emphasized harmonic and no other harmonic of comparable strength anywhere in the spectrum. Isolation causes the harmonic to pop out perceptually.

Modulation of Individual Harmonics

There are numerous techniques that can be used to cause a listener to hear out individual harmonics. Attention can be called to a particular harmonic by turning it on and off repeatedly during the presentation of a complex tone. The first demonstration on the compact disc by Houtsma *et al.* (1987) shows this effect. A 200-Hz complex tone sounds continuously. A selected harmonic is turned off for about half a second, then turned on again, and the listener hears that harmonic stand out. This demonstration is repeated for the first ten harmonics, and it could easily have gone higher.

The fact that a harmonic can be made to stand out by making it disappear and then reappear was known to Helmholtz, who further observed that after an individual harmonic has been made apparent, it continues to stand out in the complex tone for many seconds. Gradually it fades away. One can cause the emphasized harmonic to disappear immediately, however, by turning the entire complex tone off for a brief interval and turning it back on again. Somehow, this break ''resets'' the auditory system.

As an alternative to turning a harmonic on and off, amplitude modulation of a particular harmonic in a complex tone can cause that harmonic to stand out. Frequency modulation too, in the form of vibrato or jitter (random frequency modulation), of a single harmonic can cause the harmonic to stand out. A demonstration by McAdams (1984a,b) exposed the first 12 harmonics of a complex tone in sequence by giving each harmonic in turn a vibrato similar to a singer's vibrato. The individual harmonics were heard with no effort at all.

Modulation of individual harmonics must be distinguished from modulation of the complex tone as a whole. If all the harmonics of a complex tone are modulated together, in amplitude or in frequency, then it becomes more difficult to hear out individual components in the complex, not easier. When amplitude modulated together, the components of a complex tone all grow or decrease in amplitude with the same envelope so that the relative amplitudes, and the shape of the waveform, remain unchanged. When frequency modulated together, the components retain their harmonic relationship so that the waveform remains approximately periodic over the short term. This means that the frequency excursion of a harmonic at any point in time is proportional to the frequency of that harmonic, i.e.

$$\Delta f_n = n \Delta f_0 . \tag{6.1}$$

Such modulation is called ''coherent'' in the literature.

There is one circumstance in which coherent frequency modulation can cause a harmonic to stand out, and that is when the waveform is passed through a filter with a passband having a sharp edge. The filter induces an amplitude modulation of a particular harmonic as that harmonic goes in and out of the pass band. The effect sometimes appears in attempts to synthesize the singing voice, where a complex tone is frequency modulated to simulate vocal vibrato and is then passed through formant filters to simulate vowel formation by the vocal tract.

Induced amplitude modulation is also responsible for the fact that it is not hard to hear out harmonics of a complex tone that is sounding for an extended duration in a room. Moving one's head by half a meter causes harmonics to coincide momentarily with resonant peaks or valleys of the room response (a form of filtering) which causes the harmonics to stand out. The third harmonic is usually the easiest to hear in this way. It is probably because of such room effects that early acousticians, e.g. Mersenne (1636) and Sauveur (1704) became aware of the individual harmonics in complex tones. [An interesting history has been given by Plomp (1964).] Because of room resonances, listening to acoustical sources in rooms is usually inappropriate for psychoacoustical experiments. The other side of this statement is that one must be cautious when applying the results of psychoacoustical experiments to predict perception in rooms.

Vowels Again

Measuring the spectrum of the ordinary vowel in Fig. 6.6(b) led to further insight into the unifying influence of natural variations in sounds. The spectrum was made by capturing the vocal sound in a one-second digital buffer and then playing it out in a repetitive loop to create a continuous sound. This sound was then analyzed by a swept-frequency spectrum analyzer. It was discovered that one could easily hear at least one of the strong harmonics in the ordinary vocal tone of Fig. 6.6(b) while the one-second segment was endlessly repeated, even though no harmonic could be heard out in the original voice. The implication of this observation is that the formants of western vowels can be strong enough and sharp enough to cause harmonics to be individually audible if the tone is artificially frozen, but the dynamic variations of running speech or singing prevent this from happening. The normal continuous motion in running speech apparently prevents a listener from focusing on any steady-state spectrum. A second implication is that if one's goal is to cause a harmonic to pop out then there is an important advantage to stability in the production of the tone.

Cueing a Harmonic

The modulation methods that cause an individual harmonic to stand out can easily bring out the first 10 or 12 harmonics, as noted above, and they may well be able to go considerably higher. Helmholtz claimed to be able to hear the sixteenth! That individual harmonics can be perceived in this way is excellent confirmation of the physiological fact that individual harmonics are resolved at the periphery of the auditory system, but it poses a problem too because one does not expect the auditory periphery to resolve harmonics as high as the tenth, let alone the sixteenth (see Chapter 10).

This problem was addressed by Plomp (1964, 1976) and Plomp and Mimpen (1968) who used a cueing experiment: To run the experiment, the listener had a three-position switch. The center position turned on a complex tone with 12 harmonics (fundamental f_0). A second position turned on a sine tone having the frequency of one of the harmonics of the complex tone (nf_0), arbitrarily chosen.The third position turned on a sine half-way to a neighboring harmonic, either

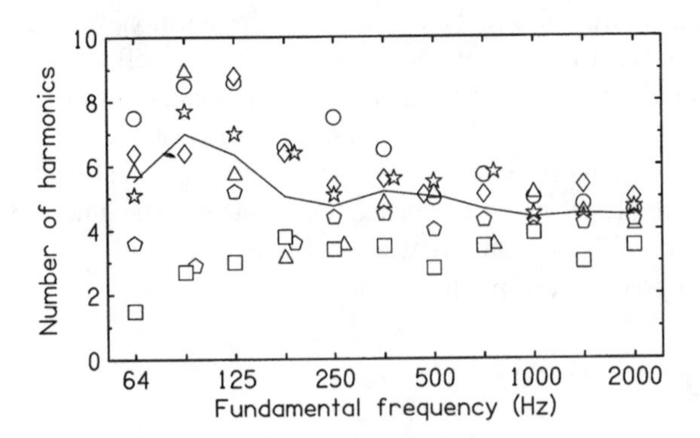

FIGURE 6.7. The number of resolvable harmonics as a function of fundamental frequency, as found by Plomp and Mimpen (1968). Data for six subjects are shown by six different symbols. The median is shown by a solid line.

$(n+1/2)f_0$ or $(n-1/2)f_0$, chosen randomly. The listener's task was to choose the switch position with the sine correctly tuned to a harmonic.

In this experiment, the sine tones serve as both test and cue. A harmonic was said to be resolvable if the listener chose the correctly tuned sine on at least 75% of the trials. Generally, listeners found it easy to identify low-numbered harmonics, but progressively harder as the cued harmonic number increased. The harmonic numbers of the highest harmonics that could be resolved, for six listeners, are given in Fig. 6.7.

The data show that listeners can resolve about five harmonics, which is the answer that one expects from forward masking or pulsation threshold measurements. Plomp (1964) went on to show that his measurements truly reflected frequency resolution, unrelated to the periodic character of his complex tones. When he substituted an aperiodic complex tone with compressed partials, the psychophysical results were essentially unchanged.

Isolating a Harmonic by Unmasking

If one listens to a periodic complex tone with the fourth harmonic pulsing on and off repeatedly, one easily hears out the fourth harmonic as a separate entity when it comes on. This effect was noted at the beginning of the modulation section above. However, there is more. After listening to this sequence for a while, one becomes aware that the fourth harmonic seems to be alternating with the fifth harmonic. The fifth harmonic is heard out as an entity just when the fourth harmonic is turned off. In general, turning off the n-th harmonic tends to pop out harmonic number $n+1$. Informal experiments show that for a 200-Hz fundamental, this effect can be heard at least for $1 \leq n \leq 9$.

Although this effect has not been the subject of formal research, it is not hard to

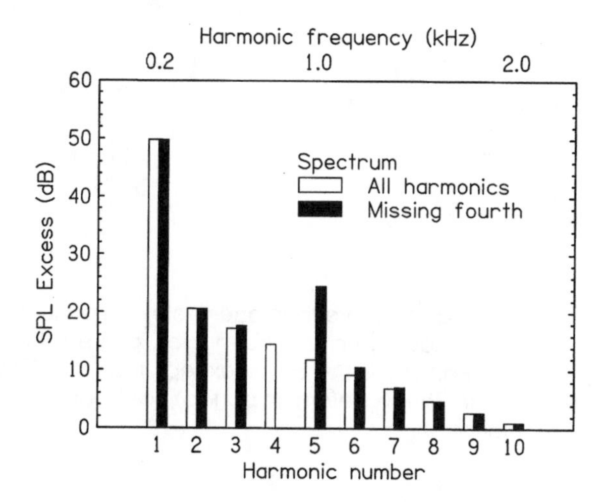

FIGURE 6.8. The SPL excess, calculated from the model of Terhardt, Stoll, and Seewann, for a complex tone where all harmonics have equal amplitudes (open bars) and where the fourth harmonic is missing (filled bars). The unmasking of the fifth harmonic when the fourth harmonic disappears is responsible for the fact that the fifth harmonic can be heard out separately.

find a reasonable explanation for it. The excitations along the tonotopic coordinates of the auditory system caused by the individual harmonics of a complex tone tend to mask one another. Because of the upward spread of masking, excitation from low harmonics particularly tends to mask excitation from higher harmonics. For example, excitation from the fourth harmonic is most effective in partially masking excitation from the fifth. When the fourth harmonic is removed, the excitation from the fifth harmonic is unmasked. The result is like a sudden boost in the amplitude of the fifth harmonic. Just as in the case of AM, this amplitude increase causes the fifth harmonic to stand out.

The magnitude of the unmasking effect can be calculated from loudness models that take into account mutual masking. Here we use a simple model by Terhardt, Stoll, and Seewann (1982a,b), which represents the effective excitation of a harmonic by a quantity called the "SPL excess." The SPL excess of harmonic n is measured in dB. It begins with the physical level of harmonic n in dB SPL, and is then reduced by three effects: (1) audiometric threshold for the frequency of harmonic n, (2) background noise in a critical band around harmonic n, and (3) masking by all the other harmonics in the spectrum. The masking effect is estimated by triangular auditory filters with linear skirts on the scale of (level)/(critical band rate) or dB/Bark (see Chapter 10). The upper skirt is level independent; the lower skirt becomes flatter with increasing level.

Figure 6.8 shows the SPL excess calculated from that model for a 200-Hz

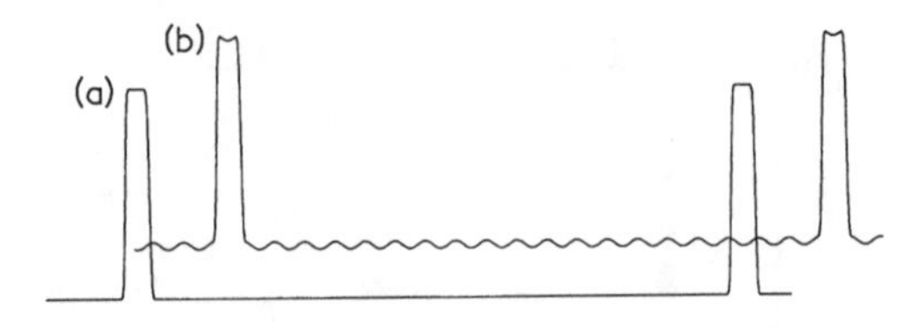

FIGURE 6.9. Part (a) shows a little more than one cycle of a pulse train with a duty factor of 1/25, made by adding up 200 harmonics, passed through a gentle Gaussian lowpass filter. Part (b) is the same except that the 20th harmonic is missing from the spectrum. The waveform in part (b) has been displaced upward and to the right for clarity.

complex tone having 12 harmonics. Only the first ten are shown. The open bars are for a tone where all harmonics have a level of 70 dB SPL. The closed bars are for the same tone except that the fourth harmonic has been eliminated. What is particularly striking is the way that the SPL excess for the fifth harmonic is boosted when the fourth disappears. It is equivalent to a 13-dB increment in the physical level of the fifth harmonic, and it is not surprising that this effect causes the fifth harmonic to become prominent.

Conspicuous by Its Absence

One of the more peculiar effects in psychoacoustics is the ''Pitch of the harmonic that is not there.'' Duifhuis (1970a,b) showed that if one listens to a low-frequency pulse train that is missing a single high-numbered harmonic then one hears out the missing harmonic as an isolated tone. Seen from a spectral point of view, this effect is indeed ''hard to reconcile with the settled order of nature.'' It suggests a unique figure-ground effect with a characteristic spectral template playing the role of ground. To confound matters further, the effect takes place for high harmonic numbers where figure and ground would not even be resolvable. A graph of the waveform as a function of time, however, suggests a more plausible point of view.

Before calculating the waveform, we can imagine what it must look like. A pulse train with a missing harmonic ought to be just what one gets if one starts with a perfect pulse train and then adds a sine tone to cancel the harmonic that one wants to eliminate. The added sine must have the right frequency and amplitude and a phase that is opposite to the phase of the harmonic in the pulse train. Thus, we expect the waveform to look like the usual two-state pulse train plus a weak sine tone. It now seems less surprising that one hears the pitch of the sine tone.

An example to illustrate this idea is shown in Fig. 6.9. Part (a) shows a segment of a pulse train (Chapter 3, Example 2) with a duty factor of 1/25. The figure was made by adding up the first 200 cosine harmonics, according to formula (5.49), so that

$$x(t)=\sum_{n=1}^{200} A_n \cos(2\pi nt/T). \tag{6.2}$$

To make the rectangular pulse, the amplitudes A_n should have the values

$$A_n=\frac{2}{\pi}\frac{1}{n}\sin(n\pi p), \tag{6.3}$$

where $p=1/25$. A waveform made with these harmonic amplitudes has Gibbs phenomena, which are regions of rapid oscillation at the discontinuities in the pulse. They can be suppressed by a smooth lowpass filtering. A Gaussian filter does a good job. Therefore, the harmonic amplitudes in Fig. 6.9(a) as actually calculated were

$$A_n=\frac{2}{\pi}\frac{1}{n}\sin(n\pi p)e^{-(2n/N)^2}, \tag{6.4}$$

where N is the maximum harmonic number, $N=200$. The Gaussian factor is monotonically decreasing. At the highest frequency, where $n=N$, it is down by a factor of e^{-4}, equivalent to 35 dB. The filtering eliminates the Gibbs oscillations at the expense of rounding the pulse somewhat. Part (b) of Fig. 6.9 was calculated with the same cosine series except that the 20th harmonic was omitted from the sum.

Duifhuis explained this pitch effect as the result of a temporal division of the signal appearing in an auditory filter around the missing harmonic. According to this explanation the sine tone is heard during the flat (low) segments of the waveform. This predicts that the effect should be harder to hear as the pulse rate increases, because the flat segments become shorter. This prediction was borne out by experiment, though Duifhuis (1970b) did find some remaining effect at a pulse rate as high as 400 Hz.

Experiments by Jian-Yu Lin (Lin and Hartmann 1995b) using the waveform of Fig. 6.9, found that it is very easy to hear the pitch of the missing 20th for a fundamental of 50 Hz, but not possible for a fundamental of 100 Hz. The level of the missing harmonic (some oxymoron that!) could be found by the method of best beats (see Chapter 17), or, with smaller variance, by a level matching task. Levels found in this way were systematically several dB below the mathematical level of the missing harmonic, probably due to partial masking. Lin's experiments also showed that when random phases were used the pitch disappeared completely. Random phases removed the flat portions of the waveform, which are required by the Duifhuis model.

Further experiments by Lin (1996) studied the detection of a missing harmonic as the frequency band around it was made narrower. As the bandwidth became smaller the pitch disappeared, even when the bandwidth was still appreciably larger than the critical band. This result challenges the auditory filter assumption of the model. It appears that the perception of a missing harmonic depends upon the existence of components outside the critical band.

METHOD: BÉKÉSY TRACKING

In traditional pure-tone audiometry the listener indicates whether or not the tone can be heard while the experimenter changes the level of the tone. In 1947, von Békésy introduced a modification wherein the listener controls the level of the tone. When the listener can hear the tone he pushes a button, which causes the level of the tone to descend at a regular rate. When the listener can no longer hear the tone he releases the button so that the level of the tone increases. It is typical for the frequency of the tone to increase slowly with time to trace out the audiometric threshold as a function of frequency. In the original Békésy audiometer, a chart recorder tracked the frequency and the level of the tone. Now, the procedure is usually implemented on a general purpose microcomputer.

The parameters of a Békésy tracking experiment depend upon the experimenter's purpose. If the goal is to make a rapid assessment of possible hearing loss, the frequency might increase at a rate of one octave per minute. If the goal is to measure threshold microstructure, one might spend 10 minutes sweeping a single octave, and the sweep would not necessarily be exponential in frequency. Typically, the level changes at a rate of about ± 2 dB per second. Reliability is increased if the test tone is pulsed. A simple procedure is to pulse the tone on for 250 ms, off for 250 ms, and to change the level by ± 1 dB on each pulse.

When the tracking run is over, the data appear as a set of upper and lower turnarounds, as shown by the dashed line in Fig. 6.10. To find the threshold, one draws the midline, which passes through the middle of each ascending and descending line segment. The midline is smoother than the line connecting turnarounds. The midline may be subjected to further smoothing by spline fit or other curve fitting procedure.

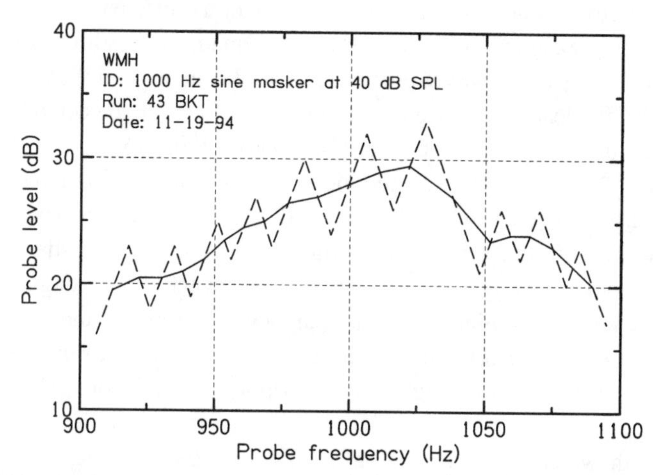

FIGURE 6.10. One-minute segment of a Békésy track in an experiment to determine the threshold for a sine tone in the presence of a 40-dB 1000-Hz sine tone masker. The midline, shown by the solid line, averages upper and lower turnarounds to estimate the threshold level.

SEGREGATION AND INTEGRATION

When two or three sustained-tone instruments play together, the individual sounds are combined in the air by linear addition of the waveforms. The sounds are also reflected from the surfaces of the room, usually many times, before they fade away. All this activity results in an acoustic field in which the sounds of the individual instruments are thoroughly entangled. Nevertheless, a listener probing this field with two ears can successfully disentangle the tones and recreate the images of the individual instruments. Further, the listener can appreciate the subtle nuances of timing and intonation that the performers give to the tones. It is truly an impressive bit of signal processing! The operations, as Helmholtz put it, ''first in separating the musical tones arising from different sources and secondly in keeping together the partial tones of each separate source'' are those of segregation and integration.

It is evident that the processes of segregation and integration cannot be entirely based upon auditory filtering. The harmonics of any one instrument fall into different auditory channels, and some harmonics from different instruments fall into any one auditory channel. Two important alternative mechanisms arise from temporal features of the signals, onset synchrony and waveform periodicity.

Onset Synchrony

It is likely that the most important acoustical mediator of perceptual segregation and integration is in the onsets of tones. If two sets of partials have different onset times, the auditory system tends to segregate them as different entities. The tone-on-tone masking experiments by Rasch (1978) found that a 30-ms advance in the onset time of a target tone was equivalent to a 40-dB increase in level. Although the advanced onset clearly produced an enormous effect on detectability in these experiments, listeners were not otherwise aware of the onset asynchrony. In a second study, Rasch (1979) measured the onset asynchrony that actually occurs in performed ensemble music when the score indicates simultaneous onsets. The RMS asynchrony varied from 27 ms for a woodwind trio to 49 ms for a string trio. Taken together, the two papers by Rasch show that onset asynchrony among accomplished ensemble players leads to increased transparency in performed music, while maintaining the impression that the musicians are playing in correct time. Transparency allows a listener to "hear through" layers of polyphonic music, even to the inner voices.

The way in which onset asynchrony tends to segregate different sources, and onset synchrony tends to unify the harmonics of individual sources, is not completely understood. Little attention has been paid to the fact that harmonics of a single musical instrument tone do not really rise simultaneously. For example, the upper harmonics of a brass instrument tone rise more slowly than the lower harmonics. Measurements of harmonic amplitudes vs time after first onset, by Risset and Mathews (1969) and Beauchamp (1975), show that high harmonics lag low harmonics by considerably more than 30 ms. The evolution of the spectrum of a tone is known to be an important contribution to timbre and to the identification of the instrument (Grey, 1977). It would be most interesting to see a psychoacoustical study of onset asynchrony among voices that had realistically evolving spectra like this.

Waveform Periodicity

Periodicity is an important element in the segregation and integration of the steady-state portion of tones in the absence of spectral anomalies. It plays a major role in segregating the sounds of different musical voices. It is always a consideration in the identification of vowel sounds in noise and the segregation of different talkers (Brokx and Nooteboom, 1982; Darwin, 1984).

To give the study of periodicity a specific focus, we imagine two periodic complex tones. One has harmonic frequencies 100, 200, 300, 400, 500, and 600 Hz. The other has harmonic frequencies 220, 440, and 660 Hz. These tones might be synthesized male and female voices. The spectra are clearly interleaved; no simple filter can pull them apart. However, the two voices are perceived as separate entities. The question of interest is, "How does the auditory system perform this segregation?" There are essentially two answers, one based upon template fitting, the other based upon neural synchrony.

Spectral Template Fitting Models

Modern theories of pitch perception (Goldstein, 1973; Terhardt, 1974b) are foremost pattern matching theories. They assume that the brain has stored a template for the spectrum of a harmonic tone, and that it attempts to fit the template to the neurally resolved harmonics of a tone. Given the ubiquity of periodic tones in the everyday environment, a template for the harmonic spectrum is a reasonable hypothesis. Applied to the present problem, it is clear that the harmonics of the 100-Hz complex tone form a pattern that fits a template that does not include any of the listed harmonics of 220 Hz. The harmonics of 220 Hz fit a template of their own, and this is one possible explanation for why the two tones are segregated. The model can be made quantitative by establishing tolerances for the template for accepting harmonics into an entity. Because the template fitting models begin with resolved partials, they look like place-theory models, and therefore, one would expect some correspondence between the tolerances and auditory frequency resolution.

This is the spirit of the DWS pitch meter (Duifhuis *et al.*, 1982; Scheffers, 1983) which sifts the resolved components through a harmonic sieve with slots of given bandwidth (4% of the center frequency) centered on harmonic values. The pitch meter was designed to find the pitch of a signal in a noisy background. Those components that do not pass through the slots in the sieve are rejected from the computation of the tone pitch and are attributed to noise. In its present form, the pitch meter does not try to find more than a single harmonic series, but there would seem to be no reason in principle why it could not be programmed to find several.

Similarly, Terhardt *et al.* (1982a,b) have extended Terhardt's virtual pitch model to allow for the perception of two or more virtual (low) pitches. As in the case of the pitch meter, the decision criterion is in the frequency domain, but it operates on the subharmonic cues generated by the components rather than on the components themselves.

Neural Synchrony Models

An alternative model for segregation and integration is based upon neural timing. It regards as crucial the fact that harmonics of a complex periodic tone are synchronized with one another, but the harmonics of two different tones are not normally synchronized. Whereas place-based models find their physiological inspiration in auditory tonotopic analysis, timing models are inspired by the synchronization of neural firing with waveform features. The reliance on neural synchrony limits the timing models to low frequency, not higher than 5 kHz.

Timing models can take various forms. The most peripheral is a waveform model, which directly exploits the synchrony of auditory nerve spikes with waveform features. This model depends upon an incomplete tonotopic analysis of the spectral components. It assumes the existence of some peripheral neurons that synchronize to at least two harmonics so that synchrony anomalies associated with

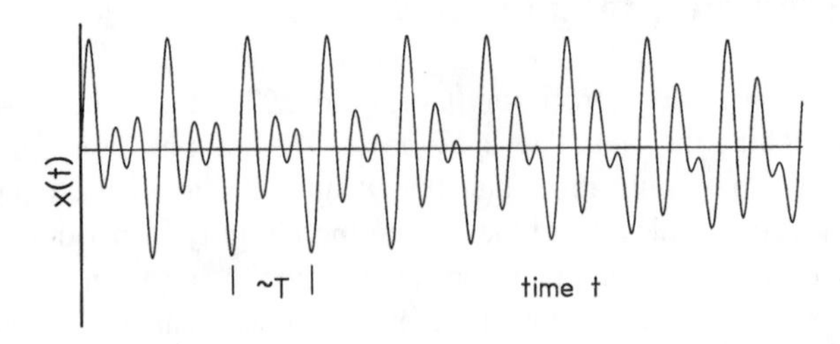

FIGURE 6.11. A tone composed of harmonics one, two, and three in which the second harmonic is mistuned by 1%. The mistuning causes the waveform to roll when viewed on an oscilloscope and causes a "rolling" auditory sensation known as the beats of mistuned consonances.

inharmonicities can be detected. To show the origin of synchrony anomalies we consider the perfectly periodic tone of Fig. 5.1 in Chapter 5 having three exact harmonics. If the second harmonic is mistuned by 1 percent, the waveform acquires the evolving character shown in Fig. 6.11. The evolution of the waveform ultimately leads to a corresponding evolution of the neural firing pattern in any neuron that is excited by both the second harmonic and one of the other harmonics. Multiple synchronies are known to occur in the eighth nerve (Javel, 1980). Other timing models, Meddis and Hewitt (1991a,b) or Patterson *et al.* (1992), are not based upon the waveform *per se*. They compare the neural activity patterns in different tonotopic channels after complete tonotopic analysis.

Mistuned Harmonic Experiments

A simple but effective way to study the segregation and integration of steady-state tones is the mistuned harmonic experiment. Here there is a complex tone in which all the partials are exact harmonics except for one. The goal of the experiment is to determine the conditions in which the mistuned harmonic is heard as a separate entity.

The most efficient experimental technique is the mistuned harmonic discrimination (MHD) experiment. It is a two-interval, forced-choice experiment where the listener hears two tones. One of them (either the first or the second, chosen randomly), is perfectly periodic; the other has a mistuned harmonic. The listener's task is to say which of the two tones includes the mistuned component.

Although much useful information can be obtained from this experiment, there are potential difficulties in interpretation. The problem is that the detection of inharmonicity is not the same as hearing out the mistuned harmonic as a separate entity. For short duration tones there are pitch shift effects. For long durations there are beats of mistuned consonances.

For short tones (50 ms or less), the most salient effect of mistuning a harmonic

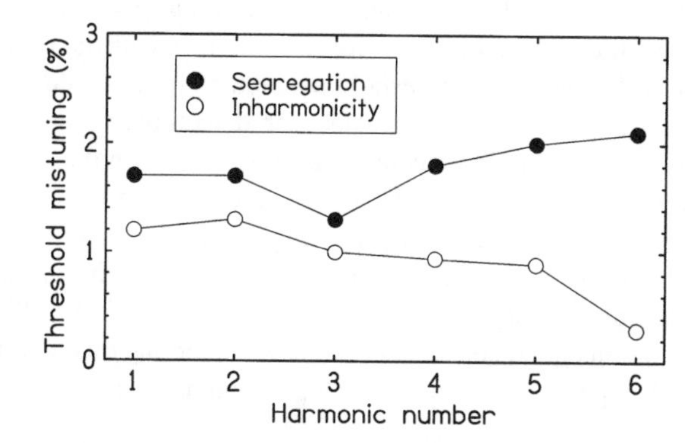

FIGURE 6.12. Thresholds for detecting the mistuning of single harmonics from Moore, Peters, and Glasberg (1985c,1986). Open symbols show the amount of mistuning required to distinguish between a periodic tone and a tone with one mistuned harmonic. Thresholds found in this experiment continue to decrease for mistuned harmonics 7 through 12. Filled symbols show the amount of mistuning required to hear out the mistuned harmonic as a separate entity. The tones consisted of the first 12 harmonics, all of equal amplitude. The fundamental frequency was 200 Hz, and the duration was 410 ms.

is to change the virtual (low) pitch of the complex tone (Hartmann, 1988). Moore *et al.* (1985a) showed that the virtual pitch shift has the same sign as the mistuned harmonic frequency shift and grows monotonically with frequency shift until the frequency shift has reached about 8%. Thereafter, the mistuned harmonic appears to be excluded from the pitch computation.

At the other end of the duration scale, the most salient effect of a mistuned harmonic for long tones is the rolling of the "beats of mistuned consonances." Literally, "beats" is a misnomer here—there are no spectral amplitude variations. But the effect resembles beats in that it imparts a sensation of "action" to the tone. Plomp (1967) proposed that the beats of mistuned consonances were caused by temporal variations in the excitation in an auditory filter where two or more tones are combined. Hall, Mathews, and Roberts (1983) disagreed, and attributed the effect to actual beats between low-frequency harmonics and difference tones caused by nonlinear distortion involving the mistuned component.

Whatever the cause of the beats of mistuned consonances, the rolling sensation is prominent when tone durations are long enough for beats to be heard, and it complicates the interpretation of MHD experiments. The effect can be seen if one compares two experiments by Moore, Peters, and Glasberg (1985c vs 1986). In 1985, listeners were asked to detect the inharmonicity caused by a mistuned harmonic. In 1986 they were asked to hear out the mistuned harmonic. As shown in Fig. 6.12, hearing out the mistuned harmonic turned out to require larger mistunings. Further, there is a different frequency dependence. Threshold mistunings for detecting inharmonicity, as a percentage of the mistuned harmonic frequency, decrease almost monotonically with increasing harmonic number. Thresholds for hearing out the mistuned harmonic are constant or rise with increasing mistuned harmonic number. The decreasing threshold percentage is consistent with a constant frequency difference, as might be expected for detecting the beats of mistuned consonances. Constant threshold percentage is consistent with the slot widths in the DWS pitch meter for segregating a partial from a complex tone.

The MHD experiment is capable of providing useful data about segregation and integration if precautions are taken. To defeat the effect of virtual pitch shift, one can randomize the fundamental frequency on each and every tone. To try to defeat the effect of beating mistuned consonances, one can keep the tone duration short, 50 ms or less. MHD experiments carried out in the author's lab (1984–1994) have shown features as follows:

(1) MHD, as measured by d', is a nonmonotonic function of signal level, being optimum near 40 dB SPL per component. This is consistent with Javel's (1980) physiological conditions for multiple synchrony.

(2) MHD depends nonmonotonically upon both signal duration, T_D, and mistuning, Δf. The nonmonotonic structure is found to scale as a function of the dimensionless de-phasing factor $T_D\Delta f$. (After sounding for duration T_D, a "harmonic" mistuned by Δf has become $T_D\Delta f$ cycles out of phase with its neighbors.) This dependence might be added to the DWS pitch meter by making the slot width inversely proportional to tone duration for short tones. It is already consistent with a neural timing model as implemented with an autocorrelation function (Hartmann, 1988).

(3) Overall, MHD depends upon duration and mistuning according to

$$d' \propto (T_D\Delta f)^{\alpha}\sqrt{T_D}, \tag{6.5}$$

where α is between 1 and 2. The extra root T_D dependence can be understood as a signal-to-noise ratio that improves, as usual, with the square root of the measuring time.

(4) MHD is seriously impaired if the mistuned harmonic is in a spectral gap. For instance, a mistuned fourth harmonic becomes much harder to detect if the third and fifth harmonics are missing from the spectrum; the third is especially important. This observation is also consistent with a waveform-based temporal model of segregation and integration.

An alternative to the mistuned harmonic discrimination experiment is the mistuned harmonic matching (MHM) experiment. Here the listener hears a complex tone with one mistuned harmonic. The mistuned harmonic number is randomized from trial to trial. The listener's task is to hear out the mistuned harmonic and to indicate which one it is by tuning a sine tone to match its pitch. The matching tone

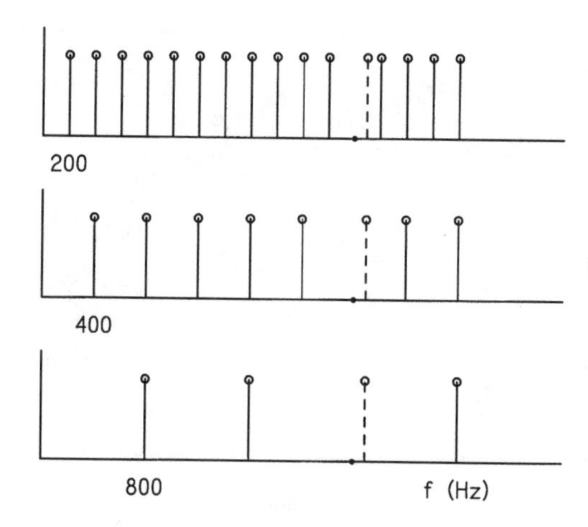

FIGURE 6.13. Three complex tones, with fundamental frequencies of 200, 400, and 800 Hz. Each complex tone has one mistuned harmonic, mistuned by +4%, shown dashed. Even though harmonics are denser for lower frequency fundamentals, the three mistuned harmonics are found to be equally detectable in an MHM experiment.

is not simultaneous with the complex tone. In fact, there is a gap (e.g. 300 ms) between the tones to avoid continuity effects.

The advantage of the MHM experiment is that there is only a small probability of a correct response unless the listener really has heard out the mistuned harmonic. There are several disadvantages: First, there are pitch shift effects that complicate the evaluation of the matches. Ultimately the experimenter needs to set criteria for a correct response. Second, the experiment is time consuming.

MHM experiments by Hartmann, McAdams, and Smith (1990) used complex tones with different fundamental frequencies, 200, 400, and 800 Hz. It was found that matching performance deteriorated for higher mistuned harmonic numbers, but mistuned harmonic number was not the important variable. Instead, what was important was the frequency of the mistuned harmonic. Thus, the ability to match a mistuned 3rd harmonic of 800 Hz resembled the ability to match a mistuned 6th harmonic of 400 Hz, or the mistuned 12th harmonic of 200 Hz. This result provided dramatic evidence that the segregation of a mistuned harmonic is not determined by spectral resolution.

At a mistuned harmonic frequency near 2400 Hz, as illustrated in Fig. 6.13, the MHM experiment found about 50% correct matches. For higher mistuned harmonic frequencies the percentage dropped rapidly. What was particularly surprising was to discover that these results were insensitive to the amount of mistuning. At 2400 Hz, it hardly mattered whether the mistuning was 4%, or 2%, or 1%. The fact that detection depends upon mistuned harmonic frequency and not harmonic number is consistent with the constant-Q assumption of the DWS pitch meter. The fact that

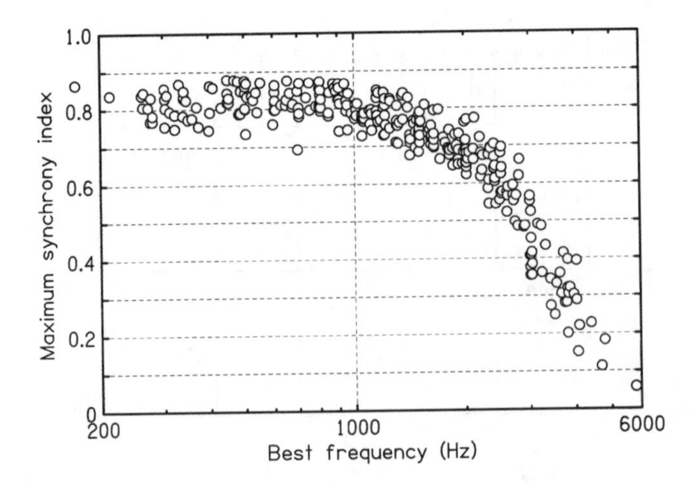

FIGURE 6.14. Synchrony indices for auditory nerve fibers measured at their characteristic frequencies by D. H. Johnson who provided the data for this plot. The decrease in synchrony with increasing frequency can account for the decreasing ability to segregate mistuned tones.

detection is insensitive to the amount of mistuning is not consistent with DWS.

The explanation for the high-frequency behavior (above 2 kHz) in the MHM experiment is that mistuned harmonics are segregated on the basis of neural synchrony anomalies. At a frequency of 2400 Hz, the synchrony between neural firing and the signal is decreasing rapidly, probably because of time-jitter noise in the firing of haircells (Teich *et al.*, 1993). The loss of synchrony is shown by Johnson's 1980 (also 1974) data in Fig. 6.14. The vanishing synchrony leaves little basis for detecting an anomaly. The effect is so dominant that little else matters. In support of this hypothesis is the fact that the percentage of correct responses for matching mistuned harmonics closely tracks the frequency dependence of Johnson's measured maximum synchrony for 315 auditory nerve neurons for cat.

PITCH AND TONE COLOR

As noted above, periodic complex tones are usually heard as integrated single entities, characterized by a pitch and tone color. This perception may be regarded as a data reduction operation whereby the auditory system avoids the burden of dealing with a plethora of resolved harmonics and deals with the entire ensemble as a unit.

Normally a periodic complex tone has a pitch corresponding to the fundamental frequency. This can be seen in pitch matching experiments, where listeners match the pitch of a complex tone by adjusting the frequency of a sine tone. In fact, listeners do not match the fundamental exactly. Walliser (1969) and Terhardt (1971) found that the pitch of a complex tone is actually slightly lower than the pitch of a sine tone having the frequency of the fundamental of the complex tone.

The tone color of a periodic complex tone is determined by the spectrum. If the level is high enough that the equal-loudness contour is sensibly flat, the tone color is insensitive to the overall level of the tone, and so the relative spectrum is usually an adequate basis for description. The relative spectrum references the harmonic levels and phases to the fundamental component, or to some other large component. The relative spectrum depends upon the fundamental frequency and the wave shape. It is possible to make some general statements about tone color based on wave shape alone or on the pattern of harmonic amplitudes, so long as the fundamental frequency lies in the two-octave range from 100 to 400 Hz, a range of particular importance for both speech and music.

(1) Spectra that decrease with increasing frequency at −6 dB per octave like the sawtooth, or other waveform with a discontinuity, sound ''bright'' if the fundamental frequency is high. They sound ''buzzy'' if the fundamental is low. Analog music synthesis methods make use of such waveforms as the starting point for the sounds of bowed strings or brass instruments because the upper harmonics have a lot of strength that can be shaped by filtering.

(2) Spectra that decrease with increasing frequency less rapidly than −6 dB per octave have a ''thin'' sound. An example is a pulse waveform with a small duty factor. Although the spectral envelope decreases asymptotically at −6 dB/octave, when the duty factor is very small this asymptotic behavior does not apply until one reaches very high harmonic numbers. For small harmonic numbers, the amplitude spectrum can be almost flat. The pulse wave with small duty factor is, therefore, a good way to begin to synthesize the sound of a flying insect.

(3) Square waves, triangle waves, and other waveforms without even-numbered harmonics are said to sound ''hollow.'' Just why they are characterized by this adjective is not certain, but it may well be an association with pipes that are open at one end and closed at the other, and which therefore have resonances at only odd multiples of a ''fundamental'' frequency.

(4) A half-wave-rectified sine waveform has one odd-numbered harmonic, the fundamental, and all the other harmonics are even numbered. When the fundamental frequency is low, this waveform has the solid sound of a ''doubled'' bass. Doubled bass is a common orchestral technique wherein a cello plays a bass line, and a bass viol plays the same notes an octave lower. For the half-wave, harmonics 2, 4, 6, etc. play the role of the cello and the fundamental component plays the role of the bass viol.

(5) A tone whose spectrum has a strong formant band in the region from the third

to the sixth harmonic, stronger than the fundamental, has a nasal tone color. This kind of spectrum is a good way to begin the synthesis of a double-reed instrument tone, such as an oboe or bassoon. It also works for a banjo which, from the viewpoint of the synthesist, is a percussive oboe.

The Spectral Envelope

The tone color of a periodic complex tone does not depend sensitively on all the details of all the harmonics. For fundamentals in the range 100 to 400 Hz, one expects that details of the low harmonics, approximately the first ten, play an important role in determining the tone color. So long as they are spectrally well behaved, higher-numbered harmonics contribute *en masse* to make the tone color brighter. For these, the *spectral envelope*, which gives the general shape of the spectrum, is more important than the details of spectrum.

Representing the spectral envelope correctly is something of a problem. The usual decibel scale representation of the levels of the harmonics exaggerates the importance of the high harmonics. For example, Exercise 5 in Chapter 5 shows the spectrum of a triangle wave as drawn by a spectrum analyzer using a decibel scale. All the harmonics seem to be important in that figure, even the 29th and 31st. Perceptually, however, the triangle wave has a ''dull'' sound; the strongest overtone is the third harmonic, and its level is more than 19 dB down from the fundamental. The next harmonic is the fifth, and its level is 28 dB down. Higher harmonics are even smaller. Curiously, the log-log plot in part (b) of this figure gives a more realistic impression. Although each harmonic has the same height (level) as in part (a), the larger width of the lower harmonics, due to the constant bandwidth of the analyzing filter, gives the (correct) impression that these lower harmonics are enormously more important.

Case of the Missing Fundamental

Periodic waveforms with a missing fundamental have played an important role in the development of the theory of pitch perception. Such waveforms have harmonic spectra, but there is no actual spectral strength at the frequency of the fundamental. Figure 6.15 shows an example of a 200-Hz complex tone with no power at 200 Hz.

It is important to realize that dropping the fundamental out of the spectrum does not change the periodicity of the waveform. This becomes evident from Fig. 6.16, which shows the harmonics of Fig. 6.15 individually as well as their sum. The lowest component, 400 Hz, has a period of 1/400 s, but the next lowest, at 600 Hz, does not repeat after 1/400 s. Therefore, the period of the sum cannot be as small as 1/400 s. The smallest time interval over which all the components repeat is 1/200 s. Therefore, the period is 1/200 s or 1/(200 Hz). Summing all the components leads to the waveform shown in the bottom line. There it becomes evident that the waveform period is 1/200 s.

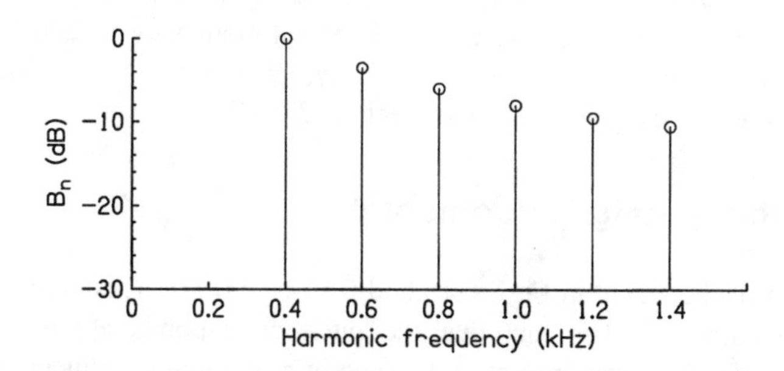

FIGURE 6.15. Spectrum of a complex tone with six harmonics of a missing 200-Hz fundamental. The harmonic levels decrease at a rate of −6 dB per octave. The fundamental is missing.

The fundamental frequency of a complex waveform (the reciprocal of the period) is the greatest common divisor of all the component frequencies. In other words, the fundamental frequency f_0 is the largest number such that each harmonic frequency can be written as an integral multiple of f_0, i.e.

$$f_n = nf_0. \tag{6.6}$$

The fundamental frequency is not changed if other harmonics, besides the funda-

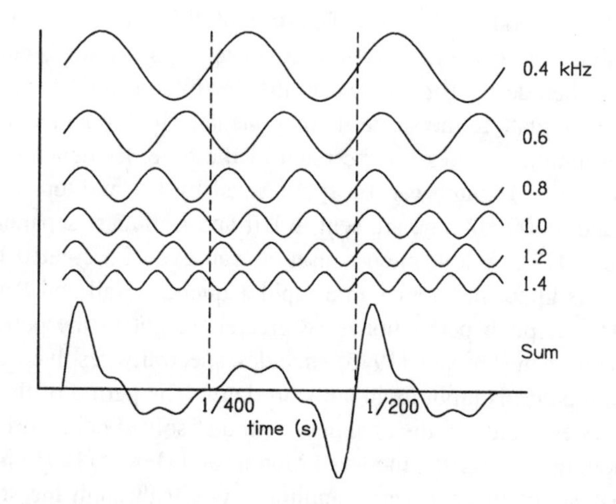

FIGURE 6.16. The six non-zero harmonics of Fig. 6.15 as a function of time, and their sum. The phases of all harmonics are "sine-phase."

mental, are missing. For example, if the spectral component frequencies are 400, 600, 1000, 1200, and 1400 Hz, the fundamental frequency is 200 Hz, and the appropriate harmonic numbers are, respectively, $n=2$, 3, 5, 6, and 7. The period of a waveform with these spectral components is $T=1/200$ s.

Pitch of the Missing Fundamental

The pitch of the signal in Figs. 6.15 and 6.16 is 200 Hz. This is the pitch of the missing fundamental. The fact that the pitch corresponds approximately to a frequency that is not even present in the spectrum was highly influential in guiding theories of pitch perception. [For a popular history of pitch theory see Wightman and Green (1974).] One obvious choice for a theory of pitch says that the perception of pitch is determined by the period of the waveform. Because dropping the fundamental from the spectrum does not change the period, it follows naturally from this theory that it should not change the pitch either, in agreement with experiment. Therefore, the pitch of such a signal is sometimes called ''periodicity pitch.'' Schouten (1940,1970) attributed periodicity pitch to the periodicity of the ''residue,'' the set of high-frequency components that are not individually resolved by auditory filtering.

Modern theories of pitch perception differ in the significance attached to periodicity itself. The spectral theories of Goldstein (1973) and Terhardt (1974b) begin with the observation that the most important harmonics in determining the pitch of a complex tone are actually individually resolved by the auditory periphery, and therefore, it does not seem appropriate to consider the period of the entire waveform. In the present example, neural channels tuned near 400 Hz see the second harmonic with a period of 1/400 s. Channels tuned near 600 Hz see the third harmonic with a period of 1/600 s. Because of this frequency analysis by the peripheral system, the period of 1/200 s never appears. In a subsequent step, the spectral models then derive (or synthesize) the low pitch from a central process that fits a harmonic template to the spectral components. The fundamental frequency of the best-fitting template codes for the pitch of the tone. Evidence in favor of this theory is that a 200-Hz pitch can be synthesized by the auditory system when a 400-Hz tone and a 600-Hz tone are sent to left and right ears separately (Houtsma and Goldstein, 1972). A low pitch sensation can even be created by presenting non-simultaneous upper harmonics in a rapid sequence (Hall and Peters, 1981).

Other models of pitch perception give greater weight to the periodicity of the waveform. Wightman's model (1973) includes spectrally resolved and spectrally unresolved harmonics as input to an autocorrelator. The period of the waveform as a whole appears explicitly in the channels with unresolved neighboring harmonics. Autocorrelation models like the model of Meddis and Hewitt (1991a,b) maintain the frequency analysis of the peripheral auditory system through the stage of neural encoding, but then add all the neural outputs together in a summary correlogram. The periodicity originally in the waveform then reappears.

GEORG SIMON OHM (1787–1854)

Georg Simon Ohm was born in Erlangen, Germany, on March 16, 1787 and was educated at the university in that town. He was a professor at the Jesuit College in Cologne in 1827 when he published the idea that became known as Ohm's Law, relating the electrical voltage and current in a resistance. The idea was so poorly received at the time that Ohm felt obliged to resign his college position. Not long afterwards, however, the significance of his work was recognized, and Ohm was made a foreign member of the Royal Society. Meanwhile, on more interesting matters, this same Ohm also said that the human auditory system operates as a spectrum analyzer that displays the power spectrum of a complex tone and is insensitive to relative phases of the components. This idea too has met with a mixed reception. In 1849 Ohm became professor at Munich, where he died on July 7, 1854.

PITCH, CHROMA, AND SHEPARD TONES

The association of pitch with frequency or with position on a musical staff is a one-dimensional view of pitch. Indeed, the American National Standards Institute (1994) definition for pitch suggests a one-dimensional continuum. It reads, in part, "Pitch is that attribute of auditory sensation in terms of which sounds may be ordered on a scale extending from low to high."

The way in which pitch is used in music, however, often makes it seem as though pitch has two dimensions, one called "chroma," the other called "pitch height." Tones with the same note name (for example, all the notes called $F\sharp$) all have the same chroma. This means that tones separated by an octave (factor of two in frequency) have the same chroma, though they have different pitch height. This concept is artfully illustrated by the pitch helix of Fig. 6.17. Here chroma is represented by angular position around the cylinder, and pitch height by elevation. Because of the helical structure, the distance between two notes that are an octave apart is smaller than the distance between two notes that are half an octave apart.

The musical validity of the two-dimensional model for pitch is frequently verified in practice. When a soprano is required to produce a note that is above her range, she will substitute a note that is an octave lower. Although she may be accused of lack of courage, it will not be said that she is singing out of tune. The difference might not even be noticed. Substituting any other note, not related to the written note by an octave, would be regarded as a more serious departure from the score, and possibly called "out-of-tune."

For musical purposes then, it seems to be true that notes with the same chroma share an equivalence, even though they may have different pitch height. The concept of equivalence has its limitations, however, even within a musical context. A melody that is played using the correct sequence of chromas, while randomly jumping octaves, is not recognizable. Thus, the musical equivalence of tones with

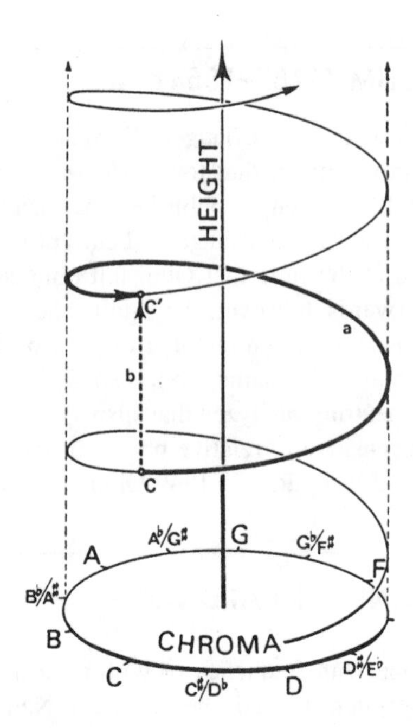

FIGURE 6.17. The pitch helix maps pitch onto two dimensions, chroma and pitch height. Tones separated by a tritone (half an octave) are on opposite sides of the chroma circle. One octave, from c to c′, is shown here with a heavy line. (From Shepard, 1982, by permission of Academic Press.)

the same chroma is not so strong as to overcome frequent violations of melodic contour (Deutsch, 1978).

An artifice that strongly emphasizes the chroma dimension of tones and deemphasizes the pitch-height dimension is the Shepard tone (Shepard, 1964). A Shepard tone is a periodic complex tone whose only harmonics are octaves of one another. Thus, the only harmonics that are allowed to have any strength are powers of two, namely, 1,2,4,8,16,32,..., and also 1/2,1/4,1/8,... . The subharmonics emphasize that, in principle, Shepard tones have no highest or lowest frequency. It is typical to impose a broad bandpass filtering, as shown by the dashed line in Fig. 6.18, to roll off the high and low frequency components.

The Shepard tones are used to create the Shepard illusion of an ever-ascending tone. To create this illusion, the frequencies of all the components of the tone are increased, either in a series of many small steps or in a glide. The frequencies change in a way that maintains the octave relationship. The frequencies continue to rise, with the highest component falling off the high end of the filter range, and a new low component entering weakly at the bottom of the range. After the frequencies have all increased by an octave, the spectrum is no different from the starting spectrum. When one listens to this sequence, one hears a tone with ascending pitch.

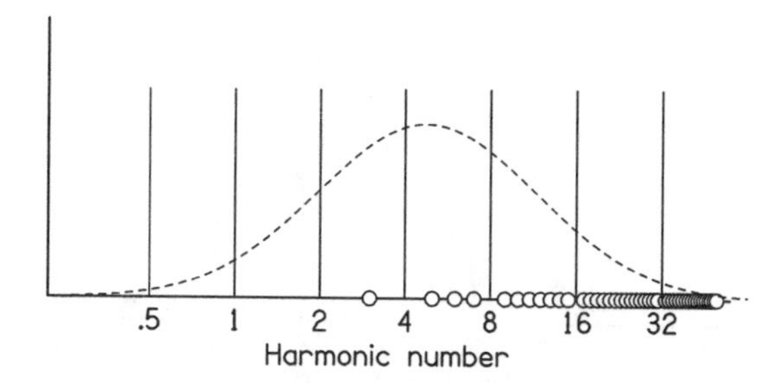

FIGURE 6.18. The spectrum of a Shepard tone. The horizontal axis is proportional to the logarithm of the frequency (i.e. to octave number) a fact that is emphasized by circles that indicate the integer numbers of harmonics that are missing from the spectrum. The dashed line shows a typical broad bandpass filter used to shape the spectrum to emphasize the central frequencies.

The illusory nature of the effect is evident after one has listened long enough to realize that although the pitch has gone up for a considerable time, the pitch range does not seem to be changing.

It is an awkward fact that the Shepard illusion does not actually require Shepard tones. Although Shepard himself (1982) explained the illusion on the basis of the special octave character of the Shepard tones, Burns (1981) showed that the illusion can be obtained with tones that do not have octave components. It seems probable that the ever-ascending tone illusion is not related to octave equivalence but resembles an Escher staircase instead. The Escher staircase is a drawing of a closed path which always seems to be going upward. Every local perspective tells the viewer that the staircase is going up, but ultimately the staircase can be followed back to its origin. Therefore, there is a conflict between local perspectives and a global understanding. That is exactly the kind of impression one gets with the illusion of the ever-ascending tone. Using the Shepard tone to generate the illusion does have the advantage that the harmonic relationship between the components of the tone leads to an integrated percept. Other choices for the spectrum of the tone might cause individual harmonics to be noticed. If a listener hears an individual harmonic disappear from the top of the spectrum or enter from the bottom, the illusion is compromised.

The Shepard tones are useful whenever an experimenter (or a composer!) wants to present chroma information and to minimize pitch height information. A feature of melodies played with such tones is that musical intervals cannot be larger than half an octave, a factor of 1.414. This can be seen by trying to make an interval (successive notes) larger than half an octave. Suppose one tries to make an ascending interval with a frequency ratio of 1.6, corresponding to an ascending minor 6th. Because Shepard tones enforce octave equivalence by their construction, this interval is physically indistinguishable from an interval with a frequency ratio of $1.6/2=0.8$, corresponding to a descending major third. (The names of musical

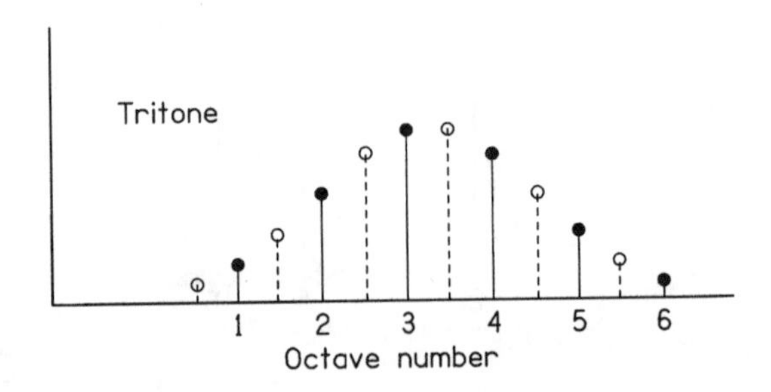

FIGURE 6.19. Spectra of two tones that are separated by half an octave, one shown by solid lines the other by dashed lines. Component amplitudes are determined by a fixed bandpass filter. Although both tones are shown on the same plot here, they actually are presented successively in the tritone paradox.

intervals are given in Chapter 11.) The descending major third is a smaller interval than the ascending minor sixth [$1/0.8=1.25<1.6$], and in the absence of bias, the brain resolves this ambiguity by perceiving the smaller interval, the descending major third.

The Tritone Paradox

What then happens when the melodic interval between two Shepard tones is exactly half an octave (factor of 1.414)? Half an octave is the interval called the "tritone," and one might expect that listeners would be completely confused about whether the interval was ascending or descending. The spectra for the two tones are given in Fig. 6.19.

Some listeners do demonstrate complete confusion. Given a large number of tritone intervals to judge, they choose "up" as often as "down." Other listeners exhibit a peculiar form of absolute pitch. They judge tritone intervals as ascending if the first tone is one of approximately half the possible chromas of the octave, and they judge tritone intervals as descending if the first tone is among the other half (Deutsch, 1986). Deutsch has found that individual judgements are independent of the spectral envelope formed by the bandpass filter, and has conjectured that this apparent innate pitch reference is related to frequency ranges of speech heard in individual environments (Deutsch, 1990). On the other hand, Repp (1994) concludes that the spectral envelope plays an important role after all. Experiments by Hartmann and Lin (1994) showed both kinds of behavior depending upon individuals. Listening to the same set of stimuli, in the same order, some individuals maintained their pitch reference despite a one-half octave shift in the bandpass filter; other individuals reversed their judgements when the filter was shifted.

EXERCISES

Exercise 1: Hearing out harmonics

Compare the boost of the fifth harmonic in the sung vowel of Fig. 6.6a and the synthesized tone in Fig. 6.5b

Exercise 2: Vibrato of high harmonics

If a complex musical tone has a vibrato rate of 5.5 Hz, what is the frequency modulation rate of the 10th harmonic? Think twice!

Exercise 3: Gibbs phenomena

Using a computer with graphical output, add up 200 cosines, according to Eq. (6.2) with amplitudes given by Eq. (6.3). Compare the result with Fig. 6.9. The oscillations at the pulse edges are the Gibbs phenomena. As the number of components increases, the oscillations become narrower, but so long as the number of components remains finite they never disappear.

Exercise 4: Mistuned harmonic limits

There are limits to the amount of mistuning that one should give to a harmonic in a mistuned harmonic experiment. Describe the special character of a tone in which the 12th harmonic is mistuned by 8.3%.

Exercise 5: Near-periodicity pitch

Consider a tone with spectral components at frequencies 601, 801, 1001, 1201, and 1401 Hz. What is the fundamental frequency? The pitch of such a tone is slightly greater than 200 Hz. This result is support for a template model of complex-tone pitch perception.

Exercise 6: Frequencies of the Shepard tone

Show that the frequencies of the Shepard tone are given by the equation

$$f_n = f_0 2^n,$$

where f_0 sets the scale note and n runs through all positive *and negative* integers.

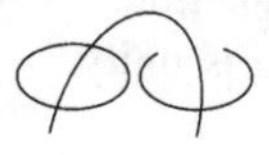

CHAPTER 7

Delta Functions

$\delta(x)$ is not a function of x according to the usual mathematical definition of a function.

P. A. M. DIRAC

BASIC DEFINITION

The delta function, sometimes called the Dirac delta function in honor of British theoretical physicist P. A. M. Dirac (1902–1984), is a useful concept in practical mathematics. The delta function is defined formally by an integral. In fact, whenever the delta function appears it is, at least implicitly, always in a context of integration.

The definition of the delta function is in two parts. First, the delta function of an argument t is a function that is zero for all values of t except when $t=0$. Second, the integral of the delta function over any interval of t that includes $t=0$ is one. In equations this definition is

$$\delta(t)=0 \quad (t\neq 0), \tag{7.1}$$

and

$$\int_{T_1}^{T_2} dt\,\delta(t)=1 \quad (T_1<0<T_2).$$

The limits of integration can be anything that we want, so long as they include the point $t=0$. If the point at $t=0$ is not included, the integral must be zero.

In a context where the argument of the delta function has dimensions, the delta function itself has dimensions that are the reciprocal of the dimensions of the argument. For instance, if one assumes that argument t above represents time, then $\delta(t)$ has the dimensions of inverse time, i.e. it has the dimensions of frequency. That is why an integral over t of the delta function is the dimensionless number one.

In order to fit the above definition, the delta function must be essentially infinitely large at $t=0$, but by necessity, this infinity is an integrable one. The delta function is often called the "unit impulse function," where the term "unit" is appropriate because the area under the function is unity.

RELATION TO THE UNIT STEP FUNCTION

The integral of the delta function is the unit step function, or theta function. It is also sometimes called the Heaviside function, after Oliver Heaviside (1850–1925). We write

$$\theta(t)=\int_{T_1}^{t} dt'\,\delta(t') \text{ (for } T_1<0), \tag{7.2}$$

where the theta function is defined by

$$\theta(t)=0 \quad (\text{for } t<0)$$

$$\theta(t)=1 \quad (\text{for } t>0)$$

$$\theta(t)=1/2 \quad (\text{for } t=0). \tag{7.3}$$

It follows that the delta function is the derivative of the theta function, i.e.

$$\frac{d}{dt}\,\theta(t)=\delta(t). \tag{7.4}$$

TRANSLATION AND SELECTION PROPERTIES

The delta function gives a spike where its argument is zero. In practice, the spike can be put anywhere. For instance, the function

$$g(t)=\delta(t-t_0) \tag{7.5}$$

is zero for all values of t except for $t=t_0$, where the spike occurs, as shown in Fig. 7.1.

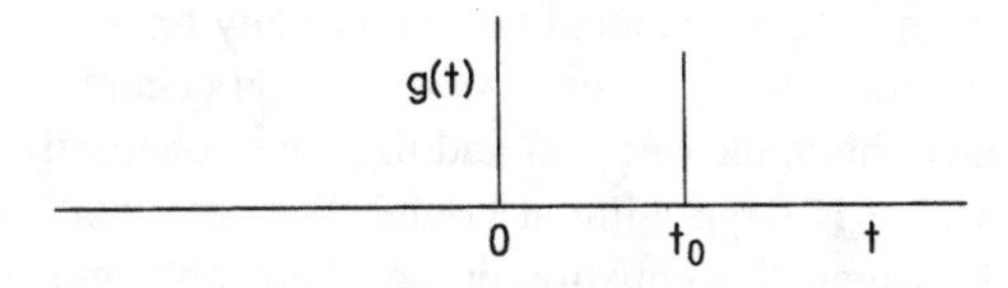

FIGURE 7.1. The translated delta function from Eq. (7.5) is zero except at the point $t=t_0$ where there is an infinite spike.

The most important feature of the delta function is its ''selection'' property. The selection property comes into play when the delta function is integrated together with another function. Suppose that the delta function multiplies a function $f(t)$, and that this product is integrated over some region that includes the spike. Then, selection says:

$$\int_{T_1}^{T_2} dt\ f(t)\,\delta(t-t_0)=f(t_0), \quad (\text{for } T_1<t_0<T_2). \tag{7.6}$$

The interpretation of this equation is that it does not matter how big $f(t)$ becomes, the fact that $\delta(t-t_0)=0$ is enough to ensure that $f(t)$ never makes any contribution to the integral. The exception occurs where $t=t_0$ and the delta function allows f to come through the integral. The delta function is so narrow that only the value of $f(t)$ right at $t=t_0$ is allowed to contribute. Therefore, the integral is $f(t_0)$. Thus, the delta function has selected from the entire function $f(t)$ its value at the single point t_0.

APPLICATION TO FOURIER TRANSFORMS

The delta function is of practical importance in the study of the Fourier integral. The key equation is this:

$$\frac{1}{(2\pi)}\int_{-\infty}^{\infty} d\omega\ e^{i\omega t}=\delta(t). \tag{7.7}$$

Some understanding of this important equation can be gained by the following reasoning: By Euler's identity, Eq. (2.8), the exponential function may be written as a sum of a sine and a cosine. Therefore, Eq. (7.7) can be rewritten as

$$\frac{1}{(2\pi)}\int_{-\infty}^{\infty} d\omega\ \cos\,\omega t\ +\ i\,\frac{1}{(2\pi)}\int_{-\infty}^{\infty} d\omega\ \sin\,\omega t=\delta(t). \tag{7.8}$$

Because of the even limits on the integral, the sine function (odd symmetry) integrates to zero. Therefore, it must be the integral of the cosine function that leads to the delta function. We consider this integral to be a sum of an infinite number of cosines of all frequencies. Because each cosine is periodic, contributions to the integral are positive and negative equally often. For any finite value of t they cancel perfectly and the integral is simply zero. An exception occurs when $t=0$, where all the cosine terms have the value one and add together coherently. Then the cosines add to give a result that is large without bound. We note that because the integral consists entirely of cosines, this equation proves the even symmetry property of the delta function, namely

$$\delta(t)=\delta(-t). \tag{7.9}$$

In Eq. (7.7) the delta function of time was written as an integral over angular frequency, but the relationship is completely symmetrical. One can write the delta function of frequency as an integral over time,

$$\frac{1}{(2\pi)}\int_{-\infty}^{\infty} dt\ e^{i\omega t} = \delta(\omega). \tag{7.10}$$

RELATION TO THE KRONECKER DELTA

The Kronecker delta is a function of two integer variables. It is defined by the equation

$$\delta_{n,n'} = 1 \quad \text{for } n = n' \tag{7.11}$$

and

$$\delta_{n,n'} = 0 \quad \text{for } n \neq n'.$$

The Kronecker delta has a selection property similar to the selection property of the Dirac delta. The difference is that the Dirac delta selects from an integral whereas the Kronecker delta selects from a sum. That is, if f_n is a function that depends upon integer n, then

$$\sum_{n=-\infty}^{\infty} f_n \delta_{n,n_0} = f_{n_0}. \tag{7.12}$$

Whereas the Dirac delta has dimensions that are the inverse of the dimensions of its argument, the Kronecker delta is dimensionless.

In dealing with periodic functions it is very helpful to have a relationship between the Dirac delta and the Kronecker delta. The relationship is established by the context in which those two functions are used. For example, Eq. (7.10) for the delta function of frequency can be understood as the limit

$$\lim_{T_D \to \infty} \int_{-T_D}^{T_D} dt\ e^{i\omega t} = 2\pi\delta(\omega). \tag{7.13}$$

The fact that the delta function (r.h.s.) is infinite when $\omega = 0$ is the direct result of the fact that the integration limits $\pm T_D$ become infinite (l.h.s.). This can easily be seen by evaluating the integral above,

$$\lim_{T_D \to \infty} \int_{-T_D}^{T_D} dt\ e^{i\omega t} = \lim_{T_D \to \infty} \frac{2\sin(\omega T_D)}{\omega}. \tag{7.14}$$

When $\omega=0$ the expression on the right can be evaluated by first taking the limit as $\omega\to0$, and then taking the limit as $T_D\to\infty$. In the first limit the right-hand side becomes $2\omega T_D/\omega$, so that in the second limit the expression becomes large proportional to $2T_D$.

It therefore follows that we can write an expression for the Kronecker delta of n and n' by simply dividing the above expression by $2T_D$, and letting ω be a linear function of the difference $n-n'$, i.e.

$$\lim_{T_D\to\infty} \frac{1}{2T_D} \int_{-T_D}^{T_D} dt\, e^{i\omega_0(n-n')t} = \delta_{n,n'} \quad (\omega_0 \neq 0), \tag{7.15}$$

or, substituting from Eq. (7.13),

$$\lim_{T_D\to\infty} \frac{\pi}{T_D} \delta[\omega_0(n-n')] = \delta_{n,n'} \quad (\omega_0 \neq 0). \tag{7.16}$$

The actual value of the arbitrary parameter ω_0 is unimportant, so long as it it finite. Its main role is to keep the dimensions right.

These elementary properties of the delta function, including the representation as an integral of oscillating functions, are enough background to allow a good start on the subject of the Fourier integral. The present chapter continues with further properties of the delta function.

FURTHER PROPERTIES OF THE DELTA FUNCTION

Some useful properties of delta functions can be expressed by simple equations, given below without proof. The arguments of the functions below have symbols t, but the equations are not restricted to cases where the argument is a time variable. Symbols b, t_0, t_1, t_2, and ρ below are real constants. *A priori* it would seem that these equations only give relationships among zeros and among infinities, given that the delta function can only be zero or infinity. In fact the equations are more useful than that because delta functions are frequently defined as the limits of certain other functions (see the representations later in the chapter). To be a legitimate contender for delta function status, these other functions need to satisfy these equations in the limit.

Symmetry and Scaling

$$\delta(t) = \delta(-t) \tag{7.17}$$

$$\delta(bt) = \frac{1}{|b|}\,\delta(t) \tag{7.18}$$

Product with a Function

$$g(t)\,\delta(t)=g(0)\,\delta(t) \tag{7.19}$$

$$t\,\delta(t)=0 \tag{7.20}$$

Product of Delta Functions or Arguments

$$\delta(t-t_1)\,\delta(t-t_2)=0 \quad \text{for} \quad t_1\neq t_2 \tag{7.21}$$

$$\int_{-\infty}^{\infty} dt\,\delta(t-t_1)\,\delta(t-t_2)=\delta(t_1-t_2) \tag{7.22}$$

$$\delta[(t-t_1)(t-t_2)]=\frac{1}{|t_1-t_2|}\,[\delta(t-t_1)+\delta(t-t_2)] \quad \text{for} \quad t_1\neq t_2 \tag{7.23}$$

Derivatives

$$\left.\frac{d}{dt}\,\delta(t)\right|_{t=t_0}=-\left.\frac{d}{dt}\,\delta(t)\right|_{t=-t_0} \qquad (\text{for all } t_0) \tag{7.24}$$

$$t\,\frac{d}{dt}\,\delta(t)=-\,\delta(t) \tag{7.25}$$

Implicit Dependence on Real Function g

$$\delta[g(t)]=\frac{\delta(t-t_0)}{\left|\dfrac{d}{dt}\,g(t)|_{t_0}\right|}, \tag{7.26}$$

where $g(t_0)=0$, and $g(t)\neq 0$ for $t\neq t_0$.

If there are several values of t where $g(t)=0$ then $\delta[g(t)]$ becomes a sum of delta functions, one at each of these special values of t, and the strength of each delta function is given by Eq. (7.26).

LATTICE SUM

See Chapter 8 for the derivation of the lattice sum. The result is:

$$\sum_{m=-\infty}^{\infty} e^{i\omega mT}=\frac{2\pi}{T}\sum_{n=-\infty}^{\infty}\delta(\omega-n\omega_0) \qquad (\omega_0=2\pi/T). \tag{7.27}$$

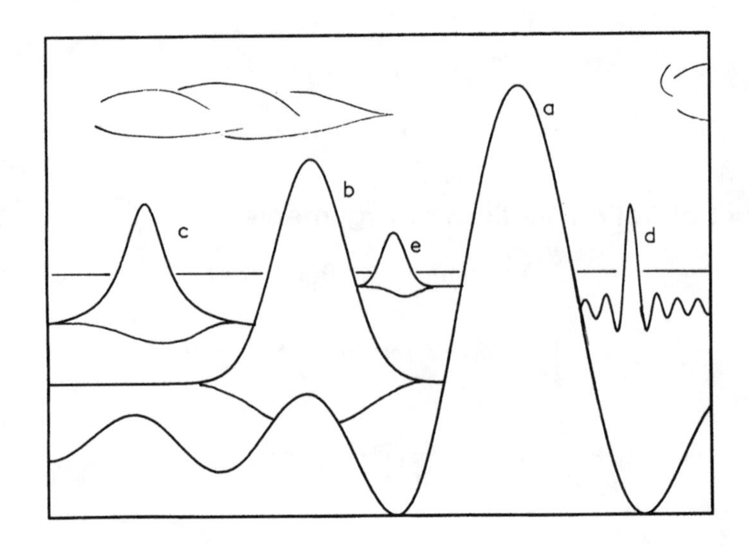

FIGURE 7.2. A congregation of peaked functions that aspire to achieve delta function status by becoming taller and thinner. (a) The sinc function; (b) the Gaussian; (c) the Lorentzian; (d) the sum of the first six cosine harmonics; (e) another Gaussian.

Because the sine function is odd, only the cosine part of the exponential survives the sum over even and odd values of m, and the sum is equivalent to a sum of cosines,

$$\sum_{m=-\infty}^{\infty} \cos(\omega m T) = \frac{2\pi}{T} \sum_{n=-\infty}^{\infty} \delta(\omega - n\omega_0) \quad (\omega_0 = 2\pi/T). \tag{7.28}$$

REPRESENTATIONS OF THE DELTA FUNCTION

Delta functions are limiting cases of ordinary peaked functions, as shown in Fig. 7.2.

$$\text{Two poles:} \quad \delta(t) = \frac{1}{2\pi i} \lim_{\rho \to +0} \left[\frac{1}{t - i\rho} - \frac{1}{t + i\rho} \right] \tag{7.29}$$

$$\text{Sinc:} \quad \delta(t) = \frac{1}{\pi t} \lim_{\omega \to \infty} \sin(\omega t) \tag{7.30}$$

$$\text{Lorentzian:} \quad \delta(t) = \lim_{\omega \to \infty} \frac{\omega}{\pi(\omega^2 t^2 + 1)} \tag{7.31}$$

$$\text{Gaussian:} \quad \delta(t) = \lim_{\omega\to\infty} \frac{\omega}{\sqrt{\pi}} e^{-\omega^2 t^2} \tag{7.32}$$

$$\text{Rectangle:} \quad \delta(t) = \lim_{t_0\to 0} R(t,t_0)$$

where

$$R(t,t_0) = 1/t_0 \quad |t| < t_0/2$$

$$R(t,t_0) = 0 \quad |t| > t_0/2 \tag{7.33}$$

EXERCISES

Exercise 1: Derivative of the sawtooth

The purpose of this exercise is to evaluate the Fourier transform of the derivative of a sawtooth by two separate ways and to show that the results agree. Figure E.1 shows a sawtooth waveform.

In Chapter 5 we showed that the Fourier series for the sawtooth waveform is

$$x(t) = \frac{-2}{\pi} \sum_{n=1}^{\infty} \frac{1}{n} \sin(2\pi nt/T). \tag{7.34}$$

(a) Differentiate this series with respect to time to show that the Fourier series for the derivative of the sawtooth is

$$(dx/dt) = \frac{-4}{T} \sum_{n=1}^{\infty} \cos(2\pi nt/T). \tag{7.35}$$

(b) From Fig. E.1 it is evident that the derivative of the sawtooth is given by a constant term, $2/T$, representing the common slope, plus a train of negative-going delta functions of strength 2, i.e. the derivative is

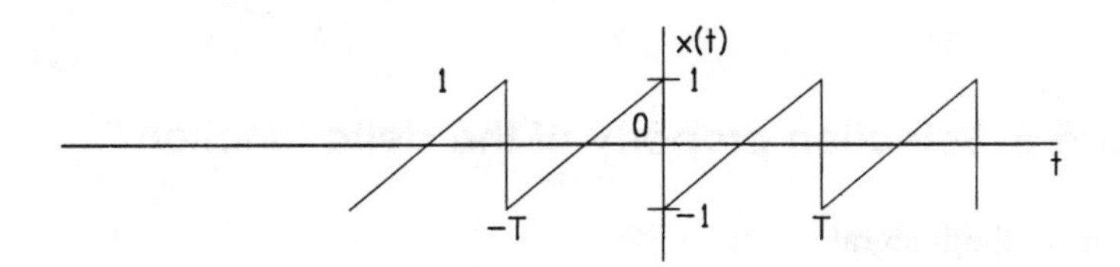

FIGURE E.1. The sawtooth has some slopes that are infinitely steep.

$$(dx/dt)=\frac{2}{T}-2\sum_{m=-\infty}^{\infty}\delta(t-mT). \tag{7.36}$$

Find the Fourier series to represent the function given in Eq. (7.36). If the world makes sense then your answer must agree with Eq. (7.35).
[Note: Because the delta function is an even function of its argument, we expect that the Fourier series for dx/dt is a sum of cosine terms. Therefore, you only need to find coefficients A_n that are the Fourier transform of Eq. (7.36). In particular, note that A_0 must turn out to be zero.

Exercise 2: A prefactor

What is the sense of having a prefactor in front of a delta function, as in the expression $x(t)=2\,\delta(t)$? Consider the awkward notion that if t is not zero then $x(t)$ is zero, with or without the factor of 2, and if t is zero then $x(t)$ is infinite, again with or without the factor of 2. Explain how a prefactor makes sense.

Exercise 3: Representations of the delta function

(a) Show that the representations of the delta function in Eqs. (7.29–7.33) are dimensionally correct.

(b) Show that each of them is an even function.

(c) Show that each of them has unit area.

(d) Show that the "sinc" representation of the delta function follows directly from the fundamental formula in Eq. (7.7).

Exercise 4: Multiplicative constant

Show that for real constant b,

$$\delta(bt)=\frac{1}{|b|}\,\delta(t).$$

Exercise 5: Selection property of the delta function

(a) Evaluate the integral

$$\int_{-\infty}^{\infty} d\zeta\,\sin(\zeta)\,\delta(\zeta-\pi/2).$$

(b) Evaluate the integral

$$\int_{-\infty}^{\infty} dx \exp(-x^2/a^2)\delta(x).$$

[Answer: (a) 1. (b) 1.]

Exercise 6: Product of delta functions

Explain in words why $\delta(x-a)\delta(x-b)=0$ for $a \neq b$.

Exercise 7: Integrals of trains of delta functions

Consider the trains of equally spaced spikes, called functions of time $f(t)$, shown in Fig. E.7. Suppose that the spikes are delta functions and draw figures to show the integrals

$$\text{Integral} = \int_0^t dt' f(t').$$

You should find a staircase and a square wave.

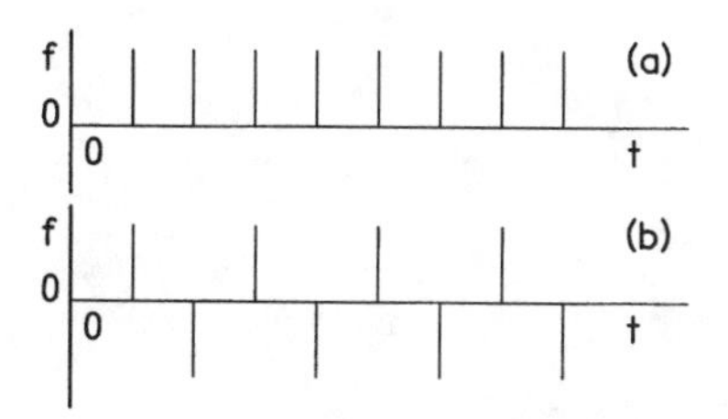

FIGURE E.7. Two different trains of delta functions.

Exercise 8: Half a delta function

Explain why

$$\int_{-\infty}^{t_0} dt\, f(t)\delta(t-t_0) = \frac{1}{2} f(t_0).$$

Exercise 9: Delta function symmetry

A popular textbook in electrical engineering represents the delta function as

$$\delta(t) = \lim_{\epsilon \to 0} \frac{1}{\epsilon} e^{-t/\epsilon} \quad \text{for } t > 0$$

and

$$\delta(t) = 0 \quad \text{for } t < 0.$$

Explain why this representation is incorrect.

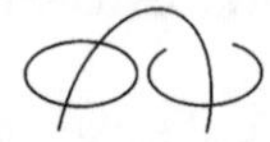

CHAPTER 8

Fourier Integral

The problem to be solved in distinguishing the partials of a compound tone is that of analyzing a given aggregate of sensations into elements which no longer admit of analysis.

HERMANN L. F. HELMHOLTZ

The Fourier transformation is a link between two different representations of an object. If the object is a signal, then the transformation links the time-domain representation and the frequency-domain representation. If the object is a pattern in three-dimensional space, for instance the points on a crystal lattice, then the transformation links the real-space representation to a reciprocal space (k-space) representation. If the object is a probability density, then its Fourier transform is the characteristic function, which links an outcome space to a conjugate space.

There are two kinds of Fourier transformation: the Fourier series, described in Chapter 5, and the Fourier integral. The Fourier series gives a representation of a signal that is restricted to cases where the signal is periodic. A periodic signal repeats itself indefinitely, into the infinite future and into the infinite past. The Fourier integral is a more general formulation. It can describe any function, periodic or not, and for a periodic function it reduces to the Fourier series. The Fourier integral gains enormous power from its generality. It is perhaps true that there is no problem in the Fourier series that cannot be done more elegantly if not more simply in terms of the Fourier integral, once one has mastered the techniques of that representation. Further, the Fourier integral finds a natural place within the larger context of integral transformations, Hilbert spaces, and linear algebra. By contrast, the Fourier series is a dead-end street. The present chapter presents the Fourier integral and relates it to the Fourier series.

Like the Fourier series, the Fourier integral is used in acoustics to find both the amplitude spectrum and the phase spectrum. It describes the frequency content of a signal in a formal way that makes it possible to predict the effects of frequency-selective processes and devices like filters.

THE TRANSFORMS DEFINED

We suppose that there is a function of time, $x(t)$, that represents a signal. There is no implication here that $x(t)$ is a periodic function. We define the Fourier transform $X(\omega)$ of $x(t)$ in terms of the Fourier integral,

$$X(\omega)=\int_{-\infty}^{\infty} dt\ e^{-i\omega t}x(t). \tag{8.1a}$$

Schematically we may write this as

$$X(\omega)=\mathcal{F}x(t). \tag{8.1b}$$

where symbol $\mathcal{F}$ stands for the operation of Fourier transforming. The integral equation shows that the Fourier transform of a function of time is a function of frequency, or, to be more precise, angular frequency ω. Normally in this chapter we will refer to functions of time by lower case letters, such as x, and their Fourier transforms by upper case letters, such as X.

The inverse Fourier transform is defined by the Fourier integral,

$$x(t)=\frac{1}{2\pi}\int_{-\infty}^{\infty} d\omega\ e^{i\omega t}X(\omega), \tag{8.2}$$

which expresses the original function of time, $x(t)$, as an integral over frequency.

Apart from a sign change in the exponent and a factor of 2π, the transform and the inverse transform are the same. This makes the Fourier transform easier to use than unsymmetrical transforms such as the Laplace transform. Because the transform and its inverse are almost identical, it may seem rather arbitrary to call one of them the "transform" and the other the "inverse." The main difference is one of approach. We regard function $x(t)$ as a physical signal, one that might be observed on an oscilloscope. Mathematically then, it is a *real* quantity. The Fourier transform, $X(\omega)$, however, is generally complex, with both real and imaginary parts.

Self-Consistency

Having defined the transform of x to X, and the transform of X to x, the first thing we ought to do is to verify that these definitions are self-consistent. If we begin with a function of time, $x(t)$, find its Fourier transform, $X(\omega)$, and then find the inverse Fourier transform, we must end up with function $x(t)$ again. We prove that as follows:

First, we simplify the notation somewhat with the convention that all integrals in this chapter are from minus infinity to plus infinity, unless explicitly stated otherwise. This is true of integrals on time and integrals on angular frequency. Therefore, we drop the limits on the integral signs below.

We begin by writing x in terms of X using Eq. (8.2)

$$x(t)=\frac{1}{2\pi}\int d\omega\ e^{i\omega t}X(\omega), \tag{8.3}$$

Next, we substitute for $X(\omega)$ by using the defining expression for the Fourier transform in Eq. (8.1),

$$x(t)=\frac{1}{2\pi}\int d\omega\ e^{i\omega t}\int dt' e^{-i\omega t'}x(t'). \tag{8.4}$$

Note the use of dummy integration variable t'. If we had used variable t as an integration variable, there would have been hopeless confusion with the main variable t.

We make the assumption that we can exchange the order of integration on time t' and frequency ω. Without this crucial step further progress is not possible. Therefore,

$$x(t)=\frac{1}{2\pi}\int dt' x(t')\int d\omega\ e^{i\omega(t-t')}. \tag{8.5}$$

The second integral in Eq. (8.5) is simply a delta function on time. Recall Eq. (7.7) which says:

$$\int d\omega\ e^{i\omega(t-t')}=2\pi\delta(t-t'). \tag{8.6}$$

Therefore,

$$x(t)=\int dt' x(t')\delta(t-t'). \tag{8.7}$$

By the selection property of the delta function, the right-hand side is equal to $x(t)$. Therefore, left and right sides are equal, and the transformations are shown to be self-consistent. The above proof of self-consistency has also employed an important mathematical technique whereby an integral is reduced to a delta function. This technique will be used repeatedly below.

TRANSFORMS OF SINE AND COSINE

The Fourier transform of a cosine with angular frequency ω_0 is

$$X(\omega)=\int dt\ e^{-i\omega t}\cos(\omega_0 t). \tag{8.8}$$

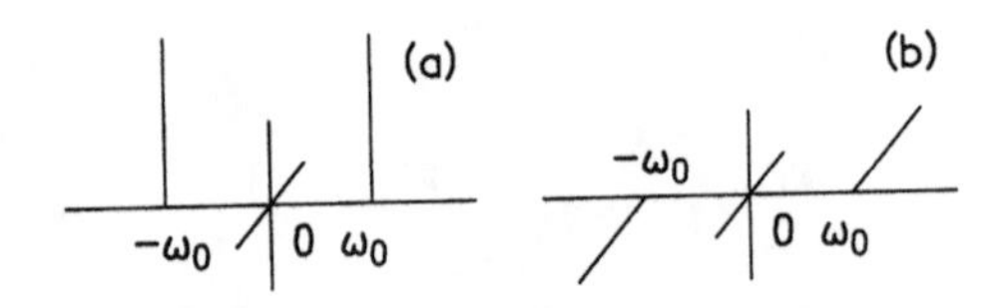

FIGURE 8.1. The Fourier transforms for (a) cos $\omega_0 t$ and (b) sin $\omega_0 t$ are shown on a three-dimensional graph with the positive imaginary axis coming out of the page.

We write the cosine as the sum of two complex exponentials, so that Eq. (8.8) becomes

$$X(\omega)=\int dt\ e^{-i\omega t}[e^{i\omega_0 t}+e^{-i\omega_0 t}]/2 \tag{8.9}$$

or

$$X(\omega)=\int dt[e^{-i(\omega-\omega_0)t}+e^{-i(\omega+\omega_0)t}]/2. \tag{8.10}$$

The integrals of exponentials are delta functions [recall from Eq. (8.6)], and so the Fourier transform of the cosine is

$$\mathcal{F}\cos\omega_0 t=X(\omega)=\pi[\delta(\omega+\omega_0)+\delta(\omega-\omega_0)]. \tag{8.11a}$$

A similar argument leads to the transform of $\sin(\omega_0 t)$,

$$\mathcal{F}\sin\omega_0 t=i\pi[\delta(\omega+\omega_0)-\delta(\omega-\omega_0)]. \tag{8.11b}$$

Both transforms have spikes at positive and negative frequencies as shown in Fig. 8.1

The Fourier transformations of phase-shifted cosines and sines can be done by combining the results above. In order to calculate the Fourier transform of $\cos(\omega_0 t+\phi)$ we begin by writing it as $\cos\phi\cos\omega_0 t-\sin\phi\sin\omega_0 t$ and then use the transforms of sine and cosine functions, together with the fact that the transformation is linear, to find that

$$\mathcal{F}\cos(\omega_0 t+\phi)=\pi(\cos\phi-i\sin\phi)\delta(\omega+\omega_0)+\pi(\cos\phi+i\sin\phi)\delta(\omega-\omega_0). \tag{8.12a}$$

In order to calculate the Fourier transform of $\sin(\omega_0 t+\phi)$ we begin by writing it as $\cos\phi\sin\omega_0 t+\sin\phi\cos\omega_0 t$ and then use the transforms of sine and cosine functions, together with the fact that the transformation is linear, to find that

$$\mathcal{F}\sin(\omega_0 t+\phi)=\pi(\sin\phi+i\cos\phi)\delta(\omega+\omega_0)+\pi(\sin\phi-i\cos\phi)\delta(\omega-\omega_0). \tag{8.12b}$$

REAL FUNCTIONS, EVEN AND ODD FUNCTIONS

We work in a context where function $x(t)$ is real. For example, the functions $\cos(\omega_0 t)$ and $\sin(\omega_0 t)$ from the above calculation are real. That calculation illustrates several important points about Fourier transforms:

> *Reality*: If $x(t)$ is real then its Fourier transform $X(\omega)$ has the property that $X(-\omega)=X^*(\omega)$, where X^* is the complex conjugate of X. (The complex conjugate of a complex number X has a real part that is the same as the real part of X and an imaginary part that is the negative of the imaginary part of X.)
> *Even symmetry*: If $x(t)$ is an even function [i.e. $x(-t)=x(t)$] then its Fourier transform $X(\omega)$ is not complex but is entirely real.
> *Odd symmetry*: If $x(t)$ is an odd function [i.e. $x(-t)=-x(t)$] then its Fourier transform $X(\omega)$ is entirely imaginary.
> *No symmetry*: If $x(t)$ is neither an even nor an odd function of t then its Fourier transform $X(\omega)$ has both real and imaginary parts.

We shall prove all four assertions by studying a more general problem. We begin with the fact that any real function can be written as the sum of an even function and an odd function, i.e. for any function x,

$$x(t)=x_e(t)+x_o(t), \tag{8.13}$$

where $x_e(t)$ is an even function of t, and $x_o(t)$ is an odd function of t.

Then, with the fact that $\exp(-i\omega t)=\cos \omega t-i \sin \omega t$, the Fourier transform $X(\omega)$ can be written as

$$X(\omega)=\int dt[\cos(\omega t)-i \sin(\omega t)][x_e(t)+x_o(t)]. \tag{8.14}$$

The product inside the integral can be expanded; some terms will be even in t and others will be odd. The integral of the odd terms over the symmetrical limits of integration will be zero, as proved in Appendix D on the integrals of even and odd functions. Therefore, we obtain a simpler form,

$$X(\omega)=\int dt \cos(\omega t)x_e(t)-i\int dt \sin(\omega t)x_o(t). \tag{8.15}$$

With this formula we have written X in terms of its real and imaginary parts, displayed separately. Clearly the imaginary part of X comes from the odd part of x, namely $x_o(t)$, and if $x_o(t)$ is zero for all time then X is purely real. Similarly the real part of X is seen to come from the even part of x, namely $x_e(t)$, and if $x_e(t)$ is zero for all time then X is purely imaginary. This proves the statements in the above box about symmetry.

It is also clear that if we change the sign of the running variable ω then

$$X(-\omega)=\int dt\ \cos(-\omega t)x_e(t)-i\int dt\ \sin(-\omega t)x_o(t) \tag{8.16}$$

or, using the symmetry of the cosine and sine functions,

$$X(-\omega)=\int dt\ \cos(\omega t)x_e(t)+i\int dt\ \sin(\omega t)x_o(t). \tag{8.17}$$

Comparing with Eq. (8.15) we see that replacing ω with $-\omega$ has made no change in the real part of X, but the sign of the imaginary part has been reversed. In other words, it has produced X^*, the complex conjugate of X. This proves the statement above about the Fourier transform of a real function.

EXAMPLE 1: *The unit rectangle*

One of the most useful transformations is the transform of the unit rectangle, defined by the equations

$$x(t)=1/T_0 \quad \text{for } |t|<T_0/2 \tag{8.18}$$

and

$$x(t)=0 \quad \text{for } |t|>T_0/2, \tag{8.19}$$

and shown in Fig. 8.2.

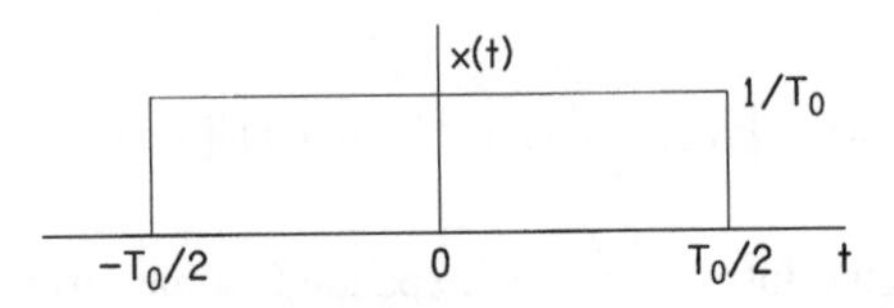

FIGURE 8.2. The unit rectangle has area of unity.

According to the definition of the Fourier transform we need to integrate on time from negative infinity to positive infinity. However, the function $x(t)$ is zero except on the interval from $-T_0/2$ to $T_0/2$, which limits the range of integration. Further, over that range function $x(t)$ is simply a constant. Therefore, we find

$$X(\omega)=\int_{-T_0/2}^{T_0/2} dt\left(\frac{1}{T_0}\right)e^{-i\omega t} \tag{8.20}$$

$$= \frac{e^{i\omega T_0/2} - e^{-i\omega T_0/2}}{i\omega T_0} \tag{8.21}$$

$$= [\sin(\omega T_0/2)]/(\omega T_0/2) \tag{8.22}$$

$$= \mathrm{sinc}(fT_0). \tag{8.23}$$

The last step comes from the definition of the sinc function: for any variable r, $\mathrm{sinc}(r) \equiv \sin(\pi r)/(\pi r)$. This function is shown in Fig. 8.3.

There is a reciprocal relation between the width of the pulse in time and the width of its Fourier transform. The figure of the pulse shows that the pulse width is T_0. The width of the Fourier transform shown in Fig. (8.3) is not so obvious. Normally it is taken to be the span between the first zero crossings, when $\pi r = \pm\pi$. Equation (8.22) shows that the zero crossings occur when $\omega T_0/2 = \pm\pi$. The span between zero crossings is therefore given by $\Delta\omega = 4\pi/T_0$, or $\Delta\omega T_0 = 4\pi$. The frequency width Δf is, therefore, given by $\Delta f T_0 = 2$. The reciprocal relationship between the width of the pulse T_0 and the frequency width Δf illustrates an important principle, sometimes called the ''uncertainty principle,'' which says that if a function is narrow in the time domain (e.g. small T_0) then its Fourier transform will be wide in the frequency domain (e.g. large $\Delta\omega$), and vice versa.

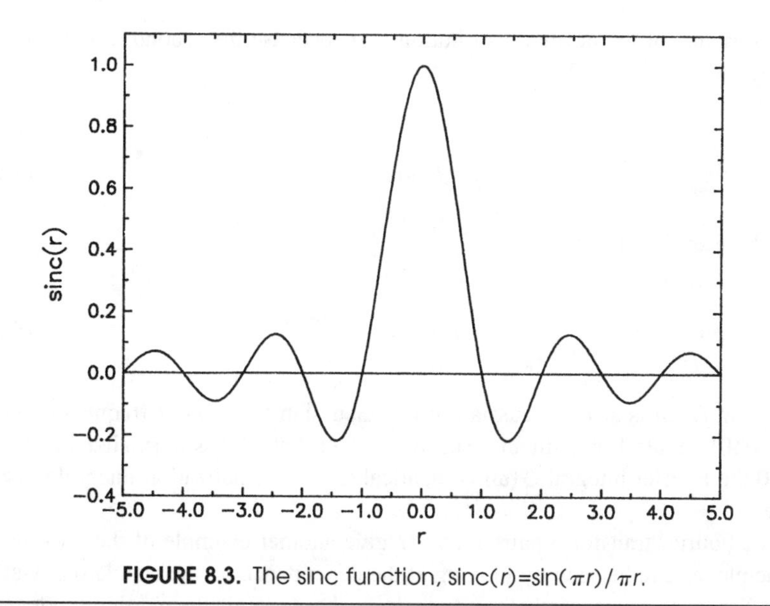

FIGURE 8.3. The sinc function, $\mathrm{sinc}(r) = \sin(\pi r)/\pi r$.

The unit rectangle from Example 1 is a useful function because it can be used as a window to allow some input values to appear in an output, while excluding other input values. A window function simply multiplies the function from which input values are to be selected. Another function that is frequently used as a window is the Gaussian, perhaps better known as the probability density for a normal distribution.

EXAMPLE 2: *The Gaussian*

In this example we examine the statement, "The Fourier transform of a Gaussian is a Gaussian."

The Gaussian function of time is

$$g(t)=\frac{1}{\sigma\sqrt{2\pi}}\,e^{-t^2/(2\sigma^2)}. \tag{8.24}$$

The function is centered on time 0. It is normalized to unit area, in that

$$\int dt\; g(t)=1. \tag{8.25}$$

The variance of the normal distribution is σ^2 because the second moment of the Gaussian is

$$\int dt\; g(t)t^2=\sigma^2. \tag{8.26}$$

The Fourier transform is

$$G(\omega)=e^{-\omega^2\sigma^2/2}. \tag{8.27}$$

Function $G(\omega)$ is also a Gaussian; it is a Gaussian function of frequency. Function $G(0)$ equals 1 as a direct result of the fact that $g(t)$ is normalized to 1. For $\omega=0$ the Fourier integral $G(\omega)$ is identical to the normalization integral over all time.

The Fourier transform pairs g and G give another example of the uncertainty principle, as can be seen by looking at the dependence on the standard deviation σ in Fig. 8.4. The parameter σ appears in the denominator of the exponential dependence on time in $g(t)$ so that $g(t)$ is narrow when σ is small. But σ appears in the numerator of the exponential dependence on frequency in $G(\omega)$ so that $G(\omega)$ is wide when σ is small.

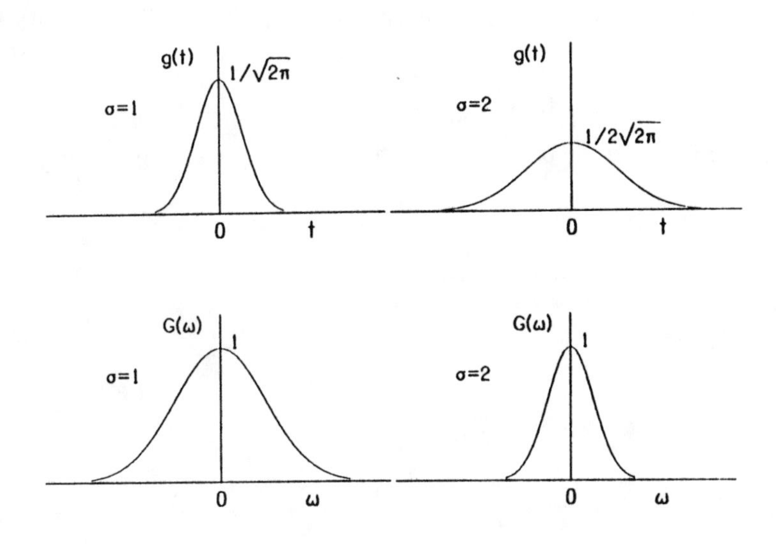

FIGURE 8.4. The Gaussian function of time for two values of the standard deviation and the corresponding Fourier transforms.

TIME SHIFTING

A general rule that is useful in practical computations relates the Fourier transform of a function of time, $x(t)$, to the Fourier transform of the same function that has been shifted along the time axis, $x(t-t_0)$, as shown in Fig. 8.5.

If the Fourier transform of $x(t)$ is $X(\omega)$ then the Fourier transform of $x(t-t_0)$ is given by

$$\mathcal{F}x(t-t_0)=e^{-i\omega t_0}X(\omega). \tag{8.28}$$

This equation says that the effect of time shifting a function by time t_0 is simply to multiply the Fourier transform of that function by a phase factor $\exp(-i\omega t_0)$.

The proof is straightforward: By definition,

$$\mathcal{F}x(t-t_0)=\int dt\ e^{-i\omega t}x(t-t_0). \tag{8.29}$$

Changing time variables, $t'=t-t_0$, we have

$$\mathcal{F}x(t-t_0)=\int dt' e^{-i\omega t_0}e^{-i\omega t'}x(t') \tag{8.30}$$

$$=e^{-i\omega t_0}X(\omega) \tag{8.31a}$$

$$=e^{-i\omega t_0}\mathcal{F}x(t). \tag{8.31b}$$

End of proof.

Time shifting is the basis of "delay-and-add filtering," treated below in the exercises. It is also at the basis of digital filtering.

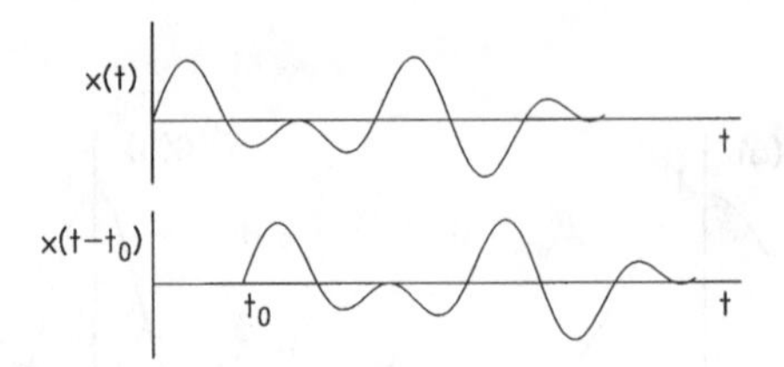

FIGURE 8.5. Function $x(t)$ and its time-shifted version.

DERIVATIVES AND INTEGRALS

The Fourier transform of the derivative of a function is closely related to the Fourier transform of the function itself. If $v(t)$ is the derivative of $x(t)[v(t)=dx/dt]$ then

$$\mathcal{F}v(t)\equiv V(\omega)=i\omega X(\omega), \tag{8.32}$$

where V is the Fourier transform of v and X is the transform of x. This formula says that the Fourier transform of the derivative is just the same as the Fourier transform of the function itself except that more weight is given to the high frequency part. The high frequency enhancement increases linearly with frequency, as in an idealized high-pass filter.

The formula can be repeatedly applied to find the Fourier transform of the n-th derivative,

$$\int dt\ e^{-i\omega t}\left[\frac{d^n}{dt^n}x(t)\right]=(i\omega)^n X(\omega). \tag{8.33}$$

For each order of derivative there is another factor of $i\omega$ in the transform.

A similar relationship holds for the integral. If $w(t)$ is the integral of $x(t)$

$$w(t)=\int_{-\infty}^{t} dt' x(t'), \tag{8.34}$$

and the average value of $x(t)$ is zero, then

$$W(\omega)=X(\omega)/(i\omega), \tag{8.35}$$

where W is the Fourier transform of $w(t)$, and X is the Fourier transform of $x(t)$.

Equation (8.35) follows directly from the formula for derivatives. Multiple integrations simply result in multiple factors of $1/(i\omega)$.

$$W_n(\omega)=X(\omega)/(i\omega)^n. \tag{8.36}$$

An obvious generalization is that differentiation and integration both produce factors of $i\omega$, with a positive exponent n for differentiation and a negative exponent n for integration.

If the average of $x(t)$ is not equal to zero, another term must be added to the integral formula, making a contribution only for zero frequency:

$$W(\omega)=X(\omega)/(i\omega)+\pi X(0)\delta(\omega). \tag{8.37}$$

Recall that the average value of $x(t)$ is $X(0)$.

The derivative and integral formulas are easy to use. However, they do not guarantee that the desired transform exists. The only thing we know is that if the transform does exist then the formulas tell us what the answer is.

One can entertain the idea that parameter n is not an integer, leading to a fractional derivative or fractional integral. It is a little difficult to say what a fractional derivative or integral might be, but it is not hard to imagine that Eqs. (8.33) and (8.36) can be well defined for fractional n. This idea is one approach to the advanced mathematical topic of fractals.

CONVOLUTION

The convolution of two functions $y(t)$ and $w(t)$, is defined by the convolution integral,

$$x(t)=\int_{-\infty}^{\infty} dt' y(t')w(t-t'). \tag{8.38a}$$

Convolution is so important that the integral is given a short-hand notation, whereby Eq. (8.38a) is rewritten

$$x(t)=y(t)*w(t). \tag{8.38b}$$

(Note that the ''star'' symbol as used here has nothing to do with complex conjugate. It indicates a convolution integral. For complex conjugate the star appears as a superscript; for convolution, the star appears on the main line.)

The convolution theorem says that if $x(t)$ is the convolution of $y(t)$ and $w(t)$, then the Fourier transform of $x(t)$ is simply the product of the Fourier transforms of $y(t)$ and $w(t)$,

$$X(\omega)=Y(\omega)W(\omega). \tag{8.39}$$

To prove this result, we write the convolution integral in Eq. (8.38) in terms of the inverse Fourier transforms of Y and W,

$$x(t)=\frac{1}{(2\pi)^2}\int dt' \int d\omega\; e^{i\omega t'}Y(\omega)\int d\omega' e^{i\omega'(t-t')}W(\omega'). \tag{8.40}$$

The integral on t' can be done to give the delta function $2\pi\delta(\omega-\omega')$. Therefore,

$$x(t)=\frac{1}{2\pi}\int d\omega\; Y(\omega)W(\omega)e^{i\omega t}. \tag{8.41}$$

But by the definition of the inverse Fourier transform,

$$x(t)=\frac{1}{2\pi}\int d\omega\; X(\omega)e^{i\omega t}. \tag{8.42}$$

Therefore, comparing Eq. (8.41) and Eq. (8.42) we see that

$$X(\omega)=Y(\omega)W(\omega), \tag{8.43}$$

which is what we set out to prove.

The convolution theorem works in reverse too. If $X(\omega)$ is the convolution of two Fourier transforms, i.e.

$$X(\omega)=\frac{1}{2\pi}\int d\omega' Y(\omega')W(\omega-\omega'), \tag{8.44a}$$

or

$$X(\omega)=\frac{1}{2\pi}\,Y(\omega)*W(\omega) \tag{8.44b}$$

then

$$x(t)=y(t)w(t), \tag{8.45}$$

and conversely. It is easy to forget the factor of 2π in Eq. (8.44). Because of it, the two convolution formulas, Eqs. (8.38) and (8.44), are not quite symmetrical.

EXAMPLE 3: *Modulation and convolution*

Balanced modulation is an example of a process that involves multiplying two functions of time. Suppose that a modulated signal is given by

$$x(t)=y(t)\cos(\omega_c t), \tag{8.46}$$

where $\cos(\omega_c t)$ is called a "carrier" signal, and $y(t)$ is a complicated modulating signal. For definiteness we suppose that $y(t)$ includes only low frequencies so that $Y(\omega)$ has vanished well before ω becomes as large as ω_c. The spectrum $X(\omega)$ can be calculated from the convolution formula in Eq. (8.44), and from Eq. (8.11a) for the Fourier transform of the cosine:

$$X(\omega)=\frac{1}{2}\int d\omega' Y(\omega')[\delta(\omega-\omega'-\omega_c)+\delta(\omega-\omega'+\omega_c)]. \tag{8.47}$$

Therefore,

$$X(\omega)=\frac{1}{2}[Y(\omega-\omega_c)+Y(\omega+\omega_c)]. \tag{8.48}$$

Figure 8.6 below shows function Y and how it is displaced along the frequency axis to make function X. To make the figure easy to draw we assumed that Y was real.

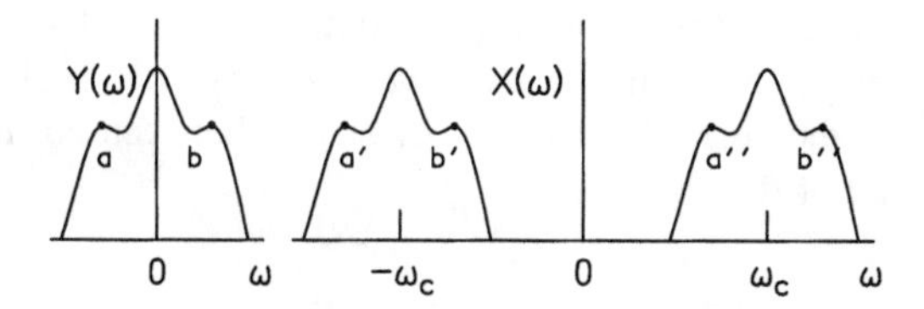

FIGURE 8.6. On the left is the spectrum of the modulating signal, $Y(\omega)$. On the right is the spectrum $X(\omega)$ of a signal that is the product of a cosine with frequency ω_c and the modulation. Points a and b in the modulation are translated to a' and b' and to a'' and b'' in the product signal, according to Eq. (8.48).

More about balanced modulation appears in Chapter 17 on beats and amplitude modulation.

INTRODUCTION TO CORRELATION FUNCTIONS

Suppose that we put a signal $x(t)$ into a system. Somewhere else we measure a signal $y(t)$. Both $x(t)$ and $y(t)$ have zero average value. We suspect that $y(t)$ is actually the same signal as $x(t)$, but unfortunately $y(t)$ has been corrupted by so much noise that we aren't sure. A single number that provides a sensible statistic deciding whether $x(t)$ and $y(t)$ are the same is the overlap integral.

$$c_0 = \frac{1}{T_D} \int_0^{T_D} dt \; x(t) y(t), \tag{8.49}$$

where T_D is an arbitrary measuring time and does not indicate period. It is easy to see why c_0 is a good statistic. If $y(t)$ tends to be positive when $x(t)$ is positive, and $y(t)$ tends to be negative when $x(t)$ is negative, then positive contributions to the overlap integral will occur. On the other hand, if $x(t)$ and $y(t)$ are not correlated, then each is positive or negative independent of the other. Then, because there is no DC component, the overlap integral is zero over the long run.

The overlap integral is a useful idea, but it is certainly limited. For instance, if $x(t) = \cos(\omega_0 t)$, and $y(t) = \sin(\omega_0 t)$, then the two functions are clearly closely related; one is just a phase shifted (or time delayed) version of the other. But the similarity would not appear in the calculated value of c_0. The value of c_0 would be zero because of the orthogonality of cosine and sine functions. Obviously there is room for improvement in our choice of statistic.

A better indicator of correlation between x and y is the cross-correlation function,

$$\gamma(t) = \lim_{T_D \to \infty} \frac{1}{T_D} \int_0^{T_D} dt' x(t') y(t' + t). \tag{8.50}$$

Here $\gamma(t)$ is the overlap of the input $x(t')$ with the output y at a later time $t' + t$. If y is actually a delayed version of x then we expect $\gamma(t)$ to be large and positive when t is equal to the delay. (The cross-correlation between cosine and sine is an exercise at the end of the chapter.)

Finite Signal Cross-Correlation Function

The cross-correlation function of Eq. (8.50) is ideal for the case in which functions $x(t)$ and $y(t)$ are periodic, or otherwise exist for all time. However, if x and y are of finite duration, then taking the limit that T_D becomes infinite will lead to zero. This is because the factor of T_D in the denominator will become large without bound while the numerator stays finite. Therefore, the correlation function $\gamma(t)$ is of no use to us. To deal with functions of finite duration, we define a different correlation function, called the *finite* signal cross-correlation function,

$$\gamma_f(t) = \int_{-\infty}^{\infty} dt' x(t') y(t' + t). \tag{8.51}$$

We now wish to find the Fourier transform of the cross-correlation function $\gamma_f(t)$ and to prove an important relation between it and the Fourier transforms of x and

y themselves. We observe that the cross-correlation integral resembles a convolution. However, it is not in quite the right form because the signs in the argument of y are wrong. Therefore, we have to make some changes in order to use the convolution formula. We can make progress by substituting $t' \to -t'$. Then we have

$$\gamma_f(t) = \int_{+\infty}^{-\infty} (-dt')x(-t')y(t-t') = \int_{-\infty}^{\infty} dt' x(-t')y(t-t'). \tag{8.52}$$

We have now achieved the convolution form: $y(t)$ is convolved with $x(-t)$. Because x is real, the Fourier transform of $x(-t)$ is $X^*(\omega)$, and therefore, from the convolution theorem,

$$\Gamma_f(\omega) = X^*(\omega)Y(\omega). \tag{8.53}$$

This is the result that we are seeking. It says that the Fourier transform of the cross-correlation function is equal to the product of the Fourier transforms of the two functions involved, with the small complication that we must take the complex conjugate of the Fourier transform of the function that is not delayed in time. More on the cross-correlation function can be found in Chapter 14 on autocorrelation and the power spectrum.

INTRODUCTION TO FILTERING

A filter has an input signal and an output signal, as shown in Fig. 8.7

Filtering is an operation that is simply described as multiplication in the frequency domain. It therefore follows that filtering becomes a convolution in the time domain. To be more specific, if we pass a signal $x(t)$ through a filter and obtain an output $y(t)$ then

$$Y(\omega) = H(\omega)X(\omega), \tag{8.54}$$

where $H(\omega)$ represents the filter and is called the "transfer function" of the filter. The equation says that the Fourier transform of the output is equal to the Fourier transform of the input except that certain frequencies, or bands of frequencies, are boosted or attenuated according to function H. This description of filtering is essentially correct, though it is not complete. In fact, functions X, Y, and H are all generally complex. Therefore, the transfer function H not only increases or reduces

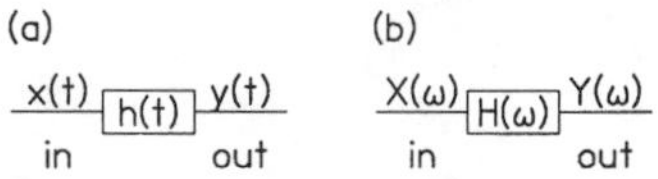

FIGURE 8.7. The filter, with its inputs and outputs, can be described (a) in time or (b) in frequency.

the strengths of some frequency components, it can also shift the phases of the components, and it generally does.

From the convolution formula Eq. (8.38) we immediately have an expression for the time dependence of the output,

$$y(t)=\int_{-\infty}^{\infty} dt' h(t-t')x(t'). \tag{8.55}$$

Here $h(t)$ is the inverse Fourier transform of $H(\omega)$. Function $h(t)$ is called the "impulse response" of the filter. The name is easy to understand because $h(t)$ is the output (the response) of the filter if the input to the filter is an impulse, i.e. if $x(t)=\delta(t)$. In symbols, we write

$$y(t)=\int_{-\infty}^{\infty} dt' h(t-t')\delta(t')=h(t). \tag{8.56}$$

Thus, the transfer function is the Fourier transform of the impulse response, and the impulse response is the inverse Fourier transform of the transfer function, i.e.

$$H(\omega)=\int_{-\infty}^{\infty} dt\ e^{-i\omega t}h(t), \tag{8.57a}$$

and

$$h(t)=\frac{1}{2\pi}\int_{-\infty}^{\infty} d\omega\ e^{i\omega t}H(\omega). \tag{8.57b}$$

More on the topic of filtering can be found in Chapter 9 on filters.

PERIODIC FUNCTIONS—FOURIER SERIES

If $x(t)$ is a periodic function with period T, then its Fourier transform consists of a sum of harmonics, each of which is a delta function in the frequency domain. In mathematical terms the above sentence can be written:
If

$$x(t)=x(t+mT) \quad \text{for all } m \tag{8.58}$$

then

$$X(\omega)=\sum_{n=-\infty}^{\infty} X_n\delta(\omega-n\omega_0), \tag{8.59}$$

where $\omega_0 = 2\pi/T$. The parameters X_n are the Fourier coefficients. They are generally complex numbers.

To show that Eq. (8.59) is correct and to find out how to calculate the X_n, we write function $x(t)$ in terms of the Fourier series from Eq. (5.2):

$$x(t) = A_0 + \sum_{n=1}^{\infty} [A_n \cos(n\omega_0 t) + B_n \sin(n\omega_0 t)]. \tag{8.60}$$

We can write the cosine and sine functions in terms of $\exp(in\omega_0 t)$ to obtain

$$x(t) = A_0 + \frac{1}{2} \sum_{n=1}^{\infty} [(A_n - iB_n)e^{in\omega_0 t} + (A_n + iB_n)e^{-in\omega_0 t}] \tag{8.61}$$

or

$$x(t) = \frac{1}{2\pi} \sum_{n=-\infty}^{\infty} X_n e^{in\omega_0 t} \tag{8.62}$$

where

$$X_n = \pi(A_n - iB_n) \quad \text{for } n>0 \tag{8.62a}$$

$$X_n = \pi(A_n + iB_n) \quad \text{for } n<0 \tag{8.62b}$$

$$X_n = 2\pi A_0 \quad \text{for } n=0. \tag{8.62c}$$

It is now easy to take the Fourier transform of $x(t)$ using Eq. (8.1),

$$X(\omega) = \frac{1}{2\pi} \sum_n X_n \int dt\ e^{-i\omega t} e^{in\omega_0 t} \tag{8.63}$$

or

$$X(\omega) = \sum_{n=-\infty}^{\infty} X_n \delta(\omega - n\omega_0), \tag{8.64}$$

which is Eq. (8.59) as advertised. This proves that $X(\omega)$ is a string of equally spaced delta functions, namely a string of harmonics. The coefficients are easily calculated if we recall from Eq. (5.15) that

$$A_n = \frac{2}{T} \int dt\ x(t)\cos(n\omega_0 t) \quad (n>0) \tag{8.65}$$

and

$$B_n = \frac{2}{T}\int dt\ x(t)\sin(n\omega_0 t) \quad (n>0). \tag{8.66}$$

Substituting into Eq. (8.62a) we find

$$X_n = \frac{2\pi}{T}\int_{-T/2}^{T/2} dt\ x(t)[\cos(n\omega_0 t) - i\ \sin(n\omega_0 t)]. \tag{8.67}$$

Having written Eq. (8.67) for positive n we realize that it holds good for negative n as well in that it also satisfies Eq. (8.62b), and, for that matter, Eq. (8.62c) as well. Therefore, for all n,

$$X_n = \frac{2\pi}{T}\int_{-T/2}^{T/2} dt\ x(t)e^{-in\omega_0 t}. \tag{8.68}$$

Getting the Fourier Coefficients from the Fourier Series

Chapter 5 includes the Fourier series for many periodic functions. To obtain the Fourier coefficients, $\{X_n\}$, for these periodic functions, one first identifies the coefficients A_n and B_n in the series and then uses Eqs. (8.62) to find the values of X_n. The coefficients A_n and B_n can be identified by associating each term in the Fourier series with a term in the defining equation, Eq. (8.59).

EXAMPLE 4: *Full-wave rectified sine*

The Fourier series for the full-wave rectified sine is given in Eq. (5.48)

$$x(t) = \frac{2}{\pi} - \frac{4}{\pi}\sum_{n=1}^{\infty}\frac{(-1)^n}{(2n)^2-1}\cos(2n\omega_0 t). \tag{8.69}$$

The term A_0 is $2/\pi$. The only other terms are $A_n = -(4/\pi)(-1)^n/[(2n)^2-1]$. From Eqs. (8.62...) it is easy to obtain coefficients X_n: $X_0=4$, and $X_n = -4(-1)^n/[(2n)^2-1]$.

Therefore the Fourier transform of the full-wave is

$$X(\omega) = 4\delta(\omega) - 4\sum_{\substack{n=-\infty \\ n\neq 0}}^{\infty}\frac{(-1)^n}{(2n)^2-1}\delta(\omega - n\omega_0), \tag{8.70}$$

or

$$X(\omega) = -4 \sum_{n=-\infty}^{\infty} \frac{(-1)^n}{(2n)^2 - 1} \delta(\omega - n\omega_0). \qquad (8.71)$$

PERIODIC FUNCTIONS AND THE LATTICE SUM

Above we showed that the Fourier transform of a periodic function $x(t)$ is a string of delta functions that we call harmonics. The proof made important use of the Fourier series representation of $x(t)$. Here we present an alternative proof that has several advantages. First, it does not make use of the Fourier series and thus avoids the appearance of a circular argument. Second, it gives us the chance to introduce the useful idea of the lattice sum.

First we define the lattice sum. Consider the function $F(\omega)$ given by a sum of exponentials,

$$F(\omega) = \lim_{M \to \infty} \sum_{m=-M}^{M} e^{i\omega mT}, \qquad (8.72)$$

where T is a time period, and m runs through all positive and negative integers. Particularly we want to compare $F(\omega)$ with the delta function, which is defined by an integral,

$$\delta(\omega) = \lim_{T_D \to \infty} \frac{1}{2\pi} \int_{-T_D}^{T_D} dt\, e^{i\omega t}. \qquad (8.73)$$

These two functions have a lot in common. For most values of ω the right-hand sides are sums of sines and cosines that are as often positive as they are negative and tend to cancel. Because the sums are infinite, the cancellation is very good and the functions are zero. For special values of ω, however, the summands or integrands are all $+1$, and then the value of the function grows without bound. For the delta function this occurs only when $\omega = 0$. For the lattice sum F this occurs whenever the argument ωmT is equal to an integer multiple of 2π, in other words, ω must satisfy the condition

$$\omega = n 2\pi / T, \qquad (8.74)$$

or

$$\omega = n\omega_0,$$

where n is an integer, and $\omega_0 = 2\pi/T$.

But although both the lattice sum and the delta function integral take on large values, they are not the same large values. The integral becomes large proportional to the value of $2T_D$. The lattice sum is over a set of discrete points given by $t = mT$,

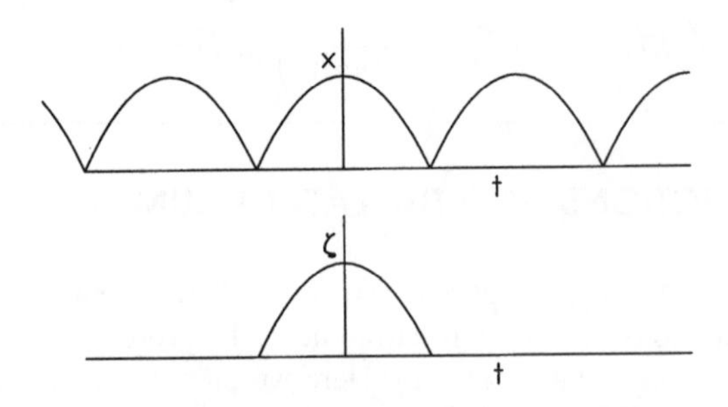

FIGURE 8.8. The periodic function $x(t)$ is obtained by successive translations of the function $\zeta(t)$.

and the lattice sum is proportional to $2M$. To put the lattice sum and the delta function on the same basis, we note that the integration duration T_D is just equal to the number of lattice points multiplied by the time interval between lattice points, which is the period, i.e. $T_D = MT$. Putting all these ideas together, we have an expression for the lattice sum,

$$F(\omega) = \sum_{m=-\infty}^{\infty} e^{i\omega mT} = \frac{2\pi}{T} \sum_{n=-\infty}^{\infty} \delta(\omega - n\omega_0). \tag{8.75}$$

This expression is a string of delta functions in frequency space. It becomes very useful in a variety of calculations with periodic functions. It makes sense dimensionally: The sum on the left of Eq. (8.75) has dimensions of unity. On the right, the factor of $1/T$ cancels the dimensions of $1/\omega$ from the delta function to lead to dimensions of unity.

We continue with the derivation of the Fourier transform of a periodic function. If $x(t)$ is periodic with period T, then it can be constructed as in Fig. 8.8 and written as

$$x(t) = \sum_{m=-\infty}^{\infty} \zeta(t - mT), \tag{8.76}$$

where function ζ is a single period of x, i.e.

$$\begin{aligned} \zeta(t) &= x(t) \quad \text{for } -T/2 < t < T/2 \\ \zeta(t) &= 0 \quad \text{for } |t| > T/2. \end{aligned} \tag{8.77}$$

By the definition of the Fourier transform,

$$X(\omega) = \int_{-\infty}^{\infty} dt\, e^{-i\omega t} \sum_{m=-\infty}^{\infty} \zeta(t - mT). \tag{8.78}$$

Rearranging the order of sum and integral, and replacing $t+mT$ with t', we find

$$X(\omega)=\sum_{m=-\infty}^{\infty} e^{-i\omega mT}\int_{-T/2}^{T/2} dt' e^{-i\omega t'}\zeta(t'). \tag{8.79}$$

In this form, the integral and the sum can be done separately. On the interval from $-T/2$ to $T/2$ the function ζ is simply the function x and therefore the integral is essentially the Fourier coefficient X_n as given by Eq. (8.68). The sum is the lattice sum above. Therefore,

$$X(\omega)=\frac{2\pi}{T}\sum_{n=-\infty}^{\infty} \delta(\omega-n\omega_0)\frac{T}{2\pi}X_n, \tag{8.80}$$

or

$$X(\omega)=\sum_{n=-\infty}^{\infty} \delta(\omega-n\omega_0)X_n, \tag{8.81}$$

which is Eq. (8.59). This concludes the alternative proof that the Fourier transform of a periodic function is a string of delta functions.

EXAMPLE 5: *Show that the Fourier transform of a train of delta functions is a train of delta functions*

This example problem illustrates the lattice sum. If the function of time is a train of delta functions equally spaced by period T,

$$x(t)=\sum_{m=-\infty}^{\infty} \delta(t-mT), \tag{8.82}$$

then the Fourier integral is easy to do by the sampling property of the delta function,

$$X(\omega)=\sum_{m=-\infty}^{\infty} e^{i\omega mT}. \tag{8.83}$$

This function is just the same as Eq. (8.72) for the lattice sum, and, as shown in Eq. (8.75), this is

$$X(\omega)=\frac{2\pi}{T}\sum_{n=-\infty}^{\infty} \delta(\omega-n\omega_0). \tag{8.84}$$

Thus the Fourier transform of the delta function train $x(t)$ is the delta function train $X(\omega)$, as shown in Fig. 8.9. The latter extends to infinitely high frequencies unattenuated because the Fourier transform of each individual pulse extends to infinity.

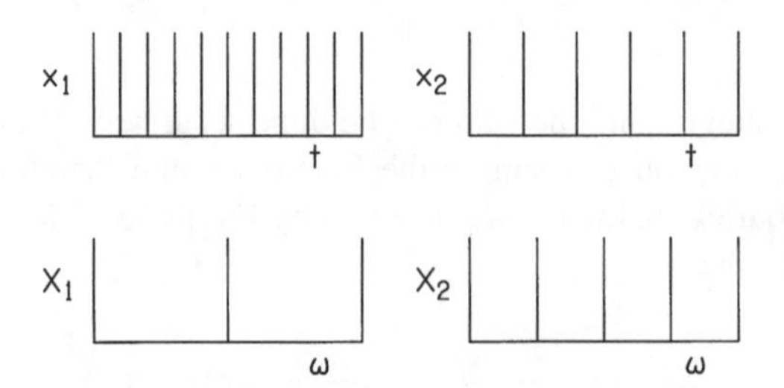

FIGURE 8.9. Trains of delta functions. On the left, the function of time $x_1(t)$ has closely spaced delta functions; its Fourier transform $X_1(\omega)$ has widely spaced delta functions. On the right, the function of time $x_2(t)$ has a longer period, and its Fourier transform $X_2(\omega)$ has a correspondingly closer spacing of delta functions.

LARGE AND SMALL SCALES

An important property of the Fourier transform is that the fine details of a periodic waveform are represented by the large-scale features of its Fourier transform. Conversely, large-scale features of the waveform are represented by fine details of the Fourier transform. In a sense, the Fourier transform turns a waveform inside out. We can illustrate this idea by using the periodic pulse waveform as an example.

The first step is to find the Fourier transform of the pulse train in Fig. 8.10(a). There, the time origin was chosen to give the function even symmetry, and so the Fourier transform will be real. As in Eq. (8.68), the Fourier coefficients are given by the Fourier transform of a single pulse:

$$X_n = \frac{2\pi}{T} \int_{-pT/2}^{pT/2} dt \, \frac{1}{p} \, e^{-in\omega_0 t} \tag{8.85}$$

$$= 2\pi \, \frac{\sin(n\omega_0 pT/2)}{n\omega_0 pT/2}. \tag{8.86}$$

But $\omega_0 T = 2\pi$, and therefore,

$$X_n = 2 \sin(\pi np)/(np) = 2\pi \, \mathrm{sinc}(np). \tag{8.87}$$

The Fourier transform is then the product of a sinc function and a train of delta functions in frequency, from Eq. (8.59),

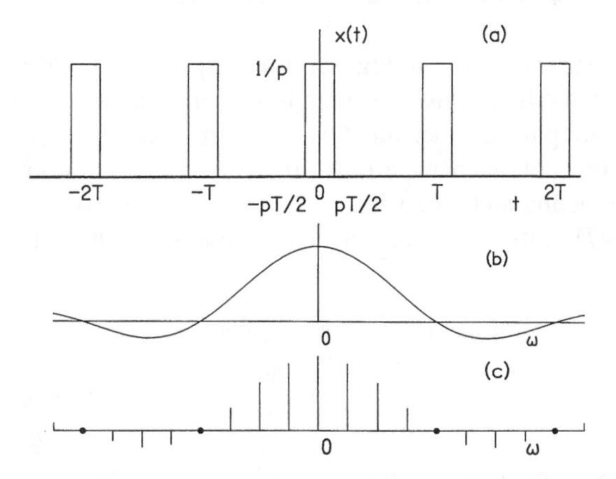

FIGURE 8.10. A periodic pulse train appears in (a) as a function of time. The Fourier transform of a **single** pulse appears in (b) as a function of frequency. The Fourier transform of the infinite pulse train in (a) is given in Fig. (c).

$$X(\omega)=2\pi\sum_{n=-\infty}^{\infty}\operatorname{sinc}(np)\,\delta(\omega-n\omega_0). \tag{8.88}$$

Because of the delta function we can rewrite Eq. (8.88) with a sinc function that is a function of a continuous argument,

$$X(\omega)=2\pi\,\operatorname{sinc}(p\omega/\omega_0)\sum_{n=-\infty}^{\infty}\delta(\omega-n\omega_0). \tag{8.89}$$

The product of the sinc function and the train of delta functions is shown graphically in Fig. 8.10(c).

To see how small-scale structure in time is represented by large-scale structure in frequency, and vice versa, we begin by noting that the small-scale structure in time is in the details of each individual pulse. This is ultimately represented by the broad sweep (extending from minus infinity to plus infinity) of the sinc function shown in Fig. 8.10(b). If we change a tiny detail of the pulses, for example if the corners were rounded somewhat, then this would be reflected by changes spread over a wide range of high frequencies in Fig. 8.10(b).

On the other hand, the large-scale structure of the pulse train is the periodicity, extending over all time. This periodicity in time is represented by that fact that the Fourier transform is a lattice function. If the large-scale structure of the pulse train were to be changed, for instance if it were long but not infinitely long, then each of the narrow spikes in Fig. 8.10(c) would be somewhat broadened. Thus, the overall character of the pulse train determines the tiny details of the Fourier transform.

FOURIER TRANSFORMS OF SGN AND THETA

The function sgn(t), shown in Fig. 8.11(a) is equal to $+1$ when t is positive and equal to -1 when t is negative. When t is zero, the function is zero (sgn(0)=0). When sgn(t) multiplies another function $x(t)$ it reverses the sign of $x(t)$ for $t<0$, but leaves the function unchanged for $t>0$. Therefore, it turns an even function of t into an odd function and vice versa.

Because sgn(t) is itself an odd function, its Fourier transform is given by

$$\mathcal{F}[\text{sgn}(t)]=-i\int_{-\infty}^{\infty}dt\ \sin(\omega t)\text{sgn}(t). \tag{8.90}$$

Equation (8.90) involves a common problem, the evaluation of an ambiguous integral at time infinity. To solve this problem, we ask ourselves, "How would anything be different if the time function of interest were multiplied by $\exp(-\alpha t)$, where α is so tiny that the decay is hardly noticeable, except, of course, at infinite time?" If the answer to the question is that nothing would really be different, then we use the exponential decay to assert that the value of the integral at infinite time is zero. Therefore, we find that

$$\mathcal{F}[\text{sgn}(t)]=-i2\int_{0}^{\infty}dt\ \sin(\omega t)=\frac{2}{i\omega}. \tag{8.91}$$

The theta function, $\theta(t)$ shown in Fig. 8.11(b), is zero for negative time and equal to 1 for positive time. As a multiplier, $\theta(t)$ forces any function to be zero for negative times. The theta function can be written easily in terms of the sgn function,

$$\theta(t)=\frac{1}{2}+\frac{1}{2}\ \text{sgn}(t). \tag{8.92}$$

Remembering that the Fourier transform of the number 1 is $2\pi\delta(\omega)$, we find the Fourier transform of the theta function to be

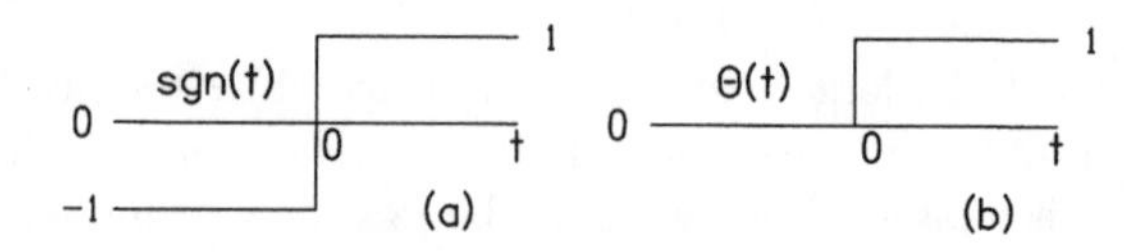

FIGURE 8.11. (a) The sign function sgn(t). (b) The theta function, or unit step function $\theta(t)$.

$$\mathcal{F}[\theta(t)] = \pi\delta(\omega) + \frac{1}{i\omega}. \tag{8.93}$$

EXERCISES

GENERAL EXERCISES

Exercise 1: Self-consistency in reverse

In Eqs. (8.3)–(8.7) we proved the self-consistency of the equations for Fourier transform pairs, x and X, by writing an equation for x in terms of X and then expressing X in terms of x. Do the reverse: show self-consistency by writing an equation for X in terms of x which you then express in terms of X.

Exercise 2: Basic Fourier transforms

(a) Prove Eq. (8.11b) for the Fourier transform of $\sin(\omega_0 t)$.

(b) Prove that if $x(t)$ is equal to a constant c then its Fourier transform $X(\omega)$ is equal to $2\pi c\,\delta(\omega)$.

Exercise 3: Even and odd functions

The text says that any real function $x(t)$ can be written as the sum of an even function and an odd function i.e.

$$x(t) = x_e(t) + x_o(t).$$

(a) Prove that this is true by showing that

$$x_e(t) = [x(t) + x(-t)]/2,$$

and

$$x_o(t) = [x(t) - x(-t)]/2.$$

(b) Suppose that $x(t) = \exp(-t/T)$. What are the even and odd functions that sum to $x(t)$?

Exercise 4: Time shifting—delay and add

One of the most common operations in signal processing is to add together a signal and a delayed version of that signal. Show that if the signal has the Fourier transform $X(\omega)$ then the sum of the signal and the same signal delayed by time t_0 has a Fourier transform given by $Y(\omega)$, where

$$Y(\omega)=[1+e^{-i\omega t_0}]X(\omega).$$

This equation has the form of a filter. The output Y is obtained from the input X by a multiplication operation. The function $[1+e^{-i\omega t_0}]$ plays the role of the transfer function $H(\omega)$. This function has a magnitude that goes to zero whenever ω is an odd multiple of π/t_0. More on delay-and-add appears in Chapter 15.

Exercise 5: Derivatives

Equation (8.32) relates the Fourier transform of a function to the Fourier transform of its derivative. Use the fact that

$$\frac{d}{dt}\cos(\omega_0 t)=-\omega_0\sin(\omega_0 t)$$

to derive a formula for the Fourier transform of $\sin(\omega_0 t)$, starting with the Fourier transform for $\cos(\omega_0 t)$.

Exercise 6: Uncertainty principle

There is a value of t for which the Gaussian $g(t)$ is equal to half of its maximum value, $g(0)$. This value of t is the half-width at half-height, Δt. There is also a value of ω for which the Fourier transform, $G(\omega)$, is equal to half of its maximum value, $G(0)$. This value of ω is the half-width at half-height, $\Delta\omega$.
Prove that the uncertainty product $\Delta\omega\Delta t$ is independent of the variance of the Gaussian. Specifically, show that

$$\Delta\omega\,\Delta t=2\ln(2)=1.386$$

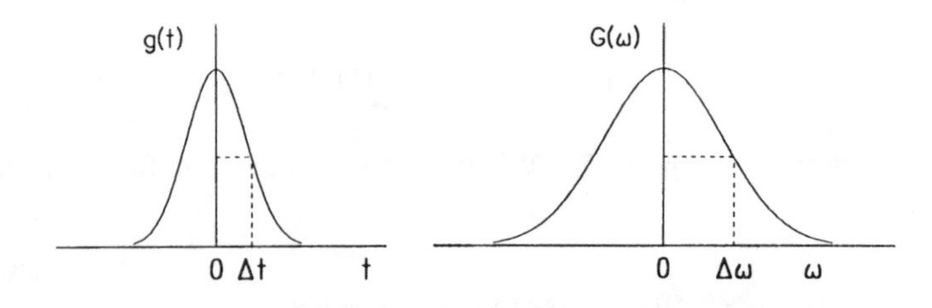

FIGURE E.6. A Gaussian distribution in time and its Fourier transform, which is a Gaussian in frequency. The figure shows the definitions of Δt and $\Delta\omega$ as half-widths at half-height.

CONVOLUTION INTEGRAL EXERCISES

Exercise 7: Convolution in frequency

Show that if $x(t)$ is the product of $y(t)$ and $z(t)$ then $X(\omega)=Y(\omega)*Z(\omega)/(2\pi)$.

Exercise 8: Which is t' and which is $t-t'$?

The order of the functions in the convolution integral does not matter, i.e. $y*z=z*y$. This means that

$$\int_{-\infty}^{\infty} dt' y(t') z(t-t') = \int_{-\infty}^{\infty} dt' z(t') y(t-t').$$

Use a change of integration variables to prove this.

Exercise 9: Fourier transform of 1

If a function $x(t)$ is multiplied by the number 1 then we still have the function $x(t)$. It follows that if $X(\omega)$ is convolved with the Fourier transform of the number 1 then the result will be the function $X(\omega)$ again. Use the fact that the Fourier transform of the number 1 is $2\pi\delta(\omega)$, to prove this necessary self-consistency.

Exercise 10: Unit step function

The unit step function $\theta(t)$ is defined as a function that is zero for negative time and is equal to one for positive time.

(a) Use the fact that $\theta(t)$ is the integral of $\delta(t)$ to verify that the Fourier transform of $\theta(t)$ is $1/(i\omega)+\pi\delta(\omega)$, as given in Eq. (8.93).

(b) Consider the function $x(t)$ that is zero for negative time and is equal to $\cos(\omega_0 t)$ for positive time. Do the Fourier integral in time to show that its Fourier transform is

$$X(\omega)=i\omega/(\omega_0^2-\omega^2).$$

(c) Calculate the Fourier transform of the function $x(t)$ in part (b) by a convolution integral. In other words, convolve the Fourier transform of the unit step function with the Fourier transform of the cosine. Compare the answer with part (b).

Exercise 11: Product of two cosines

Use the convolution formula to find the Fourier transform of $x(t)$, where $x(t)=\cos(\omega_1 t)\cos(\omega_2 t)$.

Exercise 12: Understanding modulation sidebands

The figure for the modulating signal in Fig. 8.6 showed peaks a and b. Show how these peaks are translated by the formula for the spectrum of the modulated signal to form a' and b', then also a'' and b''.

Exercise 13: Phase-shifted function

Find the cross-correlation function between the two functions, $\cos(\omega_0 t)$ and $\sin(\omega_0 t)$.

Exercise 14: Dimensions

Show that the dimensions of the Fourier transform of the cross-correlation function are time squared.

EXERCISES ON PERIODIC FUNCTIONS

Exercise 15: Series

The Fourier coefficients X_n for a periodic function were related to the coefficients A_n and B_n from Fourier series. Coefficients X_n can also be related to amplitudes and phases, C_n and ϕ_n. Show that

$$X_n = \pi C_n e^{-i\phi_n} \quad \text{for } n>0,$$

and

$$X_{-n} = \pi C_n e^{+i\phi_n} \quad \text{for } n>0.$$

Exercise 16: Representing periodicity

Consider a function $x(t)$:

(a) Explain why it is necessary and sufficient that

$$x(t)=x(t+mT), \quad \text{for all integer } m,$$

in order for $x(t)$ to be periodic with period T.

(b) Show that if x is periodic with period T, it can be written as a convolution of a function $\zeta(t)$ with a train of delta functions, where function $\zeta(t)$ is simply one cycle of function $x(t)$ and is everywhere else zero, i.e.,

$$\zeta(t)=x(t) \quad \text{for } -T/2<t\leq T/2$$

$$\zeta(t)=0 \quad \text{for } |T|>T/2.$$

In symbols, this exercise asks you to show that

$$x(t)=\sum_{m=-\infty}^{\infty}\int_{-\infty}^{\infty}dt'\,\delta(t'-mT)\zeta(t-t').$$

Exercise 17: Train of delta functions

Prove that

$$\sum_{m=-\infty}^{\infty}\delta(t-mT)=\frac{1}{T}\sum_{n=-\infty}^{\infty}e^{-in\omega_0 t},$$

where $\omega_0=2\pi/T$. This says that a lattice in time, with lattice separation T, can be represented by a Fourier series with constant amplitude $(1/T)$ and fundamental frequency ω_0.
Hint: Begin by writing

$$\delta(t-mT)=\frac{1}{2\pi}\int d\omega\; e^{-i\omega(t-mT)}.$$

TRANSFORMS OF FINITE FUNCTIONS

Exercise 18: Exponentially damped sine

Show that if $x(t)$ is an exponentially damped sine, i.e.,

$$x(t)=e^{-t/\tau}\sin(\omega_0 t) \quad (t\geq 0),$$

and

$$x(t)=0 \quad (t<0),$$

then

$$X(\omega)=\frac{\omega_0\tau^2}{1-(\omega^2-\omega_0^2)\tau^2+2i\omega\tau}.$$

Exercise 19: Raised-cosine pulse

The raised-cosine pulse is a single pulse given by the equation

$$x(t)=\frac{1}{T_D}[1+\cos(2\pi t/T_D)] \quad (|t|<T_D/2)$$

$$x(t)=0 \quad (|t|\geq T_D/2).$$

(a) Show that the Fourier transform of this pulse is given by

$$X(\omega)=\frac{\sin(\pi r)/(\pi r)}{1-r^2},$$

where $r=\omega T_D/(2\pi)$. $X(\omega)$ decreases as the cube of ω for large ω, as expected.

(b) The average value of the raised-cosine pulse is $1/T_D$. Therefore, the integral of the function in Fig. E.19a is 1. Show that this agrees with the zero-frequency limit $X(0)$ of the Fourier transform above.

The raised cosine is often used as an envelope, shaping an input signal to get a pulsed output. Therefore, the Fourier transform of the output is equal to the Fourier transform of the input convolved with the function $X(\omega)$ above.

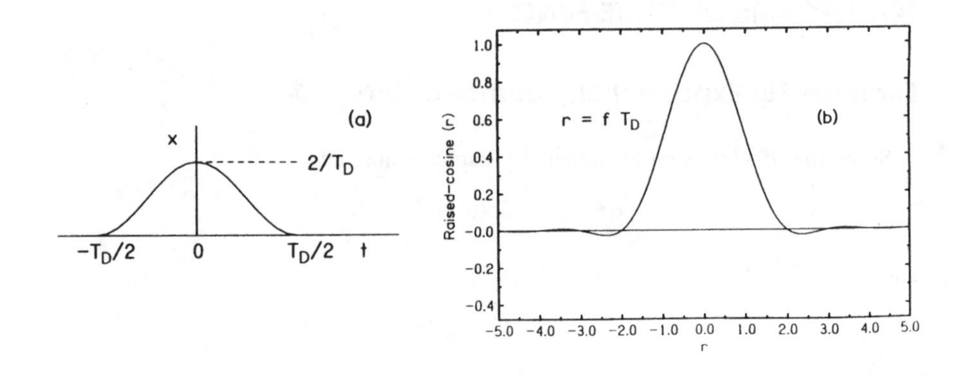

FIGURE E.19. (a) The raised-cosine pulse is a smooth function of time. We expect high-frequency components of the Fourier transform to be weak. (b) The Fourier transform of the raised cosine.

Exercise 20: Triangular and parabolic pulses

The triangular pulse (sometimes called the "hat function") is given by

$$\mathrm{tri}(t)=\frac{2}{T_D}\left(1-\frac{2|t|}{T_D}\right) \qquad -T_D/2<t<T_D/2,$$

and the parabolic pulse is given by

$$\mathrm{par}(t)=\frac{3}{2T_D}\left[1-\left(\frac{2t}{T_D}\right)^2\right] \qquad -T_D/2<t<T_D/2.$$

Both functions are zero outside the ranges given. The prefactors cause both pulses to have unit area. They are shown in Fig. E.20.

(a) Show that the Fourier transforms of triangular and parabolic pulses are, respectively:

$$\mathrm{Tri}(r)=\frac{2}{(\pi r)^2}(1-\cos \pi r),$$

and

$$\mathrm{Par}(r)=\frac{3}{(\pi r)^2}\left(\frac{\sin \pi r}{\pi r}-\cos \pi r\right),$$

where $r=\omega T_D/(2\pi)$ or, equivalently, $r=fT_D$.

Graphs of these two functions are shown below in Figs. E.20c and E.20d. They can be compared with the sinc function in Fig. 8.3, which is the Fourier transform of the rectangular pulse.

(b) The triangular pulse $\mathrm{tri}(t)$ is the convolution of a rectangular pulse with itself, where the rectangular pulse has width $T_D/2$ and height $2/T_D$. Because the Fourier transform of this rectangular pulse is $\mathrm{sinc}(r/2)$, it follows that the Fourier transform of the triangular pulse ought to be the square of the sinc function, $\mathrm{Tri}(\omega)=\mathrm{sinc}^2(r/2)$. Show that the above result for $\mathrm{Tri}(\omega)$ agrees.

Exercise 21: Trapezoidal pulse

The trapezoidal pulse, shown in Fig. E.21(a) is defined by two times, t_1 and t_2.

(a) Show that the Fourier transform of the trapezoidal pulse is given by

$$\mathrm{Trap}(r)=\frac{2}{(2\pi r)^2(t_2/t_1-1)}[\cos 2\pi r-\cos 2\pi r(t_2/t_1)],$$

where $r=\omega t_1/(2\pi)$ or $r=ft_1$.

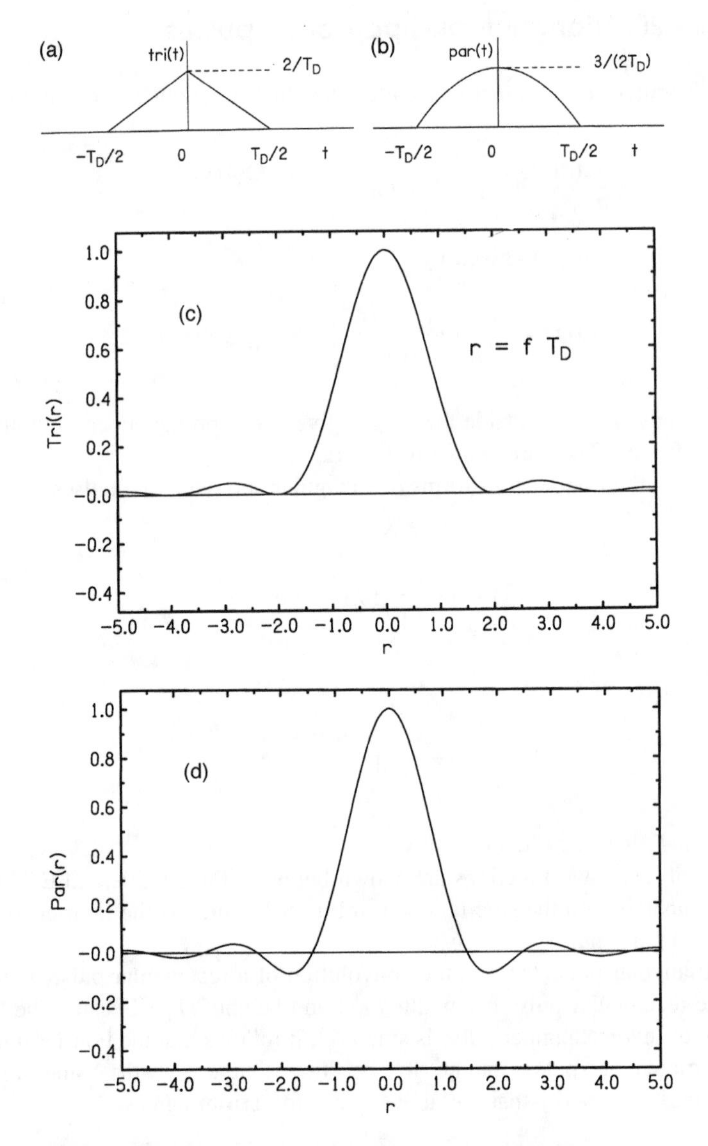

FIGURE E.20. (a) The triangular pulse. (b) The parabolic pulse. (c) and (d) Fourier transforms of the triangular pulse and the parabolic pulse. $r = fT_D$.

[Hint: Notice that the trapezoid is what you get if you integrate two rectangles, a positive rectangle between $-t_2$ and $-t_1$ and a negative rectangle between t_1 and t_2. You already know that the Fourier transform of a rectangle centered on zero is a sinc function. Therefore, you can solve this problem by dividing the sinc function by $i\omega$ (from the integral rule) and then using the time-shift rule for a rectangle translated left and right by $(t_2+t_1)/2$.]

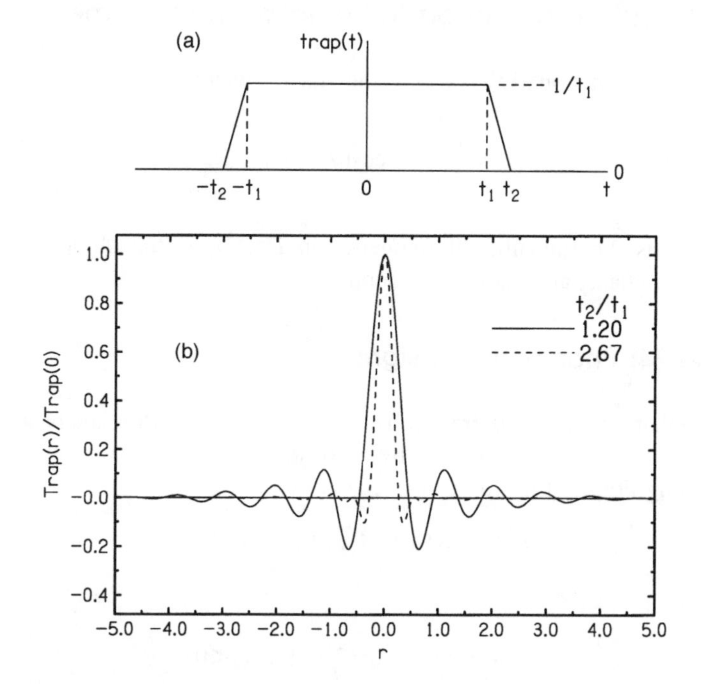

FIGURE E.21. (a) The trapezoidal pulse with t_2/t_1=1.2. (b) The Fourier transform of the trapezoidal pulse for two values of t_2/t_1.

(b) Use the series expansion for the cosine function (Appendix C) to show that the limit for zero frequency is $\text{Trap}(0)=1+t_2/t_1$. Show that this is the area under Fig. E.21(a) as expected.

(c) The trapezoid is a popular envelope for turning tones on and off. The ramped edges serve to reduce spectral splatter. Use the convolution concept to sketch the spectrum of a 1000-Hz tone that is turned on and off with a trapezoidal envelope having a total duration of $2t_2=60$ ms and rise/fall times of 5 ms. Such an envelope has $t_2/t_1=1.2$, and, as shown in Fig. E.21(b), the largest side lobe occurs at $r=\pm 0.65$. Show that these side lobes occur at 1013 and 987 Hz.

Exercise 22: The ideal amplifier

Wry wags describe an ideal voltage amplifier as a piece of wire with gain. By that they mean that the output is the same as the input except that there is more of it. Show that the impulse response of such an ideal amplifier is given by $h(t)=g\,\delta(t)$, where g is the gain.

Exercise 23: Find two errors in the following statement

Bozo says, "Function $x(t)$ is a periodic signal given by

$$x(t)=\sum_{n=1}^{\infty} C_n \cos(2\pi nt/T+\phi_n),$$

where $2\pi/T$ is the fundamental frequency in Hz, ϕ_n is the phase of the n-th harmonic in radians, and C_n is, in general, complex."

Exercise 24: Phase-shifted signal

Suppose that signal $y(t)$ is created from signal $x(t)$ by shifting the phase of each sine and cosine component of $x(t)$ by a constant phase shift $\Delta\phi$. Show that the corresponding Fourier transforms are related by

$$Y(\omega)=X(\omega)e^{i\Delta\phi} \quad (\text{for } \omega>0),$$

and

$$Y(\omega)=X(\omega)e^{-i\Delta\phi} \quad (\text{for } \omega<0).$$

CHAPTER 9

Filters

The reason why actively creative scientists seem so seldom to communicate with each other is that they do not really want to. Scientists whose work is prospering are wrapped up in it with such obsessive absorption that they want above all to be left to cultivate their gardens, and their closest approximation to neighborly behavior often amounts to little more than an inclination to borrow their neighbor's garden tools—especially the physicists'.

P. B. MEDAWAR

In the most general terms, a filter is any device that selects certain frequencies from an input and passes them through to an output while suppressing other frequencies. What is most important about a filter is its frequency response. A lowpass filter allows low frequencies to pass while attenuating high frequencies. A highpass filter passes high frequencies and attenuates the low. A bandpass filter passes only frequencies in a given band while rejecting components with frequencies above and below the band. A band-reject (or band-stop) filter passes everything except for frequencies in the stop band.

FILTER FUNDAMENTALS

In the more formal terms used in this chapter, the definition of a filter is more restrictive. In order to be a proper filter, the device must be a linear processor that modifies the spectrum of the input by multiplication. If the Fourier transform of the input signal is $X(\omega)$, then the Fourier transform of the output signal, $Y(\omega)$, is given by

$$Y(\omega)=H(\omega)X(\omega). \tag{9.1}$$

Function $H(\omega)$ characterizes the filter and is called the "transfer function." If it should happen that $H(\omega)$ is zero then the output is zero, no matter how large the input.

Because the filter is a linear processor, the output at a given frequency is directly proportional to the input at that frequency. If the input is doubled then the output will also be doubled. The constant of proportionality between input and output is the filter transfer function $H(\omega)$. If the input to the filter is the sum of two signals, then the output is the sum of the two outputs that would occur if the two input signals were processed separately by the filter. Explicitly, if input A produces output alpha and input B produces output beta, then when A and B are added together and put into the filter the output becomes alpha plus beta. An operation like this is sometimes called "superposition."

A Complex Product

Equation (9.1) has a deceptively simple appearance. In fact, all three quantities there are complex, including the transfer function. Normally it is convenient to represent the transfer function in polar form,

$$H(\omega)=|H(\omega)|e^{i\phi(\omega)}. \tag{9.2}$$

This form neatly separates two aspects of the filtering function, the amplitude response and the phase response. For any frequency ω, the amplitude of the output is equal to the amplitude of the input multiplied by the magnitude $|H(\omega)|$. The phase of the output is equal to the phase of the input plus the phase shift of the filter $\phi(\omega)$. We can illustrate these ideas with a few concrete examples.

EXAMPLE 1: *Pure tone response*

We shall shortly introduce a filter called "first-order lowpass," specified by a single parameter, the time constant. In this example we use such a filter with time constant given by $1/(200\pi)$. The amplitude and phase response of this filter to a pure tone with frequency of 100 Hz ($\omega=2\pi100$) is

$$|H(2\pi100)|=1/\sqrt{2}=0.707 \tag{9.3}$$

$$\phi(2\pi100)=-45 \text{ degrees.} \tag{9.4}$$

If a 100-Hz sine wave passes through the filter, its amplitude will be reduced by the factor $1/\sqrt{2}$ and its phase will be shifted by -45 degrees. Input and output signals are shown in Fig. 9.1.

The response at 200 Hz will be different. Here the filter transfer function is

$$|H(2\pi200)|=1/\sqrt{5} \tag{9.5}$$

$$\phi(2\pi200)=-63 \text{ degrees.} \tag{9.6}$$

At 200 Hz there is more attenuation (smaller factor $|H|$) and a larger phase shift (-63 degrees).

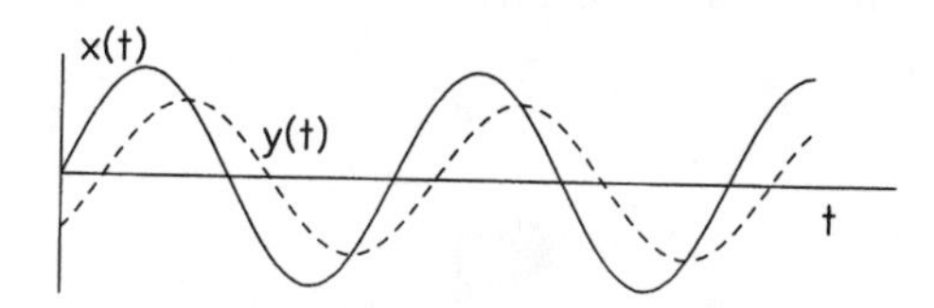

FIGURE 9.1. A 100-Hz input signal, $x(t)$, is shown by a solid line; the output signal, $y(t)$, is shown by a dashed line. The filter has reduced the signal amplitude by a factor of 0.707 and shifted its phase by −45 degrees.

EXAMPLE 2: *Complex tone response*

If the input to the first-order filter of Example 1 is the sum of unit-amplitude cosines at 100 and 200 Hz, then

$$x(t)=\cos(2\pi 100t)+\cos(2\pi 200t). \tag{9.7}$$

The spectrum is shown in Fig. 9.2(a). Because the filter is linear we can immediately find the output signal by taking over the results from Example 1. The spectrum of the output is given in Fig. 9.2(b). The amplitudes and phases of the components have been changed as in Example 1. The spectrum shows that whatever frequencies go into a filter also come out. The filter cannot change frequencies, nor can it introduce components with frequencies not present in the input.

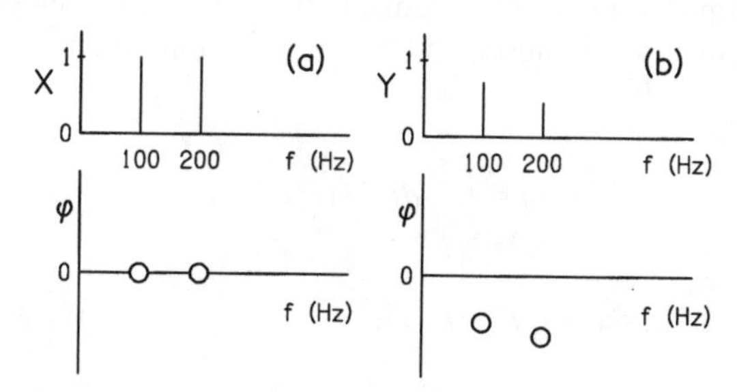

FIGURE 9.2. The amplitude and phase spectra of a tone with two components. (a) The input to the filter. (b) The output from the filter.

The output of the filter as a function of time is therefore described by

$$y(t)=\frac{1}{\sqrt{2}}\cos\left(2\pi 100t-\frac{2\pi}{360}45\right)+\frac{1}{\sqrt{5}}\cos\left(2\pi 200t-\frac{2\pi}{360}63\right). \tag{9.8}$$

Input and output signals, as functions of time, are shown in Fig. 9.3(a) and 9.3(b), respectively.

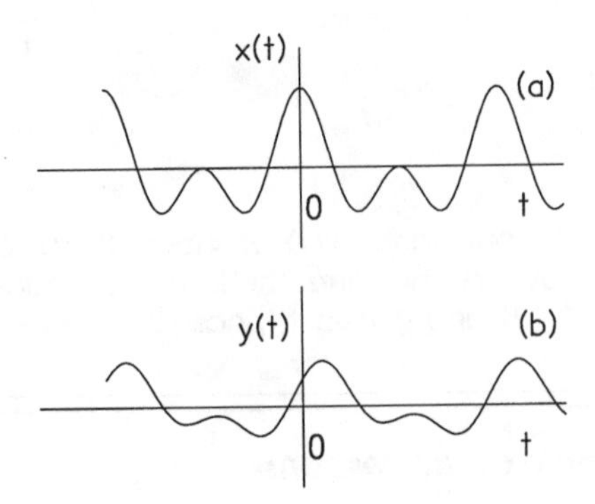

FIGURE 9.3. Signals corresponding to the spectra of Fig. 9.2. (a) The input signal. (b) The output signal. The peaks have been delayed in the output because of the phase shift of the filter. Filtering has also changed the shape of the signal because the relative amplitudes of the components have been changed and the relative phases have been changed.

Impulse Response

It is most natural to think of filters in the frequency domain where they are described by a transfer function. However, any operation in the frequency domain has a corresponding operation in the time domain. Because the frequency domain operation is multiplication, the time domain operation is convolution. Therefore, as noted in the introduction to filtering in Chapter 8, the output y is related to the input x by

$$y(t)=\int_{-\infty}^{\infty} dt' h(t-t')x(t'), \tag{9.9}$$

or with a change of variables, $t' \rightarrow t-t'$,

$$y(t)=\int_{-\infty}^{\infty} dt' h(t')x(t-t'). \tag{9.10}$$

Function $h(t)$ is called the impulse response. It is the inverse Fourier transform of $H(\omega)$,

$$h(t)=\frac{1}{2\pi}\int_{-\infty}^{\infty} d\omega \; e^{i\omega t}H(\omega). \tag{9.11}$$

Equations (9.9) and (9.10) give the output as a function of time in terms of a convolution between the input and the impulse response.

The form in Eq. (9.9) is particularly transparent. The integral equation there expresses the output at time t in terms of the input at all other times, represented by the moving variable t'. Common sense tells us that the output of a physical system cannot depend upon the input at future times. Therefore, in Eq. (9.9), whenever t' is greater than t, the contribution to the integral must be zero. This requirement is known as *causality*, and it leads to a restriction on the form of the impulse response:

$$h(t)=0 \quad t<0. \tag{9.12}$$

It is therefore possible to rewrite the integrals in Eqs. (9.9) and (9.10) with new limits

$$y(t)=\int_{-\infty}^{t} dt' h(t-t')x(t'), \tag{9.13}$$

or equivalently,

$$y(t)=\int_{0}^{\infty} dt' h(t')x(t-t'). \tag{9.14}$$

This restriction on $h(t)$ in turn implies limits on the kinds of transfer functions, $H(\omega)$, that are physically realizable.

It is not hard to see why the function $h(t)$ is called the impulse response. If the input is an impulse, i.e. $x(t)=\delta(t)$, then

$$y(t)=\int_{-\infty}^{\infty} dt' h(t-t')\,\delta(t'). \tag{9.15}$$

By the selection property of the delta function, Eq. (9.15) becomes

$$y(t)=h(t). \tag{9.16}$$

Therefore, when the input is an impulse, the output response is $h(t)$ itself.

Illustration: A Filter in Action

The process of convolving an input with a filter impulse response is shown by the cartoon in Fig. 9.4. It represents the moment when the filter is creating the output at time t, namely $y(t)$. The input $x(t')$ when $t'=t$ makes an important contribution to the output, but the inputs at previous times t', where $t'<t$, also contribute, as weighted by the impulse response $h(t-t')$. For example, the input at time $t'=t_2$ contributes as weighted by $h(t-t')$ at the open circle. Some input features, such as the large peak at $t'=t_3$ occurred so long ago that they do not make an important contribution to the output at time t. The action of this particular impulse response is to smooth the signal. For example, the input valley at time $t'=t_2$ appears in the output to be less negative than the input valley at time $t'=t_1$ because of the lingering effect of the previous peak. Because the filter smooths the signal we know that it is a lowpass filter, and this also causes features in the output to lag the input.

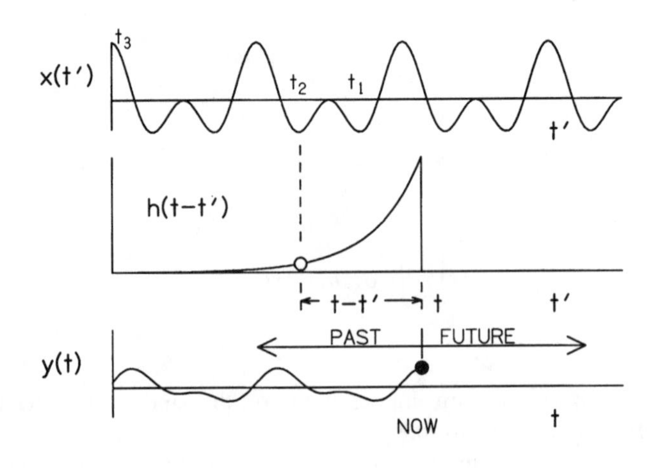

FIGURE 9.4. The cartoon shows the action of the impulse response $h(t-t')$ on the input $x(t')$ to create the output $y(t)$. The details are these: The input is the sum of two cosines, 100 Hz and 200 Hz, with equal amplitude. The filter is a first-order lowpass with time constant $\tau=1/(\pi)$ ms, corresponding to a corner frequency of 50 Hz.

Time-Invariant Systems

A filter is a linear time-invariant system. Time-invariant means that the response properties of the filter are the same now as they were a few seconds ago. It is because of the time invariance that the convolution equation applies, relating output y to input x, as given in Eqs. (9.13) and (9.14).

The key point is that the relationship between the output at time t and the input at time t' depends only upon the difference $t-t'$. It does not depend upon t and t' separately.

To illustrate this point, we consider a quite specific case. The input at time 3 ms makes a contribution to the output at a time of 5 ms. The contribution is described by

$$dy(5)=h(5-3)x(3)dt \tag{9.17}$$

or

$$dy(5)=h(2)x(3)dt. \tag{9.18}$$

A while later, at 11 ms, the output will be affected by the input at time 9 ms, given by

$$dy(11)=h(11-9)x(9)dt \tag{9.19}$$

or

$$dy(11)=h(2)x(9)dt. \tag{9.20}$$

Again the factor relating the effect of the input on the output is the number $h(2)$. It is the same number because over the time interval from 3 ms to 11 ms the filter did not change its properties. It is still the same filter.

To show how things might be otherwise, we write the equation for a linear system that is not time-invariant. Then the output is related to the input by

$$y(t)=\int_{-\infty}^{\infty} dt' h(t',t-t')y(t'), \tag{9.21}$$

Function h not only depends upon $t-t'$, it also depends absolutely on the time of the input. To relate to the previous example, $h(9,11-9)$ might be different from $h(3,5-3)$. Time-varying systems are mathematically complicated. Their transfer functions are functions of two variables, $H(\omega',\omega)$ or $H(t',\omega)$ if not completely Fourier transformed.

Awkward as it may be mathematically, the time-varying filter is actually of the utmost importance. The human vocal tract is essentially an acoustical filter, producing speech phonemes using the glottal pulse as an input. Different speech sounds result from different filter transfer functions that are made with different vocal tract shapes as the jaw, tongue, and lips move. In ordinary speech production the vocal tract is never constant. The properties of the vocal filter are always changing.

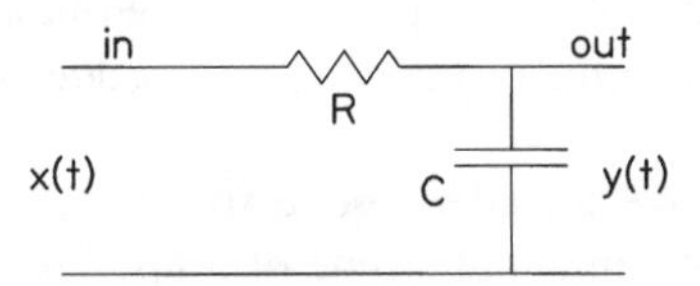

FIGURE 9.5. A first-order lowpass filter is made with a resistor (R) and a capacitor (C).

FIRST-ORDER FILTERS

To illustrate the nature of filters we consider a specific example, the first-order lowpass filter. Although it is the simplest of all filters, the first-order lowpass finds frequent application in mechanical, acoustical, and electrical domains. It is commonly used as a conceptual element in models of perception, where its role is to smooth time-varying stimuli. It leads to a continuously moving average of the recent past. An example is the Plomp-Bouman temporal integrator in Chapter 4. We choose to represent this filter in electrical form for simplicity. The schematic diagram is given in Fig. 9.5.

An explanation of the electrical elements in the filter, their impedances, and rules for combining these impedances is given in Appendix G. There it is shown that the transfer function of the filter is given by

$$H(\omega)=1/(1+i\omega\tau), \tag{9.22}$$

where τ is the time constant in seconds, given by the product $\tau=RC$. To find the real and imaginary parts of $H(\omega)$ we rationalize this complex fraction by multiplying both numerator and denominator by the complex conjugate of the denominator in order to make the denominator real. Thus, we find

$$H(\omega)=\frac{1-i\omega\tau}{1+\omega^2\tau^2}. \tag{9.23}$$

A more convenient representation of this filter is in terms of amplitude response and phase response. The amplitude response is the absolute value of the transfer function, from Eq. (9.22)

$$|H(\omega)|=\sqrt{\frac{1}{1+\omega^2\tau^2}}. \tag{9.24}$$

The phase shift is the inverse tangent of the ratio of the imaginary part to the real part, from Eq. (9.23)

$$\phi(\omega)=-\tan^{-1}(\omega\tau). \tag{9.25}$$

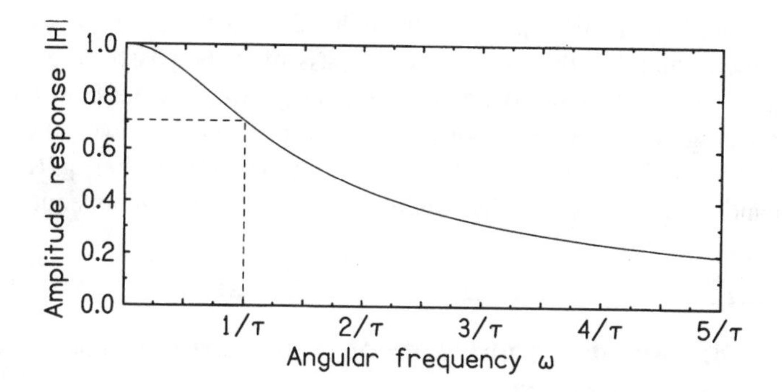

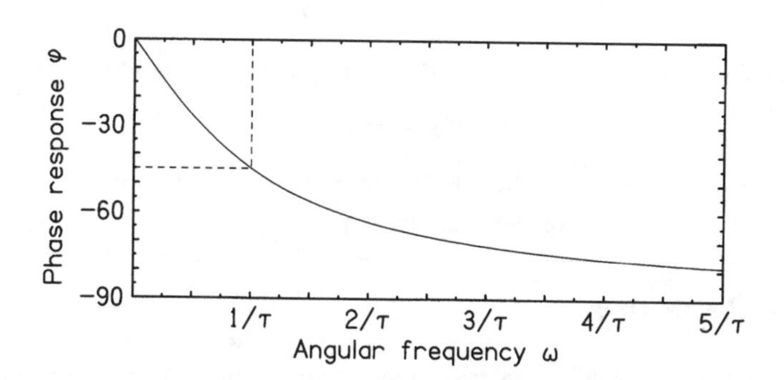

FIGURE 9.6. Amplitude response and phase response of the first-order lowpass filter. At the corner frequency, shown by a dashed line, the amplitude response is down by 3 dB, and the phase shift is −45 degrees. (See Eq. (9.27).)

These functions are plotted in Fig. 9.6.

The amplitude response function makes it obvious why this filter is called lowpass. Low frequencies are passed from input to output while high frequencies are attenuated. A lowpass filter is characterized by its corner frequency, sometimes called the "cutoff" frequency. Frequencies below the cutoff are in the passband; frequencies above the cutoff are in the stop-band. (For a filter as gentle as this first-order filter the term "cutoff" seems like an exaggeration because the attenuation in the stop-band is very modest.) In general, the cutoff frequency of a filter is defined by the 3-dB-down point, namely the frequency where the response falls 3 dB below the maximum response, which occurs at $\omega=0$.

The Bandwidth

The bandwidth of a filter is normally defined as the frequency span between three-dB-down points. In the case of the lowpass filter the passband extends down to zero frequency so that the bandwidth is simply equal to the cutoff frequency. Because a change of 3 dB corresponds to a factor of two in power, the arithmetic is particularly easy if we consider the power response of the filter, $|H(\omega)|^2$. Therefore the bandwidth $\Delta\omega$ is just that frequency ω where $|H(\omega)|^2$ is equal to half of $|H(0)|^2$.

EXAMPLE 3: *Find the bandwidth of a first-order lowpass filter in terms of the time constant*

Solution: From Eq. (9.22) we have that

$$|H(\omega)|^2 = \frac{1}{1+\omega^2\tau^2}. \tag{9.26}$$

This function is equal to half of its dc value, $|H(0)|^2$, when $\omega\tau=1$. Therefore,

$$\Delta\omega\,\tau=1. \tag{9.27}$$

This result says that the bandwidth of a first-order filter, in units of radians per second, is the reciprocal of the time constant.

Impulse Response

The impulse response of the first-order lowpass filter is given by the inverse Fourier transformation,

$$h(t) = \frac{1}{2\pi}\int_{-\infty}^{\infty} d\omega\, e^{i\omega t}\,\frac{1}{1+i\omega\tau}. \tag{9.28}$$

The integral can be done, and the answer appears in three parts,

$$h(t) = \frac{1}{\tau}\,e^{-t/\tau} \quad \text{for} \quad t>0$$

$$h(t) = 0 \quad \text{for} \quad t<0$$

$$h(t) = \frac{1}{2\tau} \quad \text{for} \quad t=0. \tag{9.29}$$

This result was obtained from Eq. (9.28) by doing a contour integral in the complex ω plane as described in Appendix E. The impulse response is a decaying exponential. It is shown as $h(t-t')$ in the center of Fig. 9.4. Because the running variable t' enters as $t-t'$ the impulse response leads to an exponentially decaying memory of things in the past.

We can check the answer for $h(t)$, given in Eq. (9.29), by finding its Fourier transform, and showing that it equals the original function $H(\omega)$ in Eq. (9.22). To do this, we have

$$\mathcal{F}[h(t)] = \frac{1}{\tau}\int_0^\infty dt\, e^{-i\omega t} e^{-t/\tau} \tag{9.30}$$

$$= \frac{1}{\tau}\left[e^{-\left(i\omega + \frac{1}{\tau}\right)t}\right] \Big/ \left[-\left(i\omega + \frac{1}{\tau}\right)\right]\Bigg|_0^\infty \tag{9.31}$$

$$= 1/(1 + i\omega\tau) \tag{9.32}$$

$$= H(\omega). \tag{9.33}$$

Because the Fourier transform is unique, we know that having got Eq. (9.32–9.33) right, we must have had the right impulse response in Eq. (9.29).

Step-Function Response

The impulse response is a way to characterize the time dependence of a filter. An alternative is the step-function response, which describes how a filter responds to the unit step function, also known as the theta function, $\theta(t)$, described in Chapter 7. Function $\theta(t)$ is defined as equal to $+1$ when $t>0$ and as equal to 0 when $t<0$. It is usual to define $\theta(0)=1/2$.

It is a simple matter to prove that the step-function response of a filter is the integral of the impulse response. If the input to a filter is a step function then, by Eq. (9.9), the output is

$$y(t) = \int_{-\infty}^{\infty} dt'\, h(t')\, \theta(t-t'). \tag{9.34}$$

Because causality requires that $h(t')=0$ for $t'<0$, the lower limit of the integral can be replaced by 0. Because $\theta(t-t')=0$ for $t'>t$, the upper limit can be replaced by t. Otherwise the θ function has the value 1, and we don't need to mention it further. Therefore, the step-function response is

$$y(t) = \int_0^t dt'\, h(t'), \tag{9.35}$$

i.e. the step-function response is the integral of the impulse response.

EXAMPLE 4: *Rise time*

The step-function response of the first-order lowpass filter is given by the integral of the impulse response in Eq. (9.29), namely

$$y(t)=\int_0^t dt' \frac{1}{\tau} e^{-t'/\tau}, \tag{9.36}$$

which integrates to

$$y(t)=1-e^{-t/\tau}, \tag{9.37}$$

as shown in Fig. 9.7.

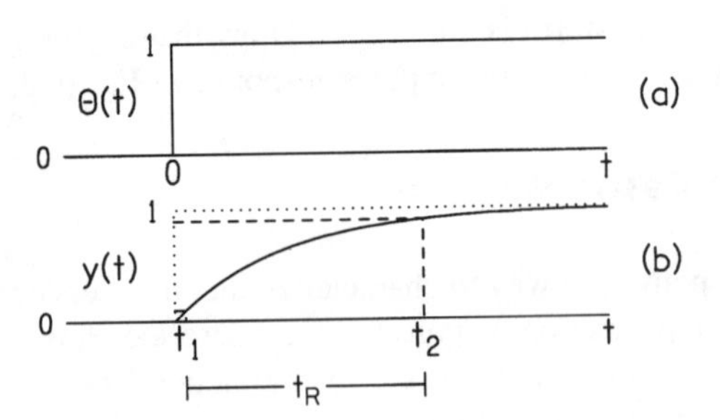

FIGURE 9.7. (a) The unit step function and (b) the response of a lowpass first-order filter to the unit step function. The rise time t_R is the time it takes the response to rise from 10% of its final value to 90% of its final value. The 10% (t_1) and 90% (t_2) points are shown by dashed lines. The long-term asymptote is shown by a dotted line.

The figure shows that the lowpass filter does not respond instantly to the unit step. All classical physical systems are that way (Nature does not make jumps), and that is why any physical system is, to some extent, a lowpass filter. The sluggishness of the response to a step function is defined by the rise time, which is a standard engineering term for the time it takes the response to rise from 10% of its final value (at time t_1) to 90% of its final value (at time t_2). Thus t_1 and t_2 are defined by $\exp(-t_1/\tau)=0.9$, and $\exp(-t_2/\tau)=0.1$. Solving these equations for t_1 and t_2 by taking the natural logs, we find the rise time,

$$t_R=t_2-t_1=2.303\tau-0.105\tau=2.2\tau. \tag{9.38}$$

We are now prepared to derive a famous equation that is widely applied in communications systems, an equation that relates the bandwidth of a system to the rise time of the system. Because $\tau = 1/\Delta\omega$ (Eq. 9.27) we have

$$\Delta \omega \, t_R = 2.2. \tag{9.39}$$

The relationship is usually stated with the bandwidth in Hz. Dividing both sides by 2π we find,

$$\Delta f \, t_R = 0.35. \tag{9.40}$$

The product of the bandwidth and the rise time is a constant; its value of 0.35. If we need to send signal pulses at a high rate, for example for digital communications, then the rise time of the system must be small so that individual pulses can be distinguished, and the bandwidth must be correspondingly large.

A Filter by Any Other Name

There are several alternative descriptions of filters that can be illustrated using the first-order lowpass. The first describes a filter by its asymptotic behavior. For high frequency, where $\omega \gg 1/\tau$, the gain of Eq. (9.24) becomes

$$\lim_{\omega \to \infty} |H(\omega)| = 1/(\omega\tau). \tag{9.41}$$

The filter gain in decibels is then

$$g = 20 \log|H(\omega)| = -20 \log(\omega\tau). \tag{9.42}$$

If the frequency increases by an octave, then the gain decreases by 6 dB ($-20 \log 2 = -6$). Therefore, the filter can be described as asymptotically -6 dB per octave, or as -20 dB per decade. The asymptotic character of a filter is particularly emphasized by a log-log plot, i.e., $\log|H(\omega)|$ vs $\log \omega$, where the high-frequency dependence becomes a straight line, as shown in Fig. 9.8.

A second way to describe a filter is in terms of the number of poles. The foundation for this representation is to imagine that angular frequency ω can have an imaginary part and is therefore a complex number that can be plotted in a complex plane like Fig. 9.9. A pole occurs for a value of ω at which the denominator of the transfer function becomes zero, making the transfer function itself infinite. Equation (9.22) for the transfer function of the first-order lowpass shows that there is only one value of ω for which this happens, namely $\omega = i/\tau$. This is the pole, represented by a symbol $\times$ in Fig. 9.9. Because there is only a single pole, the first-order filter is called a "one-pole" filter.

There is still another common name for our simple lowpass filter. It is sometimes called a lag processor. If the word "filter" makes one first think in the frequency domain, then the term "lag processor" evokes an image in the time domain. The

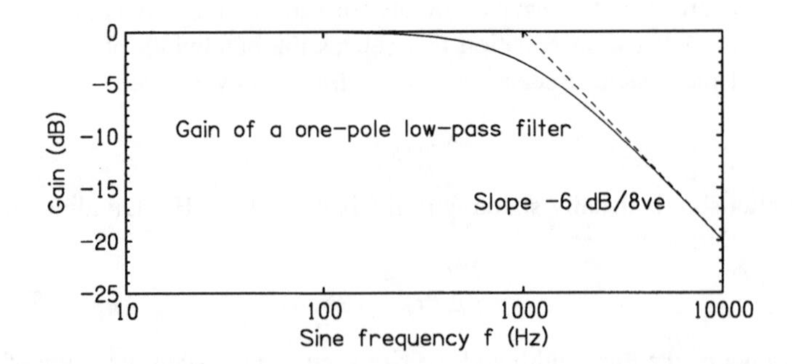

FIGURE 9.8. A log-log plot for the magnitude of the transfer function of a first-order lowpass filter plots the amplitude response from Fig. 9.6 (Gain) in decibels against the log of the frequency of the input signal. The corner frequency for the filter shown here is 1000 Hz. The high-frequency asymptotic behavior is shown by a dashed line. The asymptotic line intercepts the g=0 line at the corner frequency, where $\omega\tau=1$.

phase shift of the lowpass filter is always negative, which means that the output always lags the input. It is a classic system with inertia, the higher the frequency, the more the lag (as much as 90 degrees). Consider, for example, the response to the two negative-going bumps in the input $x(t)$ in Fig. 9.3. The first of these bumps tries to drive the output negative, and it succeeds to some extent. But the system is sluggish, and it takes the second bump to drive the output more negative.

The Highpass Filter

The one-pole highpass filter is obtained by reversing the positions of the resistor and capacitor in Fig. 9.5. Using the methods of Appendix G, we find that the transfer function is

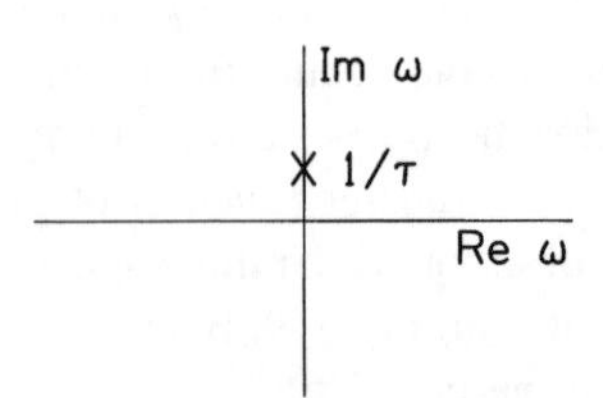

FIGURE 9.9. The complex frequency plane showing the single pole of the first-order lowpass filter. It occurs in the upper-half plane.

$$H_{HP}(\omega)=\frac{i\omega\tau}{1+i\omega\tau}. \tag{9.43}$$

It is possible to relate the transfer function for the highpass filter to the transfer function of the lowpass filter. If we multiply numerator and denominator above by $(i\omega\tau)^{-1}$ then

$$H_{HP}(\omega)=\frac{1}{1-i[1/(\omega\tau)]}. \tag{9.44}$$

Like the lowpass filter, the highpass is a function of the variable $\omega\tau$. Using the symbol H_{LP} to represent the transfer function of the lowpass filter in Eq. (9.22) we find that

$$H_{HP}(\omega\tau)=H_{LP}^{*}\left(\frac{1}{\omega\tau}\right). \tag{9.45}$$

This equation says that we can find the highpass response at any value of $\omega\tau$ by finding the lowpass response for $1/(\omega\tau)$ and then taking the complex conjugate. The complex conjugate operation reverses the sign of the phase shift. Because the phase shift of the lowpass filter was found to be always negative, the phase shift of the highpass filter is always positive.

A positive phase shift means that the output signal leads the input. At first, the idea of an output leading an input would seem to pose a crisis of causality. How can something occur at the output before it has arrived at the input? The crisis is avoided by realizing that the transfer function tells us about the response to a mathematically ideal sine wave, with no beginning or end. The peak in the output that seems to come too early is actually the response of the filter to the entire history of the sine input for all time. It is called the steady-state response.

SECOND-ORDER FILTERS

A one-pole (first-order) filter, as described above, can be either lowpass or highpass. Its transfer function has a magnitude that either decreases or increases monotonically with increasing frequency. Such a filter does not show resonance behavior, in which the filter strongly emphasizes a particular frequency band. The concept of resonance is so important that in this section we introduce a slightly more complicated filter, the second-order filter, which can show a resonance. The second-order filter also allows for much greater flexibility. Its frequency response can be lowpass, highpass, bandpass or band-reject. The resonance behavior is not limited to the bandpass configuration. Lowpass and highpass filters can also have a resonance.

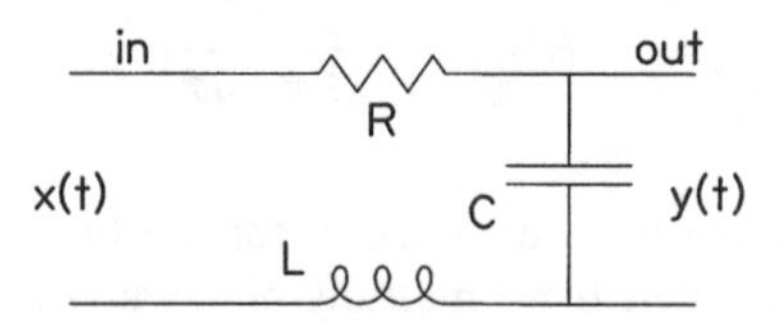

FIGURE 9.10. The second-order lowpass filter is made from three passive components: resistor (R), inductor (L), and capacitor (C).

For definiteness we study the second-order lowpass filter, with a schematic diagram shown in Fig. 9.10.

Using impedances given in Appendix G, with the equation for a voltage divider, leads to the transfer function

$$H_{LP}=\frac{1/i\omega C}{1/(i\omega C)+i\omega L+R}. \tag{9.46}$$

The transfer function can be put into more convenient form by multiplying numerator and denominator by $i\omega/L$ and defining the quantity $\omega_0=\sqrt{1/\mathrm{LC}}$. Then

$$H_{LP}(\omega)=\frac{\omega_0^2}{\omega_0^2-\omega^2+i\omega R/L}. \tag{9.47}$$

A few features of the lowpass filter are immediately evident. At low frequency (ω small) $H_{LP}(\omega)=1$, which means that at low frequency the filter acts like a simple piece of wire as expected for a lowpass filter. At high frequency, where ω^2 is so large that it dominates all the other terms in the denominator, the transfer function becomes $H(\omega)=-\omega_0^2/\omega^2$. It is left as an exercise to show that at asymptotically high frequency the gain decreases at -12 dB/octave. (Compare with a one-pole lowpass filter, which has an asymptotic dependence of -6 dB/octave.) The negative sign in $H(\omega)$ indicates a 180-degree phase shift at high frequency. Finally, it is clear that when $\omega=\omega_0$ the real part of the denominator vanishes and the transfer function can become enormous, depending on the size of resistance R. This is the phenomenon of resonance, and parameter ω_0 is called the resonance (angular) frequency.

The Q

The height and the sharpness of a resonance peak are determined by the Q of the filter, with tall, sharp resonance peaks corresponding to large values of Q. In fact, there are several different definitions of Q. All of them are approximations to the ratio of the power that oscillates back and forth among the reactive components, L and C, to the power that is dissipated in the resistive component, R. Large dissipation, corresponding to rapid damping, leads to low Q. The different definitions of Q become equivalent in the limit of large Q, and the common definition, derived below, assumes that limit.

The usual definition of Q is the ratio of the resonance frequency to the half-power bandwidth,

$$Q=\frac{\omega_0}{2\Delta\omega}=\frac{f_0}{2\Delta f}, \tag{9.48}$$

where $\Delta\omega$ (or Δf) is the amount by which the signal frequency has to differ from the resonance frequency in order for $|H(\omega)|^2$ to fall from its peak value to half of its peak value. This difference corresponds to a decrease of 3 dB.

We now find a relationship between Q and the parameters of the second-order filter by finding the half-power bandwidth of the filter. What is important in the calculation is the denominator of the transfer function in Eq. (9.47),

$$\text{Den}=\omega_0^2-\omega^2+i\omega R/L. \tag{9.49}$$

The power response of the filter depends inversely upon $|\text{Den}|^2$.

Because of the high-Q limit, the half power bandwidth $2\Delta\omega$ is small. Therefore, in our calculation we neglect terms in $(\Delta\omega)^2$ and $\Delta\omega\, R/L$.

If ω differs from ω_0 by $\Delta\omega$, then Den becomes

$$\text{Den}=2\omega_0\Delta\omega+i\omega_0 R/L, \tag{9.50}$$

and $|\text{Den}|^2$ becomes

$$|\text{Den}|^2=\omega_0^2[(2\Delta\omega)^2+(R/L)^2]. \tag{9.51}$$

To find the bandwidth we compare two cases. When $\Delta\omega=0$ the system is at resonance and the transfer function has a peak. In the other case the frequency ω is at the half-power point, and $\Delta\omega$ has a value that increases $|\text{Den}|^2$ by a factor of two. In the first case $|\text{Den}|^2=\omega_0^2(R/L)^2$. In the second case $|\text{Den}|^2$ is twice as large, which occurs when

$$(2\Delta\omega)^2=(R/L)^2. \tag{9.52}$$

There are two solutions to this equation, given by

$$\Delta\omega=\pm\frac{1}{2}\frac{R}{L}, \tag{9.53}$$

corresponding to positive and negative deviations from resonance. The difference between these two,

$$2\Delta\omega = \frac{R}{L} = BW, \tag{9.54}$$

is the half-power bandwidth, as shown in Fig. 9.11.

Finally, from Eq. (9.48) we have an equation for Q in terms of the circuit parameters,

$$Q = \frac{\omega_0}{R/L} = \frac{1}{R}\sqrt{\frac{L}{C}}. \tag{9.55}$$

It is now possible to rewrite the transfer function for the second-order lowpass filter from Eq. (9.47) in terms of resonance frequency ω_0 and Q,

$$H_{LP}(\omega) = \frac{\omega_0^2}{\omega_0^2 - \omega^2 + i\omega\omega_0/Q}. \tag{9.56}$$

When the input frequency is equal to the resonance frequency, the transfer function collapses to the simple result,

$$H_{LP}(\omega_0) = Q/i. \tag{9.57}$$

The peak amplitude response is just the number Q. The factor of i in the denominator means that at resonance the output is 90 degrees out of phase with the input. The equivalent rectangular bandwidth of the two-pole filter is the subject of Exercise 12 in Chapter 10.

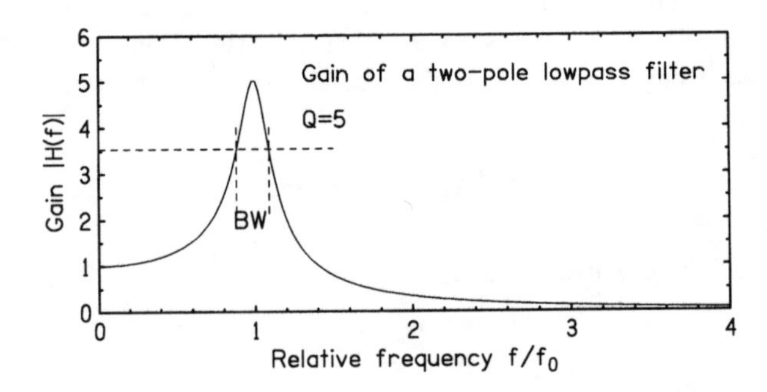

FIGURE 9.11. Amplitude response (Gain) of the second-order lowpass filter with a center frequency of f_0 and a Q of 5. The half-power bandwidth is shown by the symbol BW.

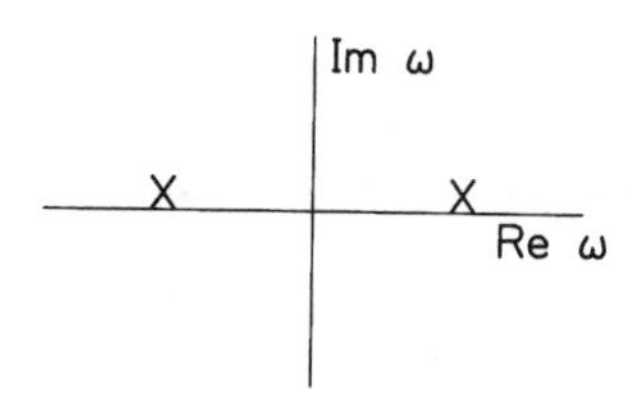

FIGURE 9.12. There are two poles for the second-order filter. Both occur in the upper-half plane. Because the imaginary coordinates of the poles are one-tenth of the real coordinates, we know that the Q of the filter is approximately 5 (see Eqs. (9.58) and (9.59)).

Poles of the Second-Order Filter

The poles of the second-order filter are found just as for the first-order filter. We assume that input frequency ω can have an imaginary part, and find the values of ω for which the denominator of the transfer function in Eq. (9.56) is zero. These values are the poles. Because the denominator is a quadratic function of ω we expect to find two poles. Writing ω as $\omega=r+ia$, and setting the imaginary part of the denominator equal to zero, we find that $a=\omega_0/(2Q)$. Setting the real part of the denominator equal to zero gives $r=\pm\omega_0\sqrt{1-[1/(2Q)]^2}$.

Therefore, the two poles occur when ω equals one of two values,

$$\omega_1=\omega_0\sqrt{1-\left(\frac{1}{2Q}\right)^2}+i\omega_0/(2Q), \tag{9.58}$$

or

$$\omega_2=-\omega_0\sqrt{1-\left(\frac{1}{2Q}\right)^2}+i\omega_0/(2Q). \tag{9.59}$$

as shown in Fig. 9.12. Not surprisingly, the second-order filter is called a "two-pole" filter.

In terms of the poles, the transfer function becomes

$$H_{LP}(\omega)=\frac{\omega_0^2}{(\omega-\omega_1)(\omega-\omega_2)}. \tag{9.60}$$

Impulse Response

The impulse response $h(t)$ of the two-pole lowpass filter is found by Fourier transforming H_{LP} from Eq. (9.56) or (9.60). This can be done with a contour

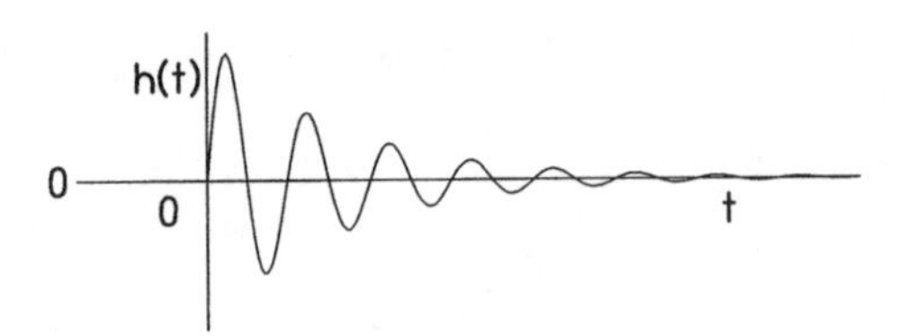

FIGURE 9.13. Impulse response for a two-pole filter having a Q of 5. Because the Q is 5, we expect to see approximately 5 oscillations as the filter rings in response to an impulse.

integral in the complex ω plane, as described in Appendix E. There are two cases of interest, one when time t is negative and the other when t is positive.

For negative t the contour at infinity diverges in the upper-half plane, so we have to do the integral in the lower-half plane. But there aren't any poles in the lower half plane, and so the integral is zero. That is what we expected to find. The impulse response has to be zero for negative values of time because of causality. For positive t, we have to do the integral in the upper-half plane. The two poles there have real parts with opposite signs, which combine together to make a sine function. The imaginary parts of the poles lead to a damping term. The final answer is

$$h_{LP}(t) = \frac{\omega_0}{\sqrt{1-[1/(2Q)]^2}} e^{-\frac{\omega_0}{2Q}t} \sin\{\omega_0 t\sqrt{1-[1/(2Q)]^2}\} \quad t \geqslant 0$$

$$h_{LP}(t) = 0 \quad t < 0. \tag{9.61}$$

The impulse response is plotted in Fig. 9.13 for a system with a Q of 5.

TRANSVERSAL FILTERS

The filters discussed so far have been standard passive filters for processing signals in analog form. We can take some steps in the direction of digital filtering by considering the transversal filter, or tapped delay line. The transversal filter consists of a delay line with taps spaced at equal values of delay. The outputs of the taps are weighted with real factors c_n and summed to create the output. Figure 9.14 gives an example for a delay line with four delays.

It is evident that the impulse response of the filter is

$$h(t) = c_0\delta(t) + c_1\delta(t-\tau) + c_2\delta(t-2\tau) + c_3\delta(t-3\tau) + c_4\delta(t-4\tau). \tag{9.62}$$

In principle, there is no limit to the number of taps on the delay line. For N taps the impulse response is

$$h(t)=\sum_{n=0}^{N} c_n\delta(t-n\tau). \tag{9.63}$$

Because the Fourier transform of $\delta(t-t_0)$ is equal to $e^{-i\omega t_0}$, it is easy to find the transfer function for the transversal filter,

$$H(\omega)=\sum_{n=0}^{N} c_n e^{-in\omega\tau}. \tag{9.64}$$

The function in Eq. (9.64) has a vaguely familiar form. It appears to be a Fourier series, but there is a difference because $H(\omega)$ is a function of frequency, and we are accustomed to seeing a Fourier series represent a signal in time. Because it is a Fourier series, we know that $H(\omega)$ is periodic in frequency, with period $\Delta\omega=2\pi/\tau$, or $\Delta f=1/\tau$. If τ is small enough the transfer function begins to repeat itself only at very high frequency, perhaps too high to be audible. Alternatively, one might want to make use of the repetition, as in creating rippled noise (see Chapter 15).

Our experience with Fourier series for functions of time was that with increasing numbers of harmonics the waveform acquired greater definition in time. With an infinite number of harmonics, one could even make the waveform into a discontinuous function of time. From this experience, we realize that the transfer function for a transversal filter can be increasingly defined as a function of frequency if we add more and more taps (greater N). With an infinite number of taps we might make a filter with an ideal rectangular transfer function.

The implementation of the transversal filter in digital signal processing is straightforward. The inputs to the summer in Fig. 9.14 are successive samples of the input signal, separated by the sampling time τ. The sampling time is small so that repetition frequency Δf is large compared to any frequency of interest.

The transversal filter is an example of a finite-impulse-response (FIR) filter. If a single impulse is put into the input, then after N time steps the output will have died

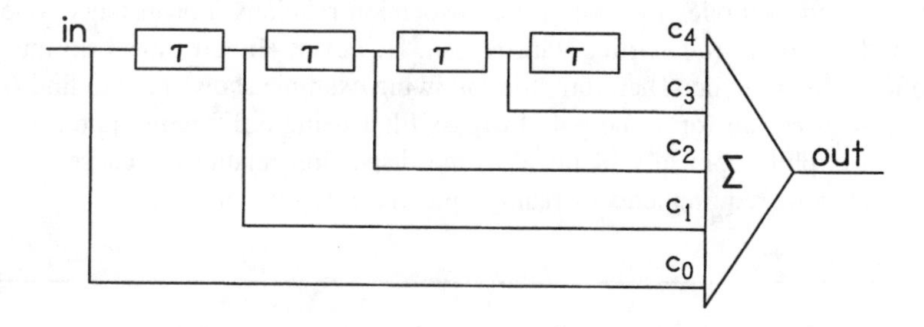

FIGURE 9.14. A transversal filter. Each element labelled τ delays the signal by time τ. Each delayed signal is weighted by a selectable constant factor c_n before summing.

away to zero. The FIR filter can be contrasted with the infinite-impulse-response (IIR) filter where some of the output is fed back into the input so that the output never dies away completely.

DISPERSION RELATIONS

Transfer functions are always complex (see Exercise 12). Therefore, a transfer function can be written as

$$H(\omega)=H_R(\omega)+iH_I(\omega), \tag{9.65}$$

where $H_R(\omega)$ and $H_I(\omega)$ are real. Functions $H_R(\omega)$ and $H_I(\omega)$ are not independent. In fact, the conditions (1) that the filter is causal and (2) that the impulse response is real, lead to strict relationships between these two functions, known as dispersion relations. In Appendix F it is shown that the real and imaginary parts of $H(\omega)$ are Hilbert transforms of one another. That means:

$$H_R(\omega)=\frac{1}{\pi}\mathcal{P}\int_{-\infty}^{\infty}d\omega'\,\frac{H_I(\omega')}{\omega-\omega'}, \tag{9.66a}$$

and

$$H_I(\omega)=\frac{-1}{\pi}\mathcal{P}\int_{-\infty}^{\infty}d\omega'\,\frac{H_R(\omega')}{\omega-\omega'}. \tag{9.66b}$$

The symbol $\mathcal{P}$ signifies that the principal value of a divergent integral should be taken. In many cases, this requires no special steps, and definite integrals from integral tables give the correct answers.

The elementary passive filters of this chapter, such as the one-pole lowpass filter, serve as good examples illustrating the dispersion relations. For instance, one can find the function $H_I(\omega)$ starting with $H_R(\omega)$. However, Hilbert transform integrals are often difficult to do. Therefore, the following example shows how to find $H_I(\omega)$ starting with $H_R(\omega)$ for a one-pole lowpass filter using a different approach. This approach is philosophically identical to the dispersion relations because it makes use of the same requirements of reality and causality.

EXAMPLE 5: *Dispersion and the one-pole lowpass filter*

Equation (9.23) shows that the real and imaginary parts of the transfer function for the one-pole lowpass filter are

$$H_R(\omega)=\frac{1}{1+\omega^2\tau^2} \tag{9.67a}$$

and

$$H_I(\omega)=\frac{-i\omega\tau}{1+\omega^2\tau^2}. \tag{9.67b}$$

The goal of this example is to show that H_I can be derived from H_R.

The first step is to find the impulse response, $h(t)$, and it will turn out that we only need to use the real part $H_R(\omega)$. The fact that the entire impulse response can be found from only the real part of the transfer function demonstrates that the imaginary part is redundant.

Like any real function, $h(t)$ can be written as the sum of an even function and an odd function:

$$h(t)=h_e(t)+h_o(t). \tag{9.68}$$

As shown in Chapter 8, the even part of $h(t)$ comes from the real part of $H(\omega)$,

$$h_e(t)=\frac{1}{2\pi}\int_{-\infty}^{\infty}d\omega\; H_R(\omega)\cos\,\omega t. \tag{9.69}$$

For the one-pole lowpass, this is

$$h_e(t)=\frac{1}{2\pi}\int_{-\infty}^{\infty}d\omega\;\frac{\cos\,\omega t}{1+\omega^2\tau^2}, \tag{9.70}$$

and this integral can be found in standard tables,

$$h_e(t)=\frac{1}{2\tau}\,e^{-|t|/\tau}. \tag{9.71}$$

Function $h_e(t)$ is shown in Fig. 9.15(a).

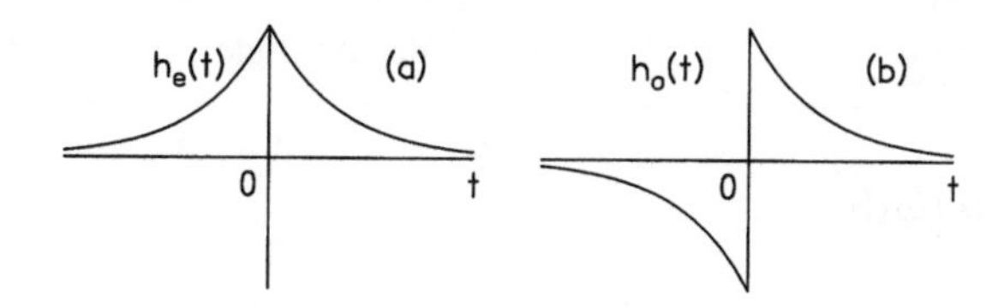

FIGURE 9.15. Even (a) and odd (b) parts of the impulse response.

Now comes the crux of the matter. Because $h(t)$ must be zero for negative t we know that $h_o(t)$ must cancel $h_e(t)$ when t is negative,

$$h_o(t)=-h_e(t) \quad (t<0). \tag{9.72}$$

But because $h_o(t)$ is odd, it must be equal to $+h_e(t)$ when t is positive. Therefore, we know that

$$h(t)=2h_e(t) \quad (t>0) \tag{9.73a}$$

and

$$h(t)=0 \quad (t<0). \tag{9.73b}$$

Because $h_e(t)$ is given by Eq. (9.71) for the one-pole lowpass,

$$h(t)=\frac{1}{\tau}\, e^{-t/\tau} \quad (t>0), \tag{9.74}$$

and this agrees with the known value of the impulse response for this filter [see Eq. (9.29)].

The next step is straightforward. We find $H_I(\omega)$ as the sine Fourier transform of the impulse response,

$$H_I(\omega)=-\int_{-\infty}^{\infty} dt\ h(t)\sin\,\omega t, \tag{9.75}$$

or

$$H_I(\omega)=-\int_{0}^{\infty} dt\ \frac{1}{\tau}\, e^{-t/\tau}\ \sin\,\omega t. \tag{9.76}$$

This integral can be found in standard integral tables. It is

$$H_I(\omega)=\frac{-i\omega\tau}{1+\omega^2\tau^2}, \tag{9.77}$$

and this agrees with Eq. (9.67b).

The example is now complete. It shows that the imaginary part of the transfer function can be derived from the real part. The steps that were done in this example for the one-pole lowpass can actually be done for any filter.

ALL-PASS FILTERS

An all-pass filter has the remarkable property that the magnitude of the transfer function is unity for all frequencies. Therefore, so far as the amplitude response is concerned the filter is invisible. However, the all-pass filter does change the phase of the signal. A simple example is the time delay. A time delay does not change any amplitudes but it changes the phases of all sine wave components by a phase shift of $\phi(\omega)=-\omega t_d$ where t_d is the delay time. The time delay does not lead to a change

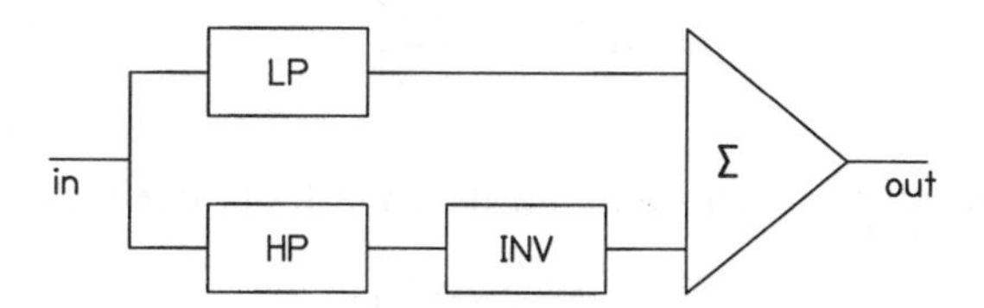

FIGURE 9.16. A one-pole all-pass filter can be made from a parallel combination of one-pole lowpass and highpass filters. The output of the highpass filter is inverted before the signals are summed. This circuit is conceptually useful, but it is not the simplest way to make a one-pole all-pass filter. See Hartmann (1979).

in the shape of a complex waveform. The general all-pass filter produces a phase shift that does change the shape of the waveform because the phase shift ϕ is a more complicated function of ω.

One-Pole All-Pass Filter

An all-pass filter can be constructed from the one-pole lowpass and one-pole highpass filters described above. The connection is shown in Fig. 9.16.

The summer adds the signals from highpass and lowpass filters. This leads to a transfer function that is the sum of the transfer functions for the two filters,

$$H(\omega)=H_{LP}(\omega)-H_{HP}(\omega), \tag{9.78}$$

where the inversion leads to the minus sign above. Therefore, from the individual transfer functions in Eqs. (9.22) and (9.43) we have

$$H(\omega)=\frac{1}{1+i\omega\tau}-\frac{i\omega\tau}{1+i\omega\tau} \tag{9.79}$$

or

$$H(\omega)=\frac{1-i\omega\tau}{1+i\omega\tau}. \tag{9.80}$$

Equation (9.80) is a transfer function that has the special form of an all-pass filter. The numerator is the complex conjugate of the denominator. The result of this special character is that when one finds the magnitude $|H(\omega)|$ one simply gets unity. For the above filter, for example,

$$|H(\omega)|^2=\frac{1+\omega^2\tau^2}{1+\omega^2\tau^2}=1. \tag{9.81}$$

To find the phase shift, we rationalize the denominator of (9.80) so that

$$H(\omega)=\frac{1-2i\omega\tau-\omega^2\tau^2}{1+\omega^2\tau^2}, \tag{9.82}$$

and

$$\phi(\omega)=\mathrm{Arg}\left(\frac{-2\omega\tau}{1-\omega^2\tau^2}\right). \tag{9.83}$$

This phase shift goes from 0 degrees to −180 degrees, as shown in Fig. 9.17.

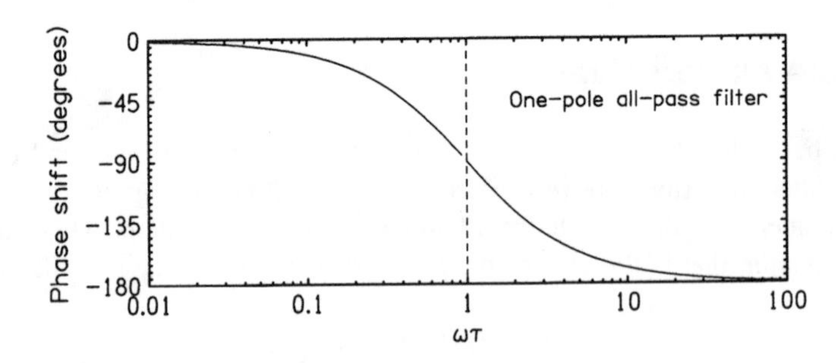

FIGURE 9.17. Phase shift of the one-pole all-pass filter.

Two-Pole All-Pass Filter

Using the general rule that an all-pass filter is constructed by arranging to have the numerator of the transfer function equal to the complex conjugate of the denominator, we can construct an all-pass filter of any order. For example, the second-order lowpass filter has a denominator that is, from Eq. (9.56).

$$\text{den}=\omega_0^2-\omega^2+i\omega\omega_0/Q. \tag{9.84}$$

Therefore, the second-order all-pass filter has a transfer function given by

$$H(\omega)=\frac{\omega_0^2-\omega^2-i\omega\omega_0/Q}{\omega_0^2-\omega^2+i\omega\omega_0/Q}. \tag{9.85}$$

It is left as an exercise to show that this filter can be constructed as in Fig. 9.18.

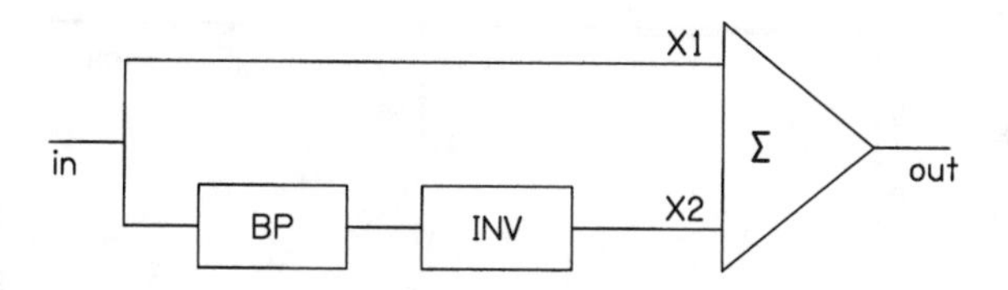

FIGURE 9.18. A two-pole all-pass filter can be made by taking twice the output of a bandpass filter and subtracting it from the original signal.

The phase shift goes through a full 360 degrees. Whereas the one-pole all-pass has a single adjustable parameter that determines the center frequency of the phase transition region (where the phase shift is 90 degrees), the two-pole all-pass has two parameters, ω_0 and Q, which control the center frequency and the sharpness of the transition. When $\omega=\omega_0$, the phase shift is 180 degrees, exactly in the middle of the phase range. The Q factor gives the width of the transition region. If we define the transition region as the frequency range where the phase shift goes from -90 degrees to -270 degrees, then, making the usual large-Q assumptions, we find that

$$2\Delta\omega/\omega_0 = 1/Q. \qquad (9.86)$$

Here $2\Delta\omega$ is the width of the transition region, in that when $\omega=\omega_0-\Delta\omega$ the phase shift is -90 degrees, and when $\omega=\omega_0+\Delta\omega$ the phase shift is -270 degrees. In words, Eq. (9.86) says that the width of the transition region is a certain percentage of the center frequency. That percentage is given by $100/Q$.

The Huggins Pitch

The all-pass filter is the experimental heart of one of the most surprising effects in psychoacoustics, namely the Huggins pitch (Cramer and Huggins, 1958). It was the first of a long line of binaural pitch effects, some of which were reviewed by Bilsen in 1977. Binaural pitch effects are created with broadband noise, too broad spectrally to produce any sense of pitch at all. The pitch that is heard is created by special interaural phase relationships.

In the case of the Huggins pitch, white noise is sent to one ear and the same noise, passed through a two-pole all-pass filter, is sent to the other. The listener hears a pitch corresponding to the center of the phase shift region where the phase shift is 180 degrees. What is fascinating about this effect is that the pitch must be centrally generated. Given a single channel of the stimulus, it is impossible to identify any frequency-specific character that would code for a pitch. One cannot even tell which of the two channels was passed through the all-pass filter. Both channels are simply white noise with random phases. In order to produce a sense of pitch both channels must be available, with phase information intact, to a central processor.

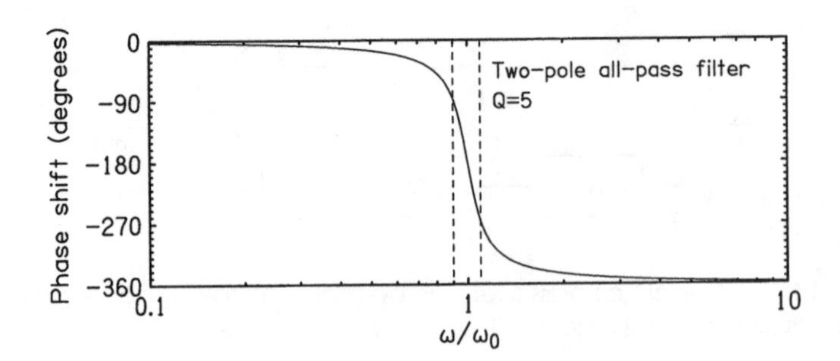

FIGURE 9.19. The phase shift of a two-pole all-pass filter with Q=5. The transition region extends from $\omega=0.9\omega_0$ to $\omega=1.1\omega_0$, as shown by dashed lines.

Because of its central character, the Huggins pitch resembles the visual effect known as "cyclopean perception" (Julesz, 1971). Here the image seen by one eye is a random-dot pattern, analogous to the random noise sent to one ear. An identical image is seen by the other eye, except for local relative deviations in dot position. When both left eye and right eye patterns are presented to a central processor, the relative deviations are interpreted as the outlines of a shape.

Huggins pitch and cyclopean perception are similar in that the signal to each peripheral sensory organ is, by itself, amorphous. The final auditory or visual image is encoded, not so much in the noise patterns, as in the relationship between the noise patterns. The final image is available only to the central nervous system, which can fuse separate inputs from the two peripheral sensory organs. This is the central cyclopean eye or the single central ear.

Because the Huggins pitch depends upon binaural synchrony, it does not exist above 1500 Hz. In this respect it resembles other binaural effects such as binaural beats, masking level difference, and lateralization due to interaural phase differences, as described in Moore (1982) Chapter 5. In fact, high-frequency signals can be localized by interaural phase if the signals are modulated (see Chapter 12 in Yost, 1994). But for these signals, the modulation itself has a low frequency.

The Huggins pitch is strongest when the center of the transition region is near 500 or 600 Hz. The Q of the all-pass filter is also important. If the transition region is too narrow (high Q), there are too few spectral components with non-zero interaural phases. If the transition region is too broad, the pitch disappears into the noise. A transition region width of about 6% of the center frequency is optimum. This corresponds to a Q of about 17, approximately three times sharper than shown in Fig. 9.19. When the Q is correctly chosen, the Huggins pitch is strong enough that listeners can match it with considerable reliability. For six listeners and five center frequencies between 300 and 700 Hz, Hartmann (1993b) found an average standard deviation of only 0.5% of the center frequency. The expected standard deviation for sine tones in this frequency range is 0.3% (see Fig. 12.2).

MINIMUM-PHASE FILTERS

The dispersion relations that relate the real and imaginary parts of a transfer function are helpful facts, but their utility is limited because the representation of a filter by its real and imaginary parts is somewhat obscure. In particular, there is no measuring process that leads to just one of these two parts. Far more useful is a representation of a filter by its amplitude response and phase response. A dispersion relation to connect those two responses would be powerful indeed. It turns out that one can find a dispersion relation that relates amplitude and phase responses if the filter has the property known as "minimum phase."

The gain and phase responses of a filter can be extracted by taking the natural log of the transfer function. Representing the transfer function by

$$H(\omega)=|H(\omega)|e^{i\phi(\omega)}, \tag{9.87}$$

we have that

$$\ln H(\omega)=n(\omega)+i\phi(\omega), \tag{9.88}$$

where $n(\omega)$ is the gain in nepers, $n(\omega)=\ln|H(\omega)|$ (Chapter 3, Exercise 8), and $\phi(\omega)$ is the phase shift in radians.

Appendix F shows that for a causal filter, $H(\omega)$ is analytic in the lower half plane; i.e. it has no poles when the imaginary part of ω is negative. This requirement led to the dispersion relations in Eqs. (9.66). But this is not enough to make $\ln H(\omega)$ analytic in the lower half plane. The log function is singular when its argument is zero. It follows that $\ln H(\omega)$ can be analytic only if there are also no *zeros* in the lower half plane. This is a new requirement. Filters that have neither poles nor zeros in the lower half plane are said to be "minimum phase" because the phase shift of such a filter can be calculated from the gain by a dispersion relation. In Appendix F it is shown that for a minimum-phase filter, where $n(\omega)$ and $\phi(\omega)$ remain finite for all ω, the dispersion relation is

$$n(\omega)=n(0)-\frac{\omega^2}{\pi}\,\mathcal{P}\int_{-\infty}^{\infty}d\omega'\,\frac{\phi(\omega')}{\omega'(\omega'^2-\omega^2)} \tag{9.89a}$$

and

$$\phi(\omega)=\frac{\omega}{\pi}\,\mathcal{P}\int_{-\infty}^{\infty}d\omega'\,\frac{n(\omega')}{\omega'^2-\omega^2}. \tag{9.89b}$$

Because $n(\omega)$ is even and $\phi(\omega)$ is odd, both integrands are even in ω', and an integral can be replaced by twice the integral from zero to infinity. Equation (9.89b) is particularly powerful. It says that if we want to find the phase shift of a system, we only have to measure the gain of the system in decibels, multiply by 0.23 to get nepers [0.23 is $1/(10 \log e)$], and do the integral. Of course, it is in the nature of the integral that in order to find the phase shift at any given frequency we need to know the gain over a wide frequency range. In some cases it may be necessary to extend the physical measurements of the gain with an analytic form that characterizes the gain at high frequency. Middlebrooks and Green (1990) calculated phase shifts in the ear canal by assuming minimum phase and using the Hilbert transform.

Equations (9.89) apply only if the filter is minimum phase. As a classic misapplication of these equations, consider an all-pass filter. The all-pass has an amplitude response of unity for all frequencies, and so the gain in nepers is $n(\omega)=0$ for all ω. It follows immediately from Eq. (9.89b) that the phase shift should be zero for all frequencies, but this is completely untrue for a real all-pass filter. By definition, every all-pass filter has at least one zero in the lower half plane and therefore an all-pass cannot be minimum phase. Another failure is the case of a pure time delay, delay t_D. Here $n(\omega)$ is again zero, but the phase shift is $\phi(\omega)=-\omega t_D$. The dispersion relations fail in this case because the phase shift does not remain finite but diverges with increasing ω. A filter with a phase shift that is pure time delay for low frequencies only, is considered by Papoulis (1962, p. 209 ff); it turns out that its amplitude response is never constant.

Filters That Are Not Minimum Phase

The dispersion relations that relate gain and phase are so useful that it is natural to ask whether there is some way they can be used if the filter is not minimum phase. The key to doing that is to separate the phase shift into two parts, one due to a minimum phase filter, and the other due to the zeros of the transfer function that prevent the filter in question from actually being minimum phase. This depends upon being able to write the filter transfer function as

$$H(\omega)=H_M(\omega)(\omega-\omega_1)(\omega-\omega_2)(\omega-\omega_3)\ldots, \tag{9.90}$$

where H_M is the part of the transfer function that is minimum phase, and ω_1 etc. are the frequencies of the zeros in the lower half plane.

Because H is a product of factors, the phase shift is the sum of phase shifts from individual factors. The total phase shift for filter H is the sum of ϕ_M, due to the minimum phase part, plus a phase shift caused by each zero, ω_1, ω_2, ω_3 etc., i.e.

$$\phi=\phi_M+\phi_1+\phi_2+\phi_3\ldots \,. \tag{9.91}$$

Phase ϕ_M obeys the dispersion relation in Eqs. (9.89) where $n(\omega)$ becomes $n_M(\omega)$, the real part of $\ln H_M(\omega)$.

The phase shift caused by a zero is not hard to calculate; we take the zero at ω_1 as an example. Complex frequency ω_1 can be written as the sum of a real part and an imaginary part. Therefore,

$$\omega-\omega_1=\omega-(\omega_r+i\omega_i). \tag{9.92}$$

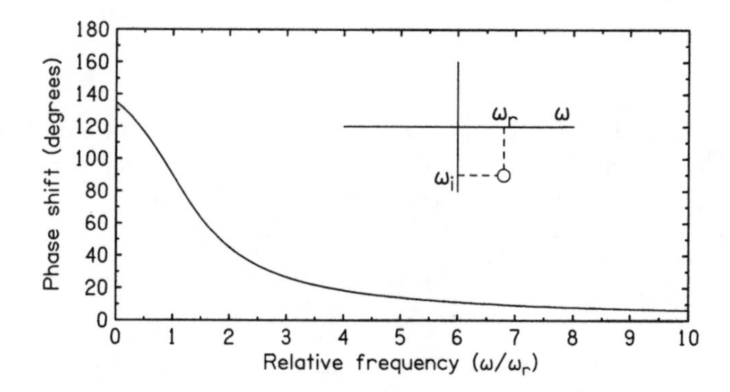

FIGURE 9.20. The inset shows a zero in the lower half plane, with $\omega_i = -\omega_r$. The curve shows the phase shift caused by the zero.

The phase shift caused by this zero is

$$\phi_1 = \text{Arg}\,\frac{-\omega_i}{\omega - \omega_r}. \tag{9.93}$$

Because the zero is in the lower half plane, ω_i is negative and the numerator is a positive number. Figure 9.20 shows a particular zero and the phase shift that it contributes to the whole. When similar phase shift calculations are done for all the zeros in the lower half plane, the sum is the difference between ϕ_M and the true phase shift ϕ.

Compensating for zeros in the lower half plane in this way requires that one know where these zeros occur. In some practical cases, such as finding an effective transfer function for a portion of the auditory system, one may not know the locations of the zeros. One may not even know whether the system is minimum phase or not. In that case, the only reasonable course is to measure both the amplitude response and the phase response.

TRANSFER FUNCTION OF THE HILBERT TRANSFORMER

Function $y(t)$ is the Hilbert transform of function $x(t)$ if $y(t)$ is related to $x(t)$ by the integral equation

$$y(t) = \frac{1}{\pi}\int_{-\infty}^{\infty} dt'\,\frac{1}{t'}\,x(t-t'). \tag{9.94}$$

Because this equation is a convolution integral we can write it in the form of a filtering operation,

$$y(t)=\int_{-\infty}^{\infty} dt' h_H(t') x(t-t'), \tag{9.95}$$

where $h_H(t)$ is the impulse response of the hypothetical filter. It follows from Eqs. (9.94) and (9.95) that taking the Hilbert transform of a signal is equivalent to passing the signal through a filter whose impulse response is

$$h_H(t)=\frac{1}{\pi}\frac{1}{t}. \tag{9.96}$$

We notice that $h_H(t)$ does not describe a causal filter; the impulse response is not zero for $t<0$. What we have is an unphysical filter, but that does not prevent us from finding the corresponding transfer function. The transfer function is the Fourier transform,

$$H_H(\omega)=\int_{-\infty}^{\infty} dt\ e^{-i\omega t}\frac{1}{\pi}\frac{1}{t}. \tag{9.97}$$

Because $1/t$ is an odd function, only the sine part of the integral survives, and

$$H_H(\omega)=\frac{-i}{\pi}\int_{-\infty}^{\infty} dt\ \frac{\sin \omega t}{t}, \tag{9.98}$$

or

$$H_H(\omega)=\frac{-i}{\pi}\int_{-\infty}^{\infty} d(\omega t)\ \frac{\sin \omega t}{\omega t}. \tag{9.99}$$

The integral of $\sin(x)/x$ is equal to π, and so function $H_H(\omega)$ is equal to $-i$ if ω is positive. If ω is negative, all signs are reversed. Therefore,

$$\begin{aligned} H_H(\omega)&=-i \quad \omega>0 \\ H_H(\omega)&=i \quad \omega<0. \end{aligned} \tag{9.100}$$

The Fourier transform here resembles the sgn function that was itself Fourier transformed in Eq. (8.91).

EXAMPLE 6: *The Hilbert transform of cos $\omega_0 t$*

The object of this example is to find the Hilbert transform of a cosine function using the filter representation of the Hilbert transformer. This is an alternative to the calculation by direct integration in Appendix E.

Let function $y(t)$ be the Hilbert transform of $\cos \omega_0 t$. We find $y(t)$ by first finding its Fourier transform $Y(\omega)$ using the transfer function $H_H(\omega)$. We recall from Chapter 8 that the Fourier transform of the cosine is given by the sum of two delta functions. Therefore,

$$Y(\omega)=H_H(\omega)\pi[\delta(\omega-\omega_0)+\delta(\omega+\omega_0)]. \tag{9.101}$$

Applying the operations of H_H, we find,

$$Y(\omega)=\pi[(-i)\delta(\omega-\omega_0)+(+i)\delta(\omega+\omega_0)], \tag{9.102}$$

or

$$Y(\omega)=\pi\frac{[\delta(\omega-\omega_0)-\delta(\omega+\omega_0)]}{i}. \tag{9.103}$$

We recognize the right-hand side as the Fourier transform of a sine. Therefore,

$$y(t)=\sin \omega_0 t. \tag{9.104}$$

It is left as an exercise to show that the Hilbert transform of $\sin \omega_0 t$ is $-\cos \omega_0 t$.

MEASURING TRANSFER FUNCTIONS AND COHERENCE

This section deals with the practical measurement of transfer functions for systems that are not ideal. It also introduces a function called the "coherence function," which provides a measure of just how far from ideal a system is. For example, we may wish to measure the transfer function of the outer ear by placing a small microphone in the ear canal. The microphone records the response to a known signal produced by a loudspeaker placed at some distant point. Such responses, made in an anechoic environment as a function of the position of the source with respect to the subject's head, are known as "head-related transfer functions."

The problem with the measurement of transfer functions is that the measuring microphone not only records the signal of the loudspeaker, it also picks up noise and some distortion too. The noise may be acoustical or electrical, and the distortion may come from the source loudspeaker or from the measuring microphone itself.

By taking the right kind of averages of the input signal to the speaker and the output signal from the microphone, it is possible to measure transfer functions in a way that is relatively immune to the noise and the distortion. This technique will be described below. The entire discussion takes place in the frequency domain. Although the process begins with input and output functions of time, it is assumed that we take the Fourier transforms of these signals, normally by computing a fast Fourier transform. The FFT leads to input and output functions evaluated at a set of

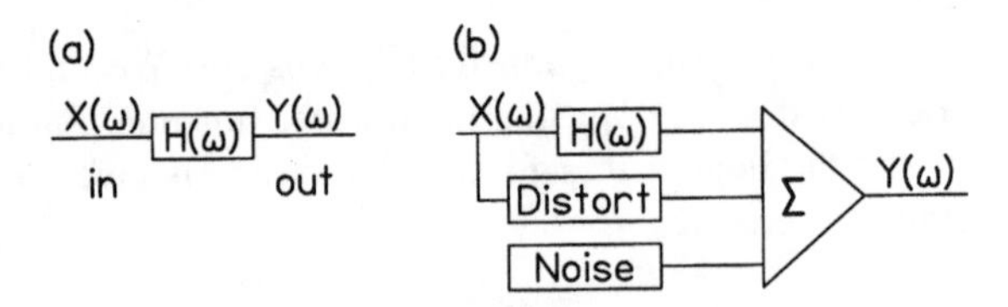

FIGURE 9.21. (a) For an ideal system, the output *Y* is equal to the product of the transfer function *H* and the input *X*. (b) For a non-ideal system the output also includes a distorted version of the input and added noise.

discrete frequencies, and these functions are complex. They have real and imaginary parts, or, equivalently, amplitudes and phases.

We begin with the idealized system, described by the transfer function $H(\omega)$ as shown in Fig. 9.21(a).

The goal of the measurement process is to find the transfer function by comparing output and input. A simple approach is to use a sine-wave input, with frequency ω, and to compare the input and output amplitudes and phases to determine the gain (or attenuation) of the filter $|H(\omega)|$ and the phase shift $\phi(\omega)$. The measurement needs to be repeated for many different values of the input frequency to determine the gain and phase shift as functions of ω.

An alternative measuring technique is to use a broadband random-phase source, like a white noise, as the input and to find the transfer function at all frequencies at once. This can be done by comparing the Fourier transforms of input X and output Y. The measured transfer function, defined as $\hat{H}$, is then given by the ratio

$$\hat{H}(\omega)=Y(\omega)/X(\omega). \tag{9.105}$$

With a fast Fourier transform, X and Y can be calculated rapidly. If input and output are described by the complex forms

$$X(\omega)=|X(\omega)|e^{i\phi_1(\omega)} \tag{9.106}$$

and

$$Y(\omega)=|Y(\omega)|e^{i\phi_2(\omega)}, \tag{9.107}$$

then the measured transfer function becomes

$$\hat{H}(\omega)=|Y(\omega)|/|X(\omega)|e^{i[\phi_2(\omega)-\phi_1(\omega)]}. \tag{9.108}$$

This technique gives a value for $\hat{H}(\omega)$ at all frequencies in the FFT except where the input has zero amplitude, a situation that would not normally be found within the band of broadband noise. Thus the measured function $\hat{H}$ is equal to the true transfer function H. This is all that needs to be said about measuring the idealized system.

In practical cases, the system under test may add noise and distortion to the output, as shown in Fig. 9.21(b). Using symbol $N(\omega)$ to represent the Fourier transform of the sum of the noise and the distortion, we express the output as

$$Y = HX + N, \tag{9.109}$$

where we drop the explicit dependence on ω to simplify the notation.

To obtain an accurate measurement of the transfer function, despite the noise and distortion, we perform ensemble averages of many independent measurements, each expressed as a function of ω by taking the FFT. The idea is to exploit the fact that HX is correlated with input X while noise plus distortion N is not. Ensemble averages of quantities over the independent measurements will be indicated by the notation $\langle ... \rangle$, as in Chapter 3.

We calculate the quantity

$$\hat{H}'(\omega) = \langle YX^* \rangle / \langle XX^* \rangle. \tag{9.110}$$

From Eq. (9.109) for Y this becomes

$$\hat{H}'(\omega) = \frac{\langle HXX^* \rangle + \langle NX^* \rangle}{\langle XX^* \rangle}. \tag{9.111}$$

Because the noise is not correlated with the signal, the average $\langle NX^* \rangle$ is zero. Because H is presumed to be stable while the averaging is going on, function H can be extracted from the averaging symbols. The average $\langle HXX^* \rangle$ becomes equal to $H\langle XX^* \rangle$, and Eq. (9.111) reduces to

$$\hat{H}'(\omega) = H(\omega). \tag{9.112}$$

Therefore, the measured transfer function is equal to the true transfer function.

The above technique depends on the lack of correlation between X and N, but it should be noted that this argument does not appeal to the cross-correlation function, because X and N are functions of frequency, not of time. What is actually uncorrelated are the values of these frequency-dependent functions obtained in different measurements. A few extra steps can make this clear: We represent X and N by amplitudes and phases,

$$\langle NX^* \rangle = \langle |N(\omega)||X(\omega)| e^{i[\phi_N(\omega) - \phi_1(\omega)]} \rangle, \tag{9.113}$$

where ϕ_N is the phase of N.

We suppose that there are a total of $\mathcal{N}_M$ different measurements, indexed by index j. Then the average can be written

$$\langle NX^*\rangle = \frac{1}{\mathcal{N}_M}\sum_{j=1}^{\mathcal{N}_M} |N_j(\omega)||X_j(\omega)|e^{i[\phi_{N,j}(\omega)-\phi_{1,j}(\omega)]}. \tag{9.114}$$

The reason that $\langle NX^*\rangle$ is zero is that the sum of the random phase factors is zero. The real part is zero because the sum of the cosines of random angles is zero. The imaginary part is zero because the sum of the sines of random angles is zero. These statements hold individually for every value of ω.

The Coherence Function

An argument similar to the above for the transfer function leads to the definition of a function known as the coherence function. The coherence function is a measure of the amount of output power that is attributable to the input, the remainder of the output being attributable to noise and distortion. The coherence function ξ^2 is defined by the following equation for averages:

$$\xi^2 = \frac{|\langle YX^*\rangle|^2}{\langle XX^*\rangle\langle YY^*\rangle}, \tag{9.115}$$

where X is the Fourier transform of the input, and Y is the Fourier transform of the output, as before. All of the averages are functions of ω and, therefore, so is ξ^2.

Writing the output as $Y = HX + N$, and using the fact that $\langle NX^*\rangle$ is zero, we find

$$\xi^2 = \frac{|\langle HXX^*\rangle|^2}{\langle XX^*\rangle(\langle |H|^2 XX^*\rangle + \langle NN^*\rangle)}, \tag{9.116}$$

or

$$\xi^2 = \frac{|H|^2\langle XX^*\rangle}{|H|^2\langle XX^*\rangle + \langle NN^*\rangle}. \tag{9.117}$$

The output power attributable to the input at frequency ω is $|H(\omega)|^2\langle |X(\omega)|^2\rangle$, that is $|H|^2\langle XX^*\rangle$, which appears in the numerator. The total output power at frequency ω is in the denominator. It follows that ξ^2 is the fraction of the output that is attributable to the input.

As in the case of the transfer function $\hat{H}'$, the key element in the calculation is the incoherence of signal and noise, specifically leading to the result that the ensemble average of signal and noise is zero. That the extraneous noise added by the system is uncorrelated with the input signal is essentially true by the definition of noise. It is less obvious that the distortion should be uncorrelated with the signal. What is important for the distortion components is the fact that the calculations of $\hat{H}'$ and ξ^2 are done independently for each frequency. Suppose, for example, that we are calculating ξ^2 for a frequency of 1000 Hz. We are therefore concerned with all distortion processes leading to components at 1000 Hz in the output. Drawing some numbers arbitrarily out of a hat, we suppose that there is a simple summation tone at $620+380=1000$ Hz, and a cubic difference tone at $2\times805-610=1000$ Hz. The phases of the distortion products depend upon the phases of the input components at 620, 380, 805, and 610 Hz, and for random noise input these phases are random across the different measurements in the averages $\langle ... \rangle$. Therefore, the ensemble average of the signal and the distortion at a particular ω is zero, as is required for the calculations of $\hat{H}'$ and ξ^2. We note that this argument requires the broadband input signal to have random phases. It would not do for the input to be a pulse train, where phases at different frequencies are locked together.

Measuring transfer functions by exploiting the correlation between input and output values can be a powerful technique. Given a long enough averaging time, it can measure the response of a concert hall during a concert with the audience present (Schroeder, 1979). The signal x is a broadband random sequence, with a Fourier transform X that is known to the analyzing computer. This signal is radiated at a low level by a loudspeaker on stage. The signal Y is recorded by a microphone at the desired location in the hall. In an amusing reversal, the music being performed plays the role of the noise and distortion which is averaged away because it is not correlated with X. The function of the music is to mask the broadband random sequence so that the audience does not hear it. The measurement process must be halted during intermission when the audience moves about, because this changes the properties of the concert hall. It is a central assumption of the method that the response H does not change during the course of the averaging. Initial experiments with this correlated averaging technique used a small general-purpose computer. The transfer function technique, and the coherence function as well, are now hardwired features of some real-time spectrum analyzers such as the Hewlett Packard 3582A. Bendat and Piersol (1971) describe the coherence function in some detail.

EXERCISES

Exercise 1: The non-filter

Consider the filter which passes a signal without change. Then the transfer function is $H(\omega)=1$. Show that the impulse response is $h(t)=\delta(t)$ and, therefore, that the convolution integral for the output leads to an output that is the same as the input.

Exercise 2: Dimensional analysis

Because input X and output Y are related by

$$Y(\omega)=H(\omega)X(\omega),$$

we know that transfer function $H(\omega)$ must have dimensions of unity. Prove, therefore, that an impulse response must have dimensions of reciprocal time.

Exercise 3: Consequences of reality

The impulse response $h(t)$ relates the output $y(t)$ to the input $x(t)$. Because x and y are both real, it follows that $h(t)$ must also be real. Show that this condition leads to the following restrictions on the transfer function, $H(\omega)$:

$$\mathrm{Re}\, H(-\omega)=\mathrm{Re}\, H(\omega)$$

$$\mathrm{Im}\, H(-\omega)=-\mathrm{Im}\, H(\omega),$$

or, in other words, $H(-\omega)=H^*(\omega)$. [Note: In a context where ω acquires an imaginary part, $H(-\omega^*)=H^*(\omega)$.]

Exercise 4: One-pole highpass

The one-pole highpass has a transfer function given by

$$H(\omega)=\frac{i\omega\tau}{1+i\omega\tau},$$

where τ is the time constant, the product of the resistance and capacitance.

(a) Show that in the limit of very low frequency the gain of the filter is $|H(\omega)|=\omega\tau$, and therefore rises at the rate of 6 dB per octave.
(b) Show that the phase shift is given by

$$\phi(\omega)=\tan^{-1}(1/\omega\tau),$$

which is always positive.

Exercise 5: Two-pole filters

The two-pole lowpass filter described in the text has transfer function given by

$$H_{LP}(\omega)=\frac{\omega_0^2}{\omega_0^2-\omega^2+i\omega\omega_0/Q}.$$

Show that the asymptotic high-frequency response is -12 dB/octave, with a 180 degree phase shift.

Exercise 6: Shift in resonance due to damping

The resonance of a two-pole filter is ω_0, where ω_0 is defined as $1/\sqrt{LC}$. When $\omega=\omega_0$, the real part of the denominator of the transfer function vanishes, causing the transfer function to be large. Interestingly, the peak of the transfer function does not occur exactly at $\omega=\omega_0$. It is possible to decrease the absolute value of the denominator further by lowering ω a little, thereby decreasing the imaginary part of the denominator while accepting a finite real part. Show that the minimum of the square of the absolute value of the resonance denominator in Eq. (9.56) occurs when

$$\omega=\omega_0\sqrt{1-\frac{1}{2}\left(\frac{1}{Q}\right)^2}.$$

In the limit of a very sharp filter the shift vanishes and $\omega=\omega_0$.

Exercise 7: Neural filters

The tuning of neurons in the auditory system is often described by a sharpness measure called Q_{10dB}. By analogy to a linear filter, Q_{10dB} is simply the ratio of the best frequency for the neuron to the bandwidth, where the bandwidth is given by that frequency deviation from the best frequency needed to make the output drop by 10 dB. Of course, for a neuron, it is not possible to do an experiment in this way. Given a sine wave input, the output of a neuron is a series of spikes. The transformation is highly nonlinear and there is no way to decide what might correspond to a 10-dB change in the output. What one can do instead is to observe changes in the input while keeping the output constant. One begins by finding the neuron's best frequency using an input stimulus of low level, and measuring the spike rate at that frequency. Then if the stimulus level is increased by 10 dB and the frequency changed to compensate, leaving the spike rate unchanged, the change in frequency gives a bandwidth that can be used to calculate Q_{10dB}.

The relationship between Q_{10dB} and Q (which might logically be called Q_{3dB}) depends upon the filter involved. Suppose that the filter is a two-pole filter as described in the text. Show that

$$Q_{10dB} = \frac{1}{3} Q.$$

Exercise 8: Three-Stage Transversal Filter

Consider the transversal filter given by Fig. E.8 and suppose that $c_2 = c_0$.

(a) Show that the transfer function is

$$H(\omega) = (c_1 + 2c_0 \cos \omega\tau) e^{-i\omega\tau}.$$

(b) Show that the phase response is $\phi = \omega\tau$. Show, therefore, that the phase shift is simply that of a pure time delay without phase distortion. The fact that it is possible to make transversal filters with no phase distortion is widely exploited in digital filtering.

(c) The transversal filter of Fig. E.8 has an impulse response given by

$$h(t) = c_0 \delta(t) + c_1 \delta(t - \tau) + c_2 \delta(t - 2\tau).$$

Show that this filter is causal.

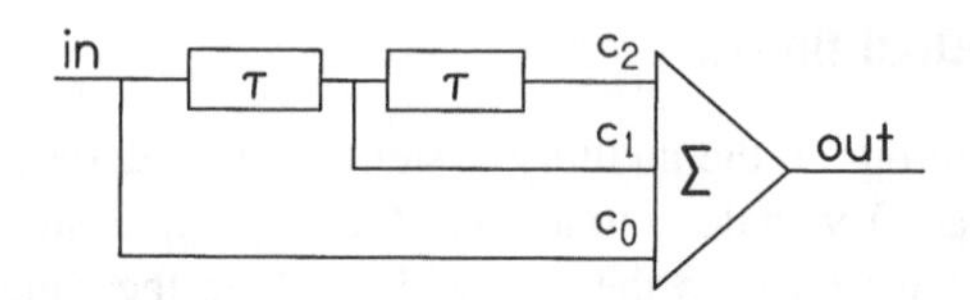

FIGURE E.8. A transversal filter with two delays.

Exercise 9: Time-domain calculation

If the input to a filter is a pure tone, $x(t) = \cos(\omega_0 t)$, then we expect that the output of the filter will also be a pure tone, with its amplitude and phase altered,

$$y(t) = |H(\omega_0)| \cos(\omega_0 t + \phi).$$

Use the impulse response for a simple one-pole lowpass filter,

$$h(t) = \frac{1}{\tau} e^{-t/\tau} \quad \text{for} \quad t>0$$

$$h(t) = 0 \quad \text{for} \quad t<0,$$

and the convolution integral,

$$y(t) = \int_{-\infty}^{\infty} dt' h(t') \cos \omega_0(t-t')$$

to find the amplitude and phase of the output.

Exercise 10: Hilbert transform of the sine

Use the filter form of the Hilbert transformer to show that the Hilbert transform of $\sin \omega_0 t$ is $-\cos \omega_0 t$.

Exercise 11: Ringing of the two-pole filter

The impulse response for the two-pole lowpass filter shows oscillations with an envelope that decays away according to the function $\exp[-\omega_0 t/(2Q)]$.

(a) Show that the decay of the oscillations, as measured by a level change in dB after n cycles, is given by

$$\Delta L = -27n/Q \quad \text{dB}.$$

(b) Analogous to reverberation time, the decay time of a ringing system may be defined as the time necessary for the ringing to decay by 60 dB. Show that the decay time (in seconds) is given by

$$T_{60} = 2.2Q/f_0.$$

(c) The decay time of a vibrational mode of a bell may be many seconds. Show that this implies a Q in the thousands.
(d) Integrate the impulse response for the two-pole lowpass filter to find the step-

function response. Show that if Q is much greater than 1, then the relationship between rise time and bandwidth is

$$\Delta f\, t_R = 0.35,$$

just the same as Eq. (9.40) for the one-pole lowpass.

Exercise 12: Necessary complexity

Filters are causal and their impulse responses are real. These two fundamental conditions lead to requirements on transfer functions. Prove that one of them is that a transfer function must be complex. Neither the real nor the imaginary part can be zero for all frequencies.

FIGURE E.12. Hobbes observes that communication systems are causal.

Exercise 13: Constructing the two-pole all-pass filter

Show that the two-pole all-pass filter of Eq. (9.85) can be constructed from a two-pole bandpass filter and a summer as shown in Fig. 9.18. For the transfer function of the bandpass, see Appendix G.

Exercise 14: Middle ear filter

Nedzelnitsky (1980) measured the pressure transfer function from the eardrum to the inner ear. The phase and amplitude data are shown in Fig. E.14. (a) Show that the phase data resemble the series connection of two bandpass filters, each of which is a two-pole filter. (b) A decade away from the resonance frequency the phase shift is about 135 degrees. Show that this corresponds to two identical bandpass filters each with a Q of about 0.2. (c) A decade below the resonance frequency the gain has dropped by about 24 dB. Show that this corresponds to two identical bandpass filters, each with a Q of about 0.4

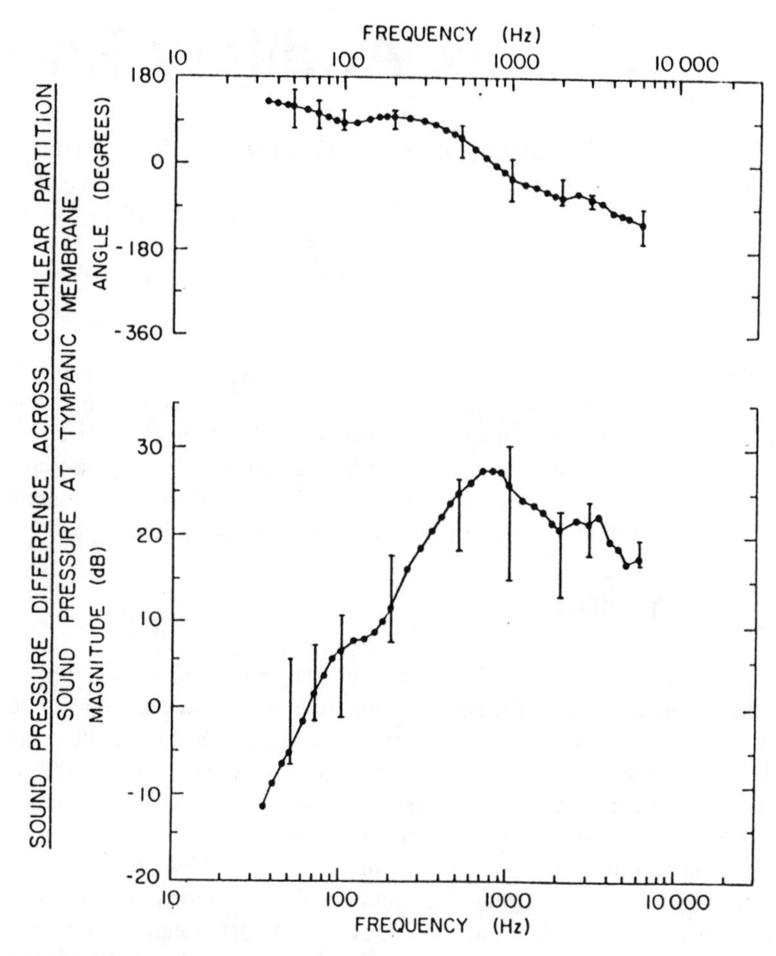

FIGURE E.14. Phase and amplitude responses of the middle ear measured by Nedzelnitsky.

CHAPTER 10

Auditory Filters

The single most salient characteristic of the auditory sense is its analytic ability.

DAVID M. GREEN

The analytic ability of the auditory sense begins at the beginning, where pressure waves in the cochlea are converted into neural impulses. Hair cells along the basilar membrane have excitation characteristics that are highly frequency selective. This selectivity, or analytic ability, can be seen in tuning curves of primary afferent fibers in the VIII-th nerve that are driven by the hair cells. The advantage of studying the system at the VIII-th nerve is that the cochlea does not have to be opened.

COCHLEAR TUNING

A tuning curve is a ''threshold of hearing'' curve for a single neuron, using a sine-tone stimulus. It shows the smallest amplitude needed to excite the nerve fiber as a function of the frequency, where a typical criterion for threshold excitation is that the presence of the signal should increase the fiber's firing rate by 10% over its spontaneous firing rate. A model tuning curve is shown in Fig. 10.1. It shows sharp bandpass tuning centered on a characteristic frequency (CF).

The tuning seen in neurons of the auditory nerve resembles filtering by an electromechanical filter. The filtering character extends also to the impulse response. If a click, approximating a delta function, is sent to the inner ear, an auditory nerve fiber shows a ringing response, so long as its characteristic frequency is not much greater than a few kHz. Figure 10.2 shows poststimulus-time (PST) histograms giving the distributions of neural spikes that result from a click at time zero for two neurons with different characteristic frequencies. The delay of the first spike is longer for the fiber with the lower characteristic frequency because it takes longer for the mechanical vibration to excite a low-frequency place in the cochlea. After the first spike, subsequent spikes are separated by a period that is the reciprocal of CF, as expected for a bandpass filter centered on frequency CF (see Fig. 9.13).

If the filter is linear then it is possible to find the magnitude of the transfer function by inverting the tuning curve of Fig. 10.1, as shown in Fig. 10.3(a). In the neighborhood of the tip, a range of about 20 dB, the transfer function can be

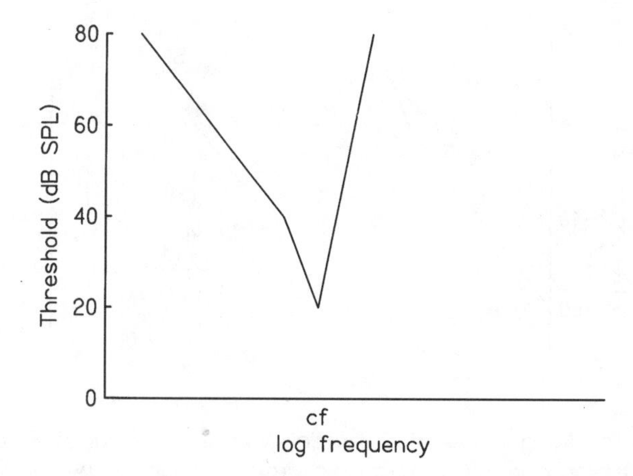

FIGURE 10.1. Model tuning curve for a neuron of the auditory nerve. The threshold at CF is 20 dB SPL. The double lower slope is observed for CF greater than 1 kHz.

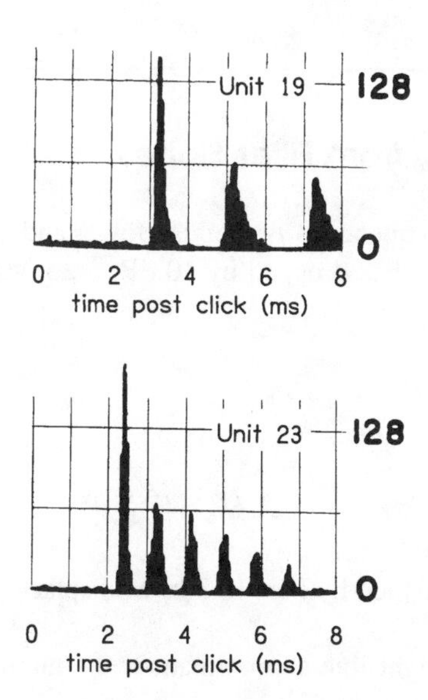

FIGURE 10.2. Poststimulus-time histograms from two auditory nerve fibers from Kiang *et al.* (1965). Unit 19 had a characteristic frequency of 540 Hz, and unit 23 of 1170 Hz. The ringing periods of units 19 and 23 appear to be 1.9 ms and 0.85 ms respectively, equal to the reciprocals of the characteristic frequencies.

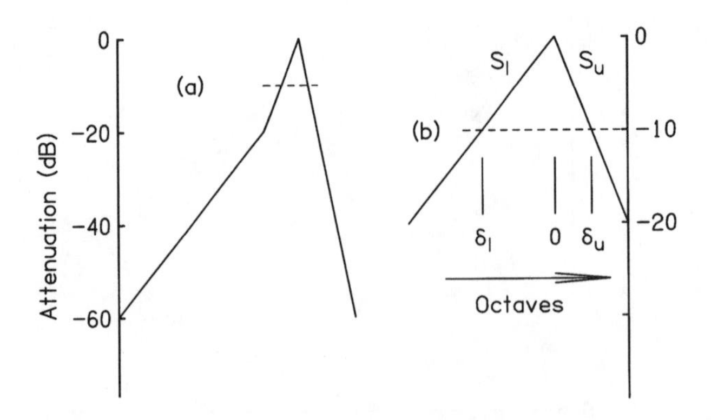

FIGURE 10.3. (a) Magnitude of a model filter transfer function obtained by inverting the tuning curve. (b) Expanded view of the tip of the tuning curve with slopes and 10-dB-down points defined. Lower and upper filter slopes, S_l and S_u, are expressed in units of dB per octave. They can be related to the Q of the auditory filter.

described by two straight lines on a log-log plot, namely attenuation in decibels vs frequency in octaves (Evans, 1968). It is possible to calculate Q_{10dB} from the slopes. The following box shows how.

Calculating Q_{10dB} from Filter Slopes

If the filter center frequency is f_c then the lower and upper frequencies, f_l and f_u, where the response has dropped by 10 dB, are given by

$$f_l/f_c = 2^{\delta_l}, \tag{10.1}$$

and

$$f_u/f_c = 2^{\delta_u}, \tag{10.2}$$

where δ_l and δ_u are defined by Fig. 10.3(b) as displacements from f_c in units of octaves.

Because of the straight-line approximation, the number of octaves by which the upper frequency exceeds f_c is

$$\delta_u = -10/S_u, \tag{10.3}$$

and similarly for the lower frequency

$$\delta_l = -10/S_l. \tag{10.4}$$

The inverse of $Q_{10\ dB}$ is given by

$$1/Q_{10dB} = \frac{f_u - f_l}{f_c}, \tag{10.5}$$

and therefore,

$$1/Q_{10dB} = 2^{(-10/S_u)} - 2^{(-10/S_l)}. \tag{10.6}$$

According to Fig. 10.3(b), slope S_u is negative, and slope S_l is positive. Equation 10.6 is summarized in convenient form in Fig. 10.4.

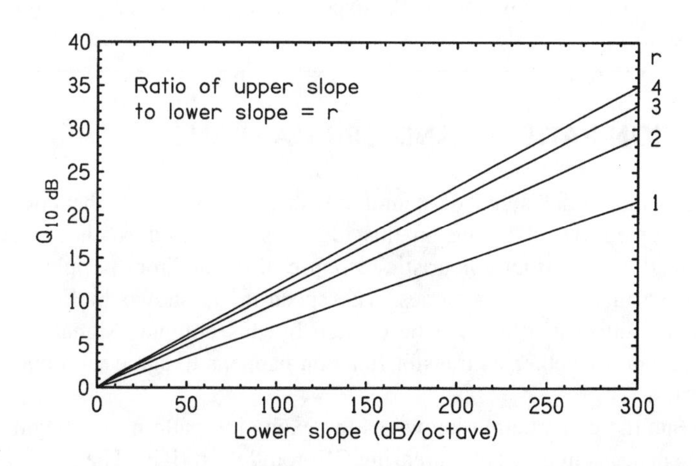

FIGURE 10.4. Equation 10.6. The value of Q_{10dB} can be read from this plot using the lower slope, given on the horizontal axis, and the ratio of the upper slope magnitude to the lower slope given by the parameter on the right.

EXAMPLE: If upper and lower slopes are −200 dB/octave and 100 dB/octave, then

$$1/Q_{10dB} = 2^{1/20} - 2^{-1/10} = 1.035 - 0.933 = 0.102, \quad (10.7)$$

so that $Q_{10dB} = 1/0.102 = 9.8$.

Filters composed of two straight segments have been used in auditory modeling. Colburn (1973) fitted physiological data to find model filter functions for a fiber labelled m with characteristic frequency f_m:

$$|H_m(f)| = (f/f_m)^{\alpha_m} \quad f \leq f_m \quad (10.8)$$

and

$$|H_m(f)| = (f/f_m)^{-2\alpha_m} \quad f > f_m,$$

where

$$\alpha_m = 4 \quad f_m < 800, \quad (10.9)$$

and

$$\alpha_m = 4 f_m/800 \quad f_m > 800.$$

The formulas give an upper slope that is twice as steep as the lower slope. Both slopes become steeper with increasing CF for CF values greater than 800 Hz. The slope of the function with exponent α_m is $S_l = 20\alpha_m \log_{10}(2) = 6\alpha_m$ (dB/octave).

EXCITATION PATTERNS AND CRITICAL BANDS

Because the initial stage of neural transduction is tuned, the entire auditory system is tuned. To elaborate on this idea, and to introduce the concept of the critical band, we consider a heuristic response of the auditory periphery to a signal that is the sum of two sine tones. The spectrum is shown in Fig. 10.5(a). An individual auditory neuron will be excited by this complex signal, more or less depending upon whether its transfer function happens to agree with one of the two frequencies.

To create the construction known as the excitation pattern, one begins by lining up all the neurons in order of increasing CF from left to right. The axis so generated is the tonotopic axis, the axis of *tonotopic place*, called z. If the axis is constructed so that distances along it are proportional to the logarithm of CF, then the neurons turn out to be approximately evenly spaced, though not exactly. Then one plots excitation as a function of z. Normally, one would plot the *driven* firing rate, which

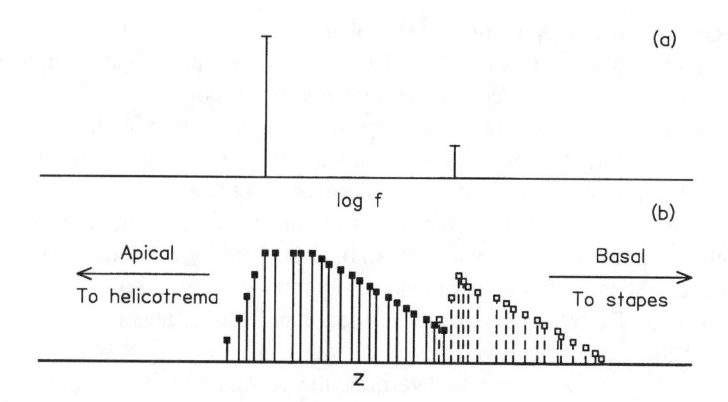

FIGURE 10.5. (a) Spectrum of an acoustical signal with two sine-tone components. (b) Cartoon showing the excitation caused by the two components. Each vertical line corresponds to one neuron of the eighth nerve. Neurons marked with a solid symbol synchronize to the low-frequency component, and those marked with an open symbol synchronize to the higher-frequency component.

subtracts off the spontaneous rate. Only excitation caused by external signals is included. Alternatively one might plot the *synchronous* firing rate, which counts only those neural spikes that are synchronized with a component of the signal.

Figure 10.5(b) shows the excitation caused by the signal in Fig. 10.5(a). Each component of the signal generates an excitation pattern centered on its CF-place. Because auditory neurons have asymmetrical filter functions, the excitation patterns are not symmetrical about the corresponding CF-places. There is an *upward-spread* of excitation. Because the low-frequency spectral component is so intense, the corresponding excitation pattern is somewhat *saturated* and has a flat top, extending to the high-frequency side. The high-frequency component of the acoustical signal is rather weaker than the low, but this does not appear to any great extent in the excitation patterns because auditory filters are *compressive*. They boost low-level signals and attenuate high-level signals.

The listener hears two separate tones from this signal because the two excitation patterns are in different places and because they are synchronized differently. Where the two excitation patterns collide, neurons might synchronize with either component of the acoustical signal, or possibly with both, but usually the low-frequency component wins, especially if the signal overall has high intensity. This causes partial masking (loudness reduction) of the high-frequency component.

Figure 10.5(b) shows a low-frequency component that is tending to overlap the place where the high-frequency component has its maximum, namely its CF-place. If the high-frequency acoustical component were so weak as to be barely detectable, then detection of this component would probably occur somewhere along the high-frequency pattern that is further away from the low-frequency excitation, somewhere higher than the CF-place for the high-frequency component. This detection strategy is known as *off-frequency listening*. As a confounding effect, off-frequency listening is so important that it raises questions for any measurement of auditory filtering based upon masking by tonal or narrowband maskers.

The idea of the critical-band filter is that the analysis along the tonotopic coordinate causes the two signal components to be processed by different channels. The channeling established at the periphery is maintained throughout the ascending auditory system. The channels are the critical-band filters, and the widths of the channels are the critical bandwidths.

The most straightforward way to determine the character of the critical-band filter is a masking experiment using a sine-tone signal and a broadband noise as a masker. It is assumed that the signal is detected in a critical-band filter centered on the signal frequency and that only the noise that comes through the filter is effective in masking the signal. Therefore, if the filter transfer function is $H(f)$, and the noise spectral density is $N_0(f)$, then the power in the filter is

$$P_N = \int_0^\infty df\, N_0(f)|H(f)|^2. \tag{10.10}$$

The signal, with power P_S, becomes barely detectable when its power gives a signal-to-noise ratio $P_S/P_N = K$. Constant K is sometimes called the "efficiency factor." It might better be called the "inefficiency factor" because larger values of K correspond to less efficient detection of the signal.

Rectangular Filters

The first determination of the critical band was made by Fletcher (1938, 1940) [see Allen (1995) for this early history]. He considered an approximation in which the auditory filter is rectangular, as shown in Fig. 10.6.

The signal power at threshold can be used to find the bandwidth of the filter. If the noise is white, N_0 is independent of frequency and can be removed from the integral. Then

$$P_N = N_0 \int_0^\infty df |H(f)|^2, \tag{10.11}$$

and for the rectangular filter with unit height and bandwidth B_R,

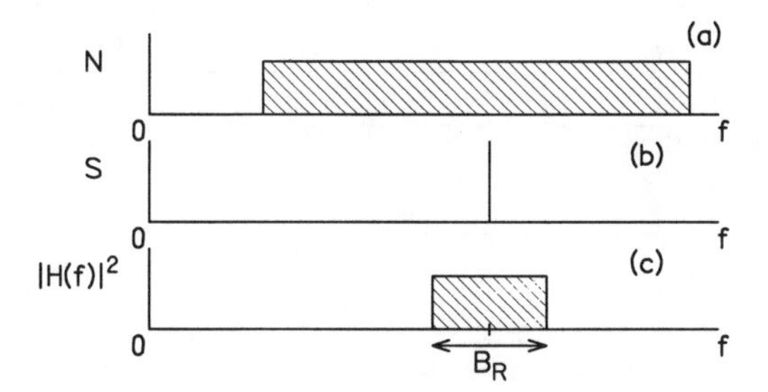

FIGURE 10.6. (a) The power spectral density for a broadband noise. (b) The power spectrum for a sine-tone signal. (c) The rectangular filter, centered on the signal frequency, passes only a part of the broadband noise. Its bandwidth is B_R.

$$P_N = N_0 B_R . \tag{10.12}$$

Therefore, the signal power at threshold becomes

$$P_S = K N_0 B_R . \tag{10.13}$$

There are two unknowns in Eq. (10.13), K and B_R, but measuring P_S gives only one data point. Therefore, Fletcher investigated the consequences of assuming that detection would occur at a signal-to-noise ratio of unity ($K=1$) and found the critical bandwidth from $B_R = P_S / N_0$. This measure of critical bandwidth is called the *critical ratio*.

Rectangular critical-band filters have obvious simplifying advantages. For continuous spectra they reduce integrals to simple products, and for line spectra, two components are either in the same critical band or they are not. If $|H(f)|^2$ is rectangular then the amplitude response $|H(f)|$ is also rectangular. In principle, we should be able to take the Hilbert transform of this amplitude response to find the phase response, under the assumption that the filter is minimum phase. In actuality, we cannot do that. There is an integral formula, known as the Paley-Wiener condition (Papoulis, 1962, p. 215), that an amplitude must obey in order for it to correspond to a causal filter. The fact that $|\log H(f)|$ is infinite outside the passband causes the rectangular filter to fail the condition. It is left as an exercise below to show that the impulse response of a rectangular filter is not causal.

Equivalent Rectangular Bandwidth—ERB

Realistic critical bands are not rectangular in shape, but the concept of a rectangular filter is so useful that it is common to describe the bandwidths of critical-band

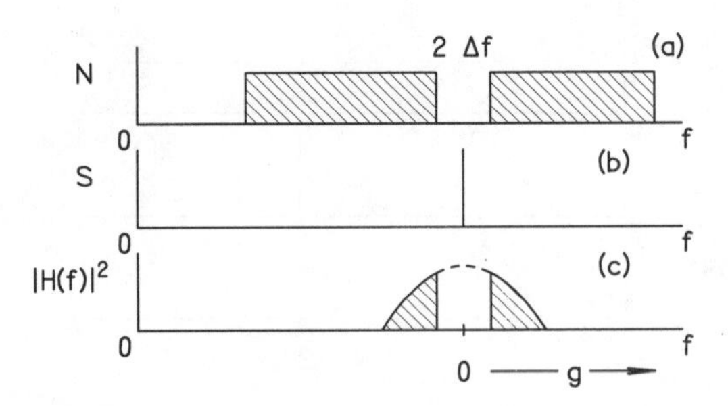

FIGURE 10.7. Masking with a notched-noise masker. The masking noise spectrum in part (a) and the signal in part (b) are the same as Fig. 10.6 except that the noise has a notch centered on the signal frequency. The auditory filter in part (c) is rounded.

filters as equivalent to a rectangular filter with a unit height and a bandwidth B_{ER}. Equivalency means that for a white noise input, the rectangular filter should pass exactly as much power as the real filter. Therefore, the equivalent rectangular bandwidth is

$$B_{ER}=\int_0^{\infty} df\,|H(f\,)|^2. \tag{10.14}$$

This definition of B_{ER} assumes that the maximum value of $|H(f\,)|$ is unity.

Notched Noise and the Roex Filter

More information about the critical-band filter can be learned by using a notched-noise masker (Patterson, 1976). The idea is shown in Fig. 10.7. The center of the notch is placed at the signal frequency, and the signal power required for detection, P_S, is measured as a function of the notch width, $2\Delta f$. An advantage of a notched masker is that it limits off-frequency listening; the listener must use an auditory filter centered somewhere in the notch.

The notched-noise experiment is capable of tracing out the shape of the critical-band filter and also giving an independent value of parameter K. The easiest way to do this is to parameterize the filter with a functional form. The rounded-exponential (roex) function was suggested by Patterson and Nimmo-Smith (1980) and Patterson *et al.* (1982),

$$|H(f\,)|^2=(1+pg)e^{-pg}, \tag{10.15}$$

where g is the normalized deviation of the frequency from the center frequency f_c, $g=|(f-f_c)/f_c|$. This filter has a single dimensionless parameter, p, to describe the

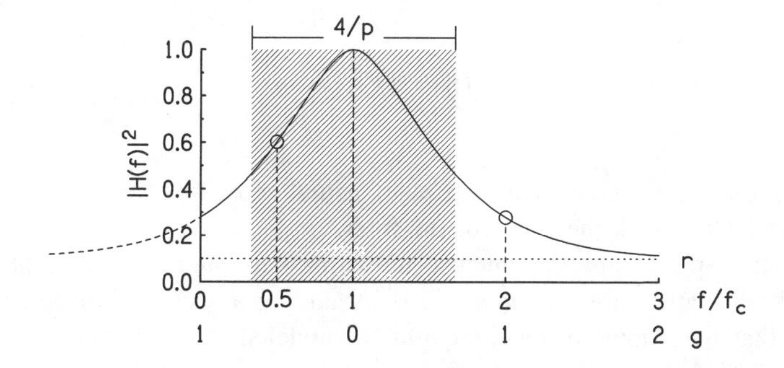

FIGURE 10.8. Cartoon showing the squared magnitude of the roex(p,r) filter with several frequency scales. Octave points are shown by circles. For purposes of illustration, upper and lower slopes are both p=3. Actual model filters are considerably sharper with values of p greater than 10. In realistic filters, constant r is also much smaller than is shown here. The dashed extension to the left indicates the symmetrizing approximation used in fitting the filter to notched-noise data. An area corresponding to the equivalent rectangular band is shaded.

critical bandwidth, and the filter is called the roex(p) filter.

Because it is an absolute value, deviation g cannot be negative. Integrals over the noise spectrum go from $g=0$ to $g=\infty$ on the right and from $g=0$ to $g=1$ on the left (see Fig. 10.8). In practice, with reasonable values of p, the exponential factor decays so rapidly that the integral can be formally continued from $g=0$ to $g=\infty$ on the left also. An exercise at the end of the chapter studies the error made by this approximation. The advantage of this symmetrizing approximation is that the filter becomes a symmetrical function, and an integral over g can be done by doubling the integral from the center of the notch to the right. The area under the symmetrized function of g is $4/p$. Therefore, the equivalent rectangular bandwidth is $B_{ER}=4f_c/p$.

Measuring the signal threshold, first with no gap ($\Delta f=0$) and then with an appropriate finite gap, is enough to determine both roex parameter p and inefficiency factor K. The ratio of the two measurements is

$$P_S(\Delta f)/P_S(0)=\int_{\Delta g}^{\infty} dg|H(g)|^2 \Big/ \int_0^{\infty} dg|H(g)|^2. \tag{10.16}$$

The ratio on the right depends only upon p, and the measured signal power ratio on the left can be used to calculate p. Then, using p to calculate H, one can find K from the equation:

$$P_S(0)=2KN_0f_c\int_0^{\infty} dg|H(g)|^2, \tag{10.17}$$

which is

$$P_S(0)=4KN_0 f_c/p. \tag{10.18}$$

Although two measurements are adequate, actual experiments are done with many values of Δf to check the shape of the filter.

Masking experiments with intense noise maskers suggest that auditory filters have a broad tail. Efforts to model this by adding a second, broader, roex filter showed that the important results could be modeled more simply by an additive constant. This led to the roex(p,r) filter (Patterson *et al.*, 1982),

$$|H(f)|^2=r+(1-r)(1+pg)e^{-pg}, \tag{10.19}$$

where again, $g=|f/f_c-1|$. To keep the area under the filter finite, the constant r can be cut off when $g>0.8$.

A schematic guide to the roex(p,r) filter is given in Fig. 10.8. The peak always occurs at $g=0$. From the form of g it is evident that the roex(p,r) filter is a symmetrical function of f about center frequency f_c. However, as a function of $\log f$ it is tilted in favor of low frequencies, partially representing the upward spread of masking. (Compare the value an octave below f_c with the value an octave above f_c.)

Asymmetrical Filters

Masking experiments using a notch that is symmetrical around the signal frequency lead to an auditory filter approximation that is symmetrical, like the roex(p) or roex(p,r). To learn about the asymmetry of auditory filters requires another experimental parametric variation, a shift of the notch center away from the signal frequency. Then the data can be fitted with a roex(p_l,p_u,r) filter, where the slope parameter above the center frequency p_u is different from the parameter below p_l (Glasberg and Moore, 1990; Moore *et al.*, 1990). The equivalent rectangular bandwidth is given by

$$B_{ER}=(2/p_l+2/p_u)f_c. \tag{10.20}$$

When the model filter is fitted to the detection data it is important to allow the filter center frequency to wander off-frequency within the notch. Results then show that auditory filters are asymmetrical with high-frequency slopes steeper than low ($p_u>p_l$). The asymmetry increases with increasing intensity. The latter is a form of nonlinear effect which is discussed in Chapter 22.

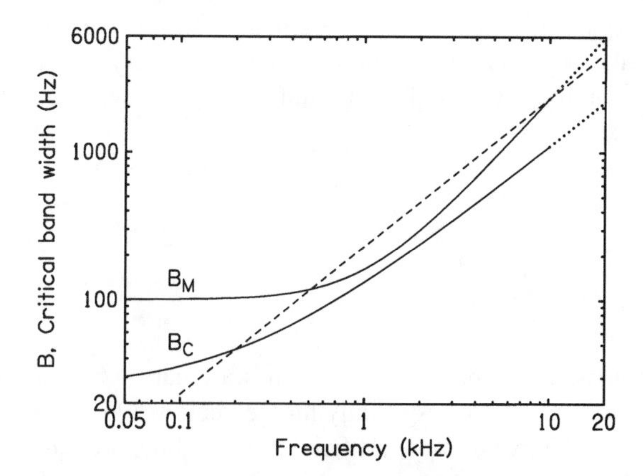

FIGURE 10.9. The critical bandwidth according to formulas 10.21 and 10.22, respectively from Munich B_M, and from Cambridge B_C. The dashed line shows the width of one-third-octave bands centered on the given frequency. Dotted lines show the extension of the formulas beyond 10 kHz.

CRITICAL BANDWIDTH

The most important thing about a critical-band filter is its width. Over the years, an enormous effort has gone into determining critical bandwidths, using a wide variety of models and experimental techniques. In 1961 Zwicker published the critical bandwidths known as the Bark scale. For several decades thereafter these bandwidths were widely accepted. They are here called "Munich" critical bands. In the 1980s, masking experiments with notched noise, as described above, led to a different set of critical bandwidths. Because the experiments were done in Cambridge, England, these critical bandwidths will be designated "Cambridge."

Figure 10.9 shows the Munich and Cambridge critical bandwidths, B_M and B_C respectively. Both critical bandwidths are equivalent rectangular bandwidths (ERBs) and the plots come from the following approximate expressions for ERBs in units of Hz:

$$B_M = 25 + 75(1 + 1.4F^2)^{0.69}, \tag{10.21}$$

and

$$B_C = 24.7(1 + 4.37F), \tag{10.22}$$

where F is the center frequency of the band in kHz. The critical bandwidth is not well known above 10 kHz, and the extensions of the formulas above that frequency are shown dotted. Critical bands are approximately 1/3 octave in width. One-third-octave filter banks are commonly used in the audio industry as a convenient idealization of human auditory filters. The one-third-octave plot in Fig. 10.9 is given by

$$B_{1/3}=2^{1/6}f-2^{-1/6}f=0.232\,f=232F. \tag{10.23}$$

At every frequency, Cambridge critical bands are narrower than Munich critical bands. The difference becomes especially marked below 500 Hz. Moore (1995) has argued that the Munich bands were artificially inflated by critical-ratio experiments that failed to take into account the increase in inefficiency factor K with decreasing frequency. This led to an overestimate of the critical bandwidth. He further argued that measurements of the critical modulation frequency gave misleading estimates at low modulation frequency (see Chapter 20).

Critical-Band Number

The critical-band number is an absolute scale for frequency based upon the ear's frequency analysis. The critical-band number for frequency f is obtained by stacking up critical bands until one has reached a height of frequency f and then counting how many critical bands are in the stack. What makes this counting tricky is that the width of the critical bands grows with increasing frequency. Therefore, to find the answer we have to do an integral.

We represent the critical-band number by the symbol z. The change in z, as the frequency changes by df, is given by

$$dz=\frac{\Delta z}{\Delta f}\,df. \tag{10.24}$$

The quantity $\Delta z/\Delta f$ is the reciprocal of the change in frequency needed to change z by one unit. But that is just the definition of the critical bandwidth, $B(f)$. Therefore,

$$dz=[1/B(f)]df. \tag{10.25}$$

It is typical to set the critical-band number equal to zero for a frequency of zero. Finally, therefore, we have

$$z(f)=\int_0^f df'\,\frac{1}{B(f')}. \tag{10.26}$$

EXAMPLE: *Find an expression giving the critical-band number for Cambridge critical bands, $B_C(f)$*

Unfortunately, the Cambridge unit has been given the name "ERB" in the literature, which stands for "Equivalent rectangular bandwidth," and therefore does not distinguish it from any other measure of the critical band since the time of Fletcher. We will call the Cambridge unit a "Cam" instead.

Solution:

From Eq. (10.22), the critical bandwidth is given by $B_C(f) = 24.7(0.00437f+1)$. Therefore,

$$z(f)=\int_0^f df'\,\frac{1}{24.7(0.00437f'+1)}. \tag{10.27}$$

It is easy to do this integral. It is

$$z(f)=\frac{1}{24.7}\,\frac{1}{0.00437}\,\ln(0.00437f+1), \tag{10.28}$$

or

$$z(f)=9.26\ln(0.00437f+1)=21.3\log(0.00437f+1). \tag{10.29}$$

For instance, 1000 Hz is equivalent to 21.3 log(5.37) = 15.5 Cams.

Equation (10.29) gives a psychoacoustical frequency scale based upon frequency resolution, as determined in notched noise experiments. Possibly it correlates with the mapping of frequency in the cochlea.

Bark Scale

The Bark scale is an absolute frequency scale based upon Munich critical bandwidths. It too is a measure of critical-band number, but inexplicably, it came to be called "critical-band rate." (It is not a rate.) The unit was named after Heinrich Barkhausen (1881–1956). The Bark scale from Zwicker's 1961 paper is given in Table 10.1.

TABLE 10.1. The Bark Scale

Bark	lower (Hz)	center (Hz)	upper (Hz)	Bark	lower (Hz)	center (Hz)	upper (Hz)
0–1	0	50	100	12–13	1720	1850	2000
1–2	100	150	200	13–14	2000	2150	2320
2–3	200	250	300	14–15	2320	2500	2700
3–4	300	350	400	15–16	2700	2900	3150
4–5	400	450	510	16–17	3150	3400	3700
5–6	510	570	630	17–18	3700	4000	4400
6–7	630	700	770	18–19	4400	4800	5300
7–8	770	840	920	19–20	5300	5800	6400
8–9	920	1000	1080	20–21	6400	7000	7700
9–10	1080	1170	1270	21–22	7700	8500	9500
10–11	1270	1370	1480	22–23	9500	10500	12000
11–12	1480	1600	1720	23–24	12000	13500	15500

On the Bark scale, integer Bark values correspond to band edges, e.g. 8 Bark is 920 Hz, and 1000 Hz is 8.5 Bark.

There is no exact analytic expression for the Bark scale. A formula to calculate Bark from frequency was given by Zwicker and Terhardt (1980), but it is not invertible. An invertible procedure by Traunmüller (1990) is given in the box, and shown in Fig. 10.10.

From Hertz to Bark and Back Again

For frequency f in Hz, calculate critical-band number z in Bark:

The formula calculates z', which is the desired value of z except for corrections that are applied below 2 Bark and above 20.1 Bark. The procedure is:

$$z' = 26.81f/(1960+f) - 0.53. \qquad (10.30)$$

$$\text{If } z' < 2.0 \text{ Bark, then } z = z' + 0.15(2.0 - z')$$

$$\text{If } z' > 20.1 \text{ Bark, then } z = z' + 0.22(z' - 20.1).$$

$$\text{Otherwise } z = z'. \qquad (10.31)$$

For critical-band number z in Bark, calculate frequency f in Hz:

The procedure is the inverse of the above. It is not approximate; it is the exact inverse. The correction, if needed, is made first:

$$\text{If } z<2.0 \text{ Bark, then } z'=2.0+(z-2.0)/0.85$$

$$\text{If } z>20.1 \text{ Bark, then } z'=20.1+(z-20.1)/1.22$$

$$\text{Otherwise, } z'=z.$$

$$\text{Then} \tag{10.32}$$

$$f=1960(z'+0.53)/(26.28-z'). \tag{10.33}$$

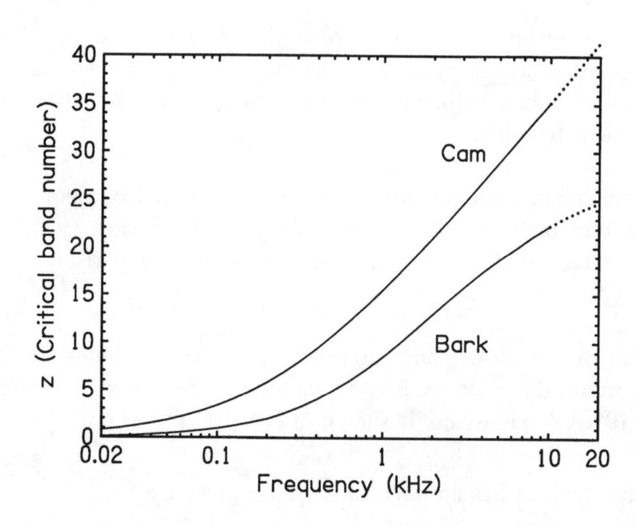

FIGURE 10.10. Plots that can be used to convert frequency in kHz to critical-band number according to Munich or Cambridge critical bands. The corresponding units are Barks and Cams, respectively.

HEINRICH BARKHAUSEN (1881–1956)

Heinrich Georg Barkhausen was born in Bremen on December 2, 1881. In 1911 he became a professor of electrical engineering at the Technische Hochschule in Dresden, where his major efforts were devoted to the development of electron tubes. He discovered the effect now called the ''Barkhausen effect,'' which is the generation of acoustical waves in a solid by the motion of domain walls when the material is magnetized. He also studied loudness and invented the ''phon.'' On February 13, 1945 his Institute of High-Frequency and Electron Tube Technology was destroyed in the bombing of Dresden. After the war he returned from West Germany to Dresden to rebuild the institute, and he remained there until his death on February 20, 1956.

GAMMATONE FILTER

The gammatone filter function is used in models of the auditory periphery to represent critical-band filters. The gammatone function has many virtues, as described by Patterson *et al.* (1991):

(1) It represents the filter by a causal impulse response. Therefore, its Fourier transform gives a transfer function with the amplitude and phase characteristics of a realizable physical filter. Further, the filter is minimum phase (see Chapter 9).

(2) The amplitude response of a gammatone filter of order 4 is very similar to the amplitude response of a roex(p) filter having the same equivalent rectangular bandwidth. The fit is even better if the equivalent rectangular bandwidth of the gammatone filter is made 10% larger.

(3) The gammatone function provides a good fit to the impulse response of auditory nerve fibers as measured with the revcor technique (Johannesma, 1972; Carney and Yin, 1988; see Chapter 14). This facilitates a systematic comparison of physiology and psychophysics.

(4) The output of a bank of gammatone filters can be calculated by an efficient computer algorithm that uses recursive infinite-impulse response filtering (Holdsworth *et al.*, 1988). Such a bank is shown in Fig. 10.11.

The gammatone filter has an impulse response given by

$$h(t)=b^{\eta}t^{(\eta-1)}e^{-2\pi bt}\cos(2\pi f_c t+\phi) \quad (t\geqslant 0)$$

$$h(t)=0 \quad (t<0). \tag{10.34}$$

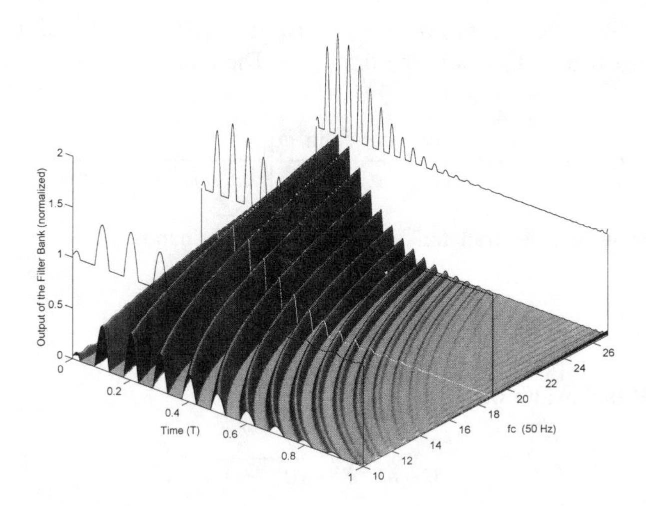

FIGURE 10.11. Pulse-train response of a quasi-continuum of gammatone filters with center frequencies f_c given as multiples of 50 Hz from 10×50=500 to 27×50 =1350. The input is a 50-Hz pulse train consisting of the first 45 harmonics in cosine phase. The time on the horizontal axis is given in units of the period (20 ms). The gammatone filters were chosen to fit the Cambridge critical bandwidths. The output of the filters is half-wave rectified for clarity. A white line emphasizes the filter with center frequency of 950 Hz. The figure shows the faster rise time, faster ringing, and faster decay of the response for filters with high f_c. It is possible to see a hint of some resolution of individual harmonics at the lowest values of f_c (from Lin, 1996).

Parameter η is the order of the filter. The name ''gammatone'' arises because the cosine function looks like a tone, and the prefactor that multiplies it is related to the gamma function. (The gamma function is the integral of the prefactor.) The impulse response of the gammatone resembles the impulse response of the two-pole filter in Eq. (9.61), in that both have an oscillating factor multiplied by a decaying exponential. The rate of decay is given by parameter b. The difference is the powers of t in the gammatone prefactor.

An exercise below shows that the corresponding transfer function is given by

$$H(f)=(\eta-1)!\left\{\left[\frac{\cos\phi-i\sin\phi}{2}\right]\left[\frac{b}{2\pi i(f+f_c)+2\pi b}\right]^{\eta}+\left[\frac{\cos\phi+i\sin\phi}{2}\right]\right.$$

$$\left.\times\left[\frac{b}{2\pi i(f-f_c)+2\pi b}\right]^{\eta}\right\}. \qquad (10.35)$$

When the power response $|H(\omega)|^2$ is used in an integral to calculate the power in a critical band filter, the integral is over positive frequencies only, and it is normally a good approximation to ignore the first term. Therefore,

$$H(f)=(\eta-1)!\left[\frac{\cos\phi+i\sin\phi}{2}\right]\left[\frac{b}{2\pi i(f-f_c)+2\pi b}\right]^{\eta}. \tag{10.36}$$

The gammatone filter transfer function has a 3-dB bandwidth of

$$B_{3dB}=2b\sqrt{2^{1/\eta}-1} \tag{10.37}$$

and a 10-dB bandwidth of

$$B_{10dB}=2b\sqrt{10^{1/\eta}-1}, \tag{10.38}$$

where B and b are in Hertz. For $\eta=4$ these are respectively $0.870b$ and $1.76b$.

The equivalent rectangular bandwidth of the gammatone filter is

$$B_{ER}=\pi(2\eta-2)!\left[\frac{2^{-(\eta-1)}}{(\eta-1)!}\right]^2 b. \tag{10.39}$$

For $\eta=4$, $B_{ER}=0.982b$.

THE UBIQUITOUS CRITICAL BAND

The concept of auditory filtering, or critical band, is the single most dominant concept in auditory theory. Mainly, it is defined by masking experiments, but it is also said that the critical band is that frequency bandwidth or frequency separation where perceptual properties change suddenly (Scharf, 1970). Critical bands are channels for processing. Spectral components that are separated in frequency so as to fall into different channels are processed separately. This idea has numerous implications:

(1) An implication for masking of a narrow-band signal is that noise energy outside the critical band of the signal has no effect on the signal threshold or loudness. Therefore, when a tone is masked by broadband noise, it is only the noise components close to the tone that raise the detection threshold. The rest of the noise can be eliminated with no effect on the signal even though the masking noise becomes less powerful and less loud. A ''band-narrowing'' experiment like this was first done by Fletcher and is a demonstration on the compact disk by Houtsma *et al.* (1987).

(2) An implication of the separate-channels picture for loudness perception was presented in Chapter 4. For tones in separate critical bands the loudnesses add. For tones in the same critical band the loudness is determined by the summed power. Because the loudness is a compressive function of power, the separated tones are louder. It was actually this aspect of loudness that first gave Fletcher the idea of the critical band.

(3) One would not expect two components in different auditory filters to produce difference tones through nonlinear distortion. The cubic difference tone, known to have a cochlear origin, obeys this rule (Goldstein, 1967a,b; see Chapter 22). Two components in different auditory filters should not produce beats either (a linear effect), because they are kept separate in the system. Therefore, roughness, which is caused by rapid beats, can only be caused by tonal components in the same critical band. Plomp and Levelt (1965) suggested that maximum roughness occurs when two components are separated by 1/4 critical band.

(4) Scaling rules, like the rule for roughness, have also been proposed for frequency discrimination (Plomp, 1964) and for pitch perception. According to the mel scale of pitch distances, 100 mels equals one bark (see Chapter 12). In general, if a critical band scale, such as the bark scale, is the correct measure of the tonotopic axis, and if a perceptual effect has a tonotopic origin, then the perception ought to follow the critical band scale.

(5) The ear can analyze a complex tone into its individual harmonics. As described in Chapter 6, Plomp (1964) and Plomp and Mimpen (1968) found that the first five to seven harmonics could be heard out. Their analysis suggested that harmonics can be heard out if, and only if, successive harmonics are separated by a Munich critical bandwidth.

(6) The binaural system is composed of neural elements that compare signals in left and right ears. These elements operate in tuned channels so that they compare left-ear features with right-ear features in the same frequency range (Scharf *et al.*, 1976). The widths of these channels agree approximately with the critical bandwidths determined in monaural listening.

(7) In general the relative phase between two signal components should be irrelevant if the two components are separated by more than a critical bandwidth. Phase should not matter then because there is no neural element that sees both components. The model therefore predicts Ohm's law of phase deafness for the low harmonics of a complex tone while allowing for phase sensitivity among the higher harmonics that are not resolved (Plomp and Steeneken, 1969). The model of phase insensitivity also applies to the indistinguishability of amplitude modulation and quasi-frequency modulation (Zwicker, 1952), as discussed in Chapter 20.

Historically, the critical band has been a unifying element in the explanation of many psychoacoustical effects, particularly their dependences on bandwidths or frequency separations. In recent years, however, the emphasis in auditory theory has changed somewhat. According to the contemporary view, most psychoacoustical effects are actually rather complicated, and although auditory filtering is an important part of any explanatory model, it is rarely enough by itself. Many aspects of hearing appear to depend upon the fact that neural firings are synchronized to the periodicities of the signal (Rose, Brugge, Anderson, and Hind, 1967; Ohgushi, 1978). This synchrony is frequency dependent, but auditory filtering models do not include considerations of synchrony.

There are also temporal effects, such as modulation transfer functions, that have a frequency dependence that can be confused with auditory filtering. For example, the perception of roughness appears to involve several frequency-dependent mechanisms beyond auditory filtering, both linear and nonlinear (Zwicker and Fastl, 1990; Lin and Hartmann, 1995a). What is more, the perception of phase is not limited to a critical band (Patterson, 1987). Even in the area of masking, recent research has identified effects that are outside the scope of auditory filtering models. Studies in profile analysis show that the ability to detect the increment in a sine tone can be improved by adding energy to frequency regions that are several critical bands removed (Green, Mason, and Kidd, 1984). Effects based on envelope fine structure are seen in co-modulation masking release (Hall *et al.*, 1984) or in the dependence of masking on noise power fluctuations (Hartmann and Pumplin, 1988; van der Heijden and Kohlrausch, 1995).

EXERCISES

Exercise 1: Rectangular filter

A filter has a rectangular transfer function with $H(\omega)=1$ inside the passband and $H(\omega)=0$ outside. The passband is centered on frequency ω_c and has a bandwidth of $\delta\omega$. Show that the impulse response is

$$h(t)=\frac{2}{\pi}\,\frac{\sin(\delta\omega\, t/2)}{t}\,\cos\,\omega_c t.$$

[Hint: Include negative frequencies by remembering that $|H(\omega)|=|H(-\omega)|$. Note that the filter is not causal; the impulse response does not vanish for negative time. The problem is that H is entirely real.]

Exercise 2: Critical ratio calculation

Consider a masking experiment where the masker is broadband noise with a spectrum level of 30 dB-SPL. A barely detectable 1000-Hz tone has a level of 46 dB-SPL.

(a) Show that a critical ratio calculation gives B_R=40 Hz.

(b) Suppose that the noise spectral level is given in illogical units as 30 dB-SPL/Hz and show that this leads to a nonsensical value of the critical ratio, B_R=1.5 Hz.

Exercise 3: Roex(p) filter

The roex(p) filter is defined by

$$|H(f)|^2=(1+pg)e^{-pg},$$

where g is the normalized deviation from the center frequency, $g=|(f-f_c)/f_c|$, and p is the slope parameter.

(a) Show that the filter Q for the 10-dB down point is given by $Q_{10dB}=p/7.78$.

(b) Show that the filter Q for the 3-dB down point is given by $Q=p/3.36$. [Hint: Trial and error shows that the approximate solution to the equation $(1+x)\exp(-x)=0.1$ is $x=3.89$, and the approximate solution to the equation $(1+x)\exp(-x)=0.5$ is $x=1.68$.]

(c) Show that the equivalent rectangular bandwidth of the roex(p) filter is

$$B_{ER}=f_c[4/p-(1+2/p)e^{-p}],$$

and show that the second term in the square brackets can usually be ignored. Therefore, another measure of Q for this filter is $p/4$, to be compared with parts (a) and (b).

[Note: Assmann and Summerfield (1990) have used the roex filter in an auditory model, with a causal impulse response made from minimum-phase transfer function.]

Exercise 4: Notched-noise experiment

(a) Show that the area under the roex(p) filter, from Δg to infinity, is given by

$$\text{Area}=\frac{1}{p}(2+p\Delta g)e^{-p\Delta g}.$$

This function can be computed to find p from a notched-noise experiment.

(b) A few data points from an experiment by Moore *et al.* (1990) can be used to calculate p and K for the roex(p) filter model: The noise had a spectrum level of 59 dB SPL, and the signal frequency was 800 Hz. When there was no gap, the signal threshold was 81 dB SPL. When the gap parameter was $\Delta g=0.3$, the signal threshold was 61 dB SPL. Show that $p=20$ and $K=1$. (Note: The value of K is unusually high because the experiment used a conservative psychophysical procedure: three-alternative forced-choice, three-down and one-up; see Chapter 24.)

Exercise 5: Slope of the roex filter

The slope of the roex(p) filter skirts, away from the peak, is dominated by the exponential decay. Show that the slope is given by

$$S=3p(f/f_c) \quad \text{dB/octave}$$

for $f<f_c$ and by the negative of this expression for $f>f_c$.
[Hint: Begin by expressing the frequency ratio as an octave variable $f/f_c=2^\rho$, and differentiate with respect to ρ. Notice that the magnitude of the slope is greater above the peak than below it.]

Exercise 6: Roex filter parameter

(a) Using Eq. (10.22) for the equivalent rectangular bandwidth of a critical band filter, and the equation for the p parameter in the roex filter, $B_{ER}=4/p$, show that p can be calculated from the equation

$$p=37/(1+0.23/F),$$

where F is the center frequency of the filter in kHz. Show that p varies from 11 to 37 as the center frequency varies from 100 Hz to 20 kHz.

(b) Use the results of Exercise 5 to calculate the slope of the critical band filter an octave away from the peak. Show that as the center frequency varies from 100 Hz to 20 kHz the slope varies from 22 to 74 dB per octave. [Note: Slopes of this order are obtained in experiments using simultaneous masking. Experiments using forward masking or pulsation threshold, as described in Chapter 6, lead to slopes that are several times steeper.]

Exercise 7: Q of the third-octave filter

A rectangular one-third-octave filter has a passband time that extends from a frequency that is 1/6-th of an octave below the center frequency to a frequency that is 1/6-th of an octave above. Show that its Q is 4.32. (Note: Fractional octaves are defined in Eq. 11.2.)

Exercise 8: Critical-band number

Suppose that critical bands are one-third of an octave in width, *per* Eq. (10.23). Use the integral formula in Eq. (10.26) to show that the critical-band number becomes

$$z=9.9(\log f-\log 20),$$

where the zero of the z scale has been set at 20 Hz. Show that this function is a straight line on Fig. 10.10.

Exercise 9: The place theory of hearing

Greenwood's (1961) mapping of frequency onto place along the basilar membrane is

$$f(z)=165(e^{az}-1),$$

where z is the place measured from the apex in mm and f is the frequency in Hertz. Parameter a is $0.06\times\ln(10)$ (approximately 0.14) mm^{-1}.

(a) Invert $f(z)$ to find $z(f)$.

(b) Show that for high frequency, $z(f)$ is a straight line when z is plotted against $\log(f)$.

Show that this straight line plot makes an error of less than 10% when $f>1000$ Hz.

(c) Show that the straight line plot corresponds to about 5 mm per octave.

(d) Suppose that a critical band corresponds to a constant distance of 0.8 mm along the basilar membrane. Show that the critical bandwidth in Hertz is given by

$$B_G=0.8\times0.06\times\ln(10)\times(f+165).$$

$$\approx 0.11(f+165).$$

[Hint: Show that the derivative is $df/dz=a(f+165)$. Then let df be the critical bandwidth B_G when $dz=0.8$.]

Exercise 10: Gammatone transfer function

The gammatone filter has an impulse response given by

$$h(t)=b^{\eta}t^{(\eta-1)}e^{-2\pi bt}\cos(2\pi f_c t+\phi) \quad (t\geq 0)$$

$$h(t)=0 \quad (t<0).$$

Fourier transform $h(t)$ to show that the transfer function $H(\omega)$ is given by Eq. (10.35).
[Hints: Consider $h(t)$ to be the product of a cosine and a prefactor. Then the transfer function is the convolution between the Fourier transform of the cosine and the Fourier transform of the prefactor. The convolution integral is easy to do because the Fourier transform of the cosine is just two delta functions. The Fourier transform of the prefactor can be done by recognizing that every time we differentiate $e^{-2\pi bt}$ with respect to $(-2\pi b)$ we bring down another factor of t. Therefore, the Fourier transform of the prefactor is obtained by differentiating the Fourier transform of $e^{-2\pi bt}$.]

Exercise 11: Gammatone bandwidth

Show that the 3-dB bandwidth and 10-dB bandwidth of a gammatone filter are given by Eqs. (10.37) and (10.38).

Exercise 12: ERB for a two-pole filter

The goal of this exercise is to prove that the equivalent rectangular bandwidth for a two-pole filter is $(\pi/2)$ times the full bandwidth at half power. The proof makes use of an advanced topic, the calculus of residues.

Equation (9.56) gives the transfer function for the two-pole lowpass filter. Dividing by Q leads to a transfer function that is unity at resonance. Therefore, the equivalent rectangular bandwidth construction leads to an ERB that is just given by the integral of the squared magnitude of the transfer function.

(a) First show that the ERB is given by the integral

$$B_{ER}=\int_{-\infty}^{\infty} d\omega \, \frac{\omega_0^2/(2Q)}{\omega^2-\omega_0^2-i\omega\omega_0/Q} \, \frac{\omega_0^2/(2Q)}{\omega^2-\omega_0^2+i\omega\omega_0/Q}$$

This integral has four poles. In the limit of large Q, these poles occur at $\omega=\pm\omega_0\pm\omega_0/2Q$.

(b) Do the contour integral including the real axis and a semicircle in the upper half plane (the lower would work too). You will find that $2\pi i$ times the sum of the residues at the two poles in the upper half plane is $(\pi/2)\omega_0/Q$. By the definition of Q this becomes $\pi/2$ times the full bandwidth at the half power point.

(c) Note that for a bandpass filter, the numerator in the integral would be $\omega\omega_0/(2Q)$, and for a highpass it would be $\omega^2/(2Q)$. Show that in the limit of large Q any of these numerators leads to the same result for the ERB.

CHAPTER 11

Musical Measures of Frequency

Music stands in a much closer connection with pure sensation than any of the other arts.

H. L. F. VON HELMHOLTZ

The scientific measure of frequency is simple. Frequency is measured in Hertz, which is the number of cycles per second. In addition, scientific and technical work make common use of two simple frequency ratios, the octave and the decade. Frequency f_2 is an *octave* higher than frequency f_1 if $f_2/f_1=2$. Frequency f_2 is a *decade* higher than f_1 if $f_2/f_1=10$.

One doesn't have to read very far in the acoustical literature before encountering *musical* measures of frequency (Helmholtz couldn't resist them). These exist because of the paramount role that frequency plays in determining the musical pitch of tones. The purpose of this chapter is to present facts that will enable the reader to cope with musical measures as they appear in the literature without delving into the musical details that are responsible for them.

INTERVALS

Unlike the scientific measures, with frequencies given in absolute Hertz, the musical measures are given by relative frequencies, i.e. by frequency ratios. Both melody and harmony are established by frequency ratios, melody by the ratios between successive tones, and harmony by the ratios between simultaneous tones. Absolute frequencies are perceptually important in music only for individuals who possess a sense of absolute pitch (sometimes called "perfect pitch").

Frequency ratios are known as musical "intervals." Thus the frequency ratio of 2:1 is the interval of an octave. The sequence 200, 400, 800, 1600, 3200, 6400 Hz corresponds to stepping up by an octave five times. The ratio 6400/200 is the interval of 5 octaves, a ratio of 2^5.

Mathematically, musical intervals are a form of exponent. Therefore, the addition of intervals corresponds to the multiplication of frequency ratios. The case of octave intervals can serve as an example: From 200 Hz to 800 Hz is 2 octaves. From 800 Hz to 6400 Hz is 3 octaves. Two octaves plus 3 octaves is 5 octaves, and for the frequency ratios,

$$2^2 \times 2^3 = 2^5. \tag{11.1}$$

Octave Notation

Octave notation is a logarithmic scale used for musical intervals. As a logarithmic scale, it is a description of frequency that resembles the way that the decibel scale describes intensity.

If there are two tones, with frequencies f_1 and f_2, then the interval between these tones is f_2/f_1, and the interval expressed in octave numbers is

$$\text{Octaves} = \text{lt}(f_2/f_1), \tag{11.2}$$

where the function "lt" means $\log_2$, i.e. log-base-two.

It is not hard to find the log-base-two of a number because one only needs to take the log to some other convenient base and then convert. For example, using log-base-10, for any number x, the "lt" is given by

$$\text{lt}\, x = (\log_{10} x)/(\log_{10} 2), \tag{11.3}$$

or, replacing the symbol $\log_{10}$ by the common symbol log, we have

$$\text{lt}\, x = (\log x)/(\log 2) = (\log x)/0.301030. \tag{11.4}$$

For example, lt 3=log 3/log 2=0.477121/0.301030=1.584963.

EXAMPLE 1

Find the base-two logs of the integers 1 through 8, which will give the relative frequencies of the harmonics of a complex tone in octave notation.

Solution:

TABLE 11.1. Octave notation for the harmonics of a complex tone

Harmonic number	Octaves removed from the fundamental $\rho(n)$
1	0
2	1
3	1.584963
4	2
5	2.321928
6	2.584963
7	2.807355
8	3

DIVIDING THE OCTAVE

The octave is the most important of all musical intervals. It appears to be a special interval in all musical cultures (Burns and Ward, 1982). However, music consisting only of octave intervals would be monotonous (literally). The octave must be divided into smaller intervals, and different cultures have different ways of doing this. Even within the western musical tradition there is a wide variety of tunings and temperaments. These have emerged from the old problem of tuning instruments with fixed frequencies like the fretted string instruments (guitar) and keyboard instruments (organ or piano). The attention devoted to this problem historically can be appreciated by examining the tome by Jorgensen (1977) or the history of renaissance acoustics by M. H. Cohen (1984b).

Although some of the approaches to the tuning of keyboard instruments have attempted to solve the problem in hardware by constructing keyboards with a large number of keys, interlaced and stacked in wondrous ways, most of the approaches have adhered to the rule that the octave shall be spanned by twelve keys, and that the musical intervals between these keys shall be approximately equal in size. Specifically, the octave is divided into six whole-tone intervals and each whole tone

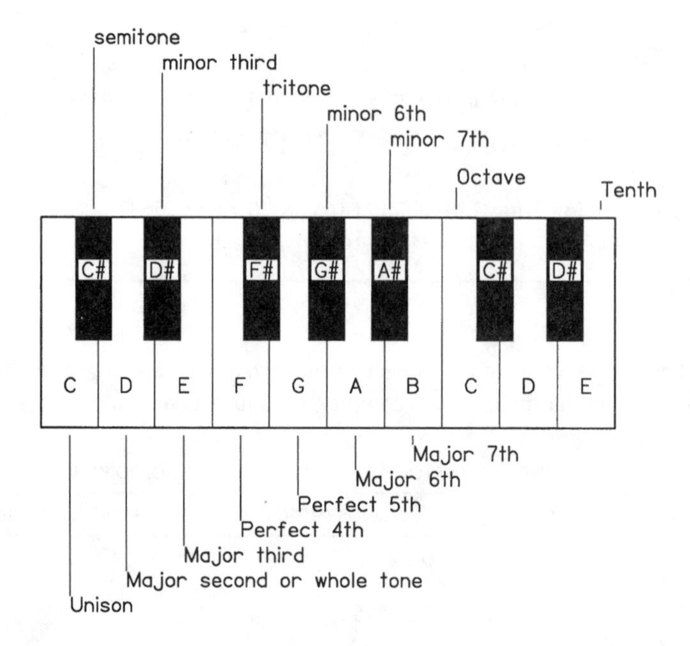

FIGURE 11.1. Section of a piano keyboard, slightly more than an octave, with the names of notes *C*, *C*$^\#$, etc. In this figure, the intervals "unison," "semitone," etc. are defined assuming that the low *C* is the reference. For example, if the *C* is followed by the nearest *E*, one would say that the musical pitch has gone up by the interval of a major third.

TABLE 11.2. Musical interval ratios for three tunings

Interval	Tuning		
	Just	Pythagorean	Equal Temperament
Unison	1.	1.	2^0=1.
minor Second	16/15=1.067	256/243=1.053	$2^{1/12}$=1.059
Major Second	10/9=1.111 or 9/8=1.125	9/8=1.125	$2^{2/12}$=1.122
minor Third	6/5=1.200	32/27=1.185	$2^{3/12}$=1.189
Major Third	5/4=1.250	81/64=1.266	$2^{4/12}$=1.260
Fourth	4/3=1.333	4/3=1.333	$2^{5/12}$=1.335
Tritone	45/32=1.406 or 64/45=1.422	1024/729=1.405 or 729/512=1.424	$2^{6/12}$=1.414
Fifth	3/2=1.500	3/2=1.500	$2^{7/12}$=1.498
minor Sixth	8/5=1.600	128/81=1.580	$2^{8/12}$=1.587
Major Sixth	5/3=1.667	27/16=1.688	$2^{9/12}$=1.682
minor Seventh	7/4=1.750 or 16/19=1.778 or 9/5=1.800	16/9=1.778	$2^{10/12}$=1.782
Major Seventh	15/8=1.875	243/128=1.898	$2^{11/12}$=1.888
Octave	2/1=2.000	2/1=2.000	$2^{12/12}$=2.000

The table gives the frequency ratios for musical intervals within the octave in three tunings, Just, Pythagorean, and Equal Tempered, per ANSI standard S1.1-1994. There are 12 intervals besides the unison.

is divided into two semitones for a total of 12 semitone intervals in the octave. Why should there be 12, and not 11 or 13? There are answers to this question (Longuet-Higgens, 1978; Balzano, 1980; Burns and Ward, 1982; Hall, 1991). One or more of them may actually be right, but none of them is compelling. For the present we take the number 12 to have fallen from heaven and not to be questioned.

The piano keyboard in Fig. 11.1 shows how keyboard instruments physically get 12 keys into the octave. The keyboard is periodic, with each octave geographically identical to any other octave. The keyboard octave includes seven white keys, called C, D, E, F, G, A, and B, and five black keys. The black keys are given names that include "sharp (♯)" or "flat (♭)". The names of the keys (and musical notes) are likewise periodic. Thus an octave above the note called C is another note called C.

The interval between successive keys is a semitone. Thus from C to $C^\sharp$ (called "C sharp") is a semitone and from $C^\sharp$ to D is another semitone. From E to F is also a semitone, though no black key is involved. Figure 11.1 also gives the musical names of intervals assuming that one uses the tone called C as a starting point.

The different western tunings correspond to different frequency ratios for the musical intervals. In what follows, we describe only three such tunings: Just, Pythagorean, and Equal Temperament. Important intervals for these tunings are given in Table 11.2.

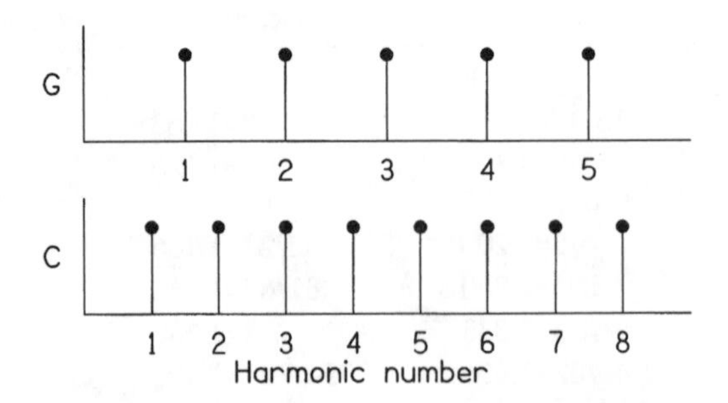

FIGURE 11.2. The notes *C* and *G* are separated by a Fifth. In Just and Pythagorean tuning the ratio of the fundamental frequencies is 3/2. As a result, the harmonics of the two notes either coincide in frequency or else are well separated.

Just Tuning

The tuning known as ''Just'' chooses ratios of small integers to represent the intervals. In doing so, the Just tuning leads to a coincidence of the frequencies of the harmonics of complex tones which avoids beats between low-order harmonics. The intervals with the greatest coincidence of harmonics correspond to the smallest integers. In order, they are: Octave=2/1, Fifth=3/2, Fourth=4/3, Major sixth=5/3, Major third=5/4, minor third=6/5, and minor sixth=8/5. (The intervals of the Fifth and Fourth are sometimes called ''Perfect Fifth'' and ''Perfect Fourth.'') A little thought shows that this sequence is constructed from the smallest possible integers that make rational fractions between 1 and 2, with the exception of the fraction 7/4. Figure 11.2 shows how the harmonics of two tones, separated by the musical interval of a Fifth, have harmonics that either coincide or are well separated from each other. As a result, the tones should sound smooth when heard together because rapid beats will not occur.

Pythagorean Tuning

The Pythagorean tuning is based upon the interval of the Fifth. All ratios in the Pythagorean column of Table 11.2 are some power of 3/2 or 2/3, appropriately translated by enough octaves so as to be in the range from 1 to 2.

EXAMPLE 2: ***Pythagorean construction***

Show how the Pythagorean intervals in Table 11.2 have been generated by going up or down by intervals of the Fifth.
Solution:
Intervals of the Fifth correspond to seven semitones on the keyboard, shown in Fig. 11.3, and to a frequency ratio of 3/2. Starting at the note called C we move seven keys to get to G, which is the Fifth, a ratio of 3/2. Moving up seven more semitones gets us to the note called D and a frequency ratio of 9/4 with respect to our starting place. This note is outside the original octave, so we drop down an octave (-12 semitones), which makes the frequency ratio 9/8. From Fig. 11.1 we find that this note is the major second (whole tone) with respect to the starting point. That agrees with the ratio 9/8 in Table 11.2. Seven more semitones get us to A and a ratio of 27/16 for the interval of the major sixth. Seven more get us to E and a ratio of 81/32, and we drop it down an octave, for a ratio of 81/64 to make the major third. Finally, seven more semitones put us at B, a ratio of 243/128, which is the major seventh.

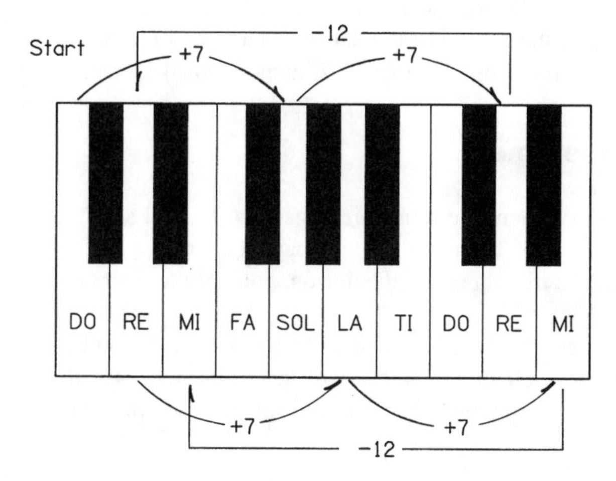

FIGURE 11.3. Moving up by four intervals of the Fifth on the keyboard to generate tones of the Pythagorean scale. To introduce another notation, the names of the notes have been changed from *C, D, E*, etc. to *DO, RE, MI*, etc. The Pythagorean sequence begins on *DO* and continues with *SOL, RE, LA*, and *MI*.

The other tones of the Pythagorean scale have been generated by moving down by intervals of the Fifth (ratio of 2/3), or, what is the same thing, moving up by an interval of the Fourth, a ratio of 4/3. This generates in turn 4/3 (the Perfect Fourth), 16/9 (the minor seventh), 64/27, which translates to 32/27 (the minor third), 128/81 (the minor sixth), and 256/243 (the minor second).

So far we have found five intervals by successive ratios of 3/2 (they happen to be major) and five intervals by ratios of 2/3 (they happen to be minor). That leaves one more interval to find, the tritone, which is in the exact center of the octave. There are two possibilities, either $3^6/2^9=1.424$ or $2^{10}/3^6=1.405$. The former is called the Augmented Fourth, the latter the Diminished Fifth.

Although it may seem as though Pythagorean tuning is significantly inferior to Just tuning because the integers in the ratios are much larger, it was actually the Pythagorean tuning that proved to be the more effective starting point for generating the musically useful temperaments known as "meantone" temperaments. The interested reader can read about the musical considerations in the texts by Hall (1991), Rossing (1990), or Strong and Plitnik (1992).

Choosing intervals by the rules for Just or Pythagorean tunings leads to semitone intervals of different size, and that presents a problem. From the keyboard in Fig. 11.1 we expect the interval between a major and minor third to be a semitone. We also expect the interval between a minor sixth and a Fifth to be a semitone. However, in Just tuning these intervals are respectively 25/24 and 16/15, which are not at all the same. In Pythagorean tuning these intervals are respectively 2187/2048 and 256/243, and these are not the same either. The problem of the semitones is solved by the third tuning in Table 11.2, namely equal temperament.

Equal Temperament

In equal temperament the ratios are chosen so that all semitone intervals are of exactly the same size. The price that one pays for this equalizing is that, except for the octave and the trivial case of the unison, none of the intervals is a ratio of small integers.

The requirements that the octave shall be a ratio of 2:1, and that there shall be 12 semitone intervals, all of the same size, determine the size of the equitempered semitone. The semitone interval must correspond to the following frequency ratio:

$$\text{Equitempered semitone} = 2^{1/12} = 1.059463094. \qquad (11.5)$$

The correctness of the ratio $2^{1/12}$ is not hard to prove: Starting at a tone with any given frequency, the above formula says that we go up by a semitone by multiplying the frequency by the ratio $2^{1/12}$. To go up by two semitones we multiply by $2^{1/12}\times 2^{1/12}=2^{2/12}$. Similarly, to go up by 12 semitones we multiply the original frequency by $2^{12/12}$, which is a factor of 2. But an interval of 12 semitones corresponds to an octave, and a factor of 2 is exactly what we need to go up by an octave. Therefore, the ratio $2^{1/12}$ is seen to be the correct size of an equal-tempered semitone.

UNIT OF CENTS

The concept of equal temperament is at the basis of another important measure of musical intervals known as the ''cent.'' The unit of cents, like the unit of octave number or equal-tempered semitone number, is based upon a ratio scale and is therefore musically and perceptually sensible. (This is not a pun.)

The size of one cent is determined by the simple statement that an equal-tempered semitone is divided into 100 cents, the division being on a ratio scale. Because there are 12 semitones per octave there are 1200 cents per octave. The frequency ratio corresponding to one cent is therefore,

$$\text{One cent} = 2^{1/1200} = 1.00057779. \tag{11.6}$$

This number is very close to unity, and so the interval of one cent is an extremely small frequency change. It is only 0.057779 percent. This change is less than half the just noticeable difference for frequency for human listeners.

It follows that if the interval between frequencies f_2 and f_1 is c cents then c satisfies the equation

$$f_2/f_1 = 2^{c/1200}. \tag{11.7}$$

This equation can be solved for c by taking the logarithms of both sides. We choose base-ten logs,

$$\log_{10}(f_2/f_1) = \frac{c}{1200}\log_{10} 2, \tag{11.8}$$

so that

$$c = \frac{1200}{\log_{10}(2)}\log_{10}(f_2/f_1) \tag{11.9}$$

$$c = 3986.313714 \log_{10}(f_2/f_1). \tag{11.10}$$

A similar expression can be written in terms of the *natural* logarithm of the frequency ratio,

$$c = 1731.234049 \ln(f_2/f_1). \tag{11.11}$$

Because cents notation and octave notation are both logarithmic, the number of cents in an interval is simply proportional to the octave number of the interval.

$$c(\text{Cents}) = 1200 \text{ Octaves}. \tag{11.12}$$

For example, using Table 11.1 we find that the third harmonic of a complex tone is above the fundamental by an interval of 1.584963 octaves, which corresponds to $1200 \times 1.584963 = 1901.96$ cents.

EXAMPLE 3: *The Fifth*

In Just or Pythagorean tuning, the interval of a Fifth is equal to 3/2. Calculate the size of this interval in cents.

Solution: From Eq. (11.10) we have

$$c = 3986 \log_{10}(3/2) = 702 \text{ cents.} \tag{11.13}$$

The value of 702 cents can be compared with the equal-tempered Fifth, by definition, precisely 700 cents. The reader may find it strange that an integer ratio like 3/2 (1.500...) translates into 702 cents, whereas an irrational number like 1.498... translates into an even 700 cents. The mathematics of tuning is sometimes strange.

EXAMPLE 4: *Cents to ratio*

A piano tuner mistunes a string by 10 cents. What is the percentage of mistuning?

Solution: From the defining equation for cents, we have

$$f_2/f_1 = 2^{c/1200}. \tag{11.14}$$

For $c = 10$ cents

$$f_2/f_1 = 2^{10/1200} = 1.0058. \tag{11.15}$$

The percentage of mistuning is

$$\frac{f_2 - f_1}{f_1} = 0.0058 = 0.58\%. \tag{11.16}$$

The natural log formula in Eq. (11.11) above leads to a convenient approximation whereby percentage changes in frequency can be converted to cents and vice versa. The approximation uses the fact from Appendix C that for small x, $\ln(1+x) \approx x$.

If Δf is the frequency change, such that $f_2 = f_1 + \Delta f$, then

$$c \approx 1731 \Delta f / f_1, \tag{11.17}$$

or

$$\text{Percentage frequency change} \approx \frac{c}{17.31} = 0.05777c, \qquad (11.18)$$

where c is the interval in cents. For percentage changes less than 3% (corresponding to 51 cents) this approximation makes an error of less than one cent.

ABSOLUTE MUSICAL SCALE

Up to now, everything that has been said about musical tuning has been relative. The frequencies of tones have been defined only with respect to other tones, through ratios which are musical intervals (octave numbers, equal-tempered semitone numbers or cents). Particular tunings (Just, Pythagorean, Equal Temperament) are rules for determining the frequencies of all the notes in an octave with respect to some starting note. Further, once the notes have been chosen for one octave they are effectively chosen for all octaves, because translation by an octave is always a factor of 2 in frequency.

At this point then, we only need to specify the frequency of a single tone, and the frequencies of all other tones within a particular tuning become uniquely defined. Traditionally, the reference is specified by assigning a frequency to the tone called "Concert A," which is $A4$, the fourth A from the bottom of the piano keyboard (where the bottom note is $A0$). It is also the third string (starting from the lowest) on the violin. Throughout the history of western music a wide variety of reference frequencies have been used for $A4$. An extensive history, starting with the 14th century, is given by Ellis (1885/1954) (see the exercises at the end of the chapter for Handel's tuning fork and Mozart's piano).

The contemporary standard is that concert A shall be given the frequency of 440 Hz. The frequencies of all other tones are then determined by the temperament. A portion of the piano keyboard with frequencies of the tones determined in this way is given in Fig. 11.4.

Although $A4=440$ is a standard recognized throughout the world, not all musicians adhere to this standard in performance. Even some major orchestras tune off the standard. These orchestras tune "sharp," setting the frequency of $A4$ to be higher than 440 Hz, hoping that the higher frequencies will make the music sound brighter and more exciting (Allman, 1989).

One occasionally encounters a standard known as the "scientific musical scale." Buried deep within the piles of accumulated junk in science departments in ancient institutions of higher learning, the amateur archeologist can find sets of old tuning forks that, when the dust is blown off, reveal such strange markings as "C 256" or "A 427" ($=5/3\times256$) or "A 430" ($=2^{9/12}\times256$). These forks represent the scientific scale, where the reference is set by letting the frequency of $C4$ be equal to 256 Hz (compare with 261.6 Hz in Fig. 11.4). Despite the appeal of a reference wherein middle C has a frequency that is 2^8 (so that all other tones called C, up and down

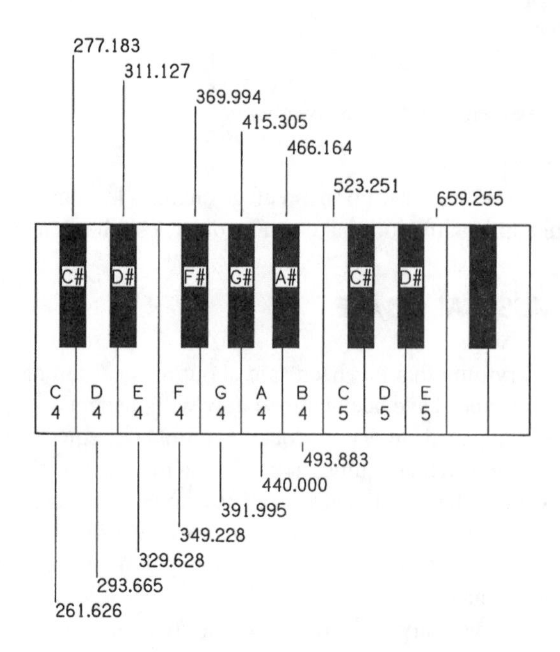

FIGURE 11.4. Section of a piano keyboard including the central octave starting on *C4* (known as "Middle *C*") and including *A4* ("Concert *A*"). The frequency of Concert *A* is 440 Hz by convention. The frequencies of the other notes are determined by equal temperament. Numbers beneath the note names give the standard octave number. Three unlabelled keys have been added on the right in honor of Russian pianist Sergei Rachmaninov (1873–1943) who, according to the Guinness Book of World Records, could span all the keys shown in this figure.

the keyboard, also have frequencies that are powers of 2) the scientific scale is not accepted musical practice at this time, and it has been banished from polite society as an unnecessary source of confusion in standards. Polite, however, is a relative term. A recent edition of the EMS catalog of scientific apparatus for schools shows that it is still possible to buy new forks tuned to the scientific scale, even though this tuning is contrary to United States and international standards.

H. L. F. VON HELMHOLTZ (1821–1894)

Hermann Ludwig Ferdinand von Helmholtz was born in Potsdam on August 31, 1821. He studied with Johannes Müller in Berlin, who inspired him to take a physical approach to physiology, a mode of investigation that was to be the hallmark of his early career. This career started brilliantly in 1847 when Helmholtz was able to show that the law of conservation of energy held universally, including the physiological domain. He elucidated the geometrical optics of the visual system, invented the ophthalmoscope, and promoted a theory of color vision. He made the first measurement of the speed of nerve impulses. Turning his attention to physiological acoustics, he explained the action of the middle ear and made surprisingly prescient conjectures on the operation of the cochlea. In his best-known acoustical work, *On the Sensations of Tone*, he related physical and physiological acoustics to harmony and music with special reference to tone color and combination tones. His latter years are more notable for contributions to physics, particularly to electrodynamics, theoretical mechanics, fluid flow, and thermodynamics. He died of a stroke on September 8, 1884. In terms of the breadth of his important contributions, Helmholtz is in a league that includes Leonardo da Vinci, and not many others.

STRETCHED AND COMPRESSED TUNING

The traditional problem of musical tuning, as described above, is that one begins with the octave, an interval of 2:1, and tries to divide it into smaller intervals that sound good when used in chords. Equal temperament is a solution to that problem, perhaps not the best solution for all kinds of music, but it is the solution that is accepted almost universally in the western world in our day. Given the rate at which western music is propagating internationally, one might dare say that it is globally accepted.

It is worth considering what happens if we turn the situation on its head, accepting equal temperament and 12 tones per octave, but taking the unusual step of allowing the octave interval to be something other than 2:1. This was the idea of Mathews and Pierce (1980), who studied stretched and compressed musical tuning. For a stretched tuning the octave is greater than a factor of 2; for a compressed tuning the octave is smaller than a factor of 2. For example, in a stretched tuning where the octave is a ratio of (2.1):1, the note which is an octave above $A440$ is $2.1\times440=A924$. For the rest of this discussion we will specialize to stretched tuning, but the case of compressed tuning is not different in principle.

It is important to distinguish two kinds of stretched musical tuning: melodic and harmonic. For *melodic* stretch the fundamental frequencies of the notes of the scale are altered. The harmonics of the notes, however, remain equal to integral multiples of the fundamental frequencies. For *harmonic* stretch, the fundamental frequencies of the scale remain those of standard equal temperament. The harmonics of each tone, however, are mistuned so that they are only approximately harmonic. The separation of neighboring harmonics is stretched.

Melodic stretch is something that one could imagine doing with a standard sustained-tone musical instrument. For instance, one could begin with blank clarinet tubes and drill holes in the wrong places so that the instrument would play the wrong notes. Eventually one could learn to place the holes in such a way as to make a stretched tuning, so that when the musician played notes an octave apart the fundamental frequencies would be separated by a factor of 2.1. The instrument would still be a clarinet, albeit a weird one.

Harmonic stretch is another matter. The physics of wind instruments, brass and woodwinds, actually enforces harmonic partials. There is no way that one can stretch the harmonics of a clarinet and still have a clarinet. The only reasonable way to get a sustained tone with stretched "harmonics" is to use computer synthesis of the tone. See for example Mathews and Pierce (1980) or Cohen (1984a). The mathematics of stretched and compressed tuning is a little tricky, and we present that below.

Melodic Stretch Only

First, for the melodic stretch, we need to define the fundamental frequencies of all the tones in the scale. Suppose that the tuning is defined by the octave ratio r. For normal tuning $r=2$. For equal temperament and general r, the frequency ratio between two tones that are separated by c cents is

$$f_2/f_1 = r^{c/1200}, \tag{11.19}$$

where there are still 1200 cents in an octave, regardless of the value of r. For example, the interval of a Fifth corresponds to $c=700$. Therefore, in stretched tuning where $r=2.1$, the frequency ratio of the Fifth is $f_2/f_1 = 2.1^{7/12} = 1.54156$. This ratio can be compared with 1.49831 for the normal equitempered Fifth interval.

In order to determine all the fundamental frequencies of the scale, we must fix a reference frequency. If we agree to keep the frequency of $A4$ equal to 440 Hz, the fundamental frequency of any note $f_0(m)$ is given by a special case of Eq. (11.19),

$$f_0(m) = 440 r^{m/12}, \tag{11.20}$$

where m is an integer semitone number, positive or negative, with respect to $A440$.

Because there is no harmonic stretch in the present case, the frequency of the n-th harmonic of note number m is simply given by

$$f_n(m) = 440 n r^{m/12}. \tag{11.21}$$

Harmonic Stretch Only

We next need a way to find the frequencies of the pseudoharmonics in the case of harmonic stretch or compression. In the case of harmonic stretch, the second "harmonic" is more than a factor of 2 above the fundamental, the third "harmonic" is more than a factor of 3 above the fundamental, etc. The actual "harmonic" frequencies in a stretched tone can be found by using the octave numbers of the harmonics from Table 11.1 as exponents and using the stretched octave factor as a base. For example, for a stretched scale with an octave of 2.1, the third "harmonic" corresponds to an interval of $2.1^{1.584963}=3.2412$. Therefore, if the fundamental frequency is $f_0=440$ Hz, the third "harmonic" has a frequency of 1426.1 Hz.

There is a simpler way to calculate the frequencies of the "harmonics" of a stretched or compressed tone. We define the octave interval as r and the octave exponent as ϵ, (epsilon) such that

$$\epsilon = \mathrm{lt}\ r, \tag{11.22}$$

i.e. $2^{\epsilon}=r$. (Notice that if $r=2$ then $\epsilon=1$.) Then the frequency of the n-th "harmonic" is given by

$$f_n = f_0 n^{\epsilon}. \tag{11.23}$$

For example, for a stretched octave of 2.1, we have $r=2.1$ so that $\epsilon=1.07039$. Then the frequency of the third "harmonic" of a 440 Hz fundamental is $440\times 3^{1.07039}=1426.1$ Hz. What makes this formula simple to use is that ϵ is a constant for a given stretch or compression and allows us to convert frequencies for all harmonic numbers.

Proof of Eq. (11.23)

To prove Eq. (11.23), we define the octave number in Table 11.1 as $\rho(n)$. It is simply the base-two log of the harmonic number. By the definition of a stretched or compressed scale, we have

$$f_n = f_0 r^{\rho(n)}. \tag{11.24}$$

Taking the base-two log, we find

$$\mathrm{lt}(f_n/f_0) = \rho(n)\,\mathrm{lt}\ r, \tag{11.25}$$

or

$$\mathrm{lt}(f_n/f_0) = \epsilon\rho(n). \tag{11.26}$$

Therefore,

$$f_n/f_0 = 2^{\epsilon\rho(n)} \tag{11.27}$$

or, because $2^{\rho(n)}=n$,

$$f_n = f_0 n^{\epsilon}, \tag{11.28}$$

which is Eq. (11.23). (End of proof.)

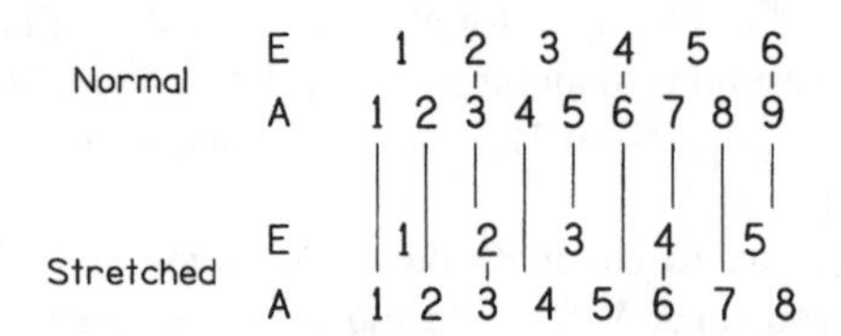

FIGURE 11.5. "Harmonic" numbers for two tones, separated by a Fifth. On top: tones *A*440 and *E*659 played normally with no stretch. The harmonic numbers appear at the correct frequencies. There is frequent coincidence between the harmonics of the two tones. On the bottom: tones *A*440 and *E*678 with both melodic and harmonic stretch (r=2.1). The interval is badly out of tune, and yet there is the same coincidence of the frequencies of the partials.

Both Melodic and Harmonic Stretch

In the case when there is both melodic and harmonic stretch (by the same value r), the frequency of the n-th harmonic of the m-th semitone is

$$f_n(m) = 440 r^{m/12} r^{p(n)}, \tag{11.29}$$

where we have again taken $A440$ as the reference. By the same logic as above, this can be calculated as

$$f_n(m) = 440 (2^{m/12} n)^{\epsilon}. \tag{11.30}$$

The Stretched Chorale

The compact disc by Houtsma, Rossing and Wagenaars (1987) (Demo 31) includes a Bach chorale synthesized in four different ways: (1) no stretch or compression, (2) both melodic and harmonic stretch, (3) melodic stretch but not harmonic stretch, and (4) harmonic stretch but not melodic stretch. When stretch occurs it is $r = 2.1$.

The reason that these different conditions are interesting is that they probe the basis of musical consonance. This can be understood with the help of Fig. 11.5. At the top is a spectrum of the tones A and E in the case of no stretch or compression. These tones are separated by the interval of a Fifth and sound smooth together. (As a musical interval the Fifth has been around since Pythagoras and it played a particularly prominent role in medieval music. We are well accustomed to listening to this interval in music.)

Because the Fifth is very close to a ratio of 3/2, the low-numbered harmonics of the two tones either coincide or else they are well separated. Therefore, the harmonics do not produce beats, and that is why the tones sound smooth together. According to Helmholtz (1885/1954) and Plomp and Levelt (1965) the smoothness is an important contribution to the sense of consonance.

If the harmonics are mistuned [condition (4)] beats develop between the harmonics. Similarly, if the fundamentals depart significantly from small integer ratios [condition (3)] there are beats. The beats lead to roughness and dissonance.

What is particularly interesting is condition (2) where both melodic and harmonic mistuning occur, by the same stretch factor. This condition is illustrated by the interval of the stretched Fifth at the bottom of Fig. 11.5. The stretched Fifth is not a familiar interval. Musicians would call it out of tune. And yet, because of the common melodic and harmonic stretch, the partials of the tones either coincide or else are well separated, and there is no roughness. The aim of the demonstration is to discover whether listeners are willing to regard foreign, but smooth, intervals like this as consonant. Based upon the judgements of five musicians and five nonmusicians, Mathews and Pierce concluded that both familiarity with the interval and smoothness contribute to the sense of consonance.

EXERCISES

Exercise 1: Major and minor thirds

(a) Table 11.2 says that the interval of the equal-tempered major third is $2^{4/12}$. Explain why this interval corresponds to 400 cents. A Just major third is an interval of 5/4. Use the conversion formula to show that this is 386 cents.

(b) Similarly show that the equal-tempered minor third is 300 cents whereas the Just minor third is 316 cents.

(c) Observe that compared to the simple ratios of the Just intervals, the equal-tempered major thirds are too large (by 14 cents) and the equal-tempered minor thirds are too small (by 16 cents). Show that the Pythagorean major and minor thirds err in the same way, only more so.

Exercise 2: Complementary intervals

The mathematics of tuning include some entertaining number facts, such as complementarity. Two intervals are complementary if they sum up to be an exact octave.

(a) Fourths and Fifths: A Fourth plus a Fifth equals an octave. This can be seen in Just or Pythagorean tuning where $4/3 \times 3/2 = 2$. Use Table 11.2 to show that this complementarity holds for equal temperament also.

(b) Major thirds (M3) and minor sixths (m6): Use Table 11.2 to show that M3+m6 =Octave for all three tunings.

(c) Major sixths (M6) and minor thirds (m3): Use Table 11.2 to show that M6+m3 =Octave for all three tunings.

Exercise 3: More interval relationships

Use Table 11.2 to show that for Just, Pythagorean, and equal temperament tunings the following rules apply:

(a) minor third+Fourth=minor sixth

(b) Major third+Fourth=Major sixth

(c) minor third+Major third=Fifth

Exercise 4: Violins

The four open strings of a violin are tuned to the notes $G3$, $D4$, $A4$, and $E4$, which means that there is an interval of a Fifth between successive strings. Show that the frequency ratio of the highest open string to the lowest open string is 27/8, if the tuning is Pythagorean.

Exercise 5: Frequency difference limen

The difference limen for the frequency of a 1000 Hz sine tone is about 2 Hz. Show that this corresponds to a difference limen of 3.5 cents.

Exercise 6: Are scientists flat?

In the scientific scale, middle C has a frequency of 256 Hz. In standard tuning, middle C has a frequency of 261.6 Hz.

(a) Show that the scientific tone is 37 cents flat.

(b) Prove that any other tone on the scientific scale is also 37 cents flat compared to its equivalent in standard tuning. Explain why this statement is true regardless of the tuning system chosen (Just, Pythagorean, equal tempered).

Exercise 7: Or are contemporary performers sharp?

According to Ellis, Handel's tuning fork was tuned to A 422.5 Hz, and Mozart's piano was tuned to A 421.6 Hz. When works by Handel or Mozart are played on modern instruments, tuned to A 440 Hz, the frequencies of the notes turn out to be considerably higher than the composers intended. Given that the interval of a semitone is six percent, show that the historical frequencies would be better approximated by playing Handel or Mozart in a key that is a semitone lower than the score.

Exercise 8: The tritone paradox

The tritone paradox, invented by Diana Deutsch (1986), presents listeners with successive tones made with harmonics that are all octaves of the fundamental (see Shepard's algorithm in Chapter 6). Listeners disagree about whether the pitch of the second tone is higher or lower than the first. Using the frequency ratios in Table 11.2, explain why the interval of the tritone might be particularly confusing.

Exercise 9: Savarts

In France, musical intervals are sometimes given in units called savarts (Felix Savart: born in the Ardennes in 1791, studied the vibrations of strings, died 1841). The interval between frequencies f_2 and f_1 in units of savarts is

$$s = 1000 \log_{10}(f_2/f_1) \quad \text{(savarts)}.$$

This can be compared with Eq. (11.10) for the interval in cents,

$$c = 3986 \log_{10}(f_2/f_1) \quad \text{(cents)}.$$

(a) Show that one octave contains 301 savarts.

(b) The French unit may reflect more than mere chauvinism. The smallest frequency JND, which occurs in the range 400–2000 Hz, is about 0.2%. Show that this change is very nearly 1 savart.

CHAPTER 12

Pitch of Sine Tones

Some sounds are higher pitched, being composed of more frequent and more numerous motions.

EUCLID (330–275 BC)

Pitch perception has played an important role in the development of auditory theory. In part this is due to the excellent experimental precision that one can get in pitch measurements. In pitch, and in no other aspect of auditory perception, one can achieve four-significant-figure reproducibility. The precision of pitch comparisons has made it possible for psychoacousticians to explore a wide variety of pitch-shift and pitch-variability effects, and these form the basis of the present chapter.

On the other hand, pitch has a reputation for being idiosyncratic. There are enormous individual differences. Whereas subjects with good pitch perception may consistently match pitches with a variability of only 0.1%, other people identify themselves as "tone deaf" and won't even attempt a pitch experiment. General Ulysses S. Grant, eighteenth President of the United States, said, "I know only two tunes; One of them is 'Yankee Doodle' and the other isn't." The listeners in pitch perception experiments are usually selected for a good pitch matching or frequency discrimination ability, whereas listeners in other psychoacoustical experiments, such as signal detection, are usually selected only for normal audiograms.

Given the considerable differences in listeners, it is not entirely clear what conclusions one can draw about the implications that pitch research has for general models of hearing. Specialists in pitch believe that the effects seen in precision pitch experiments reveal details of low- to mid-level physiological processes that occur for all normal listeners. Listeners with a normal sense of pitch are able to report those effects reliably. Listeners with a poor sense of pitch do not report these effects reliably, but, unless the listeners exhibit other auditory deficits, the low- and mid-level processes are presumed to be operative nonetheless.

Apart from individual differences, the very variety of pitch-shift effects can be disconcerting. The pitch of a tone can be increased by clenching one's jaw (Corey, 1950). The famous pitch scholar J. F. Schouten once remarked that given two successive identical sine-tone pulses one could hear the pair as ascending or descending in pitch merely by willing it to be so. Some psychoacousticians consider these effects and decide that pitch perception is too hard to study. Specialists in

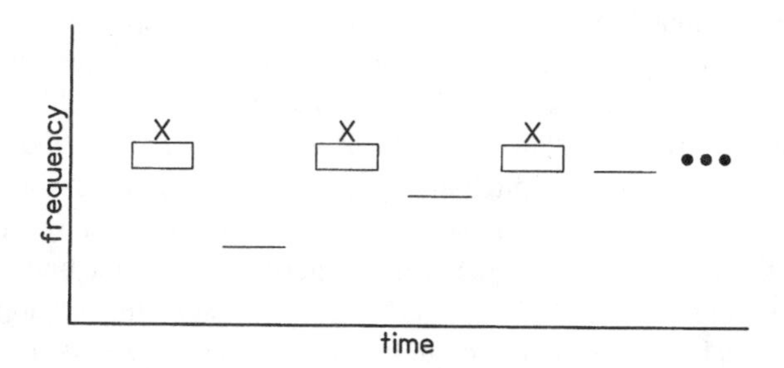

FIGURE 12.1. Sequence of sounds to determine the pitch of X. The matching sines are indicated by horizontal lines.

pitch, considering the same effects, tell their listeners not to clench their jaws and to adopt a neutral position on the tones to be judged.

DEFINITION OF PITCH

According to the American National Standards (ANSI) standard of 1994, ''Pitch is that attribute of auditory sensation in terms of which sounds may be ordered on a scale extending from low to high. Pitch depends mainly on the frequency content of the sound stimulus, but it also depends on the sound pressure and the waveform of the stimulus.''

This definition includes an emphasis on frequency content that distinguishes the pitch sensation from the loudness sensation, but it is otherwise rather unrestrictive. Psychoacousticians normally further restrict the use of the word ''pitch'' to refer to a low-to-high ordering on the scale that is used for melody in music. The ANSI 1960 definition of pitch actually includes such a musical orientation. This does not imply that pitch is restricted to musical sounds, but it does say that the psychological dimension implied by the term ''pitch'' is the same as the psychological dimension of melody. The distinction made by this additional restriction is that it separates pitch from timbre. The sound /S/ is different from the sound /SH/. They differ because /S/ is higher than /SH/, and yet one would not normally say that /S/ has a higher pitch than /SH/. Instead, it has a brighter timbre. The dimension of highness and lowness for these broadband-noise sibilant sounds is not the dimension of melody.

Better than a verbal definition of pitch is an operational definition. A sound is said to have a certain pitch if it can be reliably matched by adjusting the frequency of a sine tone of arbitrary amplitude. Figure 12.1 shows how this works. The sound, called X, whose pitch is in question alternates in a sequence with a sine tone. It is the task of the listener to adjust the frequency of the sine to match sound X in the melodic sense. This operational definition leads to a unit for measuring pitch; pitch is measured in units of Hertz, though that does not imply that the pitch sensation scales linearly with frequency.

The level of the matching sine tone has to be chosen. Depending on the task, it might be different values, or the listener might be allowed to adjust the level along with the frequency. According to convention, however, the matching tone should have a level of 40 dB SPL (Fletcher, 1934; Zwicker and Fastl, 1990, p. 105).

This operational definition of pitch allows for an estimate of variability. A typical experiment requires a listener to match the pitch of a sound many times. From a histogram of the matching frequencies the experimenter might conclude that the pitch is, for example, 400 Hz plus or minus 2 Hz. It sometimes happens that the distribution of pitch matches is bimodal, and in that case one says that the pitch is ambiguous or that there are two pitches. Naturally, this operational definition of pitch allows for individual differences. An experimenter can conclude that a given stimulus has a certain pitch only if a great majority of listeners make similar matches.

An alternative operational definition of pitch is based on melody recognition. For example, Burns and Viemeister (1976) decided that sine-wave-amplitude-modulated noise leads to a sense of pitch because listeners could recognize melodies that were created by changing the modulation rate. The definition in terms of melody is attractive because it is consistent with the melodic emphasis in the verbal definition of pitch. It suffers from the risk that listeners may be able to recognize the intervals of a melody by deciphering variations along some other perceptual coordinate—roughness for example. In any case, the melody recognition experiment must avoid rhythmic information (a long list of melodies without rhythm appears in Hartmann and Johnson, 1991). The evidence for the existence of pitch is more persuasive if the melodies come from an open set, and the evidence is strongest if listeners can recognize individual musical intervals correctly (Burns and Viemeister, 1981). The definition based on melody recognition can indicate the presence of pitch; it does not provide a measure of pitch.

Frequency Difference Limen

The frequency difference limen (DL), or just-noticeable difference in frequency, is the smallest difference between the frequencies of two tones for which the two can be discriminated. In a typical DL experiment, the listener hears two sine tones of equal amplitude in succession and must say which of the two has the higher frequency. The frequency separation for which the judgements are 75% correct is the frequency difference limen. It is commonly assumed that the basis on which listeners make their judgments is the perception of pitch, but that is not a necessary assumption. The frequency difference limen measured in this way is typically somewhat larger than the standard deviation of the matches as measured in a pitch matching experiment.

Frequency difference limens were measured by Wier, Jesteadt, and Green (1977b) using 500-ms sine-tone bursts, at different levels. Eight different frequencies were tested, f_0=200, 400, 600, 800, 1000, 2000, 4000, and 8000 Hz. They used five different levels, referenced to each subject's threshold for f_0, i.e. sensation

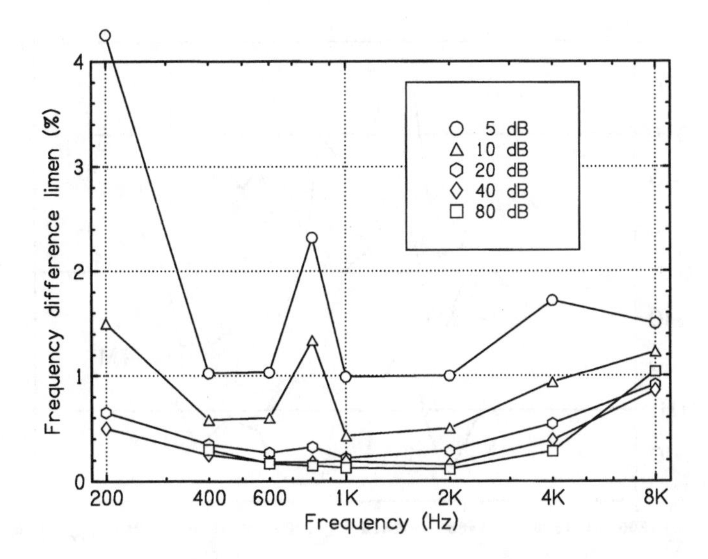

FIGURE 12.2. Frequency difference limens measured by Wier *et al.* (1977b) for sine tones with eight frequencies at five sensation levels. The peak at 800 Hz for low levels is thought to have no significance.

levels (SL). Their difference limens, converted to a percentage of f_0, are given in Fig. 12.2. The purpose of the percentage plot is that it allows for a convenient comparison with a constant-Q analyzing system, for which the percentage would be constant. The figure shows that if the level is more than about 10 dB above threshold, the DL is less than 0.5% over a wide frequency range. Half a percent translates to about 9 cents (see Chapter 11) or (9/100)-ths of a semitone.

DEPENDENCE OF PITCH ON INTENSITY

The pitch of a sine tone depends on the intensity. Although pitch-intensity effects had been known for centuries, quantitative work only began when electronics was applied to psychoacoustics (Fletcher, 1934). In a typical measurement of the pitch-intensity effect, the listener adjusts the frequency of a tone to match the pitch of another tone having a different intensity. On the average, the result of such experiments is that the pitch of low-frequency sine tones decreases with increasing intensity and the pitch of high-frequency sine tones increases with increasing intensity. This behavior is popularly known as *Stevens rule*. The boundary between low and high, where pitches change little with intensity, is between 1000 and 3000 Hz. There is rather good agreement among different experiments on this boundary (Fletcher, 1934; Stevens, 1935; Morgan, Garner, and Galambos 1951; Gulick, Gescheider, and Frisina 1989, p. 251). The pitch-intensity effect, averaged over

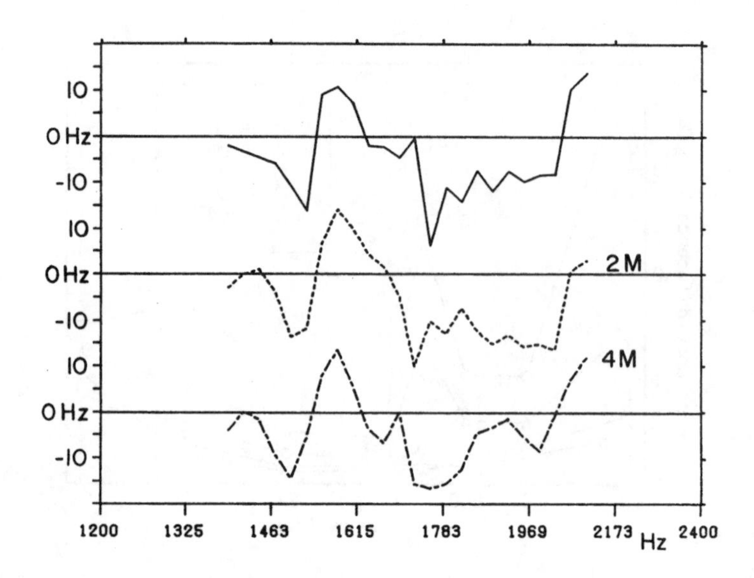

FIGURE 12.3. Pitch shift as a function of sine tone frequency in one ear for a 30-dB increase as measured by Klein (1981). Each of the 24 points on a curve is the average of five matches. The experiment was done three times over a four-month interval.

several studies, has been summarized by Terhardt, Stoll, and Seewann (1982b) in an equation for the percentage shift of the pitch, p, from the sine tone frequency, f,

$$100\times\frac{p-f}{f}=0.02(L-60)(F-2), \tag{12.1}$$

where L is the level of the test tone in dB SPL, and F is its frequency in kHz. Pitch p is the frequency of a matching sine with a reference level chosen to be 60 dB SPL.

Although Stevens rule appears to hold on the average, all experimenters who have studied more than one ear remark on the considerable individual variation in the pitch-intensity relation. Different ears for the same listener are also different, and therefore pitch-intensity experiments are normally done monaurally. Whereas Eq. (12.1) suggests that tones of very low frequency should be shifted by two percent or less, it is common to find much larger shifts at particular low frequencies and particular levels for some ears.

The observed variability of the pitch-intensity relation is not due to random error. There is considerable reproducibility. Figure 12.3 shows the pitch shifts measured by Klein (1981) in the author's right ear. Klein measured the pitch shift at 24 frequencies between 1392 Hz and 2101 Hz as the level increased from 10 dB SPL to 40 dB SPL. The top panel shows an initial experiment; the next two panels show the results of the same experiment two months later and four months later. The similarity of the functions suggests that the variation is real. Although the experiment was done near 2 kHz, where the pitch-intensity effect is expected to be small, shifts for individual frequencies show that the effect can be as large as one percent. The average over the 24 frequencies includes some cancellation between positive and negative shifts and is considerably less.

Non-monotonic variations of the pitch-intensity relation are not confined to cases in which one of the two intensities is very low, nor is it restricted to any particular frequency range (Verschuure and van Meeteren, 1975). It seems evident that there is something quite important at the origin of the pitch-intensity effect. If we could understand it we would know a great deal more about hearing, or at least about pitch.

Auditory Models and the Pitch-Intensity Effect

A priori it would seem that the pitch-intensity effect would find a natural explanation in a "place" theory of pitch perception where pitch is determined by the locus of maximum excitation on a tonotopic coordinate. It is known that the excitation pattern moves on the basilar membrane as the level of a tone increases. For a high-frequency sine tone, cochlear microphonic data (Honrubia and Ward, 1968), eighth-nerve data (Rose *et al.*, 1971) and haircell recordings (Zwislocki, 1991) show that the motion is toward the stapes (basalward) with increasing intensity. This is the direction that would correspond to an increase in pitch with increasing intensity in agreement with Stevens rule.

Psychoacoustical evidence for this motion can be found in the intensity dependence of the maximum-masking frequency. Vogten (1978) showed that a sine-tone probe of fixed frequency is maximally masked by a sine-tone masker of decreasing frequency as the intensity of the masker increases. His observations were consistent with a model in which the maximum excitation caused by a 1-kHz masker moved to a higher-frequency place by a distance equivalent to about 40 Hz for each 10-dB increase in level. A maximum masking shift actually appears in Fig. 6.10.

The problem with this cochlear account of the pitch-intensity effect is that it tends to predict pitch shifts that are far too large, as much as several octaves for a 60-dB change. Pitch shifts observed psychophysically are only a few percent. However, although the peak of the excitation pattern may shift dramatically with increasing intensity, the location of the lower frequency tail apparently changes very little, as shown in Fig. 12.4. Therefore, a place theory of pitch perception might be consistent with the facts if pitch is determined by some aspect of the lower frequency tail, perhaps where it crosses a threshold.

An alternative explanation of the pitch-intensity effect is entirely within the context of a timing model of pitch perception. Jones, Tubis, and Burns (1983) showed that a model of pitch perception based on the peaks in the inter-spike interval histogram could be made to agree with Stevens rule by introducing a simple refractory mechanism.

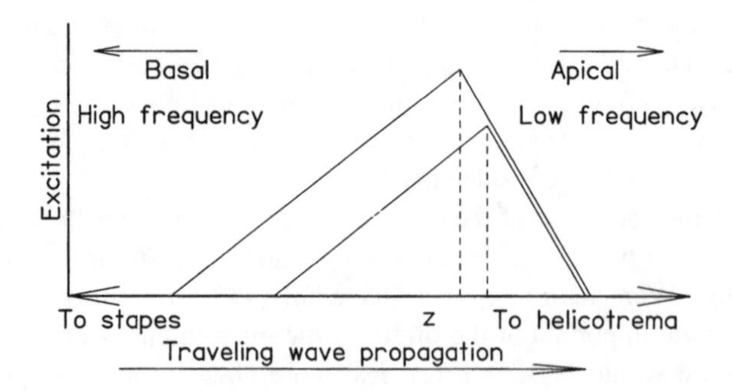

FIGURE 12.4. Schematic view of the excitation patterns for two sine tones having the same frequency, but levels differing by 30 dB. The peak moves considerably with changing intensity but the lower tail does not.

Envelope and Pitch

In 1978(b) Hartmann reported that a sine tone with a rapid exponentially decaying envelope (−1 dB/ms) had a higher pitch than a rectangularly gated (20-ms) sine tone having the same frequency, the same energy, and the same subjective duration. The effect occurred for frequencies between 400 and 3200 Hz, and it did not show the kind of individual variation expected for a pitch-intensity effect. In 1986 Rossing and Houtsma found that the envelope effect occurred for exponentially rising as well as exponentially decaying envelopes, thereby showing that the temporal course *per se* was not the crucial element. They then showed that the pitch-intensity effect becomes remarkably stable for tones of short duration, in their case 40 ms. They found that weaker tones had higher pitch over the frequency range from 200 to 3200 Hz. Thus the high-frequency part of Stevens rule appears to be violated for short-duration tones. By taking a suitable average of signal level, above a threshold, for the decaying tone, Rossing and Houtsma were able to fit their pitch-envelope data using their pitch-intensity data.

DIPLACUSIS, THRESHOLD MICROSTRUCTURE, AND EMISSIONS

As suggested by the results of pitch-intensity experiments, there are individual differences in the encoding of sine tones by different ears which lead to different pitches. One consequence is the phenomenon of *binaural diplacusis*, in which a tone of given frequency and intensity leads to different pitches in a listener's left and right ears. Diplacusis does not normally exceed four percent, though there are anecdotal reports of pathological cases where the diplacusis is several octaves. Normal diplacusis is of the same order of magnitude as the pitch-intensity effect. Possibly diplacusis is somewhat smaller because, in his study of diplacusis and level, Burns (1982) discovered that for every frequency and every ear he could always find some signal level for which the binaural diplacusis could be brought to zero. Like the pitch-intensity effect, binaural diplacusis can be a stable function of time. Brink (1970) successfully tracked it over six years, though he also found day-to-day variations (Brink, 1975a,b).

A somewhat different effect is *tonal monaural diplacusis* (Ward, 1955), where a listener hears two tones when a single sine tone is presented to a single ear. Often the sensation is of a mixture of many tones, or a very rough single tone. Tonal monaural diplacusis often occurs for tones that are near threshold; 10 dB SL is typical. Some of the complexity of the sensation appears to be the result of a difference tone, as though the sine-tone stimulus were being combined in a nonlinear way with a tone that originates within the inner ear. This inner tone, or "idiotone," can sometimes be heard as an after-tone when a stimulus tone with similar frequency is turned off.

There is evidence that the pitch-shift effects, such as diplacusis, are all related and that they are caused by inhomogeneities in the cochlea that also give rise to spontaneous otoacoustic emissions (SOAEs). The train of connections is somewhat tenuous, but it can be found. First, from the work of Burns (1982) one can regard binaural diplacusis as simply the result of a pitch-intensity effect that is different in the two ears. Next, it appears that the microstructure in the pitch-intensity function may well be related to the microstructure in the absolute threshold of hearing. Although the careful experiments by Klein (1981) and early studies by Burns found no connection, measurements on a finer scale by Burns *et al.* (1986) discovered a particularly strong correlation between microstructures in the pitch-intensity function of frequency and the absolute threshold function of frequency. Microstructure in absolute threshold, in turn, often correlates well with the spectrum of SOAEs. Zwicker and Schloth (1984) and Furst *et al.* (1992) found pronounced threshold minima centered on the frequencies of emission lines. Within a range of 15 Hz on either side of the SOAE frequency the threshold increased by 15 dB. Tonal monaural diplacusis was clearly linked to the existence of a nearby SOAE by Zurek (1981, 1985). Thomas (1975) and Kemp (1979a,b) found that frequency DLs for low-intensity tones were at a minimum at the frequency of a threshold minimum, rising by more than an order of magnitude on either side. This could have been caused by monaural diplacusis for frequencies above and below a spontaneous emission, or it could have been the result of the fact that weak tones that coincide with emissions are also louder than tones at neighboring frequencies, sometimes equivalent to an increase in intensity of 15 or 20 dB.

The effects of emissions can be shifted slightly by mechanical changes at the middle ear. Both spontaneous and evoked emissions can be shifted by ear canal pressure (Schloth and Zwicker, 1983; Naeve, Margolis, Levine, and Fournier, 1992). Kemp (1979a,b) found that by pressurizing the ear canal the frequencies of loudness maxima could be shifted by about 5%. This figure agrees with the semitone shifts in pitch that can be produced by clenching the jaw (Corey, 1950) or with the difference that Brink (1980) found in binaural diplacusis for listeners lying down or seated upright.

NOISE-INDUCED PITCH SHIFT

The pitch of a pure tone is shifted by a tone or noise band at a nearby frequency. Egan and Meyer (1950) studied the pitch shift caused by a band of noise centered at 410 Hz and 90 Hz wide. They found that sine tones above the noise band were shifted upward in pitch by the noise whereas sine tones below the noise band were shifted downward. Upward shifts were larger, as large as 1.5%, and more robust. Shifts were larger when the sine tone frequency was closer to the masking noise band.

These trends were confirmed in other studies (Webster and Schubert, 1954; Webster and Muerdter, 1965; Terhardt and Fastl, 1971), though it appears that downward shifts below the noise may occur only for low frequencies. When the experiment is transposed to higher frequencies the shift percentages increase; it is possible to obtain shifts as large as 3 or 4% for 4000 Hz. In broadband noise the pitch of a sine tone is shifted upward (Schubert, 1950) consistent with the tendency for upward shifts to be predominant.

The upward pitch shift of a sine-tone with frequency above a low-passed noise band is a demonstration on the compact disc by Houtsma *et al.* (1987). To be effective, the demonstration should be played at rather high levels (Hartmann, 1993a) where, in fact, most of the experimental work has been done.

The tendency for pitch to be repelled from narrow bands of noise is qualitatively consistent with a place model in which pitch is determined by the centroid (or other moment) of the excitation pattern, as the pattern is eroded, on one side or the other, by the noise. However, the positive pitch shift due to broadband noise is inconsistent with this model. An excitation pattern, as shown in Fig. 12.4 has a shallow high-frequency tail. Raising the threshold with masking noise therefore erodes more excitation at the high-frequency end of the pattern than at the low-frequency end, thus decreasing the moment values. This result predicts a negative pitch shift contrary to experiment.

The pitch shifts from both the narrow-band noise and the broadband noise are consistent with a place model in which only the lower-frequency tail of the excitation pattern contributes to pitch. Noise can easily erode the low-frequency end of this tail because the excitation is weak there, suggesting that positive shifts are the norm. It is more difficult to erode the high-frequency end of this tail because that is where the excitation pattern has a peak, and therefore negative pitch shifts should be exceptional.

POST-STIMULATORY PITCH SHIFTS

The pitch of a sine tone can be shifted by prior excitation (Békésy, 1960, p. 366). Christman and Williams (1963) studied the shift in the pitches of 575-Hz tones and

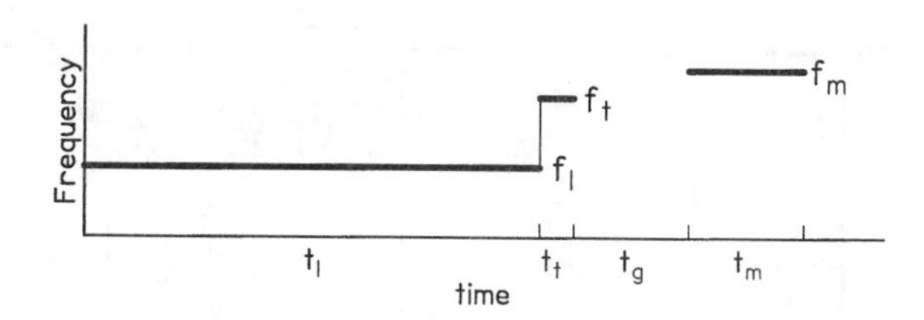

FIGURE 12.5. Frequency-step experiment showing the leading tone (*l*), target tone (*t*), gap (*g*), and matching tone (*m*). There is no waveform discontinuity between the leading tone and target. The pitch shift caused by the leading tone is the frequency difference $f_m - f_t$.

635-Hz tones that followed a 600-Hz adapting tone. The adapting tone had a level of 85 dB SPL and lasted for 60 s. They were able to observe pitch shifts for as long as a minute after the end of the adapting tone. As would be expected from an excitation pattern model, the pitch of the 635-Hz tone was shifted upward and the pitch of the 575-Hz tone was shifted downward, but by a lesser amount. Similar results, but with equal positive and negative shifts, were found by Larkin (1978) using a 3-minute 70-dB adapter. When Larkin tested with contralateral adapters he found no significant shift, suggesting that the pitch shifts have a peripheral origin.

Similar shifts in the pitch of a brief sine tone can be induced by leading bands of noise that are only a few seconds in duration (Rakowski and Jaroszewski, 1974; Jaroszewski and Rakowski, 1976). Again the pitch shift caused by low-pass and high-pass noise is toward the place of less masking. Also consistent with a place model based on the lower-frequency region of the excitation pattern is the frequent observation (e.g. Elliott, Sheposh, and Frazier, 1964) that pitch shifts associated with significant temporary threshold shifts are positive. If the TTS acts to reduce the excitation uniformly at all frequencies then it is the low-frequency tail that has its characteristic features moved towards the higher frequency place. The 1974 paper by Rakowski and Jaroszewski showed an unexpected effect whereby prior white-noise stimulation led to negative pitch shifts for sine tones below 1.5 kHz and positive pitch shifts for sine tones above this frequency, the same turning point found in the pitch-intensity effect.

Particularly reproducible pitch shifts can be obtained in experiments that use a leading sine tone having a duration of about a second (Hartmann and Blumenstock, 1976; Rakowski and Hirsh, 1980). Because the pitch shifts induced by the leading tone last for less than 300 ms, pitch matching experiments can be done with an ipsilateral matching tone that follows the target tone pulse by as little as 1 s. This technique avoids the complexity of the contralateral matching techniques that are used in the study of longer-term adaptation.

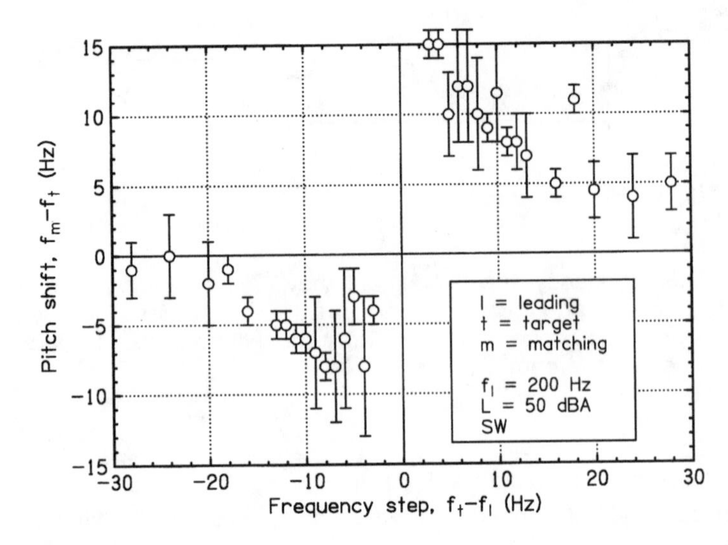

FIGURE 12.6. Pitch shifts, as measured by the difference between matching and target frequencies as a function of the frequency step between target and leading tone frequencies.

Figure 12.5 shows the stimulus of the "frequency-step" experiment, where the target tone is phase-continuous with the prior leading tone, as used by Hartmann and Blumenstock (1976) and Hartmann and Kanistanaux (1979). This stimulus allows for a brief target tone while minimizing spectral splatter.

Figure 12.6 shows typical data from the frequency-step experiment when the leading tone frequency was 200 Hz and all tones had levels of 50 dBA. The durations in milliseconds were: t_l=2000, t_t=40, t_g=1000, and t_m=500. The data show that the largest shifts occurred for the targets closest to the leading tone, although the experiment did not permit f_t-f_l to become extremely small. As expected for an adapting stimulus, the pitch shift tends to have the same sign as the difference between the target and leading tone frequencies. In a similar experiment, with a brief gap between leading tone and target, Ebata, Tsumura, and Okda (1983) found that the leading tone must be at least 300 ms in duration to produce pitch shifts of this kind. For shorter leading tones the sign of the effect was reversed: the pitch of target tones moved in the direction of the leading tones.

PLACE THEORIES AND TIMING THEORIES

Place theories of pitch perception hold that pitch originates in some spatial aspect of the initial pattern of excitation on the basilar membrane. Because primary auditory cells are distributed along the length of the membrane, and because the tonotopic organization established by this distribution is maintained throughout the entire auditory pathway, there is good anatomical justification for the place theory. Early measurements of the basilar membrane motion employed very high sound levels and damaged cochleas and found excitation patterns that were so broad that it was hard to believe that our exquisite sense of pitch could originate in such a pattern. Place theorists were required to develop elaborate models of excitation pattern sharpening to try to account for the precision of pitch perception in terms of place (e.g. Huggins, 1952). After half a century of refinements in methods, physiological experiments are now producing excitation patterns that are so precise that there seems to be little need for secondary processes in viable place theories of pitch.

Timing theories of pitch perception hold that pitch originates in some aspect of the periodicity of neural firings. Timing theories are particularly suited to complex tone perception, where the fundamental may be missing (see Chapter 6), but they are expected to apply for sine tones too. Timing models in which pitch is determined by the periodicity of the envelope of the signal, or the periodicity of the fine structure, have difficulty accounting for the many pitch-shift effects that are observed experimentally. The problem is that these periodicities follow the periodicity of the stimulus too rigidly. A model based on the inter-spike interval (ISI) histogram does not have this kind of logical problem. The peaks of the ISI histogram might be shifted to times that are longer or shorter than the period, or multiples of the period. In fact, it appears that there are only two restrictions on the freedom of ISI peaks: the area under the peaks must equal the total number of counts in the experiment, and the first moment of the histogram must equal the duration of the experiment. Autocorrelation functions are similar; they too can reflect cochlear features by shifts in their peaks. Ohgushi (1983) developed an explanation for the phenomenon known as octave stretch that appealed to the prolongation of short ISIs by refractory effects.

The principal limitation of the timing model is that neural timing cannot be a viable mechanism for perception for tones above 5 kHz because neurons do not maintain their synchrony with the stimulus above 5 kHz. The period histogram for firing becomes flat, indicating that a neuron is equally likely to fire at any phase of the input signal. This limitation, however, turns out to be one of the strengths of the model because pitch perception too deteriorates at high frequency. Several studies have shown that the sense of musical pitch disappears for high frequencies, at about 5 kHz (e.g. Semal and Demany, 1990), although Burns and Feth (1983) showed that some melody identification is possible above 10 kHz. The frequency difference limen shown in Fig. 12.2 begins to increase above 2 kHz, and that correlates well with the decrease in neural synchrony observed by Johnson (Fig. 6.14).

The high-frequency increase in difference limen is particularly pronounced for targets of short duration (Henning, 1966; Moore, 1973). Place models of pitch have great difficulty in explaining the surprisingly good perception of the pitch of tones that are 15–50 ms in duration because the spectral splatter of these short tones is reflected in the width of excitation patterns. Timing models do not necessarily suffer from this limitation. The abrupt transition in frequency difference limen for short tones that occurs near 5 kHz argues for a timing model below 4 or 5 kHz and a place model above (Moore, 1982).

Recent work (e.g. Meddis and Hewitt, 1991a,b) has shown that timing models based on neural autocorrelators offer attractive explanations for a number of pitch effects, including effects due to mistuning the harmonics of complex tones. Model neural autocorrelators employ a mechanism suggested by Licklider (1951, 1959) in which a train of spikes is delayed by a neural delay line and then combined with the undelayed train in a coincidence detector. If $s(t)$ is the train of neural spikes, then the coincidence between this train and the train delayed by τ is $c_\tau(t)$,

$$c_\tau(t)=s(t)s(t-\tau). \tag{12.2}$$

If the coincidence detector accumulates coincidences over a duration T_D, then its output becomes the integral

$$a(\tau)=\int_0^{T_D} dt\ s(t)s(t-\tau), \tag{12.3}$$

and function $a(\tau)$ is essentially an autocorrelation function, as will be described in Chapter 14. Activity in the channel with delay τ could be considered evidence for periodicity at frequency $1/\tau$ by a central processor. Additional evidence for this frequency would be provided by activity in a channel with delay of 2τ or 3τ etc., because of the periodicity of the stimulus. A problem with this model is that it requires neural delay lines with long delays. Although there is physiological evidence for delays of the order of tenths of milliseconds in the binaural system (Yin, Chan, and Carney, 1987; Yin and Chan, 1990), there is no physiological evidence for delays that are as long as the periods of low-frequency tones where the timing model is supposed to operate (see Exercise 8).

Curious too is the fact that the unusual microstructure effects seen in diplacusis and in the pitch-intensity function are particularly prominent in the range from 1 to 2 kHz, firmly in the frequency range where the timing model presumably applies. These effects, connected—as they appear to be—to threshold microstructure, would find a more natural explanation in terms of a place model, where the magnitude of excitation rather than timing is relevant. Shift effects due to microstructure and partial masking would be easier to understand within a timing model if the firing pattern of a neuron included some signature of the neuron's best frequency. If the firing of a neuron at the basal end of an excitation pattern were different from the firing at the apical end, then there might be an explanation entirely within a timing model for the pitch shift that occurs when the apical end is disabled, for instance by lowpassed masking noise. This is apparently not the case. The timing pattern of a neuron does not include a special signature that would indicate its best frequency.

At this point, the most viable alternative appears to be that the pitches of sine tones of all frequencies are determined partly by place principles and partly by timing principles, with the weighting of these two principles changing in favor of place as the frequency increases (Ohgushi, 1978). Meanwhile, back in the lab, the competition between place and timing models continues. The reader is invited to stay tuned.

GEORG VON BÉKÉSY (1899–1972)

Georg von Békésy was born in Budapest on June 3, 1899. Because his parents were in the diplomatic service, they traveled a great deal and Békésy received a diverse education, ending with a doctorate in chemistry. In 1923 he began to work for the Royal Hungarian Institute for Research in Telegraphy. After 1940 he held an additional appointment as professor of physics at the University of Budapest. His research in hearing began with a search for an improved earphone for telephones. He studied the mechanical impedance of the ear in order to match it better. His experiments on cochlear mechanics, for which he received the 1961 Nobel Prize in Physiology or Medicine, were described in a paper published in 1928. In 1949 he came to Harvard University, where he worked at the Psychoacoustics Laboratory run by S.S. Stevens. In 1960, a collection of his papers was published as a book, *Experiments in Hearing*, an eclectic mix of physiological experiments, psychoacoustics observations, introspective listening, opinions, and gadgets. (In 1989 it was reprinted by the Acoustical Society of America.) Békésy's wider scientific influence derived from his work, inspired by Mach, on inhibitory mechanisms in nervous systems. After retirement from Harvard he went to the University of Hawaii; he died in Honolulu on June 13, 1972. His art collection, assembled with scholarly dedication throughout his career, is at the Nobel Foundation in Stockholm.

MEL SCALE

The mel scale is an attempt to find a measure of the psychological sensation of pitch similar to the sone scale for loudness (see Chapter 4). There is not just one mel scale, there are several. In 1937 Stevens *et al.* asked listeners to adjust the frequency of a tone so that its pitch was half as high as a reference pitch. This led to a relative mel scale for pitch, which was then made absolute by saying that 1000 Hz should correspond to 1000 mels. In 1940 Stevens and Volkman asked listeners to divide the range from 200 to 6500 Hz into four equal intervals and found a considerable compression of the perceptual range above 1000 Hz compared to the 1937 scale. The 1940 scale was also referenced to 1000 Hz and is shown in Fig. 12.7.

The mel scale has a slope that is always less than the slope of the musical scale. For instance the three-octave range from 200 Hz to 1600 Hz (ratio of 8) is compressed to a perceptual ratio of only 4.3. The mel scale slope increases monotonically with frequency up to about 5000 Hz. It therefore says that an interval of one octave at low frequency is perceptually much smaller than an interval of one octave at high frequency. For instance, from 100 to 200 Hz, the difference on the mel scale is $300-160=140$ mels, whereas the octave 1000 to 2000 Hz corresponds

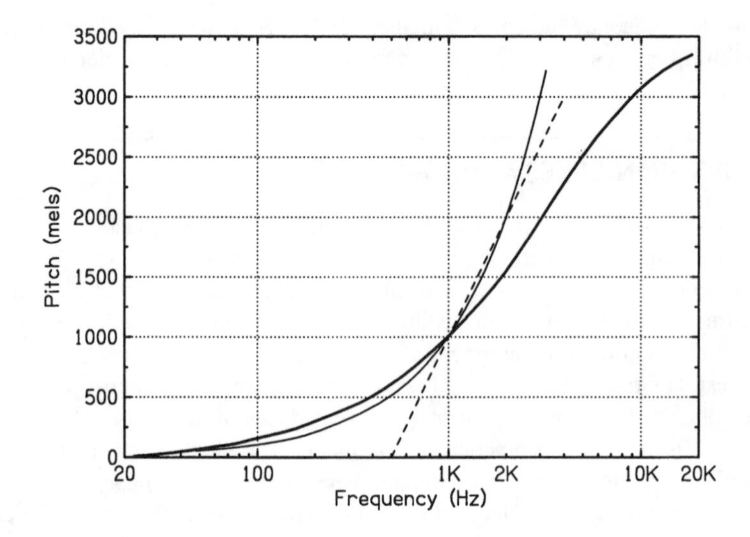

FIGURE 12.7. The heavy line is the mel scale from Stevens and Volkman (1940). The thinner line corresponds to pitch given by frequency in Hz, and the dashed line shows pitch proportional to octave number (the musical scale).

to a difference of 1550−1000=550 mels.

Zwicker and his colleagues (Zwicker and Fastl, 1990) invented a different mel scale, with 125 Hz set equal to 125 mels and spanning about 2400 mels so that 100 mels corresponded to one Bark. A comparison with physiological observations suggested that both scales might correspond to distances along the basilar membrane for points of maximum excitation, such that 100 mels=1 Bark≈1 mm from apex to stapes. A similar calculation by Zwislocki (1965) concluded that one mel corresponded to 12 primary auditory neurons and that one critical band corresponded to 1300.

Because the critical band number (Bark) is a measure of auditory resolution, usually measured by masking in some form, the correspondence between critical band and cochlear distances seems natural. Masking occurs when excitation patterns collide. The correspondence between pitch and place, while appealing, is less necessary on logical grounds; it is more a matter of faith.

The mel scale has frequently been criticized because it contradicts the use of pitch in musical practice (e.g. Attneave and Olson, 1971). The review chapter on pitch scales by Burns and Ward (1982) does not even mention the mel scale. It has also been criticized on psychoacoustical and physiological grounds by Greenwood (1990, 1991, 1995) who argued that the cochlear map, the critical band function, and the scale of pitch should all be closer to logarithmic, with less flattening at low frequency. The Cam scale comes closer to Greenwood's (1961) critical band than does the Bark scale (see Chapter 10).

The scaling of pitch that leads to the mel scale is sensitive to many experimental variables. The slope is increased if there is a standard frequency near the center of the experimental range, compared to a standard at an extreme, and the slope increases as listeners make more judgements (Beck and Shaw, 1961, 1963). Scaling closer to logarithmic, like the musical scale, is obtained if the overall range of frequency is less than four semitones (Harris, 1960).

The musical objection to the mel scale correctly observes that it is counter-intuitive for different octaves to have different perceptual sizes. The relationships among musical intervals are independent of octave height. On the other hand, the perceptual contrast between 2000 and 1000 Hz is generally found to be greater than the contrast between 200 and 100 Hz, and this fact is captured by the mel scale. The paradox might be resolved by the idea that there are two dimensions of pitch, the tone height and the chroma (see Chapter 6). The mel scale may apply to the tone-height dimension of pitch but not to the chroma dimension. The mel scale may finally turn out to be more useful as a measure of brightness for such sounds as octave bands of noise than as a measure of pitch.

PITCHES OF SINES AND COMPLEX TONES

As noted in Chapter 6, a periodic complex tone has a pitch that is equal to, or slightly less than, the frequency of the fundamental. For example, a sawtooth signal with a fundamental frequency of 200 Hz may be matched by a sine with frequency of 200 Hz. It has been widely observed, however, that if the listener matches a sawtooth using another sawtooth then the variability of matches will be smaller. A possible explanation for this result is that in the sawtooth-sawtooth match the listener uses the harmonics of the tone to do the task, possibly a single high harmonic. Because the difference limen in frequency is a smaller percentage at 800 Hz than it is at 200 Hz, a listener might reduce the variance of complex-tone matching by choosing to match the fourth harmonic rather than the fundamental. Henning and Grosberg (1968) suggested that matching for complex tones is as good as the best of the matches for all the individual harmonics (see also Faulkner, 1985). This suggests a strategy of matching by analysis, whereas the pitch of a complex tone is a synthetic percept.

Alternatively the entire set of harmonics might be used. According to the optimum processor model of the synthesized pitch of a complex tone (Goldstein, 1973), the standard deviation σ_0 for estimating the pitch of a complex tone with fundamental frequency f_0 is related to the standard deviations σ_k for the pitches of consecutive harmonics with frequencies f_k by the formula

$$(f_0/\sigma_0)^2 = \sum_{k=\text{bot}}^{\text{top}} (f_k/\sigma_k)^2. \tag{12.4}$$

Using pure-tone difference limens for the values of σ_k overestimates the precision of complex-tone discrimination. However, Moore, Glasberg, and Shailer (1984) found that Eq. (12.4) could be used to relate the difference limens for complex tones to the difference limens for the individual harmonics as measured *within* the complex tone. By contrast, assuming that complex-tone discrimination was only as good as discrimination for the best harmonic overestimated the complex-tone difference limen (see also Moore and Glasberg, 1990).

The harmonics of a complex tone go together to synthesize a pitch. Just how the harmonics are combined is a question of great interest. A promising experimental approach is to mistune one or more harmonics and study the effect on the synthesized low pitch. Ritsma (1967) mistuned harmonics in blocks; Moore, Glasberg, and Peters (1985a) mistuned individual harmonics. Both experiments led to the conclusion that the harmonics that are most important in determining the low pitch are those that are resolved by the auditory periphery. Ritsma concluded that for fundamental frequencies between 100 and 400 Hz, harmonics three, four, and five are dominant. This is the idea of *spectral dominance.* Terhardt (1974b) proposed that spectral dominance was the property of a broad frequency range, centered on 700 Hz, and not of particular harmonic numbers. Moore *et al.* (1985a) concluded that the dominant harmonics were among the first six, but individual differences prevented them from being more specific.

The observed importance of the resolved harmonics supports the idea that pitch perception takes place at high levels where excitations from different peripheral channels are recombined. Pattern matching or template fitting models are like that (Goldstein, 1973; Terhardt, 1974b). The model by Goldstein derives the low pitch from a pattern match to the frequencies of the harmonics.

The model by Terhardt differs in that the relevant input to the pattern matching model from the harmonics is not their frequencies but their pitches. As a result, any effect that shifts the pitch of a harmonic will potentially shift the low pitch of the complex tone. For example, from Stevens rule, increasing the level of a complex tone should have the effect of decreasing its pitch if the dominant harmonics have frequencies less than 1000 Hz. Terhardt's model of virtual pitch (Terhardt *et al.*, 1982a,b) calculates the spectral pitch of each harmonic based on its individual level and an estimate of partial masking by neighboring harmonics. If the pitch of a harmonic is shifted upward by adding noise, that should shift the low pitch upward as well. A review of experiments to test this idea was given by Houtsma (1981); his own experiments led to inconclusive results. Given the importance of the question, whether low pitch is established by harmonic frequencies or by harmonic pitches, it is surprising that so little work has been done on it. At this time the best evidence for the importance of component pitch, as opposed to component frequency, in the formation of complex tone pitch seems to be the diplacusis work by van den Brink (1975a,b), who showed that the diplacusis for a three-component complex tone could be predicted by averaging the diplacusis found in the harmonics.

EXERCISES

Exercise 1: Review this experiment

An article submitted to JASA says that listeners matched the pitch of a 2000-Hz sine tone with a standard deviation of 3 Hz. As a reviewer of this article would you accept that statement?

Exercise 2: Pitch versus timbre

Explain how it is possible for complex tone A to have a lower pitch but a brighter tone color than complex tone B.

Exercise 3: Pitch shifts

Pitch shifts caused by masking by lowpass or highpass noise are summarized by saying that the pitch moves in the direction of better hearing. If you suppose that the threshold of hearing is limited by masking due to internal noise, how do the high- and low-frequency behaviors of Stevens rule compare with this summary?

Exercise 4: Maximum negative pitch shift

Show that negative pitch shifts predicted by Eq. (12.1) are less than 2.5%. (Note: For particular individuals and frequencies, shifts can be more negative than that.)

Exercise 5: Pitch-intensity predictions

(a) Equation 12.1 gives pitch shifts as a function of level, from which one can calculate the expected change in pitch when a sine tone increases from a low level to a high level. Prove that it is not necessary to know both the low and high levels; the prediction can be made by knowing only the difference in levels.

(b) The data in Fig. 12.3 indicate pitch shifts of the order of $\pm 1\%$. Show that these are about twice as large as the prediction from Eq. (12.1), not to mention the alternating sign in the data.

Exercise 6: Zwicker's place model

Zwicker (1970) calculated the frequency difference limen by assuming that two frequencies were discriminable when the frequency difference resulted in a 1-dB change along the low-frequency slope of the excitation pattern (see Fig. 12.4). Based on masking studies he took the slope to be -27 dB/Bark.

(a) Show that this leads to a frequency DL of (1/27)-th of the critical bandwidth.

(b) Use Eq. (10.21) for the critical bandwidth according to the Munich (Bark) scale to calculate the frequency DLs and compare with Fig. 12.2. Some example critical bandwidths from the Bark scale are:

f(Hz)	Critical bandwidth (Hz)
200	100
1000	160
8000	1700

Exercise 7: Neural autocorrelation

Suppose that neural signal $s(t)$ is a train of Dirac delta function spikes,

$$s(t)=\sum_m \delta(t-t_m),$$

where the t_m values are the times when spikes occur. Show that the autocorrelation function is a sum of Dirac delta functions,

$$a(\tau)=\sum_m \sum_{m'} \delta(t_m-t_{m'}-\tau).$$

Exercise 8: Neural delay line

Consider a neural autocorrelator model that calculates the first five peaks of an autocorrelation function for a 200-Hz tone.

(a) Show that this calculation assumes the existence of a neural delay of at least 25 ms.

(b) Show that the same calculation for a 100-Hz tone requires a delay of at least 50 ms.

Exercise 9: Complex tone frequency difference limen

According to Eq. (12.4) the standard deviation in the pitch of a complex tone is related to the standard deviations in the pitches of the harmonics. Prove that the standard deviation, as a percentage of the frequency, for the complex tone is smaller than the smallest percentage standard deviation for any harmonic.

Exercise 10: Number of difference limens

Wier *et al.* (1977b) fitted their difference limen data to the formula

$$\log(\Delta f) = a\sqrt{f} + b,$$

where the common log is used. This is equivalent to the natural log equation

$$\ln(\Delta f) = \alpha\sqrt{f} + \beta,$$

where $\alpha = a/0.4343$ and $\beta = b/0.4343$.

The number of difference limens (NDL) between a bottom frequency and a top frequency can be calculated by integrating the reciprocal of Δf,

$$\text{NDL} = \int_{f_{\text{bot}}}^{f_{\text{top}}} \frac{df}{\Delta f}.$$

(a) Show that

$$\text{NDL} = \int_{f_{\text{bot}}}^{f_{\text{top}}} df\, e^{-(\alpha\sqrt{f}+\beta)}.$$

(b) Do the integral to show that

$$\text{NDL} = 2e^{-(\alpha\sqrt{f_{\text{bot}}}+\beta)}\left(\frac{\sqrt{f_{\text{bot}}}}{\alpha} + \frac{1}{\alpha^2}\right) - 2e^{-(\alpha\sqrt{f_{\text{top}}}+\beta)}\left(\frac{\sqrt{f_{\text{top}}}}{\alpha} + \frac{1}{\alpha^2}\right).$$

(c) For sine tones at 80 dB SL, Wier *et al.* (1977b) found that $a = 0.028$ and $b = -0.696$. Show that between their top and bottom frequencies of 8000 Hz and 200 Hz NDL=1785. By applying the same formula, show that there are 2034 difference limens in the range from 100 Hz to 10,000 Hz.

CHAPTER 13

Applications of the Fourier Transform

Where more is meant than meets the ear.
JOHN MILTON

The Fourier transform converts the representation of a signal from the time domain to the frequency domain. Often the frequency-domain representation leads to insight, and this is particularly true for the many systems, mechanical, electrical, and biological, where filtering is explicitly involved. For example, if a transmission system has a bandwidth of 10 kHz, then it is of interest to know that most of the energy in a speech signal is below 10 kHz. This fact about the speech signal is learned by Fourier analyzing it.

Because the auditory system analyzes signals into different channels specified by frequency, the Fourier transform often leads to insight about hearing. The auditory system is nonlinear, and therefore the Fourier transform, which is a linear operation, cannot capture all the tuning characteristics of the system. However, to a first approximation, the tuning of the system can usually be represented as a linear filter. For example, if the audiogram of a hearing-impaired individual exhibits a high threshold for a 4-kHz sine tone then one expects that the individual will have difficulty detecting 4 kHz when it is a component of speech or music. This expectation stems from the assumption that the system is approximately linear because the rule of superposition does not apply when nonlinearity is present.

This chapter presents several examples in which a Fourier analysis of signals leads to insight into perception. The signals are originally defined in the time domain, but the most useful representation is the energy spectrum, which comes from the Fourier transform. The chapter concludes with the spectral rake, where the signal is originally defined in a frequency representation, but insight into perception comes from the inverse Fourier transform into time.

ENERGY SPECTRUM AND POWER SPECTRUM

The energy and power spectra are introduced formally in Chapter 14, where they are related to the autocorrelation function. They are introduced informally here, building on some ideas from Chapter 3. We recall that $x(t)$ represents a signal, and $x^2(t)/R$ represents a power. The impedance R is a resistance of one ohm (acoustical, mechanical, or electrical), and we imply its existence in saying that $x^2(t)$ has units of power.

The Fourier transform of $x(t)$ is $X(\omega)$, and it is a time integral so that $X(\omega)$ has dimensions of signal×time. The transform $X(\omega)$ has both magnitude and phase. The quantity $|X(\omega)|^2$ has only magnitude information, and its dimensions are energy spectral density. (Note: signal^2×time^2=power×time^2=energy/frequency.)

The energy density $|X(\omega)|^2$ is well defined if the signal $x(t)$ is of limited duration. According to definition, the Fourier integral extends over all time, but if signal $x(t)$ has a finite duration, the integral can stop when the signal stops. Then $|X(\omega)|^2$ is finite for all ω. With the notation substitution $A_f(\omega)=|X(\omega)|^2$, the quantity $A_f(\omega)d\omega$ gives the amount of energy that was in the signal between frequency ω and $\omega+d\omega$. Therefore, $A_f(\omega)$ is a density, where subscript f stands for "finite."

If signal $x(t)$ does not have a finite duration, for instance if it is an ideal sine wave, then the range of integration never stops, and $|X(\omega)|^2$ has a non-integrable infinity at any frequency ω where a component exists. To solve this problem, Chapter 14 defines the Fourier integral as a limit,

$$X(\omega)=\lim_{T_D\to\infty}\int_{-T_D}^{T_D} dt\, e^{-i\omega t}x(t), \tag{13.1}$$

and then defines the power spectrum as

$$A(\omega)=\lim_{T_D\to\infty}\frac{1}{2T_D}|X(\omega)|^2, \tag{13.2}$$

where it is understood that the limits in Eqs. (13.1) and (13.2) are taken simultaneously.

The power spectrum $A(\omega)$ is a power spectral density. It is the energy spectral density that is accumulated over time, from $-T_D$ to T_D, divided by $2T_D$. As the average rate of energy flow, $A(\omega)d\omega$ measures the power between ω and $\omega+d\omega$. Whereas the energy density describes a signal that is in the past (otherwise, how could we have integrated all of it?) the power density has some of the character of an ongoing flow: energy per unit time equals power. These concepts, $A_f(\omega)$ and $A(\omega)$, arise frequently. They are important to hearing because if there is power at frequency ω then neurons with characteristic frequency ω are excited. The power spectrum, or energy spectrum, does not include any phase or time information, but for some signals phase and time information may be unimportant in describing what interests us about perception.

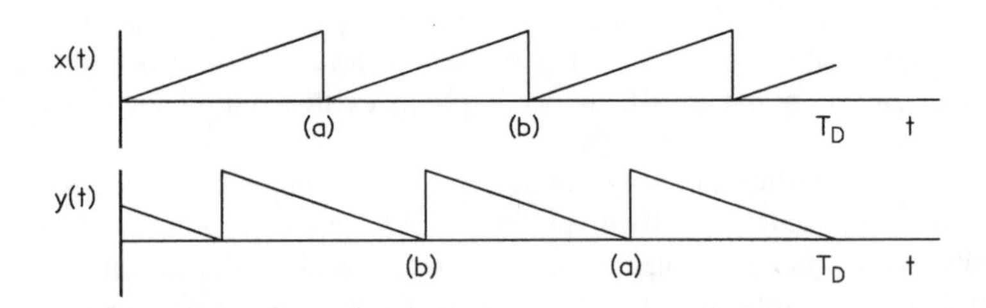

FIGURE 13.1. A sawtooth $x(t)$ and its time-reversal $y(t)$. Corresponding points on the waveform are marked with the same letter.

Time-Reversed Signals

For as long as there has been recorded sound, it has been possible for acousticians to study the effects of reversing the arrow of time by playing sounds backward. A lot can be learned by listening to sounds played backward, especially for speech sounds, where the tendency to perceive normal speech categorically often obscures acoustical features. A key point is that when a signal is played backward its spectrum is unchanged. A proof of that statement is as follows:

Consider a signal $x(t)$ which extends from time $t=0$ to time $t=T_D$. Let signal $y(t)$ be signal $x(t)$ played backward. In other words,

$$y(t)=x(T_D-t) \quad (t<T_D). \tag{13.3}$$

An example is shown in Fig. 13.1 for a sawtooth wave.

Substituting this equation into the definition of the Fourier transform, we find that the Fourier transform of $y(t)$ is

$$Y(\omega)=\int_{-\infty}^{\infty} dt\, e^{-i\omega t}x(T_D-t). \tag{13.4}$$

Changing the variable of integration to $t'=T_D-t$, we get

$$Y(\omega)=-e^{-i\omega T_D}\int_{-\infty}^{\infty} dt' e^{i\omega t'}x(t') \tag{13.5}$$

or

$$Y(\omega)=-e^{-i\omega T_D}X(-\omega). \tag{13.6}$$

Therefore, $Y(\omega)$ is just the same as $X(\omega)$ except that there is a phase factor, and the sign of ω is reversed. Because $x(t)$ is real, $X(-\omega)=X^*(\omega)$.

To find the energy spectrum of y we square the absolute value of $Y(\omega)$. The absolute value of the phase factor is 1, and the absolute value $|X(-\omega)|$ is just the same as $|X(\omega)|$ because the absolute value of a number and its complex conjugate are the same.

Therefore, we have

$$|Y(\omega)|^2=|X(\omega)|^2, \tag{13.7}$$

which says that the energy spectra of the signal and the time-reversed signal are the same, which is what we set out to prove.

If a signal sounds the same when it is played backward and forward then it is possible that the basis of the perception is in the energy spectrum. An example is the sawtooth wave in Fig. 13.1. If a signal sounds different when played backward then the energy spectrum is not an adequate basis to discuss perception. One must complicate the representation, at the least by considering a time-variant power spectrum.

An interesting example is the usual sequence of English vowels, a,e,i,o,u. As pronounced by an American talker they sound like ''AY, EE, EYE, OH, YOU.'' When these sounds are played backward the ''EE'' and the ''OH'' are only minimally changed, but the other three vowels are changed beyond all recognition. For example, ''EYE'' is turned into ''YAH.'' It follows that vowel sounds ''EE'' and ''OH'' are adequately characterized by their steady-state formants. The other three are not.

Reversing a signal can lead to unexpected results, even for perceptual characteristics that one might expect to depend only on the power spectrum. For example, Patterson (1993) studied a pulse train made from exponentially decaying sine tones. He found that the pure-tone character of the sound, especially its dependence on decay time, is markedly changed when the pulse train is played backward.

Turning on a Tone

When an audible signal is turned on abruptly there is an onset transient that is heard as a click, or pop, or thud. Often this transient is regarded as an unwanted noise, and efforts are made to reduce its effect. Unfortunately, the noise is particularly prominent if the signal is a pure tone, which happens to be the most useful signal of all in the study of hearing.

It is possible to reduce the effect of the onset transient by choosing the starting phase of the tone. A reasonable first guess is that the transient is minimized if the starting phase is chosen to make a ''sine tone'', as shown in Fig. 13.2(a). This is often described as turning on the tone in *sine* phase. By way of justification one might use a spectral argument that says that the *sine* phase leads to a waveform without a discontinuity and therefore produces minimum spectral splatter. However, as will be shown below, this qualitative spectral argument is incomplete. The choice of sine starting phase minimizes the onset noise only for low-frequency tones. The surprising fact is that for high-frequency tones the onset noise is minimized by

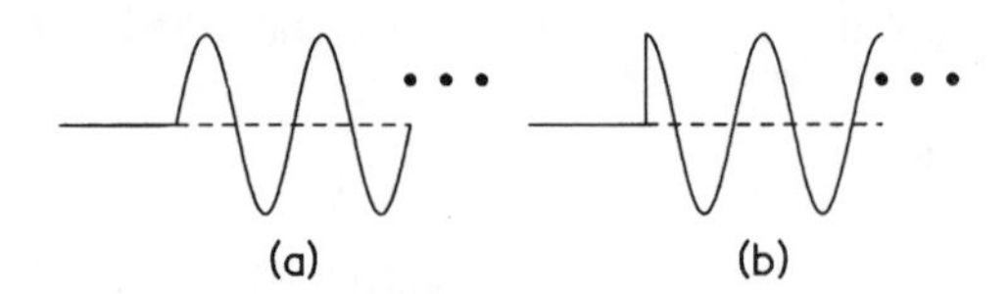

FIGURE 13.2. A waveform with sine phase is shown in part (a), and a waveform with cosine phase is shown in part (b).

turning on the tone in *cosine* phase, as shown in Fig. 13.2(b). The complete spectral approach follows:

The abrupt pure-tone signal with unit amplitude is defined as

$$x(t)=0 \quad \text{for } t<0$$

$$x(t)=\sin(\omega_0 t+\phi) \quad \text{for } t\geq 0, \tag{13.8}$$

where ω_0 is the angular frequency of the tone ($=2\pi f_0$, and f_0 is the frequency in Hz). Phase angle ϕ is the starting phase; it is zero for *sine* phase and 90 degrees for *cosine* phase.

It is left as an exercise at the end of the chapter to show that the Fourier transform of $x(t)$ is given by

$$X(\omega)=\frac{\omega_0 \cos(\phi)+i\omega \sin(\phi)}{\omega_0^2-\omega^2}. \tag{13.9}$$

The energy spectrum is the squared magnitude of X,

$$|X(\omega)|^2=\frac{\omega_0^2 \cos^2(\phi)+\omega^2 \sin^2(\phi)}{(\omega_0^2-\omega^2)^2}. \tag{13.10}$$

Equation (13.10) has two limits of special interest: *sine* phase ($\phi=0$), where

$$|X(\omega)|^2=\left[\frac{\omega_0}{(\omega_0^2-\omega^2)}\right]^2, \tag{13.11a}$$

and *cosine* phase ($\phi=\pi/2$), where

$$|X(\omega)|^2=\left[\frac{\omega}{(\omega_0^2-\omega^2)}\right]^2. \tag{13.11b}$$

The spectral functions in Eqs. (13.11a) and (13.11b) are plotted in Fig. 13.3. Both spectra have non-integrable infinities when ω is equal to ω_0 because the signal extends to infinite time. These infinities are no trouble because we are interested only in the rest of the spectrum, which represents splatter from the abrupt onset. For

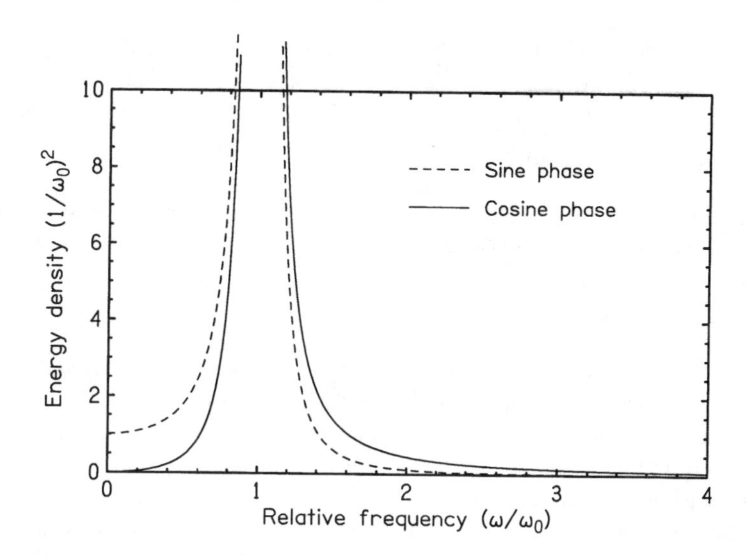

FIGURE 13.3. Energy spectra calculated for sine phase (dashed line) and cosine phase (solid line) from Eqs. (13.11a) and (13.11b) respectively. The vertical axis shows energy density, in units of $1/\omega_0^2$. The horizontal axis is the frequency parameter relative to the tone frequency.

frequencies well above ω_0, the spectrum of the abrupt *sine* tone, Eq. (13.11a), decreases as the inverse fourth power of the frequency, whereas the spectrum for the abrupt *cosine*, Eq. (13.11b), decreases only as the inverse square. As a result, the abrupt *cosine* leads to more high-frequency energy than the abrupt *sine*. The high-frequency splatter is perceived as a click sound, and it is smallest for *sine* phase. If the frequency of the tone is low, such that splatter to the higher frequencies is perceptually important, then using *sine* phase minimizes the noise.

The price that one pays for using *sine* phase is that the spectrum does not go to zero at low frequency. By contrast, the spectrum for *cosine* phase vanishes as the square of the frequency as the frequency goes to zero. As a result, the abrupt *sine* leads to more low-frequency energy than the abrupt *cosine*. The low-frequency splatter may be perceived as a thud, and it ought to be smallest for *cosine* phase. If the frequency of the tone is high, such that splatter to the lower frequencies is perceptually important, then using *cosine* phase should minimize the noise.

Listening experiments with pure tones that were turned on abruptly and turned off slowly (Hartmann and Sartor, 1991) confirmed the conclusions of these spectral arguments. Onset noise was minimized with *sine* phase for low-frequency tones and with *cosine* phase for high-frequency tones. The boundary between low- and high-frequency regions was different for transducers having different frequency responses, as would be expected.

Using high-quality headphones, four different listeners found boundaries ranging

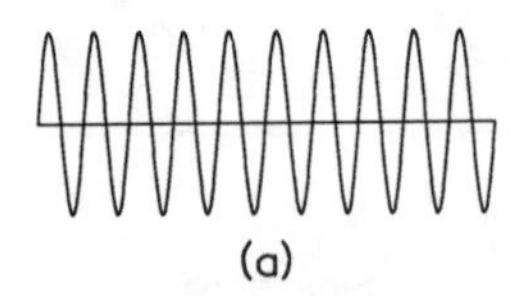

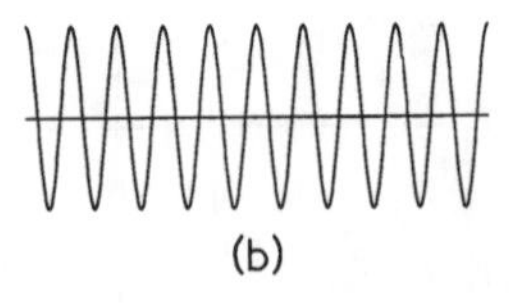

FIGURE 13.4. Pulses of a 1000-Hz tone with 10-ms duration: (a) sine phase (b) cosine phase. These pulses are used as maskers to mask an even shorter tone pulse.

from 300 Hz to 1200 Hz. It is difficult to understand why individual listeners differed so greatly about the boundary frequency, especially because the data showed that each listener was quite certain about the location of the boundary. What seems likely is that selective attention to different frequency regions was responsible. Onset noise above the frequency of the tone sounds like a click; onset noise below the frequency of the tone sounds like a thud. Listeners may have attended to one of these sounds and discounted the other until it became so intense that it was impossible to ignore.

Although the listening experiments did not find a definitive boundary frequency, they did show that for each listener a boundary did exist. For pure tones with frequency below the boundary, *sine* phase minimized the onset noise on essentially 100 percent of the trials. For pure tones above the boundary, a *cosine* phase minimized the noise essentially 100 percent of the time. At no time was an intermediate starting phase, such as 45 degrees, less noisy than the better of the extreme phases.

Sine and Cosine Masking

In 1968 Green studied the masking caused by short tonal maskers. Figure 13.4 shows two of his masking waveforms, 1000-Hz tone pulses with 10-ms duration.

Green observed that the spectrum depends upon the onset and offset phases. Relative to the *sine*-phase pulse in Fig. 13.4 part (a) the *cosine*-phase pulse in part (b) has greater high-frequency energy. The ratio of the energy density functions is, as in Eqs. (13.11),

$$\frac{|X_{\cos}(\omega)|^2}{|X_{\sin}(\omega)|^2} = \left(\frac{\omega}{\omega_0}\right)^2, \tag{13.12}$$

where ω_0 is the frequency of the tone ($2\pi 1000$ r/s in the case of Fig. 13.4). On a log-log plot this function is a straight line with a slope of 6 dB per octave. One might expect to see this phase difference in a masking experiment. Suppose that the target tone to be detected has a frequency of ω. Suppose further that the masking caused by the tone pulse with frequency ω_0 is simply proportional to the energy density at the frequency of the target. Then one expects that the difference in target

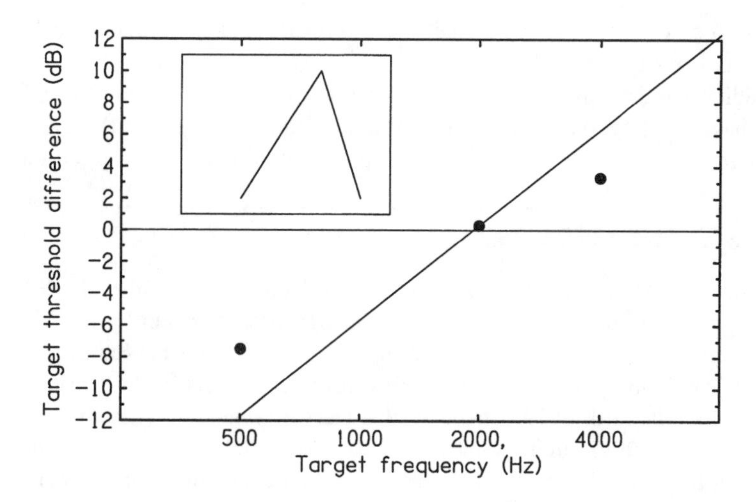

FIGURE 13.5. The three dots show the measured difference in the threshold for detection of a 5-ms pulse when masked by a cosine masker and when masked by a sine masker. The straight line shows a slope of 6 dB per octave, the expected result if the effects of the masker depend only upon the energy density at the target frequency. The inset shows the power response function of a model auditory filter that can account for the measured data.

threshold level caused by changing from a *sine* phase masker to a *cosine* phase masker will follow the 6-dB-per-octave function.

Green's experiment used a 5-ms target pulse that started at the first zero crossing of the masker. The target was passed through a filter to smooth the edges of the pulsed signal and to try to make the onset/offset phase of the target a negligible factor. The results for a 1000-Hz masker and target frequencies of 500, 2000, and 4000 Hz are shown in Fig. 13.5. Also shown in a straight line with a 6-dB-per-octave slope, the expected threshold difference.

It is evident that the experimental results show the effect of phase and that the sign of the effect is as predicted. However, the slope is much smaller than predicted. The cause of the disagreement is very likely that the effective part of the masker is not determined by an infinitely sharp filter centered at the target frequency. Instead, the effective masker is that passed by an auditory filter with a low-frequency skirt that is shallower than the high-frequency skirt. Green found that his data were best fitted by skirts of +12 dB/octave and −24 dB/octave, as shown in the inset in Fig. 13.5. The idea of an asymmetrical auditory filter such as this is not new. It has been amply confirmed by many masking experiments, including the notched-noise experiments described in Chapter 10. It is interesting to see that the asymmetrical auditory filter continues to apply to events of such short duration. However, the slopes are considerably less steep than found in usual masking experiments (see

Exercise 6 of Chapter 10), very likely as the result of the short time scale of the pulsed masker experiment.

In addition to the masking effect, the spectral differences between sine and cosine phases have an effect on the pitch of short tones. Rajcan (1972, 1974) found that the extra high-frequency energy in the cosine-phase condition led to a pitch increase.

S. S. STEVENS (1906–1973)

Stanley Smith Stevens was born in Ogden, Utah on November 4, 1906. He was orphaned at the age of 17, and shortly thereafter was sent by the Mormon Church on a mission to Liege, Belgium, knowing no French. Returning to the United States three years later he studied diverse subjects at the University of Utah and at Stanford. In 1931 he arrived at Harvard intending to study medicine, but he discovered psychology instead, and he never left Harvard. After doing a thesis on the quality of tones, he worked on the development of an operational theory of psychological measurement and simultaneously assisted Hallowell Davis in physiological studies of the peripheral auditory system. During World War II he directed the Harvard Psychoacoustics Laboratory, which improved voice communication in noisy military aircraft. Returning to research in the 1950s he and his students developed the magnitude estimation and magnitude production techniques that led to the power laws of psychophysics. An avid skier, Stevens died in Vail, Colorado on January 18, 1973.

ALTERNATING LATTICE

To study the role of periodicity in hearing, psychoacousticians have compared two kinds of sequences, ''unipolar'' and ''alternating.'' For a unipolar sequence, a brief signal segment is repeated indefinitely with repetition period T_0. The alternating sequence is the same except that odd-numbered segments are inverted (i.e. are multiplied by -1). An example is given in Fig. 13.6. The top line shows a particular signal segment, four cycles of a sine wave. The next line shows the unipolar sequence $x(t)$ and the alternating sequence $x_a(t)$ made from this segment.

A variety of different functions have been used for the brief signal segment. The sine wave burst corresponds to an experiment by Gold and Pumphrey (1948). An experiment by Flanagan and Guttman (1960) used pulses, and an experiment by Warren and Wrightson (1981) used bursts of noise.

What these experiments have in common are the underlying lattice functions. In each case, the unipolar signal is the convolution of a signal segment $\zeta(t)$ with the lattice function $l(t)$ shown in the third line of Fig. 13.6. The alternating signal is the convolution of the same segment with the alternating lattice function $l_a(t)$. A description of the signals in terms of their lattice functions is particularly useful if one wants to study the spectral properties of the signals. The key is the Fourier transforms of the lattice functions themselves.

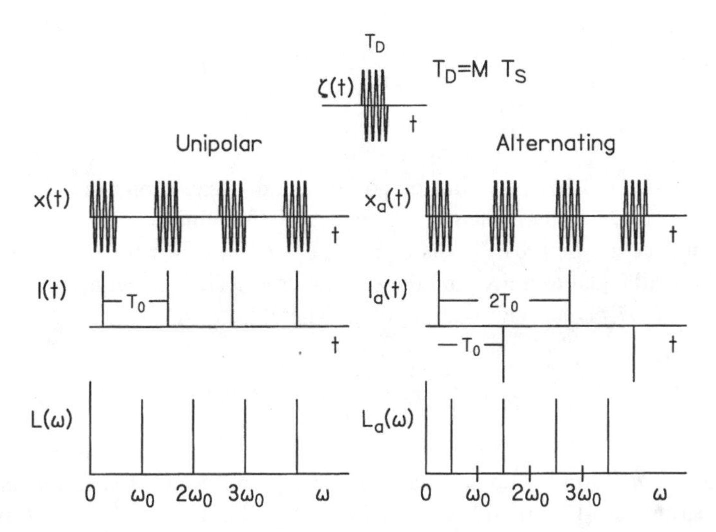

FIGURE 13.6. The top line shows a brief signal segment made from M cycles of a sine tone whose period is T_S. One can use it to make a unipolar signal $x(t)$ or an alternating signal $x_a(t)$, by convolving the segment with unipolar or alternating lattice functions, respectively $l(t)$ and $l_a(t)$. Note that the period of the alternating function is $2T_0$, where T_0 is a reference period from $l(t)$. Finally, $L(\omega)$ and $L_a(\omega)$ are Fourier transforms of the unipolar and alternating lattice functions, respectively.

The unipolar lattice $l(t)$ is a series of delta functions occurring at times that are integral multiples of a period T_0, namely $t = mT_0$. The Fourier transform of this function was derived in Chapter 8. Because of the sampling property of the delta function it was not hard to do, and the answer was:

$$L(\omega) = \frac{2\pi}{T_0} \sum_{n=-\infty}^{\infty} \delta(\omega - n\omega_0). \tag{13.13}$$

This Fourier transform is a set of a harmonics of a fundamental angular frequency $\omega_0 = 2\pi/T_0$.

We next consider the alternating lattice function, $l_a(t)$. It is the same as the unipolar function except for an alternating factor, and is given by

$$l_a(t) = \sum_{m=-\infty}^{\infty} (-1)^m \delta(t - mT_0). \tag{13.14}$$

As for the unipolar lattice function, the delta functions make it easy to find the Fourier transform, $L_a(\omega)$. Using the fact that $(-1)^m = e^{im\pi}$, we find

$$L_a(\omega)=\sum_{m=-\infty}^{\infty} e^{-im(\omega T_0+\pi)}. \tag{13.15}$$

Because Eq. (13.15) is a sum of phasors, and because phasors with arbitrary phases sum to zero, the function $L_a(\omega)$ is zero for almost all values of ω. The exceptions occur when $\omega T_0+\pi$ is equal to 0, or 2π, or 4π, or 6π, etc. Then the phasors are all equal to unity, and the sum diverges. Thus $L_a(\omega)$ has delta function spikes when ωT_0 is an odd multiple of π. Therefore,

$$L_a(\omega)=\frac{2\pi}{T_0}\sum_{n=\mathrm{odd}} \delta(\omega-n\omega_0/2), \tag{13.16}$$

where $\omega_0=2\pi/T_0$, and n ranges over all positive and negative odd integers.

Comparing Eq. (13.16) with Eq. (13.13) we see that the Fourier transform of the alternating lattice has a fundamental frequency that is half that of the unipolar lattice, and it has only odd harmonics. The latter might have been predicted because of the half-wave symmetry of the alternating lattice.

Gold and Pumphrey Experiment

In 1948, Gold and Pumphrey invented an experiment intended to find the bandwidth of auditory filters. The special character of their experiment was that it attempted to measure the temporal response of the filters, from which the width of the frequency response (bandwidth) could be inferred. The experiment used both unipolar and alternating sequences of sine-wave bursts with large gaps between the pulses, as shown in Fig. 13.6, where the pulse duration T_D and the pulse separation T_0 were integer multiples of T_S, the period of the sine. These parameters are defined as follows:

Sine angular frequency: $\omega_s=2\pi/T_s$

Duration of each pulse: $T_D=MT_s$

Separation between pulses: $T_0=NT_s$

Unipolar fundamental angular frequency: $\omega_0=2\pi/T_0$

The idea was that if the listener could distinguish between the unipolar and alternating sequences, then it must be because the response of the auditory filter was slow enough that each pulse was superimposed upon a trace of (at least some of) the previous pulse. A slow response time, in turn, corresponds to a narrow bandwidth. The experiments showed that when the frequency of the sine-tone pulses was high (e.g. $\omega_s/2\pi=2$ kHz or 5 kHz), listeners could distinguish between the sequences for values of separation T_0 that were much longer than expected. This suggested that the auditory filters were much narrower than had been previously supposed, based upon measurements made in the frequency domain.

This experiment was criticized by Hiesey and Schubert (1972) and by Green, Wier, and Wightman (1975). These authors pointed out that Gold and Pumphrey had assumed that listeners were only using information at the sine tone frequency ω_s to make their decisions. In fact, there is information available over the entire spectrum, starting at frequencies as low as the fundamental of the alternating signal, $\omega_0/2$, which is $\omega_s/2N$. Furthermore, this low-frequency information is evidently very important to listeners because the experimental results did not change when the signals were lowpass filtered so as to eliminate the regions around the supposedly critical frequency of ω_s. The narrow bandwidth found in the experiments turned out not to be inconsistent with expectations based upon measurements of bandwidth for the low frequencies that the listeners apparently used to perform the task.

The low-frequency spectra which finally proved to be the undoing of the Gold and Pumphrey experiment can be determined from the Fourier transforms of the unipolar and alternating lattice functions, as calculated above, and the Fourier transform of the brief signal segment. The final spectrum is simply the product of the Fourier transforms of lattice and segment. The lattice function depends upon the separation integer N, and the segment depends upon the duration integer M.

The Fourier transform of the segment $\zeta(t)$, consisting of M cycles of a sine wave centered on time $t=0$, is given by

$$Z(\omega)=(i)(-1)^M \sin(M\omega T_s/2)\left[\frac{1}{\omega-\omega_s}-\frac{1}{\omega+\omega_s}\right]. \tag{13.17}$$

Function $Z(\omega)$ is purely imaginary, as expected for an odd-symmetry function. Therefore, the spectrum, $|Z|^2$, looks a lot like $|Z|$ itself. Equation (13.17) makes it evident that the spectrum has a peak at the sine tone frequency ω_s. Rewriting the equation in terms of the fundamental frequency of the unipolar lattice (ω_0) makes it easier to think about the amplitudes of the harmonics in the spectrum,

$$Z(\omega)=(i)(-1)^M \sin\left(\pi\frac{M}{N}\frac{\omega}{\omega_0}\right)\left[\frac{1}{\omega-N\omega_0}-\frac{1}{\omega+N\omega_0}\right]. \tag{13.18}$$

Figure 13.7 shows $Z(\omega)$ and the transforms of both the unipolar and the alternating lattice functions for the particular case $M=4$, $N=10$. To find the amplitude spectrum one weights the delta functions of the lattice by $|Z(\omega)|$. To find the power spectrum one weights by the square, $|Z(\omega)|^2$. The former is shown in the last two lines of the figure. Although the spectrum is largest in the neighborhood of the sine wave frequency, there is a lot of spectral strength at lower frequencies and ample opportunity for listeners to observe low-frequency differences between the spectra of unipolar and alternating signals. The listeners' strategy thus became another example of off-frequency listening (Leshowitz and Wightman, 1971) as described in Chapter 10.

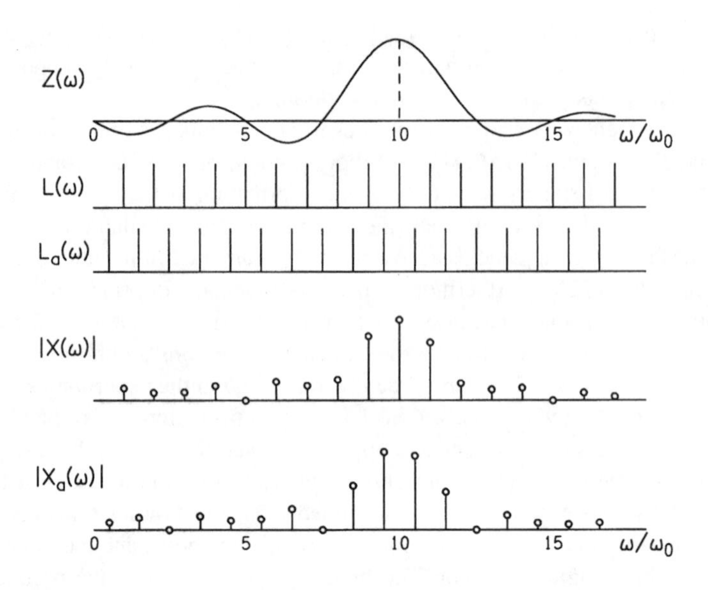

FIGURE 13.7. $Z(\omega)$ is the Fourier transform of the sine tone signal segment $\zeta(t)$. It multiplies the delta function spectra for unipolar, $L(\omega)$, and alternating, $L_a(\omega)$, lattice functions to give the spectra of the Gold-Pumphrey stimuli. These spectra are shown by the last two lines of the Figure. (M=4, N=10.)

FORGETFUL FOURIER TRANSFORMER

The comparison of unipolar and alternating pulse functions can be used to demonstrate the failure of a perceptual model based upon Fourier analysis to cope with situations where the perceptual-analysis duration becomes a critical factor. To see the nature of the problem we approach this comparison from two sides, short and long.

When the pulse separation T_0 is short, there are many pulses within the time interval over which perceptual analysis takes place. We expect that a Fourier transform of the entire pulse train is a perceptually relevant operation and that the spectrum is perceptually meaningful. The Fourier-analysis model predicts that the unipolar pulse train leads to a complex tone with a pitch near a frequency of $1/T_0$, and a tone color that depends upon the nature of the signal segment $\zeta(t)$. It also predicts that the alternating pulse train is a similar complex tone, except that its pitch is about an octave lower, near $1/(2T_0)$, because the period of the alternating pulse is twice as long. These predictions are borne out by experiment.

When the pulse separation T_0 is long, for example when the pulses are separated by a full second, then each pulse is analyzed separately by the auditory system. The energy spectrum for each individual pulse of the alternating pulse train is the same whether the pulse is right side up or inverted. Therefore, the Fourier model predicts that the train of pulses sounds the same for both the unipolar and alternating cases.

This prediction too is borne out by experiment.

The problem occurs when the pulse separation T_0 is neither short nor long. Using noise for segment ζ, Warren and Wrightson (1981) found that for T_0 between 10 and 50 ms the perceived periodicity of the alternating pulse train was intermediate between $1/(2T_0)$ and $1/T_0$. One way to approach this problem, and other problems too, is to use a combined frequency and time representation that was suggested by Flanagan in 1972. The representation involves an analysis called a ''forgetful Fourier transformer.'' The output of the transformer is a function of both frequency and the present time,

$$X(\omega,t)=\int_{-\infty}^{t} dt' x(t')e^{-(t-t')/\tau}e^{-i\omega t'}. \tag{13.19}$$

This transform resembles the Fourier transform except that instead of integrating all the way to infinity, it stops at the present time t. Also, before being transformed, the signal $x(t')$ is weighted by a function that decays exponentially into the past, causing the transformer to forget things that happened long ago. The time constant of the memory is parameter τ.

With τ set equal to 100 ms, the time-dependent spectrum $|X(\omega,t)|^2$ provides a representation of the signal that accords with the experiments of Warren and Wrightson (Hartmann, 1981): When the period, T_0, is short, this spectrum is nearly independent of time, t, but the periodicity can be seen in the dependence on ω, which shows spikes corresponding to harmonics of the signal. When the period is long, the spectral spikes are blurred, but there is a strong temporal modulation of the overall spectral level. For intermediate periods, between 10 and 50 ms, there is a smooth transition for both frequency and temporal dependences.

The forgetful Fourier transform has several interesting properties. In some ways it behaves like an ordinary Fourier transform. It is left as an exercise to show that the inverse transform is given by

$$x(t)=\frac{1}{\pi}\int_{-\infty}^{\infty} d\omega\, X(\omega,t)e^{i\omega t}, \tag{13.20}$$

which is the same as the inverse for the complete Fourier transform in Eq. (8.2), except for a factor of 2.

An alternative interpretation is that $|X(\omega,t)|$ represents the time-dependent output of a filter. It is not a rigorously correct filter, but it is physically similar to a real filter because of the way it weights past events, and it may be a reasonable approach to an auditory filter in some circumstances. The rest of this section develops that interpretation.

Suppose that ω_0 is the center frequency of an auditory filter. We define its time-dependent output, given input $x(t)$, by a convolution integral,

$$X(\omega_0,t)=e^{-i\omega_0 t}\int_{-\infty}^{\infty} dt' x(t')h_{\omega_0}(t-t'). \tag{13.21}$$

Here $h_{\omega_0}(t)$ is the impulse response for the auditory filter centered on frequency ω_0. Equation (13.21) would be exactly the right formula for filtering except for the leading phase factor $\exp(-i\omega_0 t)$ that has mysteriously been inserted. We note, however, that we are interested in $|X(\omega_0,t)|$, and that is unchanged by the leading phase factor, and so we press on. Our goal is to find an expression for $h_{\omega_0}(t)$ so that Eq. (13.21) becomes the same as Eq. (13.19), with the notational change that parameter ω_0 replaces ω.

Because of the leading phase factor, we achieve our goal if the filter impulse response is given by

$$h_{\omega_0}(t)=0 \quad (t<0), \tag{13.22}$$

and

$$h_{\omega_0}(t)=\frac{1}{\tau}e^{i\omega_0 t}e^{-t/\tau} \quad (t\geqslant 0).$$

The filter is causal, which is good, but the impulse response is complex, which is bad. The impulse response of a physical filter is real and never complex. If we press on anyway, we will find that this approach may not be ridiculous after all.

Assuming that $h_{\omega_0}(t)$ is a legitimate filter we find the corresponding transfer function, $H_{\omega_0}(\omega)$, by taking the Fourier transform,

$$H_{\omega_0}(\omega)=\frac{1/\tau}{i(\omega-\omega_0)+1/\tau}. \tag{13.23}$$

This can be compared with the transfer function for a two-pole bandpass filter centered on ω_0, from Exercise 2 in Appendix G,

$$H_{BP}(\omega)=\frac{i\omega\omega_0/Q}{\omega_0^2-\omega^2+i\omega\omega_0/Q}, \tag{13.24}$$

where Q is the Q of the filter.

This bandpass transfer function is large only when $\omega\approx\omega_0$, but in that case, $\omega_0^2-\omega^2$ is approximately $2\omega(\omega_0-\omega)$, and

$$H_{BP}(\omega)\approx\frac{\omega_0/(2Q)}{i(\omega-\omega_0)+\omega_0/(2Q)}. \tag{13.25}$$

From Eq. (9.61) we find that the Q of the two-pole bandpass filter is related to the time constant by $\tau=2Q/\omega_0$. Therefore, the bandpass transfer function becomes

$$H_{BP}(\omega)\approx\frac{1/\tau}{i(\omega-\omega_0)+1/\tau}, \tag{13.26}$$

and this equation is exactly the same as Eq. (13.23). This shows that the forgetful Fourier transform gives an approximation for the magnitude of the output of a model auditory filter, with a characteristic frequency given by the frequency parameter of the transform, and a time dependence given by the time parameter of the transform.

Because the impulse response of the filter is not real, $H_{\omega_0}(-\omega)$ is not equal to $H^*_{\omega_0}(\omega)$. We find the equivalent rectangular bandwidth by integrating over only the positive part of the frequency axis. Then, because the maximum value of $H_{\omega_0}(\omega)$ is unity, the equivalent rectangular bandwidth in Hertz is given by

$$\text{ERB}=\int_0^\infty df|H_{\omega_0}(f)|^2=\frac{1}{2\pi}\int_0^\infty d\omega|H_{\omega_0}(\omega)|^2=\frac{1}{\tau}\left[\frac{1}{4}+\frac{1}{2\pi}\arctan(2Q)\right]. \tag{13.27}$$

As expected for a filter, there is a reciprocal relationship between bandwidth and time constant. It shows a weak dependence on Q so that $\text{ERB}\cdot\tau=1/4$ for small Q and $\text{ERB}\cdot\tau=1/2$ for large Q. The 3-dB bandwidth, where $|H_{\omega_0}(f)|^2=1/2$, is given by $2(f-f_0)=1/(\pi\tau)$. An application of the forgetful Fourier transformer (called the "Fourier-t transform") to speech spectrograms, together with earlier references to the method, was described by Heinbach (1988).

UNCERTAINTY PRINCIPLE

When the Fourier integral was introduced in Chapter 8, it became evident that there was a reciprocal relationship between the duration of a signal and the width of its frequency-space representation. For the rectangular pulse of Fig. 8.2, the duration T_0 was reciprocally related to the separation between the first zero crossings of the energy spectrum, Δf, by the rule: $T_0\Delta f=2$. Also in that chapter, Exercise 6 showed that for a Gaussian pulse, the half widths of the time and frequency representations were related by $\Delta t\,\Delta f=\ln(2)/\pi$. The reader can find similar relationships for signal pulses of other shapes, exponential, raised-cosine, triangle, parabola, and trapezoid, whose spectra were calculated in Exercises 18–21 of Chapter 8. These reciprocal relationships are known as uncertainty products. They are generally of the form $\Delta t\,\Delta f=$constant, where the constant depends upon the pulse shape and upon the mathematical definitions of the widths of the distributions in time and frequency, and is generally of the order of 1.

Formally, the uncertainty principle is defined by the variance of distributions. Uncertainty $(\Delta t)^2$ is the variance of time about the mean time of the power $x^2(t)$, and uncertainty $(\Delta\omega)^2$ is the variance about the mean angular frequency of the energy density $|X(\omega)|^2$. The smallest possible uncertainty product occurs when $x(t)$ is a Gaussian function. Then $X(\omega)$ is also a Gaussian, and the uncertainty product becomes

$$\Delta t\,\Delta\omega = 1/2. \tag{13.28}$$

It can be shown that this is the smallest possible value of the product, and, therefore, for the general signal,

$$\Delta t\,\Delta\omega \geqslant 1/2, \tag{13.29a}$$

or

$$\Delta t\,\Delta f \geqslant 0.08. \tag{13.29b}$$

As applied to psychoacoustics, the principle of uncertainty says that if the duration of a tone is made short then the uncertainty about the frequency of the tone increases. It is usual to imagine that the frequency uncertainty should appear in the form of an uncertainty about the pitch of the tone. *A priori* it is not an unreasonable assumption. After all, a digital frequency counter has that kind of limitation: To get an accuracy of 0.1 Hz, it needs to count for 10s. If the tone lasts only 1s then only 1-Hz accuracy is possible. A bank of linear filters also behaves that way. A brief tone pulse causes them all to ring, with little regard for the best frequency of the individual filters.

Reasonable or not, the uncertainty principle as applied to pitch perception is an idea with very little quantitative experimental support. Whereas the uncertainty principle sets a lower bound on the uncertainty product that is approximately equal to 1 for most signal envelopes, pitch perception experiments show that human listeners regularly beat the uncertainty principle by as much as an order of magnitude. For example, in a two-interval forced-choice task using 1000-Hz tones, rectangularly gated with 10-ms duration, Moore (1973) found a frequency difference limen of 5.6 Hz for one listener and 6.2 Hz for another. The uncertainty product is therefore, $\Delta f\,\Delta t \approx 0.06$, which is smaller than would be allowed even if the envelopes had been Gaussian. Further, whereas a straightforward application of the uncertainty principle predicts a frequency-difference limen that scales inversely with the tone duration, experimental results usually follow a more complicated dependence on duration, as described in the next section.

Pitch of Short Tones

Tones of very short duration sound like a click, which does not have a strong pitch. But even a click elicits some sensation of pitch in that the click can be

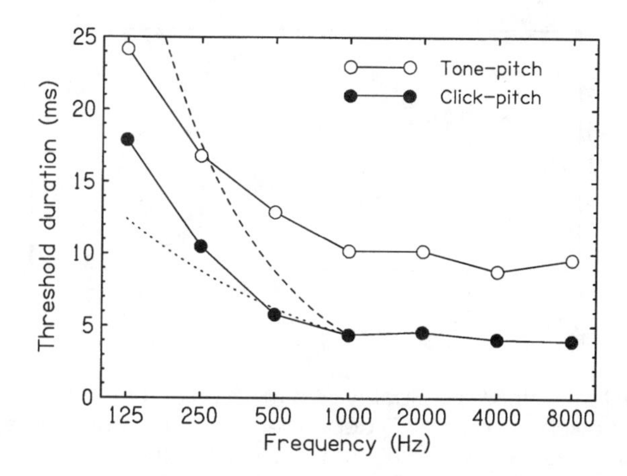

FIGURE 13.8. Click-pitch and tone-pitch thresholds from Doughty and Garner. The tone intensity was 110-dB SPL (not as loud as one might imagine for these short tones). Dashed lines starting at click-pitch threshold for 1 kHz show inverse square-root dependence (light) and inverse first-power dependence (heavy). The latter corresponds to a constant number of cycles (4.4 cycles, in fact) in the pulse.

"high" or "low." A 1-kHz tone pulse can be distinguished from a 2-kHz tone pulse from only a single cycle of each. However, the pitch character of such a pulse is very different from the pitch character of a tone.

Doughty and Garner (1947, 1948) identified two pitch thresholds, "click pitch" and "tone pitch." Click-pitch threshold is the minimum duration needed for a listener to identify the "general frequency range." A tone with click pitch mainly sounds like a click. Tone-pitch threshold is the shortest duration for which a tone burst sounds mainly like a tone. These threshold durations, as a function of tone frequency, are shown in Fig. 13.8. Figure 13.8 shows that the threshold durations are approximately constant for tones of 1 kHz or above. Below 1 kHz, the thresholds lie between a constant duration and a constant number of cycles.

The experiments of Doughty and Garner required a criterion for pitch. Difference limen experiments do not. Liang and Chistovich (1961) performed a pitch matching experiment using tones with durations from 3 to 600 ms. The standard deviation of the matches, Δf, was interpreted as the difference limen. Liang and Chistovich plotted difference limens as a function of tone duration on a log-log plot in order to search for power-law behavior. They identified three regions, a small-duration region where Δf varied inversely with the duration, a central region where Δf varied inversely as the square-root of the duration, and an asymptotically long region where Δf became constant, independent of duration. These regions are shown in Fig. 13.9.

The frequency dependence of break-point parameter T_1 was such that the increase in frequency difference limen with decreasing duration becomes more

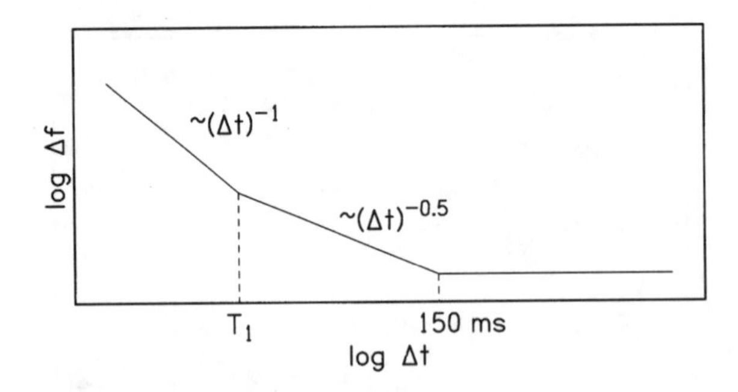

FIGURE 13.9. Pitch matching variability as a function of tone duration as found by Liang and Chistovich. Parameter T_1 depended upon the tone frequency, varying from 15 ms for a 4000-Hz tone to 40 ms for a 250-Hz tone.

dramatic for lower tone frequencies. This behavior agrees with difference limens found by Henning (1970) and Moore (1973).

There are individual possible explanations for the different regions. The reciprocal first-power dependence in the small-duration region agrees with the expectation from the uncertainty principle. The inverse-square-root dependence in the central region might be understood from a statistical argument that says that the standard deviation of a distribution of measurements varies inversely as the square root of the number of measurements. If it is presumed that the auditory system accumulates data at a constant rate, then the standard deviation should vary inversely as the square root of the duration.

There are general theoretical treatments too. Siebert (1970) calculated the frequency difference limens for short tones by applying the Cramér-Rao inequality to model neural timing patterns. This inequality gives a lower bound to the variance of a stochastic quantity (see Eadie *et al.*, 1971). The calculation predicted difference limens that were considerably smaller than observed experimentally. Goldstein and Srulovicz (1977) showed that the same statistical method, applied only to the inter-spike intervals, gave a remarkably good fit to Moore's data. Important to the success of the model was that it included a realistic representation of the decrease in neural synchrony with increasing frequency (see Fig. 6.14).

Uncertain about Uncertainty

To make an uncertainty-principle model of pitch perception starting with a spectral treatment of short tones is a two-step process. First is the mathematical step in which a tone pulse with particular envelope shape and duration is characterized by a bandwidth. Second is the interpretative step that associates the bandwidth with variability in the perception of pitch. It is possible to separate these two steps somewhat. Ronken (1971) measured the frequency difference limen for short tones with different envelope shapes: rectangular, exponential, Gaussian, and gamma. When Ronken plotted the difference limens as a function of bandwidth, the envelopes of different shape led to greatly different plots. However, there is no uniquely correct definition of bandwidth and it is possible that a different definition than the one that Ronken used would have led to better agreement.

Ronken was able to unify the frequency difference limens measured with different envelopes by avoiding the concept of bandwidth altogether. Instead, he found that difference limens scaled with an effective duration, determined as the length of time that the signal envelope power exceeded a threshold. Interestingly, Rossing and Houtsma (1986) found that a similar criterion unified the data on pitch shifts caused by duration changes or intensity changes.

The interpretative step in the spectral model of pitch can be challenged by reinterpreting the model pitch process while maintaining a spectral framework. Consider a pitch matching experiment with an alternating sequence of standard and matching tones. The analysis of this experiment from the point of view of the uncertainty principle is that the task requires the listener to align two spectral lines and that the widths of the lines, caused by the short duration, prevent the listener from doing this perfectly. This analysis may be incorrect. It may be that the true nature of the task is that the listener aligns two complete excitation patterns and not two lines. In that case the line widths would not necessarily be detrimental.

These two interpretations can be compared experimentally by a pitch matching experiment using tones of different duration. For example, in a ''same'' condition the standard tone and the matching tone may both be 12.5 ms long. In a ''different'' condition the duration of one tone remains 12.5 ms while the duration of the other increases to 50 ms. According to the uncertainty principle, the matching variability should decrease. That is because the variability in the match is a combination of the uncertainty in each of the two tones, and reducing the uncertainty of one of them is bound to decrease the uncertainty of the combination. However, an experimental test by Hartmann, Rakerd, and Packard (1985) disagreed. Going from the ''same'' condition to the ''different'' condition caused the matching variance about the mean match to *increase* 80 percent of the time. The authors explained this failure of the uncertainty principle by hypothesizing that in a pitch matching experiment, listeners attempt to align excitation patterns. But when the tones have different durations the patterns have different widths, and it is impossible to match up two patterns having different widths.

The opposite result was obtained in a two-interval forced-choice task. In that task, increasing the duration of one of the two tones led to a *decreasing* difference limen 60 percent of the time, in agreement with the uncertainty principle. It appears that the speed of the forced-choice task, with only one presentation of the tone pair, focuses the listener's attention on pitch *per se*. It does not permit a detailed match of excitation patterns.

Uncertainty Principle in Quantum Mechanics

The uncertainty principle was introduced into mechanics in a 1927 paper by Werner Heisenberg. The uncertainty principle applies to the dynamical description of atomic and subatomic particles because, according to quantum mechanics, these particles behave like waves. The fact that uncertainty relationships apply to classical waves, such as acoustical waves, had long been evident from the mathematics of the Fourier transform.

In a curious twist, the uncertainty principle was reintroduced into acoustics in 1931 by Stewart, who had become familiar with Heisenberg's application to quantum mechanics. However, Gabor (1946) notes that Mach had explored the acoustical implications of relationships like the uncertainty principle in 1871.

The acoustical uncertainty principle in Eq. (13.29) says that $\Delta t\,\Delta\omega \geqslant 1/2$. In quantum mechanics, the energy of a particle $\mathcal{E}$ is given by $\hbar\omega$, where ω is the particle angular frequency, and $\hbar$ is Planck's constant divided by 2π ($\hbar = 1.055\times 10^{-34}$ Joule·s). Therefore, the quantum mechanical principle says that

$$\Delta t\,\Delta\mathcal{E} \geqslant \hbar/2. \tag{13.30}$$

This uncertainty principle relates time and energy.

A similar argument relates position variable r and propagation constant k, where k is determined by the wavelength from $k = 2\pi/\lambda$. A particle that is described by a single propagation constant is completely delocalized. An observer hasn't a clue about where it is. In order to localize the wave, with an envelope that has a maximum at some place in space, the observer needs to sacrifice some of the certainty he has about the propagation constant. There is a reciprocal relation between the variances of the distributions in position space and k-coordinate space.

$$\Delta r\,\Delta k \geqslant 1/2, \tag{13.31}$$

where equality applies if the distributions are Gaussian.

In quantum mechanics, the momentum of a particle p is given by $\hbar k$, where the units of $\hbar$ can be written as m·kg m/s. Therefore, the uncertainties in momentum and position are related by

$$\Delta r\,\Delta p \geqslant \hbar/2. \tag{13.32}$$

SPECTRAL RAKES

The kind of signal that is of interest in this section consists of regularly spaced discrete spectral components packed into a frequency band, as shown by the spectrum in Fig. 13.10. Because the frequency separation between the components is constant and the amplitudes are all the same, the spectrum resembles the tines of a

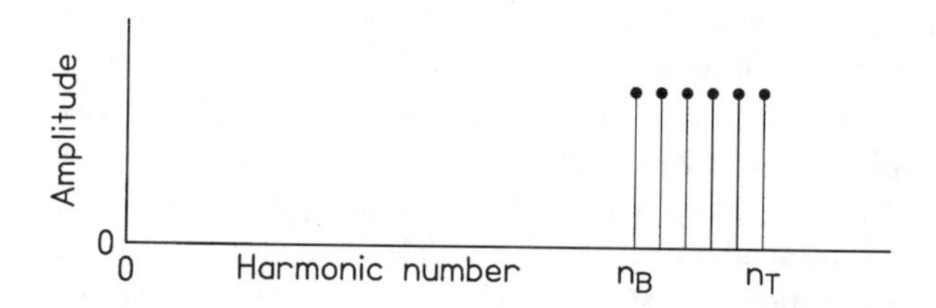

FIGURE 13.10. A "rake" spectrum consisting of discrete components equally spaced in frequency. All components have the same amplitude.

garden rake. Hence, it will be called a "spectral rake." For definiteness we shall assume that the components are all harmonics of a low fundamental frequency equal to the frequency separation.

Signals with rake spectra are not found in nature, and they are not easily generated with analog electronics either. They are examples of the many useful signals that became available with the advent of digital synthesis.

The general rake-spectrum signal is given by the equation

$$x(t)=\sum_{n=n_B}^{n_T} \cos(n\omega_0 t+\phi_n), \tag{13.33}$$

which is the sum of components, all with amplitude 1, starting with the bottom component n_B and ending with the top component n_T. The number of components is N, given by $N=n_T-n_B+1$.

It is possible to generate $x(t)$ by a computer program that adds up N cosine components for each time point t. The computed values of $x(t)$ can be stored in a file and later played out through a digital-to-analog converter. A faster way to generate $x(t)$ is to use the Fast Fourier Transform, beginning with the desired amplitude spectrum (amplitudes in the band equal to one, and amplitudes outside equal to zero) and the desired phase spectrum (the set $\{\phi_n\}$). Use of the FFT automatically generates a set of components that are equally spaced in frequency.

The common frequency spacing corresponds to angular frequency ω_0. Therefore, the fundamental frequency in Hertz is $f_0=\omega_0/2\pi$, and the signal has periodicity $T=1/f_0$. Therefore, all the components of the signal are harmonics of f_0, though it may not be particularly useful to think of them in this way if f_0 is small.

Dense Components

If f_0 is a subaudible frequency then the components are densely spaced. For example, f_0 might be 2 Hz, in which case $x(t)$ repeats itself twice per second and the period is 1/2 s. The tight spacing of the components has an important implication for the sound of the signal. Many components fall within the same auditory filter and are not separately analyzed. A consequence of this is that the relative phases are important in determining the sensation caused by this signal. Another consequence is that Eq. (13.33), summing discrete components, is never the most psychologically pertinent way to express the signal because the components are not individually heard. The actual perceptual result depends upon the phases.

If the phases are random variables, then $x(t)$ becomes what is known as ''equal-amplitude random-phase noise.'' It is usual for the random phases to be distributed with a rectangular density from 0 to 360 degrees, i.e. all possible phases are equally probable. This noise band represents the ultimate in bandpass filtering. It is as though white noise were passed through a filter with a perfectly flat top and infinitely steep skirts. For example, if there are 51 components, separated by 2 Hz and centered on 1000 Hz, then abrupt band edges occur at 950 and 1050 Hz. Further, the components in the passband are not merely of equal strength on the average as in the case of thermal noise. Instead, every sample of the noise has every component equal in amplitude to every other component. The implications of equal amplitudes are discussed in Chapter 23 on noise.

Noise constructed in this way differs from filtered noise produced by a thermal noise generator also because it is periodic. This periodicity is readily perceived by a listener. No matter what the random phases are, there always seems to be some characteristic of the noise that permits a listener to determine the periodicity. This is true of narrow-band noise and broadband noise.

The effect was first studied by Guttman and Julesz (1963) who referred to noise features such as ''whooshing,'' or ''rasping,'' or ''clanking'' as cues to the periodicity. The longest period for which the periodicity can still be detected is unknown. A period of 1s is effortless, and 2s is quite possible. The world's record appears to be 10s (Warren and Bashford, 1981). The ability to perceive the periodicity in cyclic noise is one of the minor mysteries of hearing. It is not clear what kind of spectro-temporal characteristic is responsible for the effect. It sounds as though there are particular instants in time when power sloshes from one frequency region to another, and yet no particular cues can be found in a broadband spectrogram (Kaernbach, 1992, 1993). Conceivably periodicity is not signaled by isolated events but rather by the perception of the pattern as a whole. Using a tapping experiment, Kaernbach found agreement among some listeners on the location of the characteristic features in time. In a series of elegant experiments he also studied the frequency extent of the features by time-shifting part of the spectrum. Any feature that was disrupted by the shift was regarded as spanning the particular frequency where the time-shift cut was made. Kaernbach found that listeners used both spectrally local and spectrally diffuse features as cues for recognizing the periodicity of the noise.

Constant Phase

As noted above, the spectral rake resembles the output of a filter with infinitely steep skirts. One would expect that a filter having infinitely steep skirts at given top and bottom frequencies would impress the signature of these frequencies on a time representation of the output. Indeed it does. It is particularly easy to see this result in the case where all the values of ϕ_n are the same.

If all the values of phase are zero then the sum in Eq. (13.33) becomes a sum of cosines, which leads to the sharpest possible pulse. It is effectively a delta function passed through an ideal filter with a rectangular passband and no phase shift.

Although the choice of a constant phase of zero corresponds to the sharpest pulse, other choices of a constant phase also lead to spiky waveforms. For definiteness below we consider the case where all the phases are $-\pi/2$, which turns each cosine function in the sum into a sine function; hence,

$$x(t)=\sum_{n=n_B}^{n_T} \sin n\omega_0 t. \tag{13.34}$$

What is special about the case of constant phase is that the sum can be done exactly. The first step is to write each sine function as the imaginary part of $\exp(in\omega_0 t)$,

$$x(t)=\operatorname{Im}\sum_{n=n_B}^{n_T} e^{in\omega_0 t}. \tag{13.35}$$

From this exponential form, we realize that the simplification of constant phase has given us a complex geometric series, as described in Appendix C–Geometric Series. From the appendix we extract the result for the sum of such a series as a function of the first and last terms, respectively r^{n_B} at the bottom and r^{n_T} at the top. The result is $x=(r^{n_B}-r^{(n_T+1)})/(1-r)$. Therefore,

$$x(t)=\operatorname{Im}\frac{e^{in_B\omega_0 t}-e^{i(n_T+1)\omega_0 t}}{1-e^{i\omega_0 t}}. \tag{13.36}$$

The denominator can be turned into something like a sine function if we multiply it by $e^{-i\omega_0 t/2}$. Multiplying the numerator by the same factor, we get

$$x(t)=\operatorname{Im}\frac{e^{i\left(n_B-\frac{1}{2}\right)\omega_0 t}-e^{i\left(n_T+\frac{1}{2}\right)\omega_0 t}}{e^{-i\omega_0 t/2}-e^{i\omega_0 t/2}}. \tag{13.37}$$

The denominator is actually $-2i\sin(\omega_0 t/2)$, which is purely imaginary. Therefore, to get the imaginary part of the entire expression, we must take the real part of the numerator. But the real parts of the complex exponentials in the numerator are just cosine functions. Finally, therefore, we have

$$x(t)=\frac{\cos\left(n_B-\dfrac{1}{2}\right)\omega_0 t-\cos\left(n_T+\dfrac{1}{2}\right)\omega_0 t}{2\sin\omega_0 t/2}. \tag{13.38}$$

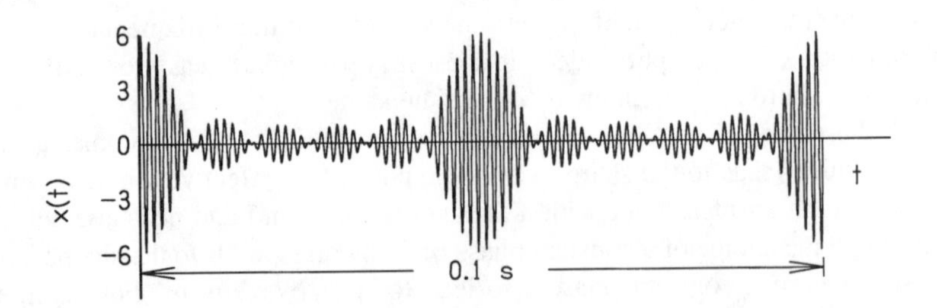

FIGURE 13.11. In an interval of time from 0 to 0.1 second, there are two complete cycles of the waveform made from harmonics 40 through 45 of a fundamental, f_0=20 Hz. For each cycle of the waveform there are six beats between frequencies of 39.5f_0 and 45.5f_0 (790 Hz and 910 Hz). The fastest oscillations represent the frequencies actually in the spectrum, in the neighborhood of 900 Hz.

This expression for the signal waveform includes three periodicities. In the numerator there is one oscillation near the frequency of the bottom of the band and another near the frequency of the top. The precise frequencies in the numerator occur at half a harmonic number below the bottom of the band and at half a harmonic number above the top. In the denominator is a sine function that oscillates at half the frequency of the fundamental. This oscillation is slow compared to the oscillations in the numerator and that is the reason that it may make sense to consider the oscillations of the numerator by themselves. During those intervals of time when the denominator is approximately constant, the periodicities found in the numerator may actually be a reasonable approximation to the periodicities of the waveform as a whole.

An example of this kind of waveform is given in Fig. 13.11, which shows the sum of harmonics 40 through 45 of a fundamental frequency f_0=20 Hz. The harmonics all have sine phase and equal amplitudes. Therefore, the waveform is given by

$$x(t)=\sum_{n=40}^{45} \sin 2\pi n f_0 t. \tag{13.39}$$

The periodicities in the numerator can be important enough to generate sensations of pitch for signals with a rake spectrum. Such pitches are called "edge pitches" for evident reasons. The signal shown in Fig. 13.11 generates edge pitches, as shown by the pitch matching data in Fig. 13.12. The variability of the matches is rather high, especially at the high end, but there is little question that the listener is matching edge pitches.

If the bottom of the spectrum extends all the way down to the fundamental frequency (n_B=1) then the lower edge pitch disappears and the remaining edge pitch is easier to hear. Using such a signal, and f_0=50 Hz, Kohlrausch and Houtsma (1992) found pitch matching variability of only 0.2%. However, the pitch matches

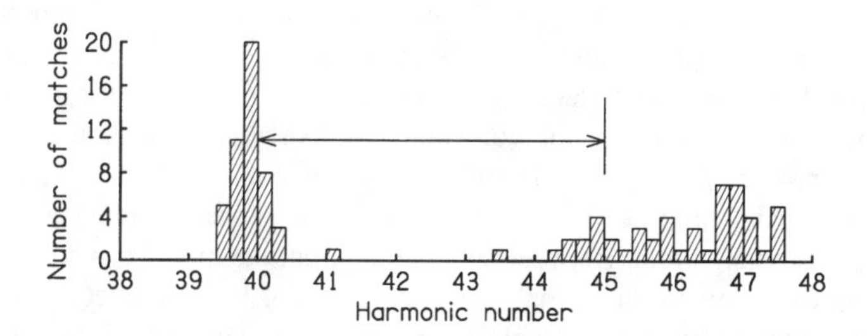

FIGURE 13.12. Histogram showing the distribution of 100 matches to the pitches generated by the signal of Fig. 13.11. Because the fundamental frequency was randomized ±10% from trial to trial, the *x*-axis shows the matching frequency on a scale of harmonic numbers. If the listener matches the periodicities in the numerator, peaks should occur at harmonic numbers 39.5 and 45.5. The listener was usually aware of both pitches in the signal and consciously attempted to match the ''high pitch'' and the ''low pitch'' equally often. The double arrow indicates the stimulus band; the level was 68 dBA.

could be systematically shifted from above the edge frequency to below the edge frequency by reducing the level from 70 dB SPL to 10 dB SPL.

It is not necessary for the starting phase of the components to be constant in order to create edge pitches. A band of equal-amplitude random-phase noise with sharp cutoff also leads to a sensation of pitch at the top and bottom of the band (Zwicker and Fastl, 1990, p. 120ff). The pitches are near the edges of the band, but tend to lie within the band, shifted by several percent towards the center (Klein and Hartmann, 1981). This shift is opposite to the shift of the periodicities in the numerator of Eq. (13.38) which are, in any case, small for noise because the components are closely spaced.

Beats and the Frequency-Domain Grating

The previous section summed equally separated sine components in a spectral rake having an appreciable bandwidth. The frequencies of the highest and lowest components were far enough apart that one could distinguish pitches at both these edges. But it might be that the bandwidth is so small that the frequencies of the two edges cannot be resolved. In that case, it makes sense to write the waveform $x(t)$ in an alternative way that exhibits beats. This topic, and its application to the frequency-domain grating, is discussed in Chapter 17, where the subject of beats is covered in some depth.

CAVEAT ON REPRESENTATIONS

This chapter has presented several examples in which insight into perception is gained by a spectral representation of the signal. It also presented an example in which a temporal representation was found to be useful, and one example where both representations were combined. It is important to realize that identifying the spectrum or the time dependence as a convenient reference for discussing a psychoacoustical effect does not constitute an explanation of the effect. True explanatory power only comes from a physiologically plausible mathematical model of a hearing process. The model accepts the physical signal as input and predicts a perceptual output. Between input and output the model can be represented either in the time domain or in the frequency domain. If the model is linear, then which representation one uses is merely a matter of convenience; it is not a fundamental issue. If the model is nonlinear, the choice assumes a larger role. Nevertheless, a representation is not yet a model. Therefore, explanations of perceptual results based upon a signal representation, such as the descriptions in the present chapter, remain incomplete at best. The advantage to an auditory scientist of a repertoire of representations for the signal is that perspicuous representations suggest successful models.

EXERCISES

Exercise 1: Fourier transform of the pure tone with an abrupt onset

The abrupt pure-tone signal with unit amplitude is defined as

$$x(t)=0 \quad \text{for } t<0$$

$$x(t)=\sin(\omega_0 t+\phi) \quad \text{for } t\geqslant 0.$$

Show that its Fourier transform is given by Eq. (13.9), namely,

$$X(\omega)=\frac{\omega_0\cos(\phi)+i\omega\sin(\phi)}{\omega_0^2-\omega^2}.$$

[Hint: You may assume that the pure tone is multiplied by an exponentially decaying envelope. The decay is so slow that we do not notice it during our lifetime, but the decay does mean that the tone has gone away before an infinite amount of time has elapsed.]

Exercise 2: Energy density of a pulse

Find the spectra for the Green pulses in Fig. 13.4. Specifically, show that if there are M periods so that the duration of the pulse is $T_D=MT_0$ then for *sine* phase,

$$|X_{\sin}(\omega)|^2=\left[\frac{\omega_0}{(\omega_0^2-\omega^2)}\right]^2[2-2\cos\,\omega M T_0],$$

and for *cosine* phase,

$$|X_{\cos}(\omega)|^2=\left[\frac{\omega}{(\omega_0^2-\omega^2)}\right]^2[2-2\cos\,\omega M T_0].$$

Note that the ratio is $(\omega/\omega_0)^2$, in agreement with Eq. (13.12).
[Hint: Consider that a tone pulse can be created by starting with a tone that is unending, as in Exercise 1, and subtracting from it a delayed version of the same tone. Then use the general result for the Fourier transform of a delayed signal from Chapter 8. This technique can work for a delay that is an integer number of periods.]

Exercise 3: Gold and Pumphrey spectrum

The Fourier transforms of the unipolar and alternating functions in Fig. 13.6 are given by the product $X(\omega)=Z(\omega)L(\omega)$ and $X_a(\omega)=Z(\omega)L_a(\omega)$. Because functions L and L_a are sums of delta functions, the values of ω are limited to harmonic frequencies. That leads to simplifications in the expressions for the X and X_a. For example, show that if z is given by Eq. 13.18 then

$$X(\omega)=(i)(-1)^M\sum_{n=-\infty}^{\infty}\delta(\omega-n\omega_0)\sin\left(\pi\frac{M}{N}n\right)\left[\frac{1}{n-N}-\frac{1}{n+N}\right].$$

Exercise 4: Inverting the forgetful Fourier transformer

The forgetful Fourier transform $X(\omega,t)$ is given by Eq. (13.19). Use the fact that the Dirac delta function is an even function to show that the inverse transform is given by

$$x(t)=\frac{1}{\pi}\int_{-\infty}^{\infty}d\omega\,X(\omega,t)e^{i\omega t},$$

as in Eq. (13.20).

Exercise 5: Uncertainty principle

Psychoacoustical experiments on frequency discrimination beat the limits of the uncertainty principle. Does this seem to violate a principle of wave physics? What has to be assumed about the encoding of pitch by the auditory system in a model that predicts that the uncertainty principle should hold?

Exercise 6: Information theory

The uncertainty principle is a central element in information theory. How then can one explain the fact that a reciprocal-reading frequency counter can measure the frequency of a sine tone with arbitrarily good precision from only a single cycle of the waveform? [Note: This kind of counter measures the time between zero crossings and computes the frequency by calculating the reciprocal.]

Exercise 7: Spectrum of a short tone

Consider a 1000-Hz tone that is turned on abruptly with cosine phase and is then turned off 50 ms later. This tone burst contains 50 cycles. From the convolution theorem of Chapter 8 it can be shown that the Fourier transform of the burst is

$$X(\omega)=\frac{\sin(\omega-\omega_0)T_D/2}{(\omega-\omega_0)T_D/2}+\frac{\sin(\omega+\omega_0)T_D/2}{(\omega+\omega_0)T_D/2}.$$

This function is shown in Fig. 13E.7, together with the limit for a tone of infinite duration. The two functions look somewhat similar. Explain why this visual similarity depends more importantly upon the number of cycles in the waveform than it does upon the burst duration itself. [Hint: Think about how $X(\omega)$ would look if the tone frequency were reduced to 100 Hz.]

Exercise 8: Sum of cosines

Use the formula for the geometric series to show that if $x(t)$ is a sum of cosines,

$$x(t)=\sum_{n=n_B}^{n_T} \cos n\omega_0 t,$$

then

$$x(t)=\frac{\sin\left(n_T+\frac{1}{2}\right)\omega_0 t-\sin\left(n_B-\frac{1}{2}\right)\omega_0 t}{2\sin \omega_0 t/2}.$$

Compare with Eq. (13.38).

Exercise 9: Sum for constant phase angle

Because one can sum sines exactly, and one can sum cosines exactly (Exercise 8), one also can do the sum for any constant phase. Use the fact that

$$x(t)=\sum_{n=n_B}^{n_T} \cos(n\omega_0 t+\phi_0)$$

is equal to

$$x(t)=\cos\phi_0 \sum_{n=n_B}^{n_T} \cos n\omega_0 t - \sin\phi_0 \sum_{n=n_B}^{n_T} \sin n\omega_0 t$$

to show that

$$x(t)=\frac{\sin\left[\left(n_T+\frac{1}{2}\right)\omega_0 t-\phi_0\right]-\sin\left[\left(n_B-\frac{1}{2}\right)\omega_0 t-\phi_0\right]}{2\sin\omega_0 t/2}.$$

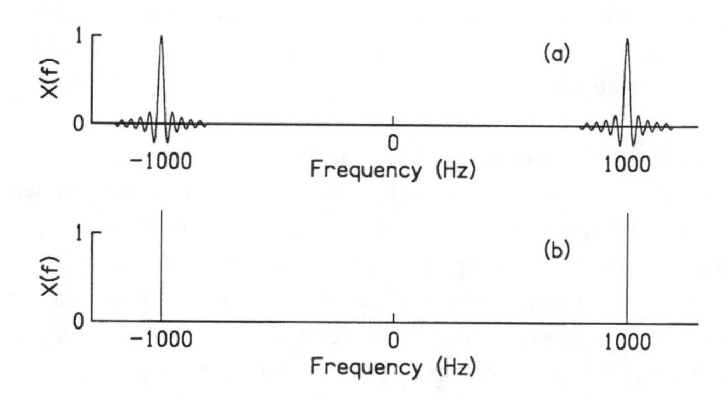

FIGURE 13E.7. The Fourier transform of a 50-ms burst of a 1000-Hz tone in (a) can be compared with the transform of an infinite tone burst in (b).

CHAPTER 14

Correlation Functions and Spectra

The attractiveness of autocorrelational analysis... lies not in revealing anything that cannot be found through frequency analysis; it lies in the fact that the operations involved in carrying out the autocorrelational analysis are quite different from those involved in making the frequency analysis.

J. C. R. LICKLIDER

Chapter 8 introduced the cross-correlation function, which computed the overlap integral between a function $x(t)$ and another function $y(t)$ that had been displaced in time. The time displacement, called the ''lag'' and given the symbol τ, was the independent variable in the cross-correlation function. The autocorrelation function is the cross-correlation function when the two functions involved are the same, i.e., when function y is actually function x.

There are two definitions for the autocorrelation function, depending on whether the signal, described by function $x(t)$, extends to infinite time or whether it is limited in time. Frequently it is possible to regard signals as limited in time. Such signals will be called ''finite signals.'' Even if a duration of a signal is somewhat vague, extending to the indefinite future or indefinite past, it is normally practical to regard it as finite. Perhaps the signal can be multiplied by an exponentially decaying function to cause it to go to zero for values of time in the distant past or far future, well removed from times of interest. The finite autocorrelation can be used in such cases, and along with it, the finite power spectrum.

By contrast, the classic example of a non-finite function is a periodic signal. Here, the definition of periodicity requires that the signal extend to the infinite future and the infinite past. The Fourier transforms of such signals are made up of delta functions which arise precisely because of the infinite extent in time of these periodic functions. For such signals a special autocorrelation function must be used, which leads to a correspondingly unusual power spectrum. We begin with finite signals.

FINITE-DURATION SIGNAL AUTOCORRELATION FUNCTION

The finite-signal autocorrelation function is defined as function $a_f(\tau)$,

$$a_f(\tau)=\int_{-\infty}^{\infty} dt\ x(t)x(t+\tau). \tag{14.1}$$

It is understood that signal $x(t)$ does not extend to infinite time in either direction. If it did then $x(t)$ would have finite values for unlimited time and the integral above would diverge. To keep the integral finite we are restricted here to time-limited signals.

For $\tau=0$, function a_f $(\tau=0)$ is simply the total energy in the signal,

$$\text{Energy}=a_f(0)=\int_{-\infty}^{\infty} dt\ x(t)x(t). \tag{14.2}$$

This is evident from the fact that $x^2(t)$ is the instantaneous power in the signal. (As usual we are assuming a load with unit resistive impedance.) Integrating the power over time leads to the energy.

For finite values of τ, $a_f(\tau)$ will generally be less than $a_f(0)$. Autocorrelation tells us how similar a signal is to a time-delayed version of itself, and this similarity is greatest when the delay is zero. Only if the function is strictly periodic can the autocorrelation function at some non-zero lag be as large as the autocorrelation function for zero lag. But strictly periodic functions are excluded by our restriction that the signals be of finite duration. It follows that the autocorrelation function at finite lag must be smaller than its value for zero lag. But if the function repeats itself often, $a(\tau)$ for $\tau>0$ may be only slightly smaller than $a(0)$. The autocorrelation function is, therefore, a useful indicator of approximate periodicities.

As an example, we consider rush-hour traffic. Suppose that we observe that westbound automobile traffic is unusually heavy at 5 pm on Wednesday. We also observe that it is heavy at 5 pm on Thursday. We conjecture from these two observations that westbound traffic density has a 24-hour period. Mathematically we describe this by letting x be the traffic density. Then if t=Wednesday at 5 pm, and τ=24 hours, the function $x(t)x(t+\tau)$ multiplies together the densities at 5 pm on Wednesday and Thursday. The autocorrelation function prescription tells us that in order to find out whether 24 hours really is a special time interval we ought to integrate the product $x(t)x(t+\tau)$ over a long span of time, perhaps letting t extend over a month, while keeping the lag τ equal to 24 hours. We will conclude that 24 hours is a special time interval for traffic if the autocorrelation function has a value for τ=24 hours that is larger than for other values of τ. In general, if $x(t)$ is nearly periodic, then $a_f(\tau)$ has a peak when τ is equal to the period of $x(t)$.

Wiener-Khintchine Relation

A famous theorem, known as the Wiener-Khintchine relation, says that the Fourier transform of the autocorrelation function is the energy spectral density. It is not hard to prove this fact. First we note that the autocorrelation function in Eq. (14.1) resembles a convolution integral. If the autocorrelation function for $x(t)$ were exactly a convolution integral then its Fourier transform would be exactly the square of $X(\omega)$, because of the convolution theorem. However, the autocorrelation function is different from a convolution integral because a sign in the time argument is reversed [compare Eq. (14.1) with Eq. (8.38a)]. The result is that the Fourier transform of the autocorrelation of x is actually $X(\omega)$ times its complex conjugate. The next paragraph proves this fact.

By the definition of the Fourier transform, the transform of $a_f(\tau)$ from Eq. (14.1) is

$$A_f(\omega)=\int_{-\infty}^{\infty} d\tau\, e^{-i\omega\tau}\int_{-\infty}^{\infty} dt\, x(t)x(t+\tau). \tag{14.3}$$

We change variables so that $t+\tau$ is t'. Then, because $i\omega\tau=i\omega t'-i\omega t$, we have

$$A_f(\omega)=\int_{-\infty}^{\infty} dt'\int_{-\infty}^{\infty} dt\, e^{i\omega t}e^{-i\omega t'}x(t)x(t'). \tag{14.4}$$

The substitution has succeeded in reducing the problem to two independent integrals, whence

$$A_f(\omega)=X(\omega)X(-\omega)=X(\omega)X^*(\omega)=|X(\omega)|^2. \tag{14.5}$$

The last step follows from the fact that the Fourier transform of a real function has the property that $X(-\omega)=X^*(\omega)$.

The spectrum $A_f(\omega)$ is a function that measures the amount of energy in the signal that happens to be between frequencies of ω and $\omega+d\omega$. It is therefore an energy density, i.e. energy per unit frequency.

Thus, the Fourier transform of the autocorrelation function, $a_f(\tau)$, is the energy density $|X(\omega)|^2$, and the inverse Fourier transform of the energy density is $a_f(\tau)$,

$$a_f(\omega)=\frac{1}{2\pi}\int_{-\infty}^{\infty} d\omega\ e^{i\omega\tau}|X(\omega)|^2. \tag{14.6}$$

INFINITE-DURATION SIGNAL AUTOCORRELATION FUNCTION

The autocorrelation function for signals of infinite duration is defined by analogy to the cross-correlation function for infinite-duration signals, by normalizing the overlap integral to the total integration time. Therefore,

$$a(\tau)=\lim_{T_D\to\infty}\frac{1}{2T_D}\int_{-T_D}^{T_D} dt\ x(t)x(t+\tau). \tag{14.7}$$

As an example, we calculate the autocorrelation function for a sine wave. If $x(t)=A\sin\omega_0 t$, then

$$a(\tau)=\lim_{T_D\to\infty}\frac{A^2}{2T_D}\int_{-T_D}^{T_D} dt\ \sin\omega_0 t\ \sin\omega_0(t+\tau). \tag{14.8}$$

We expand the function $\sin\omega_0(t+\tau)$ as $\sin\omega_0 t\cos\omega_0\tau+\cos\omega_0 t\sin\omega_0\tau$. Because the integral is over even limits in t, only the first of these two terms will contribute. The integral then becomes the average of the $\sin^2$ function over a time interval $2T_D$, and that average is the number 1/2. Therefore,

$$a(\tau)=\frac{A^2}{2}\cos\omega_0\tau. \tag{14.9}$$

Equation (14.9) for the autocorrelation function of a sine signal agrees completely with intuition. The maximum value of $a(\tau)$ occurs when $\tau=0$. This is always the case for an autocorrelation function because at $\tau=0$ the integral just evaluates the overlap of a signal with itself, which maximizes the integral. Next, the autocorrelation function is periodic. As shown in Fig. 14.1, it has the same periodicity as the sine wave itself (period T) because if one chooses $\tau=T$ then $x(t+\tau)$ is $x(t+T)$ which is the same as $x(t)$ by the definition of periodicity. Therefore, again the autocorrelation integral is just the overlap of a signal with itself.

Intuition also suggests that because the autocorrelation integral covers the whole time axis, it does not matter what the starting phase is. It is left as an exercise to show that the autocorrelation function for the function $x(t)=A\sin(\omega_0 t+\phi)$ is the same as in Eq. (14.9) above, independent of ϕ.

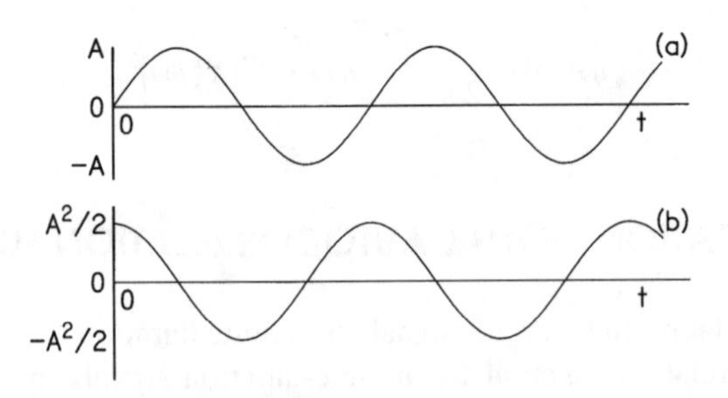

FIGURE 14.1. (a) A sine wave. (b) The autocorrelation function for the sine wave in (a).

Autocorrelation Function for any Periodic Function

The purpose of introducing the infinite-duration autocorrelation function is to enable us to deal with periodic signals. From one point of view, the exercise is merely a mathematical abstraction because all real signals have finite duration. However, it is very convenient to represent a signal as a Fourier series, with no restrictions on time, and therefore, it is definitely worthwhile to develop both an autocorrelation function and a power spectrum that allow us to cope with the mathematical fiction of a periodic signal.

Equation (8.62) gives a Fourier series form of a general periodic signal $x(t)$,

$$x(t)=\frac{1}{2\pi}\sum_{n=-\infty}^{\infty} X_n e^{i\omega_0 nt}. \tag{14.10}$$

If we plug this expression for $x(t)$ into the autocorrelation expression in Eq. (14.7), we get a double sum, one on dummy variable n, the other on dummy variable n'. Therefore,

$$a(\tau)=\lim_{T_D\to\infty}\frac{1}{4\pi^2}\sum_{n=-\infty}^{\infty}\sum_{n'=-\infty}^{\infty} X_n X_{n'} e^{i\omega_0 n\tau}\frac{1}{2T_D}\int_{-T_D}^{T_D} dt\ e^{i\omega_0(n+n')t}. \tag{14.11}$$

We next recall from Chapter 7 that the normalized time integral is the Kronecker delta function, i.e.

$$\lim_{T_D\to\infty}\frac{1}{2T_D}\int_{-T_D}^{T_D} dt\ e^{i\omega_0(n+n')t}=\delta_{n,-n'}. \tag{14.12}$$

The selection property of the Kronecker delta reduces the double sum to a single sum, and the autocorrelation function becomes

$$a(\tau)=\frac{1}{4\pi^2}\sum_{n=-\infty}^{\infty} X_n X_{-n} e^{i\omega_0 n\tau}. \tag{14.13}$$

Because X_{-n} is the complex conjugate of X_n we can write $a(\tau)$ as

$$a(\tau)=\frac{1}{4\pi^2}|X_0|^2+\frac{1}{2\pi^2}\sum_{n=1}^{\infty}|X_n|^2 \cos n\omega_0\tau. \tag{14.14}$$

Equations (8.62) and (5.21a) relate coefficients X_n to the Fourier series coefficients A_n and B_n and to the Fourier amplitude C_n. Substituting, we find

$$a(\tau)=A_0^2+\frac{1}{2}\sum_{n=1}^{\infty}(A_n^2+B_n^2)\cos n\omega_0\tau \tag{14.15a}$$

or

$$a(\tau)=A_0^2+\frac{1}{2}\sum_{n=1}^{\infty}C_n^2 \cos n\omega_0\tau. \tag{14.15b}$$

The latter form especially emphasizes the fact that the autocorrelation function depends only on the amplitudes of the harmonics of a complex periodic signal and not on the phases. The ultimate conclusion for a periodic signal is this: Regardless of the starting phases, the autocorrelation function is simply a sum of cosines, with amplitudes given by the squares of the amplitudes of the harmonics. Equation (14.15) also shows that one normally wants to remove the dc component of a signal (getting $A_0=0$) before calculating the autocorrelation function to avoid offsetting the autocorrelation function by a positive constant that has nothing to do with the flow of information.

EXAMPLE 1: *Case of the missing fundamental*

The operation of the autocorrelation function can be illustrated with a tone composed of the 4th, 5th, and 6th harmonics of a missing fundamental whose frequency is 200 Hz. The waveform is constructed by adding components, with all amplitudes equal to 1, in sine phase. The power spectrum is shown in Fig. 14.2(a). The waveform is periodic, with a period of 5 ms, as shown by the plot of the signal in Fig. 14.2(b). This figure shows two cycles exactly. Because of the sine phase, the waveform is zero at time zero and does not become quite as large as the maximum possible value of 3.0.

The autocorrelation function (for a signal of infinite duration) is shown in Fig. 14.2(c). This function was calculated by replacing the sum of sines used to calculate the signal in 1(b) by a sum of cosines and dividing by two. According to Eq. (14.15) this is all we have to do to get the autocorrelation function. As expected, the maximum value occurs at $\tau=0$, where the autocorrelation function becomes equal to the total power in the signal. This power is 3/2, obtained by getting 1/2 from each of the three components. Because the signal is periodic with period 5 ms, the autocorrelation function is too. Most important, it is the peak at $\tau=5$ ms that indicates the 5-ms periodicity of the signal.

A glance at Fig. 14.2(b) suggests that the signal has some structure on the scale of 1 ms. There is no actual periodicity at 1 ms, but there are features that seem to oscillate on approximately this time scale. That is not surprising. The average frequency of the signal is 1000 Hz, and one might expect some form of approximate periodicity corresponding to 1 ms. Therefore, it is reasonable to find that the autocorrelation function shows a secondary peak near a lag of 1 ms. (The peak does not occur exactly at 1 ms.) Because the waveform extends indefinitely in time, the same correlation peak occurs near a lag of -1 ms. In fact, because of the periodicity of the autocorrelation function, this secondary peak appears four times in Fig. 14.2(b), near 1, 4, 6, and 9 ms. [$4=5-1$; $6=5+1$; $9=10-1$.]

Figure 14.2(c) shows that there are some values of τ for which the autocorrelation function is negative. The most negative value occurs near $\tau=0.5$ ms. The explanation of this negative peak is easily found in Fig. 14.2(b) for the signal. If the signal is lagged by $\tau=0.5$ ms so that point A on function $x(t)$ coincides with point B on $x(t+\tau)$, then there is a rather strong tendency for function $x(t)$ and function $x(t+\tau)$ to have different signs. Their product, therefore, tends to be negative, and negative contributions like this lead to an autocorrelation integral that is negative.

In this example we made observations which compared a waveform with a lagged version of that waveform and showed how this comparison roughly agreed with the autocorrelation function. This example was a special case where all the components were in sine phase. It is interesting to realize that if we had been looking at a waveform made with different phases then the waveform would have had an entirely different shape, and yet it would somehow have been possible for us to make the same observations because the autocorrelation function is independent of the phases of the components.

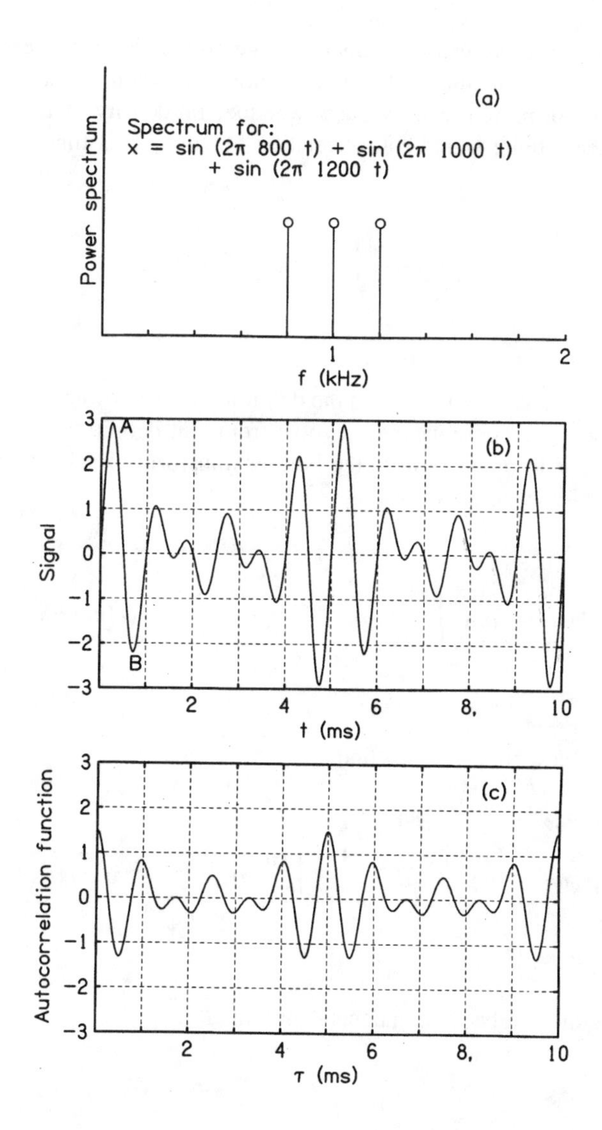

FIGURE 14.2. (a) Power spectrum for a signal with a missing fundamental. (b) Signal with a missing fundamental. (c) Autocorrelation for the signal of (a) and (b).

Wiener-Khintchine Relation for Infinite-Duration Signals

The Wiener-Khintchine relation for *finite-duration* signals says that the Fourier transform of the finite autocorrelation function is the energy spectral density. For an *infinite-duration* signal we cannot talk about energy or energy density because these are infinite. However, we can talk about average power, defined as energy per unit time. The Wiener-Khintchine relation for infinite-duration signals says that the power spectrum, or more precisely the power spectral density, $A(\omega)$, is the Fourier transform of the infinite signal autocorrelation function. This function is

$$A(\omega)=\int_{-\infty}^{\infty} d\tau\, e^{-i\omega\tau}a(\tau). \tag{14.16}$$

Our approach is to take Eq. (14.16) as the definition of the power spectrum and then to show that sensible consequences devolve from that definition.

The first step is to relate $A(\omega)$ to the Fourier transform of $x(t)$, namely to $X(\omega)$. Substituting in for $a(\tau)$, we have

$$A(\omega)=\lim_{T_D\to\infty}\int_{-\infty}^{\infty} d\tau\, e^{-i\omega\tau}\frac{1}{2T_D}\int_{-T_D}^{T_D} dt\, x(t)x(t+\tau). \tag{14.17}$$

Changing variables, $t+\tau=t'$, we find

$$A(\omega)=\lim_{T_D\to\infty}\int_{-\infty}^{\infty} dt'\,\frac{1}{2T_D}\int_{-T_D}^{T_D} dt\, e^{-i\omega t'}e^{i\omega t}x(t)x(t'). \tag{14.18}$$

The two integrals can be done independently, and

$$A(\omega)=\lim_{T_D\to\infty}\frac{1}{2T_D}|X(\omega)|^2. \tag{14.19}$$

Equation (14.19) is peculiar. Because the denominator becomes infinite, it appears that Eq. (14.19) should give zero for $A(\omega)$, unless $|X(\omega)|^2$ compensates in some way. As we shall see below, that is what happens for periodic functions, so that we end up with sensible results from the definition of $A(\omega)$. The example that follows shows that clearly.

EXAMPLE 2: ***Power spectrum of a cosine wave***

According to Eq. (8.11), the Fourier transform of a cosine wave with frequency ω_0 and unit amplitude is $X(\omega)=\pi[\delta(\omega-\omega_0)+\delta(\omega+\omega_0)]$. Substituting into Eq. (14.19) squares this function. There are cross terms, but they must be zero because the value of ω cannot be both $+\omega_0$ and $-\omega_0$. Therefore,

$$A(\omega)=\lim_{T_D\to\infty}\frac{1}{2T_D}\,\pi^2[\delta^2(\omega-\omega_0)+\delta^2(\omega+\omega_0)]. \tag{14.20}$$

The square of the Dirac delta function is one of the stranger animals in the zoo. However, the normalizing denominator $2T_D$ converts one of these delta function factors into a Kronecker delta, multiplied by $1/(2\pi)$, as shown in Eq. (7.16). Therefore,

$$A(\omega)=\frac{\pi}{2}\,[\delta(\omega-\omega_0)+\delta(\omega+\omega_0)]. \tag{14.21}$$

In the end, the power spectrum is a sum of delta functions, just like the Fourier transform. Both are made up of simple line components.

Now that we are convinced that the Fourier transform of the autocorrelation function for an infinite signal is the power spectral density, we can invert the process and write an expression for the autocorrelation function,

$$a(\tau)=\frac{1}{2\pi}\int_{-\infty}^{\infty}d\omega\, e^{i\omega\tau}A(\omega), \tag{14.22}$$

or, because $A(-\omega)=A(\omega)$,

$$a(\tau)=\frac{1}{\pi}\int_{0}^{\infty}d\omega\,\cos(\omega\tau)\,A(\omega). \tag{14.23}$$

Average Power and Parseval's Theorem

The average power in a signal can be found by averaging the instantaneous power over a long span of time,

$$\bar{P}=\lim_{T_D\to\infty}\frac{1}{2T_D}\int_{-T_D}^{T_D}dt\,x(t)x(t). \tag{14.24}$$

But from Eq. (14.7) we realize that this is just the autocorrelation function $a(\tau)$ evaluated at $\tau=0$. Therefore,

$$\bar{P}=a(0). \tag{14.25}$$

Further insight can be gained by writing the function $a(\tau=0)$ in terms of its Fourier transform, the power spectrum. Because $\tau=0$, the transform becomes

$$\bar{P}=a(0)=\frac{1}{2\pi}\int_{-\infty}^{\infty}d\omega\ A(\omega)=\frac{1}{\pi}\int_{0}^{\infty}d\omega\ A(\omega). \tag{14.26}$$

The combination of Eqs. (14.24) and (14.26) is known as Parseval's theorem. It is often written as

$$\lim_{T_D\to\infty}\frac{1}{2T_D}\int_{-T_D}^{T_D}dt\ x^2(t)=\lim_{T_D\to\infty}\frac{1}{2\pi T_D}\int_{-\infty}^{\infty}d\omega|X(\omega)|^2, \tag{14.27}$$

which has an agreeable symmetry. The theorem says that one can find the average power either by integrating the signal in time or by integrating the power spectrum, which is the power density function, over frequency. An example is the cosine function with unit amplitude. We know that its average power is 1/2. It is reassuring to find that if we integrate the power spectrum given in Eq. (14.21) then we also get a value of 1/2. [Each of the two delta functions in Eq. (14.21) contributes $\pi/2$, and dividing by 2π gives 1/2.]

AUTOCORRELATION FOR BANDS OF NOISE

Because the autocorrelation function is so closely connected to the power spectrum, a signal that has a power spectrum that is easy to describe ought to have an autocorrelation function that is also easy. A band of noise provides a good example. In the strict sense, we do not normally have enough information about a band of noise to calculate either the power spectrum or the autocorrelation function. However, we normally do know enough to describe a noise on the average. Below, the ensemble-average spectral density and the ensemble-average autocorrelation function will be called the spectral density and the autocorrelation function. We deal with two cases of interest, one with constant spectral density, the other with spectral density proportional to ω^{-1}, as shown in Fig. 14.3.

Flat-Spectrum Noise

We consider first a spectral density $A(\omega)$ which is zero for all frequencies except in a band from ω_1 to ω_2, and which is constant within that band. In other words, over the band where the noise exists the spectrum is white. We calculate the autocorrelation function from Eq. (14.23):

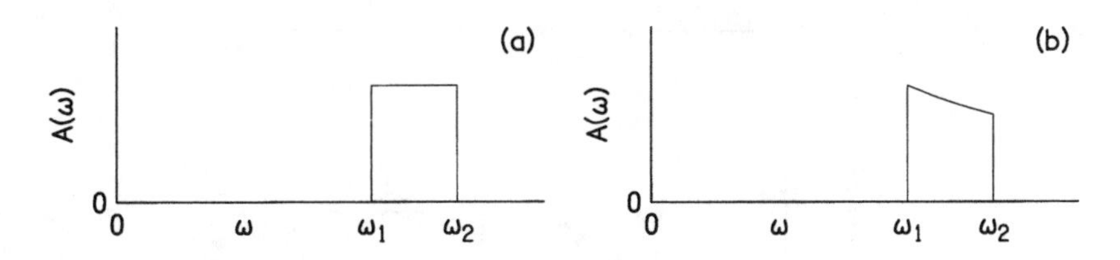

FIGURE 14.3. Average spectral density for two noise bands: (a) flat over the band, and (b) pink over the band.

$$a(\tau)=\frac{1}{\pi}\int_0^\infty d\omega\ A(\omega)\cos\omega\tau. \tag{14.28}$$

The constant value of $A(\omega)$ is related to the total power of the noise $\overline{P}$ by Eq. (14.26), so that $A(\omega)=\pi\overline{P}/(\omega_2-\omega_1)$. [Note: It is, therefore, related to the noise spectral density expressed in watts per Hertz via $A(\omega)=N_0/2$.] The autocorrelation function becomes a simple integral,

$$a(\tau)=\frac{\overline{P}}{\omega_2-\omega_1}\int_{\omega_1}^{\omega_2} d\omega\ \cos\omega\tau. \tag{14.29}$$

The definite integral of the cosine is the difference between two sines, $\sin(\omega_2\tau)$ and $\sin(\omega_1\tau)$. A more interesting expression comes out if we change the variables to be the average frequency $\overline{\omega}=(\omega_1+\omega_2)/2$ and the bandwidth $\Delta\omega=\omega_2-\omega_1$. Then we get

$$a(\tau)=\frac{2\overline{P}}{\Delta\omega\,\tau}\sin\left(\frac{\Delta\omega\,\tau}{2}\right)\cos\overline{\omega}\tau. \tag{14.30}$$

What is wonderful about this equation for the autocorrelation function is that it expresses the answer in terms of a dimensionless form of the autocorrelation lag. All we need to do is relate the bandwidth to the average frequency and we get a universal autocorrelation function in terms of the parameter $\overline{\omega}\tau$. If the band is one-third octave wide, the relationship becomes $\Delta\omega=0.23\overline{\omega}$ (see Exercise 14).

Then the autocorrelation function is

$$a(\tau)=\frac{2\overline{P}}{0.115\overline{\omega}\tau}\sin(0.115\overline{\omega}\tau)\cos\overline{\omega}\tau. \tag{14.31}$$

It is a function of a single variable $\overline{\omega}\tau$, as plotted in Fig. 14.4. Because it is a general function, it holds for any one-third octave band, regardless of center frequency. The horizontal axis in Fig. 14.4 is $\overline{\omega}\tau/(2\pi)$, or $\overline{f}\tau$, where $\overline{f}$ is the band center frequency in Hz. A convenient way to think about the horizontal axis is to regard it as the autocorrelation lag in units of the reciprocal of the center frequency. Therefore, if

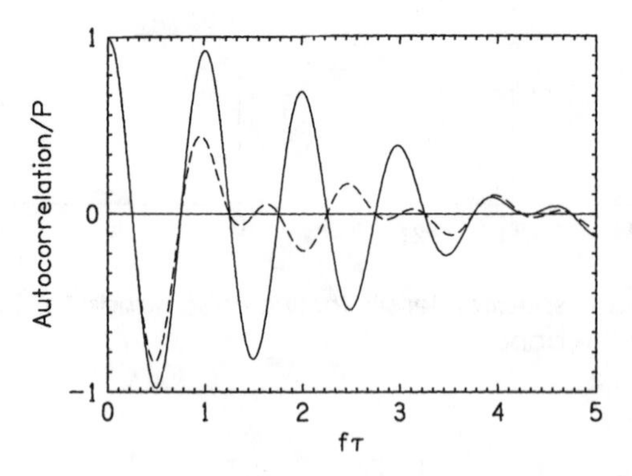

FIGURE 14.4. General autocorrelation function for two bands of noise. Solid line: one-third octave width. Dashed line: one octave width. The abscissa is the product of the band center frequency and the autocorrelation lag $\bar{f}\tau$ or $\bar{\omega}\tau/2\pi$.

the center frequency is 2000 Hz, the point $\bar{f}\tau=3$ corresponds to an autocorrelation lag of 3/2000 s or 1.5 ms.

If the band is one octave wide, the bandwidth is two-thirds of the center frequency, $\Delta\omega=2/3\bar{\omega}$, and the autocorrelation function is given by

$$a(\tau)=\frac{3\bar{P}}{\bar{\omega}\tau}\sin\left(\frac{1}{3}\bar{\omega}\tau\right)\cos\bar{\omega}\tau. \tag{14.32}$$

This function too is plotted in Fig. 14.4. The envelope of the autocorrelation function can also be studied for much narrower bands. Bordone-Sacerdote and Sacerdote (1971) used the decay of autocorrelation to measure the precision of unison choral singing. The precision of monks in Gregorian singing was evidenced by a very slow decay.

Wide-Band White Noise

In the limit that the bandwidth of white noise, Δf, becomes infinite, Eq. (14.30) shows that the autocorrelation function goes to zero. The only way to keep it from being zero is for τ to be zero. Therefore, the autocorrelation function becomes a delta function. Using the representation of the delta function in Eq. (7.30) leads to the result

$$a(\tau)=N_0\delta(\tau), \tag{14.33}$$

where N_0 is the spectral density in watts/Hz.

Pink-Spectrum Noise

The other spectral shape in Fig. 14.3 has a spectral density proportional to ω^{-1}. Over the band ω_1 to ω_2, where the noise density is not zero, the spectrum is pink. There it is given by

$$A(\omega)=\frac{\pi\overline{P}}{\ln(\omega_2/\omega_1)}\frac{1}{\omega}. \tag{14.34}$$

The natural log in the denominator makes sure that when one integrates $1/\omega$ over the range ω_1 to ω_2 one gets the correct total power, per Eq. (14.26).

From Eq. (14.28) the autocorrelation function then becomes

$$a(\tau)=\frac{\overline{P}}{\ln(\omega_2/\omega_1)}\int_{\omega_1}^{\omega_2}d\omega\,\frac{\cos\,\omega\tau}{\omega}, \tag{14.35a}$$

or, changing variables to dummy $\xi=\omega\tau$,

$$a(\tau)=\frac{\overline{P}}{\ln(\omega_2/\omega_1)}\int_{\omega_1\tau}^{\omega_2\tau}d\xi\,\frac{\cos\,\xi}{\xi}. \tag{14.35b}$$

The integral cannot be done in closed form, but it is easy to get a series solution for it using the power series for the cosine from Appendix C. We find

$$\int d\xi\,\frac{\cos\,\xi}{\xi}=\ln\,\xi-\frac{\xi^2}{2\cdot2!}+\frac{\xi^4}{4\cdot4!}-\frac{\xi^6}{6\cdot6!}+\frac{\xi^8}{8\cdot8!}-\cdots. \tag{14.36}$$

The power series is a convergent alternating series, which has the excellent property that the absolute error made by keeping only a finite number of terms is smaller than the first term that is neglected. Therefore, one only needs to calculate the series twice, once with N terms and once with $N+1$ terms to know when the calculation starts to fail. It is, nevertheless, awkward to calculate this series.

To calculate $a(\tau)$, the series needs to be evaluated at the upper and lower band edges. For a one-third octave band noise these edges occur $f_2=\overline{f}2^{1/6}$ and $f_1=\overline{f}2^{-1/6}$, where $\overline{f}$ is the geometric mean frequency of the band. Straightforward calculations show that for $\overline{f}\tau$ ranging from 0 to 1, the autocorrelation function differs by only a few percent from the autocorrelation function calculated for a one-third octave band noise with a constant spectral density in the band, which was shown by the solid line in Fig. 14.4.

SYMMETRY OF THE AUTOCORRELATION FUNCTION

The autocorrelation function is an even function of the lag, i.e. $a(\tau)=a(-\tau)$. This can be proved from the fact that the power spectrum and energy spectral density are even functions of frequency. Alternatively one can prove it from the original definition of the autocorrelation function, which is the approach we take here:

From Eq. (14.1) we have

$$a_f(\tau)=\int_{-\infty}^{\infty} dt\; x(t)x(t+\tau). \tag{14.37}$$

Therefore, for negative lag,

$$a_f(-\tau)=\int_{-\infty}^{\infty} dt\; x(t)x(t-\tau). \tag{14.38}$$

We change variables to let $t'=t-\tau$. Then

$$a_f(-\tau)=\int_{-\infty}^{\infty} dt' x(t'+\tau)x(t'). \tag{14.39}$$

But t' is a dummy variable of integration, just like t in Eq. (14.37). Therefore, Eq. (14.39) is the same as Eq. (14.37) and $a_f(-\tau)=a_f(\tau)$, which is what we set out to prove. Therefore, it is common to display the autocorrelation for only positive values of the lag. This fact is of some use in the next section.

CROSS-CORRELATION IN THREE EASY CASES

The cross-correlation between functions $x(t)$ and $y(t)$ is defined as function $\gamma(\tau)$. If function $y(t)$ is a simple transformation of function $x(t)$ then the cross-correlation function can be written as a simple variation on the autocorrelation function, $a(\tau)$, for $x(t)$ alone. Here are several examples:

(a) *Time delay:* If $y(t)$ is a time-delayed version of $x(t)$ with delay T_d [i.e. $y(t)=x(t-T_d)$] then

$$\gamma(\tau)=a(\tau-T_d). \tag{14.40}$$

This means that the cross-correlation function is a lag-shifted version of the autocorrelation function. The lag shift is the delay between x and y.

EXAMPLE 3: *Computed Cross-Correlation*

Figure 14.5 shows the normalized cross-correlation function measured between the two ears of a KEMAR (Knowles Electronics Manikin for Acoustics Research). The KEMAR was fitted with large red silicone ears and wore a cotton polo shirt with green and white stripes. The source of sound was a loudspeaker at an azimuth of 45 degrees with respect to the forward direction. The loudspeaker signal was a band of noise, about one octave wide, from 504 Hz to 1000 Hz.

The analog outputs of KEMAR's ears were digitized at a rate of 50,000 samples per second per ear. Half a second of sound was analyzed, which means that the total record was 250 times longer than the longest lag (2000 μs) to be studied. The long record was used to give a smooth-looking cross-correlation function despite the noisy input. The analysis required displacing the right-ear signal $y(t)$ with respect to the left-ear signal $x(t)$ in sample increments, $T_d = \Delta k\, \Delta t$, ($\Delta t = 20$ μs) and summing the cross products:

$$\gamma(\Delta k) = \frac{1}{\text{Norm}} \sum_{k=1}^{N-\Delta k} x(k) y(k+\Delta k). \tag{14.41}$$

The normalizing denominator is the geometric mean of the power in each ear:

$$\text{Norm} = \sqrt{\sum_{k=1}^{N-\Delta k} x^2(k) \sum_{k=1}^{N-\Delta k} y^2(k)}. \tag{14.42}$$

It is not hard to remember the formula for the normalizing denominator. It is just what is needed to make $\gamma(0)$ equal to unity if x is identical to y. It makes function $\gamma(\tau)$ independent of the amplitudes of signals x and y.

The dashed line in Fig. 14.5 is the theoretical normalized cross-correlation function for octave-band noise taken from the autocorrelation function in Eq. (14.32), or Fig. 14.4, and shifted to a lag of $T_d = -510$ μs, per Eq. (14.40) (see Exercise 15). Power $\overline{P}$ is set equal to 1 for normalization. Because of the even symmetry of the autocorrelation function, the cross-correlation function shown in Fig. 14.5 is symmetrical about $T_d = -510$ μs.

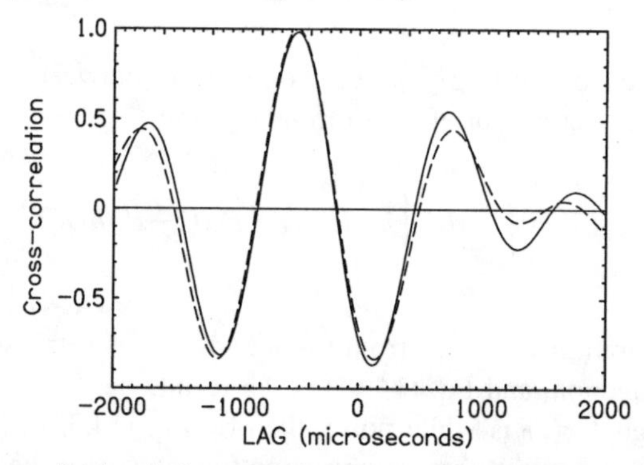

FIGURE 14.5. The solid line is the measured cross-correlation between left- and right-ear signals of a manikin exposed to an octave band of noise impinging from an azimuth of 45 degrees. The dashed line is the calculated cross-correlation function.

The calculated cross-correlation function is not in perfect agreement with the measured function because the head does not introduce a pure time delay. Instead, the head and other anatomical features of the torso perform some filtering. As a result, the spectrum in each ear is not flat over the octave band, and the spectra in different ears are different. Neither of these two effects is included in the constant delay model used in the calculations.

(b) *Common phase shift:* If $y(t)$ is a phase-shifted version of $x(t)$, where all harmonics of $x(t)$ are shifted by a common phase shift $\Delta\phi$, then the cross-correlation function is described by an equation like Eq. (14.15),

$$\gamma(\tau)=\frac{1}{2}\sum_{n=1}^{\infty} C_n^2 \cos(n\omega_0\tau+\Delta\phi). \tag{14.43}$$

Unlike the constant time delay, a constant phase shift is highly disruptive of the correlation between input and output. The phase shift effectively introduces a different time delay for each harmonic, a delay of $\Delta t_n=\Delta\phi/(n\omega_0)$. Such a system is said to be "dispersive" because the shape of the waveform is changed by the system. An exception is a common phase shift of π, where $y(t)$ is the same as $x(t)$ except for a sign change. Then the cross-correlation function is the negative of the autocorrelation function.

(c) *Filtering:* If $x(t)$ is the input to a filter and $y(t)$ is the output, then the cross-correlation between x and y is related to the autocorrelation for x alone by

$$\gamma(\tau)=\int_{-\infty}^{\infty} dt'\,a(t'+\tau)h(t'). \tag{14.44}$$

This can be proved by writing y as the convolution between x and h and then substituting in to the cross-correlation formula giving

$$\gamma(\tau)=\int_{-\infty}^{\infty} dt' \int_{-\infty}^{\infty} dt\; x(t+\tau)x(t-t')h(t'). \tag{14.45}$$

By changing integration variables, from t to t'', where $t''=t-t'$, we can extract the autocorrelation function, and Eq. (14.44) is the result.

If the input signal $x(t)$ is white noise, then by Eq. (14.33) $a(t'+\tau)$ is a delta function, $N_0\delta(t'+\tau)$, and the cross-correlation function gives the impulse response of the filter:

$$\gamma(\tau)=N_0 h(-\tau). \tag{14.46}$$

CROSS-SPECTRUM

The Wiener-Khintchine Theorem relates the autocorrelation function to the power spectrum. It is natural to look for an equivalent transformation for the cross-correlation function and to refer to the result as the cross-spectrum.

We begin with the definition of the cross-correlation function between finite-duration signals $x(t)$ and $y(t)$, namely

$$\gamma_f(\tau) = \int_{-\infty}^{\infty} dt\ x(t) y(t+\tau). \tag{14.47}$$

Then we write $x(t)$ and $y(t)$ in terms of their Fourier transforms, $X(\omega)$ and $Y(\omega)$. The integral on time t can then be done to yield

$$\gamma_f(\tau) = \frac{1}{2\pi} \int_{-\infty}^{\infty} d\omega\, X(\omega) Y(-\omega) e^{i\omega\tau}. \tag{14.48}$$

By analogy with the Wiener-Khintchine relation, the quanity $X(\omega)Y(-\omega)$ is the cross-spectrum. Unlike the energy spectrum, the cross-spectrum is complex. Because signal $x(t)$ is real, $X(-\omega) = X^*(\omega)$. Similarly $Y(-\omega) = Y^*(\omega)$. Therefore, defining the cross-spectrum as $\Gamma_f(\omega) = X(\omega) Y(-\omega)$, we have that $\Gamma_f(-\omega) = \Gamma_f^*(\omega)$. This symmetry simplifies practical calculations somewhat. Whereas the initial integral is complex,

$$\gamma_f(\tau) = \frac{1}{2\pi} \int_{-\infty}^{\infty} d\omega\, \Gamma_f(\omega) e^{i\omega\tau}, \tag{14.49}$$

the symmetry allows us to write it as the sum of two real integrals,

$$\gamma_f(\tau) = \frac{1}{\pi} \int_{0}^{\infty} d\omega [\mathrm{Re}\ \Gamma_f(\omega) \cos \omega\tau - \mathrm{Im}\ \Gamma_f(\omega) \sin \omega\tau]. \tag{14.50}$$

Expressing the cross-correlation function in terms of the cross-spectrum leads to some insight concerning the effects of filtering on cross-correlation, as shown in the example below:

EXAMPLE 4: *Binaural model*

Cross-correlation models are frequently used to represent the process of binaural interaction, particularly the processing of interaural time differences. They are continuum representations of the neural cross-correlator postulated by Jeffress (1948).

Suppose that signal $x(t)$ is the input to the left ear, and signal $y(t)$ is the input to the right. Both x and y are headphone signals. A great amount of physiological and psychoacoustical evidence indicates that the binaural interaction of these signals takes place in frequency channels, established by peripheral filtering. Models of the binaural system commonly assume that left ear and right ear peripheral filters are the same, characterized by transfer function $H_c(\omega)$. This is true of the models by Colburn, Stern, and their colleagues (e.g. Stern and Colburn, 1978) and also of the models from Blauert's school (e.g. Lindemann, 1986).

Defining $x_c(t)$ and $y_c(t)$ as the left and right filtered signals in channel c, we express the Fourier transforms by

$$X_c(\omega)=H_c(\omega)X(\omega), \tag{14.51}$$

and

$$Y_c(\omega)=H_c(\omega)Y(\omega), \tag{14.52}$$

whence,

$$Y_c^*(\omega)=H_c^*(\omega)Y^*(\omega). \tag{14.53}$$

Therefore, if $\Gamma_c(\omega)$ is the cross-spectral function for the filtered signals in channel c, then $\Gamma_c(\omega)$ is related to the unfiltered cross-spectral function by

$$\Gamma_c(\omega)=|H_c(\omega)|^2\Gamma_f(\omega). \tag{14.54}$$

It follows that the cross-correlation function within channel c is given by

$$\gamma_c(\tau)=\frac{1}{2\pi}\int_{-\infty}^{\infty} d\omega |H(\omega)|^2\Gamma_f(\omega)e^{i\omega\tau}. \tag{14.55}$$

The above results show that the cross-correlation function within a frequency channel is related to the cross-correlation function of the signal by the power response of the peripheral filter, namely by $|H(\omega)|^2$. But because the power response of a filter does not involve the phase shift, we find that the cross-correlation function within a channel is independent of the phase shift of the peripheral filter that establishes the channel. The phase-shift independence holds only if left and right auditory filters are the same.

THE REVCOR TECHNIQUE

The reversed-correlation (Revcor) technique is an experimental method used by neurophysiologists to estimate the effective filter characterizing the linear response of an auditory neuron. The idea begins with the fact, from Eq. (14.46), that when the input to a filter is white noise, the cross-correlation between input and output is the impulse response of the filter. Unfortunately, the output that is available physiologically represents both filtering and the highly nonlinear process of spike generation. The output is a series of spikes,

$$s(t)=\sum_{i=1}^{N} \delta(t-t_i), \tag{14.56}$$

where times t_i are the spike times. The cross-correlation between the input x and the output spike train is the revcor function,

$$\gamma_R(\tau)=\int_{-\infty}^{\infty} dt\; x(t)s(t+\tau) \tag{14.57}$$

or

$$\gamma_R(\tau)=\sum_{i=1}^{N} x(t_i-\tau). \tag{14.58}$$

Thus, the revcor function is just the sum of the input values at the instants of neural spikes, offset by the lag. If the spike times occur with probability density $p(t_i)$, then the ensemble-averaged revcor function is given by

$$\langle\gamma_R(\tau)\rangle=\int_{-\infty}^{\infty} dt\; x(t-\tau)p(t). \tag{14.59}$$

DeBoer and deJongh (1978) explained the revcor by supposing that neural activation takes place in three successive stages. The first is linear filtering. The second is a nonlinear transformation of the filter output. The third is a spike generation process in which the probability of a spike is a linear function of the output of the second stage. From Eq. (14.59) it is evident that if the nonlinear second stage were actually linear then the probability density would be proportional to the linearly filtered signal, and $\langle\gamma_R\rangle$ would give the impulse response of the filter (Eq. 14.46). A theorem by Price (1958) says that if the white noise input is Gaussian then the ensemble-averaged revcor remains an estimate of the impulse response despite the nonlinearity of the second stage. The Fourier transform of the average revcor function is expected to be the transfer function of the auditory filter.

DeBoer and deJongh, and Carney and Yin (1988) made extensive comparisons between predictions based on revcor filters and PST histograms within a given neuron. Such comparisons are aided by modeling the revcor impulse response as a gammatone filter (Johannesma, 1972). Revcor filters show effects expected of auditory filters, broadening with increasing signal intensity and increasing damping for increasing characteristic frequency.

An important feature of the revcor model is that the probability of a spike should vary directly with the transformed signal. This implies that neural synchrony is preserved. At high frequency, where spike times become independent of signal phase, the revcor function tends to zero. Other effects on spike probability, such as refractoriness, affect the revcor, even though one would not naturally model them as part of the auditory filter.

The comparison with PST histograms shows a systematic departure. Revcor functions generally initially build with time, like gammatone functions, whereas PST histograms usually decay monotonically. In part this discrepancy can be explained by refractoriness and adaptation. Figure 14.6 shows a hypothetical single-neuron filter, its impulse response, and firing probability for three levels of the signal. Nonlinear neural mechanisms lead to an emphasis of the first peak over the following peaks. Because the frequency analysis by the cochlea is essentially a nonlinear process, a model consisting of a linear filter followed by a nonlinearity inevitably runs up against limitations.

AUTOCORRELATION AND PITCH PERCEPTION

In 1951, J. C. R. Licklider proposed that the perception of pitch is mediated by a neural autocorrelation process. In one form or another, the concept has been a viable option ever since. An autocorrelation model predicts a pitch on the basis of a peak in an autocorrelation function. If a peak occurs at lag τ, then that is expected to cue a pitch corresponding to frequency $1/\tau$.

In several important ways, autocorrelation fits well with what we know about pitch. As shown in the missing fundamental example on p. 337, a signal composed of harmonics of f_0 leads to an autocorrelation function with its principal peak at $\tau = 1/f_0$, whether or not a fundamental is present. Thus, it can be said that an autocorrelation model predicts the pitch of the missing fundamental. Further, the pitch of a periodic complex tone is known to be independent of the phases of the resolved harmonics, and this agrees with the phase independence of the autocorrelation function.

On the other hand, as a model for pitch, autocorrelation has some peculiar properties. Because the power spectrum for a pure tone is a delta function, the autocorrelation function is a cosine with the period of the tone (see Fig. 14.1). A peak of the cosine function is very broad and it seems strange for a precise percept like the pitch of a pure tone to be encoded by such a broad peak. Stranger still is the case of rippled noise. Here the ensemble-averaged power spectrum is a cosine function. Therefore, the autocorrelation function is a delta function. On this basis we should expect rippled noise to lead to a pitch of exceptional precision, but this is not what is observed experimentally.

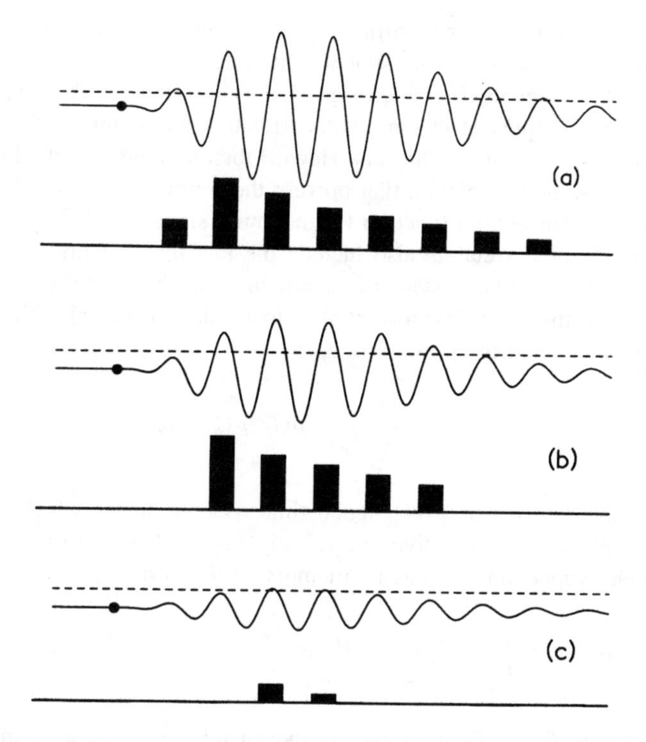

FIGURE 14.6. Cartoon showing the response of a hypothetical neuron made from a filter followed by a threshold device, noise, and other nonlinearities. The stimulus is a click at three different levels (a) high, (b) medium, (c) low. The filter has an impulse response that is a gammatone function with η=4, CF=1.17 kHz, and an equivalent rectangular bandwidth of 151 Hz, as given by the Cambridge scale (see Chapter 10). The click occurs at time zero at the far left, but the impulse response does not begin until the click has traveled to the 1.17 kHz place, a time marked by a dot. A neuron can fire when the filter output exceeds the threshold, shown dashed. Firing probabilities are shown in the idealized PST histograms.

More realistic autocorrelation models of pitch perception do not let the autocorrelator work on the signal itself. Instead, autocorrelation is preceded by other processes. All modern models begin with auditory filtering, and therefore, autocorrelation is confined to frequency channels. This has the effect of broadening the peak in the case of rippled noise. It does nothing to help the pure tone.

Auditory transduction converts the signal into a train of neural spikes. A real neural autocorrelator would work on this representation of the signal. Auditory models work on a representation that gives the probability of a neural spike $p(t)$. The probability is non-negative. Evans (1968) used an exponential rectifier to obtain a probability from a filtered signal, $p(t)=\exp[cx(t)]$, where c is a constant. Most of the recent models use half-wave rectification (Colburn, 1977; Stern and Shear, 1992). The probability may be weighted by a synchrony factor (Srulovicz and

Goldstein, 1983). It may include refractory effects (Jones, Tubis, and Burns, 1985) as well as other hair-cell effects: spontaneous rate, compression, saturation, and adaptation for brief signals (Meddis, 1988; Meddis and Hewitt, 1991a,b). It is worth noting that if the outputs of channel autocorrelators are simply added to make a summary autocorrelogram (Meddis and Hewitt) then it is only the nonlinear operations that precede autocorrelation that prevent the summary autocorrelogram from being just an autocorrelation function for the entire signal itself.

Model neural autocorrelators also include the fact that a neural system cannot integrate over infinite time. Instead of calculating a mathematically exact autocorrelation function, they calculate a running autocorrelation, $a(t,\tau)$ which is a function of both the lag and the present time, t:

$$a(t,\tau)=\int_{t-T_W}^{t} dt' p(t')p(t'-\tau), \tag{14.60}$$

where T_W is the duration of a rectangular time window during which autocorrelation can take place. An alternative to a rectangular window is a running autocorrelator with an exponentially decaying memory of the past,

$$a(t,\tau)=\int_{-\infty}^{t} dt' p(t')p(t'-\tau)e^{-(t-t')/T_C}e^{-(t-t'+\tau)/T_C}. \tag{14.61}$$

Correlation times T_W or T_C are typically two or three milliseconds, in agreement with minimum integration times found by Viemeister (1979). A memory function such as this damps the autocorrelation function for increasing τ, removing periodicity and emphasizing peaks at small τ.

Pattern-Transformation Model

An early pitch model resembling autocorrelation was the pattern-transformation model (PTM) by Wightman (1973). Unlike autocorrelation models, which Fourier transform the power spectrum, the PTM transforms the excitation pattern. For a pure tone with frequency ω, the excitation at place z is given by the auditory filter $|H_z(\omega)|$. The pattern transformation is

$$a(\tau)=\frac{1}{2\pi}\int_{-\infty}^{\infty} dz \cos(z\tau)|H_z(\omega)|. \tag{14.62}$$

But because z is not a frequency, parameter τ is not a time. Unlike autocorrelation models, the PTM does not depend on temporal synchrony. A resemblance to autocorrelation can be introduced by making the homogeneity assumption. With homogeneity, parameter z is a frequency-like variable and

$$|H_z(\omega)|=H_0(z-\omega), \tag{14.63}$$

where function H_0 is real. Homogeneity is frequently assumed in the construction of excitation patterns. It says that the frequency dependence of the transfer function for an auditory filter at one place determines the place dependence of excitation for one frequency. With this assumption, $a(\tau)$ becomes the sum of $\cos(\omega\tau)$ and $\sin(\omega\tau)$ terms, where each term has a damping factor, essentially determined by the Q of the filter-like function H_0. If H_0 is an even function the sine factor is zero.

In practice, the pattern-transformation model established trends that are characteristic of autocorrelator models, successful predictions for broad-band stimuli like rippled noise and complex tone pitch. The pitch of the missing fundamental and the ''pitch-shift'' effect (incrementing each harmonic frequency by a constant) are also predicted by the model.

Interspike-Interval Function

The interspike-interval (ISI) function, often plotted as a histogram in neural-spike timing experiments, is similar to the autocorrelation function for firing probability. The autocorrelation $a(\tau)$ can be described as the probability that if there is a spike at time t then there will be a spike at time $t+\tau$. The ISI function can be described as the probability that if there is a spike at time t then there will be a spike at time $t+\tau$ with no other spike between these two times. For small values of the lag, τ, the two functions are similar. For long τ the ISI fades away as long intervals between successive spikes become improbable, but the autocorrelation function continues unattenuated. The ISI function is difficult to calculate. It is normally approximated by an autocorrelation function with a finite memory or an exponentially decaying factor on the lag. Goldstein and Srulovicz (1977) used such a model to calculate frequency discrimination for short tones. Although the two functions are similar, an explanation for the octave enlargement effect based on neural timing (Ohgushi, 1983) is consistent with an autocorrelator model for pitch and not with an ISI model (Hartmann, 1993b). A physiological correlate of the ''pitch-shift'' effect can be seen in the ISI histogram measured in the cochlear nucleus (Rhode, 1995).

EXERCISES

FINITE-DURATION SIGNALS

Exercise 1: Reality

After Eq. (14.2) in the text, it was said that the instantaneous power is $x^2(t)$. Why did it not say $|x(t)|^2$ instead?

Exercise 2: Energy from two kinds of integral

The integral of $x^2(t)$ over all time is the energy in the signal. The integral of $|X(\omega)|^2$ over all ω is also proportional to the energy in the signal. Prove that this correspondence is dimensionally correct by starting with what you know about the dimensions of $X(\omega)$, defined by the Fourier transform.

Exercise 3: Exponentially decaying signal

Suppose that the signal is

$$x(t) = \exp(-t/\tau_0) \quad \text{for } t \geq 0$$

$$x(t) = 0 \quad \text{for } t < 0.$$

(a) Take the Fourier transform of $x(t)$, and use it to show that the energy spectrum is

$$A_f(\omega) = |X(\omega)|^2 = \frac{\tau_0^2}{1 + \omega^2 \tau_0^2}.$$

(b) Find the autocorrelation function $a_f(\tau)$, starting with $x(t)$. Be careful with the limits of the integrals! Show that the autocorrelation function is

$$a_f(\tau) = (\tau_0/2)\exp(-\tau/\tau_0) \quad \text{for } \tau \geq 0$$

$$a_f(\tau) = (\tau_0/2)\exp(+\tau/\tau_0) \quad \text{for } \tau < 0.$$

This autocorrelation function, together with the original signal, is shown in Fig. E.3

(c) Find $A_f(\omega)$ by starting with $a_f(\tau)$ and taking the Fourier transform. The result should agree with part (a).

(d) Show that the total energy is $\tau_0/2$ by integrating $x^2(t)$ over all time, and observe that this agrees with the limit $a_f(0)$.

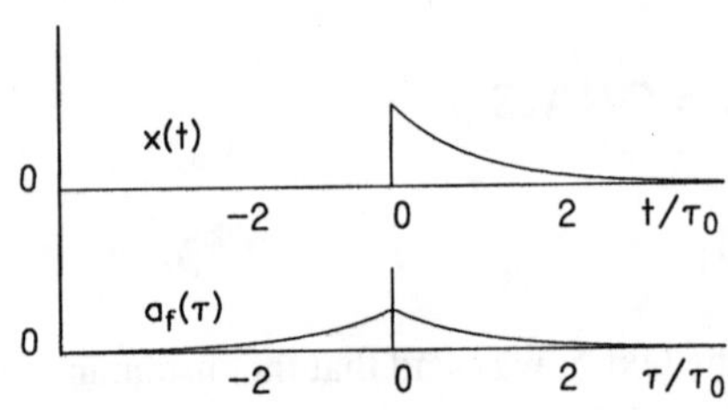

FIGURE E.3. The exponentially decaying signal and its autocorrelation.

Exercise 4: The rectangle function

Suppose that $x(t)$ is the rectangle function which is equal to 1 on the interval $0 \leqslant t \leqslant T$ and which is 0 outside that interval.

(a) Find the autocorrelation function.

(b) Find the total energy by integrating $x^2(t)$ over time.

(c) Find the energy spectrum by first finding the Fourier transform of $x(t)$ and squaring its magnitude.

(d) Find the energy spectrum by using the autocorrelation function from part (a) and taking the Fourier transform of that. The answer should agree with part (c).

(e) Find the total energy by integrating the energy spectrum from part (c) or (d) using the formula

$$\text{Energy} = \frac{1}{2\pi} \int_{-\infty}^{\infty} d\omega \, A_f(\omega).$$

Compare with the answer from part (b).

Exercise 5: The delta function

The finite-signal autocorrelation function of a delta function is a delta function, i.e. if $x(t) = \delta(t)$ then $a_f(\tau) = \delta(\tau)$.

(a) Prove the above statement by using the definition of $a_f(\tau)$ as a product of $x(t)$ and a lagged version of itself.

(b) Prove the statement by first finding the Fourier transform $X(\omega)$ and then using the Wiener-Khintchine theorem.

Exercise 6: Symmetry of the energy spectrum

Prove that the energy spectrum, $A_f(\omega)$, is even, i.e. $A_f(\omega) = A_f(-\omega)$.

INFINITE-DURATION SIGNALS

Exercise 7: An understandable mistake, but a bad one

Simplicio is of the opinion that the way to calculate the power spectrum is first to find the power, namely $x^2(t)$, and then to find its Fourier transform. That is, he thinks that

$$A(\omega)=\mathcal{F}[x^2(t)].$$

Show that he is wrong. [Hint: Using a sine wave for signal x makes a good test.]

Exercise 8: The sine

(a) Show that the autocorrelation function for a sine is independent of the starting phase, i.e. if $x(t)=A\sin(\omega_0 t+\phi)$ then $a(\tau)$ does not depend on ϕ.

(b) Show that Eq. (14.14) for the autocorrelation function (valid for any periodic signal) goes to the function in Eq. (14.9) when the signal is a sine, i.e. when $x(t)=A\sin\omega_0 t$.

(c) Use the Wiener-Khintchine relation to calculate the power spectrum of a sine and show that the result agrees with the power spectrum for a cosine in Eq. (14.21).

Exercise 9: The constant function

(a) Show that the autocorrelation function of a constant is a constant, i.e. if $x(t)=c$ for all t then $a(\tau)=c^2$.

(b) Power spectrum from the autocorrelation function: Use the Wiener-Khintchine relation to show that if $x(t)=c$ for all t, then the power spectrum $A(\omega)=2\pi c^2\delta(\omega)$.

(c) Power spectrum from the Fourier transform of the signal: Use Eq. (14.19) to show that if $x(t)=c$ for all t, then the power spectrum $A(\omega)=2\pi c^2\delta(\omega)$.

Exercise 10: The square wave

Consider the square wave, with values $+1$ and -1, in Fig. E.10.

(a) From the definition of the autocorrelation function in Eq. (14.7) explain why the autocorrelation function for the square wave is a triangle wave.

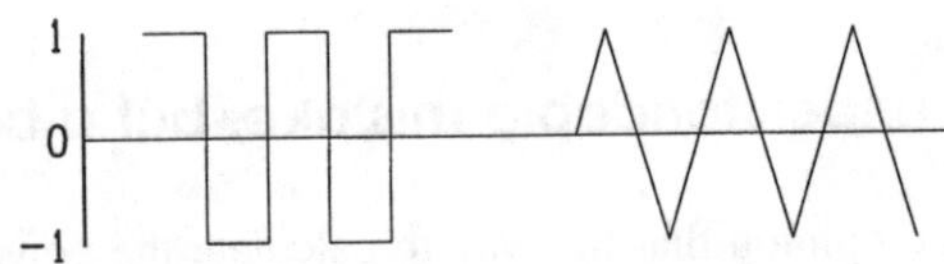

FIGURE E.10. A square wave for Exercise 10 and a triangle wave for Exercise 11.

(b) The Fourier series representation of the square wave is given in Eq. (5.25). Substitute the Fourier amplitudes for the square wave into Eq. (14.15) to find the autocorrelation function for the square wave. Compare with Eq. (5.28) which is the Fourier series for the triangle wave.

(c) The average power in the square wave is 1. (In fact, even the instantaneous power in the square wave is always 1.) Show that the average power, given by $a(0)$ from part (b) is 1, as expected. [An infinite sum from Appendix C will be helpful here.]

Exercise 11: The triangle wave

Consider the triangle wave, with maximum values $+1$ and -1, shown in Fig. E.10.

(a) The Fourier series representation of the triangle wave is given in Eq. (5.28). Substitute the Fourier amplitudes for the triangle wave into Eq. (14.15) to find the autocorrelation function for the triangle wave.

(b) Use Eq. (14.26), which averages the power in a signal, to show that the average power in the triangle wave is 1/3. Then show that the average power, given by $a(0)$ from part (a) is also 1/3, as expected. [An infinite sum from Appendix C will be helpful here.]

Exercise 12: How much of the autocorrelation function is useful?

Consider a periodic function with period T. We know that the autocorrelation function is periodic, $a(\tau)=a(\tau+T)$. Show that if we know the autocorrelation function from $\tau=0$ to $\tau=T/2$, then we know the autocorrelation function for all lag values τ, positive or negative.

Exercise 13: Long-period functions

Consider a periodic white noise created by adding together 200,000 cosine waves, equally spaced in frequency by 0.1 Hz. The cosine components have random phase and equal amplitude. This noise is white from zero to 20,000 Hz, and it repeats itself every 10 seconds. Consider also a train of clicks, created by adding together the same 200,000 cosine waves, but with zero phase. The click train is a signal which is complete silence except that every ten seconds there is a sharp click. Its temporal structure could hardly be more different from the continuous noise.

(a) Use the Wiener-Khintchine theorem to show that the noise and the click train have the same autocorrelation function.

(b) Use the integral definition of the autocorrelation function and a Fourier series representation of the noise and the click train to show that the noise and the click train have the same autocorrelation function.

Exercise 14: Noise bandwidth

The goal of this exercise is to find the relationship between the bandwidth and the average frequency for a one-third-octave band of noise, to be used in Eq. (14.31).

(a) Show that if ω_2 is one-third octave higher than ω_1 then $\omega_2 = 1.26\omega_1$. Therefore the average frequency is $1.13\omega_1$.

(b) Show that the bandwidth is 0.23 times the average frequency.

Exercise 15: Interaural delay

To fit the cross-correlation function in Fig. 14.5, an interaural delay of 510 μs was assumed. This value can be compared with the prediction of a diffraction formula in the low-frequency limit, Delay$=(3r/c)\sin\theta$, where r is the radius of a presumably spherical head, c is the speed of sound, and θ is the azimuth of the source with respect to the forward direction (Durlach and Colburn, 1978). Using the standard value of 8.75 cm for the head radius, and 34400 cm/s for the speed of sound, calculate the delay for a source at 45 degrees. Show that the 510 μs value agrees to better than 10%.

CHAPTER 15

Delay-and-Add Filtering

Directly or indirectly, all questions concerning the sensation of sound must come for decision to the ear, as the organ of hearing; and from it there can be no appeal.

LORD RAYLEIGH (J. W. STRUTT)

The delay-and-add process has appeared on several previous occasions. It was introduced in an exercise in Chapter 8, and extended to form the basis of the transversal filter in Chapter 9. The concept of delay-and-add is so important that it appears frequently in signal processing. In this section we deal with delay-and-add in detail with particular reference to the stimulus known in psychoacoustics as "rippled noise."

IMPULSE RESPONSE AND TRANSFER FUNCTION FOR DELAY AND ADD

The basic delay-and-add block diagram is shown in Fig. 15.1. The output $y(t)$ is equal to the sum of the input $x(t)$ plus a scaled and time-delayed version of the input, where the scale factor is the constant gain g and the delay time is the constant T. It is usual for the gain to be $-1 \leq g \leq 1$. From Fig. 15.1, the output is given by

$$y(t)=x(t)+gx(t-T). \tag{15.1}$$

The Spectrum

It is the spectral features of $y(t)$ that are of particular interest. Therefore, we need to find the Fourier transform $Y(\omega)$. We will do it in two different ways:

First, following the elementary techniques of Chapter 8, we realize that because the Fourier transform is linear we can find the Fourier transform of $y(t)$ by taking the transforms of $x(t)$ and $gx(t-T)$ and adding them. The transform of $x(t)$ is $X(\omega)$, and the transform of $x(t-T)$ is $e^{-i\omega T}X(\omega)$. Therefore, the sum is

$$Y(\omega)=[1+ge^{-i\omega T}]X(\omega). \tag{15.2}$$

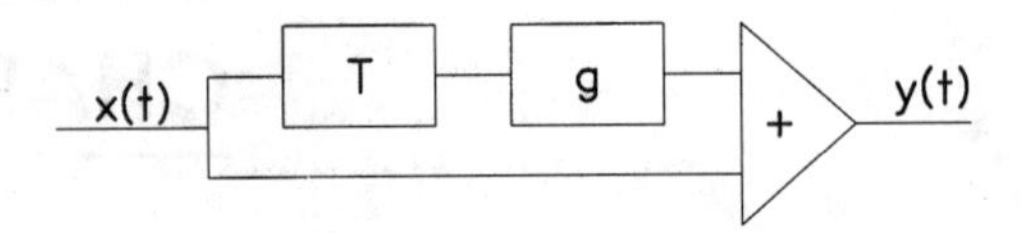

FIGURE 15.1. The delay-and-add process includes delay (T), gain scaling (g), and addition (+) operations.

This is the answer for the Fourier transform.

The second technique represents the delay-and-add process as a filter. It is evident that Eq. (15.1) can be written as a convolution integral using the delta function,

$$y(t)=\int_{-\infty}^{\infty} dt'[\delta(t-t')+g\,\delta(t-t'-T)]x(t'). \tag{15.3}$$

The reason for doing this is that it puts $y(t)$ into the form of a filtered signal,

$$y(t)=\int_{-\infty}^{\infty} dt'\,h(t-t')x(t'), \tag{15.4}$$

where the impulse response is given by

$$h(t)=\delta(t)+g\,\delta(t-T). \tag{15.5}$$

The corresponding filter transfer function is given by the Fourier transform

$$H(\omega)=\int_{-\infty}^{\infty} dt\; e^{-i\omega t}h(t)=1+ge^{-i\omega T}. \tag{15.6}$$

Because input X and output Y are related by the transfer function, we have

$$Y(\omega)=H(\omega)X(\omega), \tag{15.7}$$

or

$$Y(\omega)=[1+ge^{-i\omega T}]X(\omega). \tag{15.8}$$

This is the same answer as Eq. (15.2) obtained by adding Fourier transforms.

Representing the delay-and-add as a filtering process emphasizes that the output spectrum is a simple transformation of the input spectrum. The process is sometimes called "comb filtering," but in this book we reserve that term for delay-and-add filtering with feedback, described below.

The delay-and-add filter can be described in terms of its amplitude and phase responses,

$$H(\omega)=|H(\omega)|e^{i\phi(\omega)}. \tag{15.9}$$

The amplitude response is the square root of $|H(\omega)|^2$, namely

$$|H(\omega)|=\sqrt{(1+ge^{-i\omega T})(1+ge^{+i\omega T})}, \tag{15.10}$$

so that

$$|H(\omega)|=\sqrt{1+g^2+2g\cos\omega T}. \tag{15.11}$$

The phase shift is found by calculating the ratio of the imaginary part of $H(\omega)$ to the real part,

$$\tan\phi=\frac{\mathrm{Im}(1+ge^{-i\omega T})}{\mathrm{Re}(1+ge^{-i\omega T})} \tag{15.12}$$

i.e.

$$\tan\phi=\frac{-g\sin\omega T}{1+g\cos\omega T}. \tag{15.13}$$

Therefore, the phase shift caused by delay-and-add filtering is

$$\phi(\omega)=\arctan\left[\frac{-g\sin\omega T}{1+g\cos\omega T}\right]. \tag{15.14}$$

So long as g is in the usual range, $-1\leqslant g\leqslant 1$, the phase shift is between 90 and -90 degrees.

As a function of frequency, the transfer function is $H(f)=|H(f)|e^{i\phi(f)}$, where

$$|H(f)|=\sqrt{1+g^2+2g\cos 2\pi fT}, \tag{15.15}$$

and

$$\phi(f)=\arctan\frac{-g\sin 2\pi fT}{1+g\cos 2\pi fT}. \tag{15.16}$$

Figure 15.2 shows the transfer functions (amplitude and phase) for several different values of the delay gain g. An interesting case corresponds to $g=1$, where spectral zeros occur at frequencies that are odd multiples of $1/(2T)$, and spectral peaks occur at integer multiples of $1/T$. Thus, for a delay $T=2$ ms, spectral peaks occur at 500, 1000, 1500,...Hz. Values of gain g with magnitude less than one, or greater than one, lead to less dramatic filtering with no zeros, only minima, as shown in Fig. 15.2(c) for $g=1/2$. It is left as an exercise to show that the shape of the amplitude response is unchanged if gain g is replaced by its reciprocal g^{-1}. This

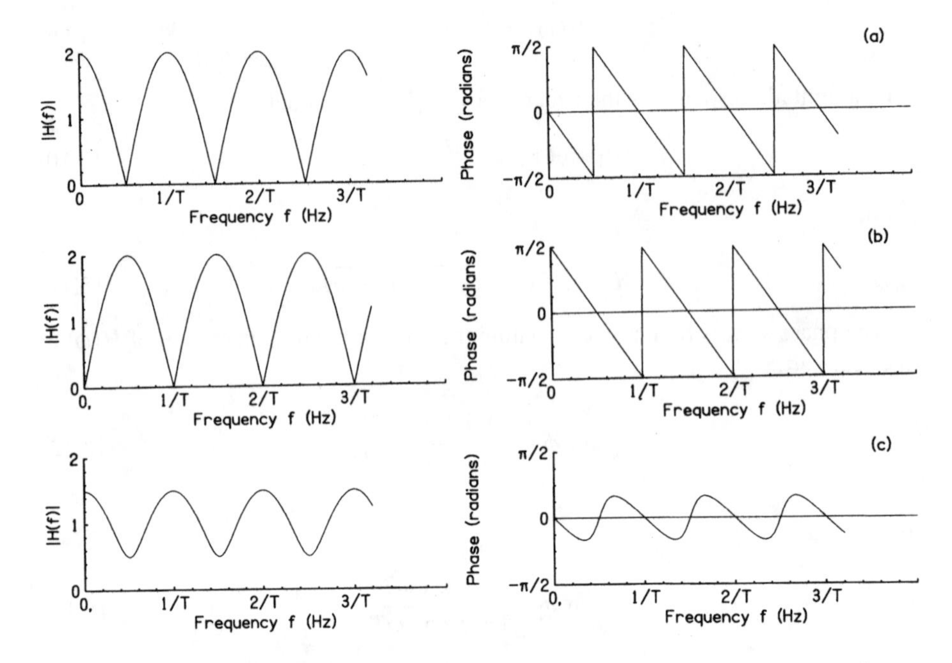

FIGURE 15.2. Amplitude response (left) and phase shifts (right) are shown as functions of frequency for different gains of the delayed signal (a) g=1, (b) g=−1, (c) g=1/2.

change leads to the same combination of relative amplitudes. For example, the amplitude response for $g=2$ looks just like the response for $g=1/2$, shown in Fig. 15.2(c).

The special case of spectral zeros points up the value of a third approach to the delay-and-add process, namely wave interference. The output of the process is zero only if there is precise cancellation between a sine component of the input and the same component after a delay. In order for complete cancellation to occur, the amplitudes of the input sine and the delayed sine must be the same ($g=1$), and the delay must introduce a 180-degree phase shift, $2\pi fT=\pi$, or $2\pi fT=\pi+2\pi$, or $2\pi fT=\pi+4\pi\ldots$, in agreement with Fig. 15.2(a).

REPETITION PITCH

If a broad-band signal is passed through a delay-and-add filter like the system described above, the filtered output creates a sensation of pitch known as repetition pitch (Bilsen and Ritsma, 1969/70). The details of the input signal are not particularly critical; the pitch effect works best if the input does not have a good pitch of its own. Both single clicks and white noise are used in Demonstration 26 on the ASA compact disc by Houtsma *et al.* (1987). The pitch can be heard for delay times T between 1 and 10 ms.

Two repetition pitch configurations have been studied in some detail. If the delayed signal is added without inversion to the undelayed (direct) signal the configuration is known as *cos+*. The spectrum resembles the amplitude response shown in Fig. 15.2(a), so that peaks of the spectrum occur at integer multiples of a fundamental frequency $1/T$, the reciprocal of the delay. The pitch of the signal, measured in pitch matching experiments, is also $1/T$.

If the delayed signal is added with inversion to the undelayed (direct) signal the configuration is known as *cos−*. The amplitude response is shown in Fig. 15.2(b). The harmonic peaks occur at frequencies $f=(2n+1)/(2T)$. These are all odd harmonics of frequency $1/(2T)$, but unlike the case of a square wave, there is no pitch at a frequency of $1/(2T)$. Instead this signal has two pitches, one approximately $0.88/T$, the other approximately $1.14/T$. The two pitches in the *cos−* condition are weaker than the pitch in the *cos+* condition. Pitch matches to broadband repetition pitch stimuli from Yost, Hill, and Perez-Falcon (1978) are shown in Fig. 15.3. The matches to the *cos+* stimulus fall on the 45-degree line. The matches to the *cos−* stimulus clearly do not.

Perceptual Models

Understanding the pitch of rippled noise poses an interesting challenge. Because the peaks in the *cos+* configuration are harmonics of $1/T$, it is, perhaps, not surprising that this configuration leads to a pitch of $1/T$. Because the peaks on the *cos−* configuration are separated by $1/T$ it is reasonable to get a pitch close to $1/T$. On the other hand, the peaks are rather broad, and the spectrum does not look much like the line spectrum of a complex periodic sound. One wonders if possibly the time delay is playing a more direct role *per se* in creating the pitch. There are interpretations of rippled-noise pitch in both the spectral domain and in the time domain.

One key to our present understanding of rippled-noise pitch has been experiments that have used bandpass filters to present only a portion of the rippled-noise spectrum to listeners. Particularly important is the *cos−* condition. There, it was found that as the center frequency of the passband increases the difference between the two pitches of the *cos−* configuration becomes smaller. Both pitches approach the value $1/T$ (Bilsen and Ritsma, 1969/70).

The results are consistent with a perception model in which the listener lines up the best consecutive-harmonic template with the available peaks of the rippled-noise spectrum. For the *cos+* condition the best fit obviously corresponds to a harmonic spectrum with a fundamental frequency of $1/T$. For *cos−*, the frequency of the best fitting fundamental depends on the part of the spectrum that needs to be fitted. The higher the harmonic numbers of the peaks, the better one can fit the available spectrum with a spectral template consisting of the harmonics of $1/T$.

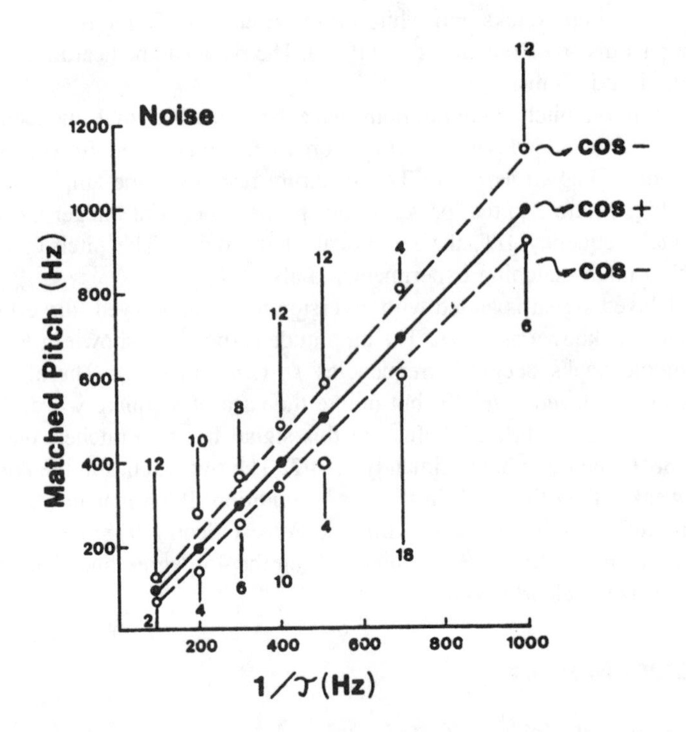

FIGURE 15.3. Matched pitches as a function of the reciprocal of the delay for *cos+* and *cos−* rippled noise. The numbers on the graph indicate the number of trials, out of a total of 24, where the two listeners in this experiment matched the high and low *cos−* pitches.

In the case of broadband noise in the *cos−*configuration, one has to supplement the idea of harmonic template matching with the idea of spectral dominance, as described in Chapter 12. According to Ritsma's (1967) spectral-dominance theory, the pitch of a complex tone is primarily determined by the frequencies of harmonics 3, 4, and 5. Thus, to understand the broadband case, where the pitch matches occur at $0.88/T$ and $1.14/T$, one would look for the harmonic series that fits best in the frequency region of these spectral peaks.

Figure 15.4 shows how this is imagined to work. We suppose that we are doing an experiment where a rippled noise in the *cos−*configuration is being matched with a rippled noise in the *cos+* configuration. The delay in the *cos−* configuration is fixed at T. The listener has the task of adjusting the delay in the *cos+* configuration to match the pitch of the *cos−* signal. From experiment we know that there are two values of the delay for the *cos+* that match the pitch of *cos−*, namely $T_1=1.14T$ and $T_2=0.88T$. The pitch in *cos+* with delay T_1 is $0.88/T$ and the pitch in *cos+* with delay T_2 is $1.14/T$. In Fig. 15.4 we see that the power spectrum for the *cos−* configuration has peaks that nearly coincide with the third and fourth

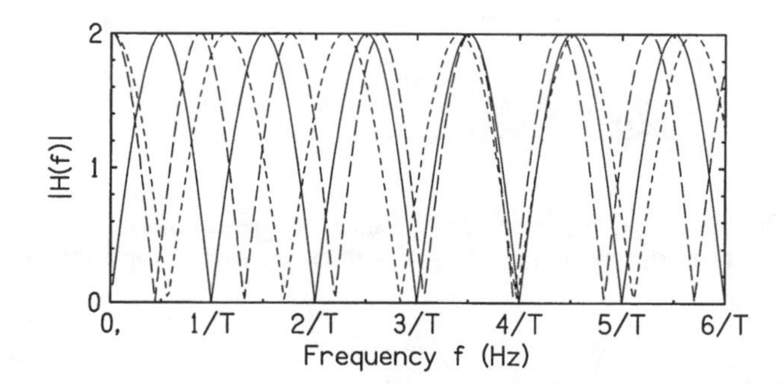

FIGURE 15.4. The spectrum for broadband rippled noise in the *cos*– configuration is shown by the solid line. The delay of the *cos*– configuration is T. Two spectra for matching sounds in the *cos*+ configuration are shown by dashed lines. The long dashes come from a delay of $T_1 = 1.14T$, the shorter dashes come from a delay of $T_2 = 0.88T$. Notice the near coincidence of the spectra in the vicinity of $f = 4/T$.

harmonic peaks for the $cos+$ configuration with delay T_2. Also, the power spectrum for the $cos-$ configuration has peaks that nearly coincide with the fourth and fifth harmonic peaks for the $cos+$ configuration with delay T_1. The significance of the coincidences for these particular values of T_1 and T_2 is that they occur in the spectral-dominance region. Because different listeners may have different dominance regions, one expects some variability in the matching values of T_1 and T_2.

The spectral matching model described above was not the first theoretical attempt to understand repetition pitch. Bilsen and Ritsma (1969/70) had initial success with a model in which pitch was determined by the time separation between the strongest peaks in the waveform fine structure, as filtered by model auditory filters in the spectral-dominance region. The spectral matching model was applied only several years later, by Bilsen and Goldstein in 1974 (see also Bilsen, 1977). Subsequently, Yost and his colleagues (e.g. Yost, 1982) attributed the rippled-noise pitch to a peak in the autocorrelation function seen in the output of filters tuned to the spectral-dominance region. A corresponding peak has been seen in the temporal response of neurons in the cochlear nucleus (Shofner, 1991).

Within the confines of mathematical models, it is not possible to distinguish between a model based on autocorrelation and a model based on the power spectrum. Because these two functions are Fourier transforms of one another they both convey the same information and lead to the same predictions. It does not follow, however, that the two models are equally good. To decide which is better requires that one first acknowledge that it is unlikely that either precisely describes what the nervous system actually does. One of the models is probably a better approximation to reality. Possibly further physiological experiments will suggest the nature of the pitch mechanism. Possibly additional psychoacoustical experiments will reveal features that are more readily accommodated by one of the models or the other. Possibly comb filtering experiments described below will be decisive.

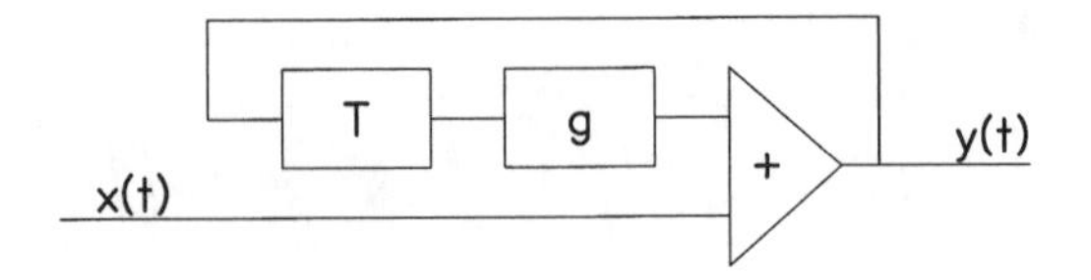

FIGURE 15.5. The comb filter is made by using the operations of delay (T), attenuation (g) and summation (+) as shown, with delay and attenuation in the feedback loop.

What all three of the mathematical models of repetition pitch have in common is that they appeal to spectral dominance in some form to define the frequency band where the model perceptual analysis occurs. Spectral dominance is an idea that emerged from pitch experiments using line spectra. We note here that the concept seems especially plausible as applied to the repetition pitch of broadband signals for the following reason: Template matching models of pitch perception suggest that the spectral peaks that are important in determining the low pitch of a complex tone are those which themselves produce a stable sense of pitch. The spectral peaks of rippled noise all have the same width in units of frequency. Therefore, on a logarithmic scale, higher frequency peaks are relatively narrower. For instance, the first peak of the $cos-$ configuration has a width of two octaves, whereas the fourth peak has a width of only 0.28 octaves. The latter leads to a much stronger sense of pitch. One therefore expects that the fourth peak will be more important than the first in determining the repetition pitch of this stimulus. To understand why peaks with even higher frequency are not even more important, one can revert to the original argument for spectral dominance which says that the spectral peaks which are resolved by the peripheral auditory system are the ones that are important for pitch. Higher frequency peaks, corresponding to higher harmonic numbers, are less well resolved, and this might be especially true when the spectral peaks are broad, as they are for rippled noise. The role of spectral dominance in repetition pitch has been challenged by the pitch matching data of Warren and Bashford (1988). However, the experimental situation is made complicated by the possibility of considerable individual differences in spectral-dominance regions.

COMB FILTERS

The peaks of the delay-and-add filter, as shown in Figs. 15.2 and 15.4, are cosine-shaped and dull. A filter with sharp peaks can be made by putting the delay and gain operations in a feedback loop, as shown in Fig. 15.5. Such a filter can be called a comb filter because the amplitude response has regularly spaced sharp peaks that resemble the teeth of a comb.

With the figure as a guide, we write the equation for the output $y(t)$ as a function of the input $x(t)$,

$$y(t)=x(t)+gy(t-T). \tag{15.17}$$

This equation includes an implied iteration. For the first iteration, we observe that if the time $t-T$ is substituted for t above then

$$y(t-T)=x(t-T)+gy(t-2T). \tag{15.18}$$

If this value for $y(t-T)$ is substituted into Eq. (15.17) then

$$y(t)=x(t)+gx(t-T)+g^2y(t-2T). \tag{15.19}$$

The next iteration requires that we substitute for $y(t-2T)$ to get

$$y(t)=x(t)+gx(t-T)+g^2x(t-2T)+g^3y(t-3T). \tag{15.20}$$

Obviously this process can be iterated indefinitely, so that

$$y(t)=x(t)+gx(t-T)+g^2x(t-2T)+g^3x(t-3T)+\ldots\,. \tag{15.21}$$

This equation says that the output y is equal to the input plus a version of the input that has been delayed and attenuated, plus a version that has been delayed and attenuated a second time, plus a version that has been delayed and attenuated a third time, plus... . There is no end to this regression and it would not seem possible to obtain a closed-form mathematical expression to describe it. The solution to this problem is to let the Fourier transform work a little magic on it.

We begin by Fourier transforming Eq. (15.17). Then, using the standard result for the Fourier transform of a time-delayed signal from Eq. (8.28), we can write the Fourier transform of Eq. (15.17) as

$$Y(\omega)=X(\omega)+ge^{-i\omega T}Y(\omega), \tag{15.22}$$

which can be solved, using only algebra, to give

$$Y(\omega)=\frac{X(\omega)}{1-ge^{-i\omega T}}. \tag{15.23}$$

Therefore the transfer function of the comb filter is

$$H(\omega)=\frac{1}{1-ge^{-i\omega T}}, \tag{15.24}$$

and the squared magnitude of the transfer function is

$$|H(\omega)|^2=\frac{1}{1+g^2-2g\cos\omega T}. \tag{15.25}$$

The amplitude response for several values of g is given in Fig. 15.6. As g approaches 1, the peaks become infinitely sharp.

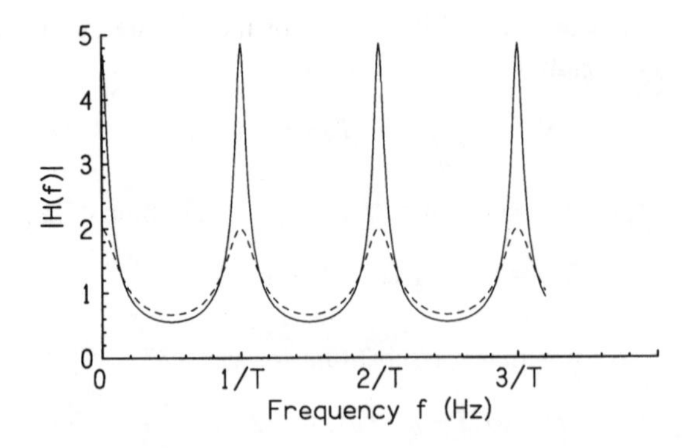

FIGURE 15.6. Amplitude response of the comb filter with two values of gain. Solid line, g=0.7943. Dashed line, g=0.5.

Bilsen (1980) studied the pitch created by comb-filtered noise. As would be expected, the sharp spectrum led to a sensation of pitch that is stronger and more reproducible than the pitch of cos+rippled noise, which has only a single delay operation. One way to study the pitch strength is to measure the just-noticeable difference for the delay time because different delay times are presumed to be recognized by pitch differences. For $g=0.7943$, corresponding to an attenuation of 2 dB in the feedback loop, Bilsen found that the JND for delay time is only one third as large for comb filtering compared to cos + rippled noise.

Sharp Teeth with a Finite Impulse Response

In the year 1693, the Dutch physicist and astronomer Christian Huygens (1629–1695) discovered a form of repetition pitch while he was visiting the castle at Chantilly de la Cour in France. Standing outdoors between a fountain and a stone staircase he noticed that the noise from the fountain was transformed by the staircase into a sound with an emergent musical tone. The situation, as imagined by F. M. M. Bilsen (Frans Bilsen's father), is shown in Fig. 15.7 taken from the paper by Bilsen and Ritsma (1969/70), which also includes the original report by Huygens written in French.

By experimenting with the resonances of paper tubes of different lengths, Huygens discovered that the pitch of the tone was determined by the common width of the steps. Each step introduced a delay T, equal to twice the step width divided by the speed of sound. From our perspective, we see that Huygens heard the results of comb filtering with delays of T, $2T$, $3T$, etc., but unlike the infinite progression in Eq. (15.17), this progression came to an end because there were only a finite number of steps. Therefore, Huygens had a finite-impulse-response comb filter.

To approximate the spectrum heard by Huygens, we consider the sum of an original reflection plus N delayed reflections,

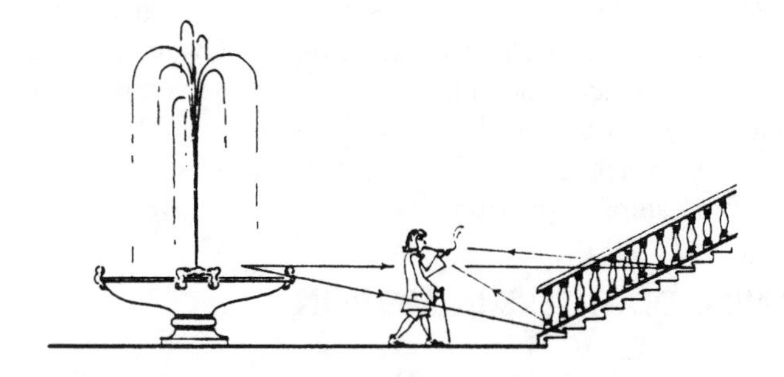

FIGURE 15.7. Huygens in the courtyard.

$$y(t)=x(t)+gx(t-T)+g^2x(t-2T)+g^3x(t-3T)+\ldots g^Nx(t-NT). \tag{15.26}$$

The Fourier transform becomes a geometric series, with a common ratio of $ge^{i\omega T}$. From the section on complex geometric series in Appendix C, we find the filter transfer function for such a series of a finite number of reflections. The input and output are related by

$$Y(\omega)=X(\omega)\,\frac{1-g^{(N+1)}e^{i(N+1)\omega T}}{1-ge^{i\omega T}}, \tag{15.27}$$

which is the infinite sum starting with the first reflection minus the infinite sum starting with reflection number $N+1$. The transfer function on the right is increasingly sharp with increasing N, and the pitch it makes is increasingly clear. Below we refer to this system as the *N-staircase*.

An alternative way to get sharp peaks in the transfer function of a rippled-noise filter is to make a series combination of numerous delay-and-add filters like Fig. 15.1. A total of N such stages makes the *N-cascade*, where output and input are related by

$$Y(\omega)=X(\omega)[1+ge^{-i\omega T}]^N. \tag{15.28}$$

Although the N-staircase and the N-cascade have transfer functions with similar principal maxima, the N-staircase has secondary maxima and the N-cascade does not.

Yost, Sheft, Shofner, and Patterson (1995) have suggested that comparing the N-staircase and N-cascade conditions can resolve the ambiguity between the power-spectrum model and the autocorrelation model in favor of autocorrelation. They observe first of all, that for given N the two conditions sound very similar. This argues against the power-spectrum perceptual model because the two power spectra differ by the secondary maxima present in N-staircase. The similarity argues in favor of a model based on the first peak in the autocorrelation function, which is similar for N-staircase and N-cascade. Second, these authors find that for both N-staircase and N-cascade, listeners are sensitive to the value of N, e.g. $N=4$ versus $N=5$. On the mathematical side, this sensitivity is observed in the first peak of the autocorrelation function but not in the power spectrum. This second observation leads to the same conclusion as the first: the autocorrelation function seems to provide the more natural explanation.

DELAY-AND-ADD FILTERS IN MOTION

As suggested by the experience of Huygens, delay-and-add filtering in various forms occurs often in ordinary life. The effects are particularly evident when sources are in motion so that the delay between the two sounds changes slowly in time, causing the pitch of the filtered noise to change. Bilsen and Ritsma describe the effect of walking toward a steam locomotive that is blowing off steam, where the path length between the direct sound and the sound reflected from the train platform changes as one walks. [Note that the pitch changes because of the change in path length, not because of Doppler shift, which is audible only for velocities faster than walking.] In our own day, the sounds of a jet airplane plus the reflection from nearby buildings, leads to a filtered broadband noise with moving delay as the airplane takes off. Indeed, delay-and-add filtering as used in the processing of rock music is often called "jet sounds." The spectral effects of moving delay-and-add seem to be especially noticeable when the sounds are loud.

Processing music with moving delay-and-add filtering may have had its origin in the Abbey Road studio of the Beatles, where the technique was known as "flanging." It works like this: The musical signal is sent to two separate but identical reel-to-reel tape recorders. Both recorders are set to record the signal at the record heads and to play back the recordings, almost immediately thereafter, from the playback heads a few centimeters downstream. The outputs from the playback heads are summed. If the spacings between record and play heads and the tape speeds are identical on both recorders, the resulting summed signal is simply twice as large as either signal alone. If one slows the tape speed of one machine by placing a thumb on the flange of the payout reel (hence the name "flanging") then a time delay is introduced between the signals from the two recorders. By changing the pressure on the reel, one can vary the time delay, which leads to filtering with peaks and valleys that move up and down in frequency slowly with time.

Now that reel-to-reel tape recorders are about as rare as steam locomotives, the flanging effect is created with digital delay lines. Feedback, leading to comb filtering, is normally included as one of many options in generalized delay-based processors. The mathematics of flanging, and a comparison with the music processing technique known as "phasing," was given by Hartmann (1978a).

An easy way to experiment with delay-and-add filtered noise informally in the laboratory is to pass the same broadband noise to two loudspeakers, one in front of the listener and the other behind. As the listener moves forward and backward the delay between transit times for the two speakers changes. D_1 is the distance from speaker 1 to the ears and D_2 is the distance from speaker 2 to the ears, then the time delay is $T=(D_1-D_2)/c$, where c is the speed of sound. If D_1 and D_2 are measured in centimeters then T, measured in ms, is $T=(D_1-D_2)/34.4$. As a quick demonstration one can use a portable FM radio, tuned off-channel to make only broadband noise. The noise from the radio plus a reflection from a hard flat wall can produce a filtered signal with a pitch.

EXERCISES

Exercise 1: Delay gain versus delay attenuation

The amplitude response of a delay-and-add filter with delay T, and delayed-signal gain g, is given by

$$|H(f)|=\sqrt{1+g^2+2g\cos 2\pi fT}.$$

Show that the shape of this response is unchanged if gain g is replaced by its reciprocal g^{-1}. The only effect of this replacement is to multiply the amplitude response by a frequency-independent constant.

Exercise 2: Delay-and-add filtering with a gain of ± 1

(a) Show that in the special case that $g=1$ the amplitude response of the delay-and-add filter in Exercise 1 becomes

$$|H(f)|=2|\cos(\pi fT)|,$$

and if $g=-1$ then

$$|H(f)|=2|\sin(\pi fT)|.$$

(b) Use the half-angle tangent formula from Appendix B to show that for $g=1$ the phase shift is given by

$$\phi=-\pi fT,$$

except that there are jumps of π radians when needed to keep the function in the range $-\pi/2$ to $\pi/2$.

Exercise 3: Attenuated delay—tiny ripples

In a delay-and-add filter of the cos+ type, the amplitude response becomes flatter as the gain of the delayed signal g is reduced. The frequencies of the peaks and valleys do not change, but the level differences between peaks and valleys become smaller.

(a) Show that the level difference between peaks and valleys is

$$L_p - L_v = 20 \log \frac{1+g}{1-g}.$$

(b) Yost and Hill (1978) found that repetition pitch can be heard distinctly for an attenuation of 20 dB, for which $g=0.1$. Show that the level difference between peaks and valleys is 1.7 dB.

Exercise 4: Generalized delay-and-add filter

Almost all the published work on repetition pitch has used the two configurations *cos+* and *cos−*. In the case of *cos+* the delayed signal (delay T) is added to the direct signal, and spectral peaks appear at frequencies that are integer multiples of $1/T$, i.e. where fT=integer n. In the case of *cos−* the delayed signal is subtracted from the direct signal, and spectral peaks appear where $fT=n+1/2$. These two configurations are the only possibilities if the signal operations are limited to delay, addition, and inversion. However, with computer processing one can create a generalized filter by applying a frequency-independent phase shift $\Delta\phi$ to the delayed signal prior to addition to the direct signal. (Signals *cos+* and *cos−* then are seen as special cases where the applied phase shift is zero or π radians.)

Show that the spectrum of white noise passed through the generalized filter has peaks where $fT=n+\Delta\phi/(2\pi)$. This means that the spectrum of rippled noise maintains its shape; the effect of the phase shift is to slide the spectrum rigidly along the frequency axis by a shift of $\Delta f=\Delta\phi/(2\pi T)$. Shifts of $\Delta\phi=\pi/2$ and $3\pi/2$ were investigated by Bilsen and Ritsma (1969/70). The pitches were found to lie in between those for *cos+* and *cos−*, in agreement with a spectral dominance model of repetition pitch.

Exercise 5: Quickie repetition pitch

A listener hears a noise source both directly and as reflected by a wall, as shown in Fig. E.5. The pitch of the noise changes as the listener moves.

(a) Explain why the pitch increases from 500 Hz to 600 Hz as the distance between

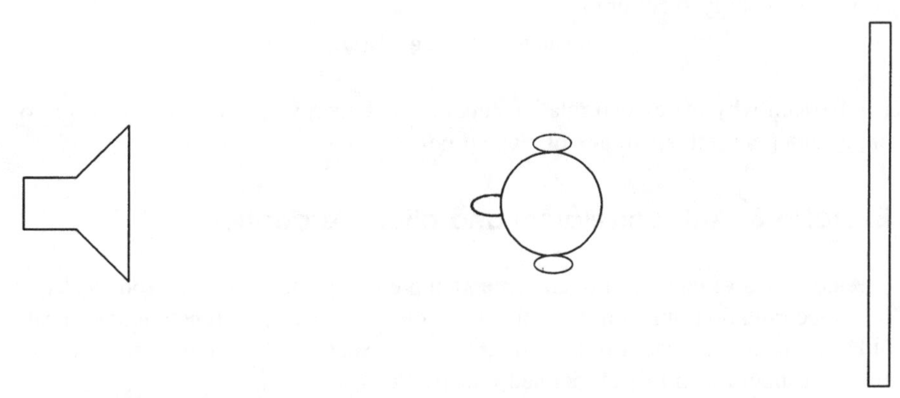

FIGURE E.5. Repetition pitch with a single source and a wall.

the listener and the wall decreases from 34.4 cm to 28.7 cm. (Assume that the speed of sound is 34400 cm/s.)

(b) Show that the pitch of the noise does not depend on the distance from the source to the listener.

Exercise 6: Staircase versus cascade

(a) Show that the power response of the N-staircase filter in Eq. (15.27) is given by

$$|H(\omega)|^2 = \frac{1+g^{2(N+1)} - 2g^{(N+1)} \cos(N+1)\omega T}{1+g^2-2g \cos \omega T}.$$

(b) Show that the power response of the N-cascade filter in Eq. (15.28) is given by

$$|H(\omega)|^2 = [1+g^2+2g \cos \omega T]^N.$$

(c) Show that the power response of the N-staircase filter has a maximum whenever ω is an odd-integer multiple of $\pi/[(N+1)T]$. The N-cascade filter does not have secondary maxima like these.

Exercise 7: Pulse pair

A signal consisting of two spikes separated by a brief delay T is perceived as a single pulse with pitch $1/T$.

(a) If the signal is given by $x(t) = \delta(t) + g\,\delta(t-T)$, show that the autocorrelation function is given by

$$a_f(\tau) = (1+g^2)\delta(\tau) + g[\delta(\tau+T) + \delta(\tau-T)],$$

so that the energy spectrum is

$$A_f(\omega)=1+g^2+2g\cos(\omega\tau).$$

(b) Explain why the autocorrelation function and energy spectrum in part (a) also apply to an ensemble average of rippled noise.

Exercise 8: Autocorrelators and pitch perception

Autocorrelator models of pitch perception predict that pitches arise from peaks in the autocorrelation function. Specifically, if the autocorrelation function has a peak at lag τ then one expects a pitch at frequency $1/\tau$. Sharp peaks in the autocorrelation function should lead to well-defined pitches.

(a) Show that on the basis of this model one would expect the pitch of rippled noise (Exercise 7) to be more salient than the pitch of a sine tone. This expectation does not agree with experiment.

(b) Suggest a way of improving the autocorrelator model so that it does not contradict experiment so badly.

CHAPTER 16

Probability Density Functions

The time is out of joint.
WILLIAM SHAKESPEARE, *HAMLET*, ACT I SC. 5.

A waveform is a function of time and this functional dependence is usually an important aspect of anything that we do with the waveform. There are occasions, however, when we are less interested in the time-dependent details of a waveform and more interested in the values themselves, or in the distribution of these values. For instance, we may be interested only in the energy of a wave accumulated over some time interval. Then we care only about what waveform values occur within the time interval; we do not care about precisely when they occur, or in what order.

In these cases a waveform can be adequately described by a "probability density function" (PDF), to which we will give the symbol $P(x)$. The PDF can be used to find the probability that the waveform has a certain value. Specifically, function $P(x)dx$ is the probability that the waveform has a value that lies between a particular value x and $x+dx$. Because it is a probability, $P(x)dx$ has dimensions of unity. Therefore, $P(x)$ itself is a density and has dimensions of $1/x$. For instance, if $x(t)$ represents the pressure in an acoustical signal then $P(x)$ has dimensions of 1/pressure.

The unimportance of time to the PDF is shown in Fig. 16.1. The wave in Fig. 16.1(a) is a sine wave. The wave in Fig. 16.1(b) is obtained by putting the time out of joint and rearranging some segments of Fig. 16.1(a). Obviously these waveforms are very different; their Fourier transforms and power spectra are very different, and they sound very different too. However, the waveforms in Figs. 16.1(a) and 16.1(b) take on the same instantaneous values, and therefore the PDF, shown in Fig. 16.1(c), is the same for both.

DERIVATION OF THE PDF

The key to deriving an expression for the PDF is that for any waveform there are some values of x that occur over longer intervals of time causing the PDF to be

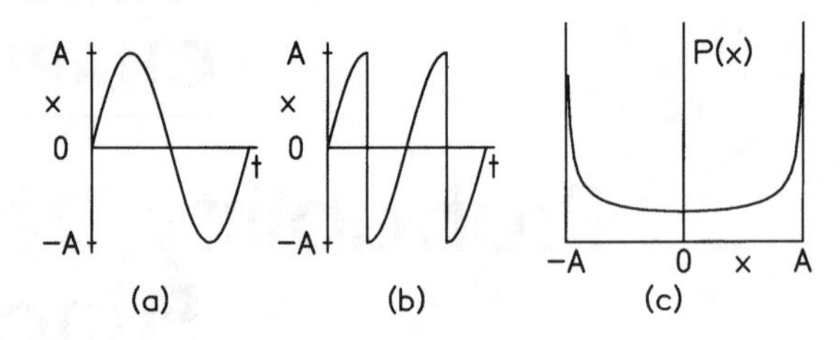

FIGURE 16.1. (a) A sine wave. (b) The sine wave with some segments interchanged. (c) The PDF for both (a) and (b).

larger for these values of x. Therefore, the PDF is proportional to $1/|dx/dt|_x$. This is the reciprocal of the derivative of the waveform, evaluated at a point where the waveform $x(t)$ has the particular value x. Because of the absolute value the function is always positive. The idea is that if the derivative has a large magnitude then the waveform moves rapidly through the point x and the chance of catching it in the vicinity of x is small. But if the derivative is small then the waveform spends more time near the value x, and there is a correspondingly large contribution to the PDF.

EXAMPLE 1: *PDF for a sine wave*

We illustrate the above idea for the calculation of a PDF by using it on the sine waveform, $x(t)=A\sin(\omega t)$. The amount of time that the sine spends in a range Δx about the value x is

$$\Delta t=\frac{1}{|dx/dt|_x}\Delta x. \tag{16.1}$$

Because the wave is periodic with period T, the probability $P(x)\Delta x$ can be calculated from the fraction of the time that the wave spends near the value x,

$$P(x)\Delta x=\Delta t/T. \tag{16.2}$$

Evaluating the derivative in Eq. (16.1) and substituting for Δt into Eq. (16.2) gives

$$P(x)\Delta x=\frac{1}{T\omega A|\cos(\omega t)|_{t=t(x)}}\Delta x, \tag{16.3}$$

or

$$P(x)=\frac{1}{2\pi A|\cos(\omega t)|_{t=t(x)}}, \tag{16.4}$$

because $T\omega = 2\pi$. The implicit dependence on x through the inversion $t=t(x)$ is awkward. Fortunately it is easy to represent $A\cos(\omega t)$ in terms of the original signal, $x(t)=A\sin(\omega t)$, because $\cos^2\omega t = 1-\sin^2\omega t$, and therefore,

$$P(x)=\frac{1}{2\pi\sqrt{A^2-x^2}}. \tag{16.5}$$

What is wrong with our evaluation of $P(x)$ so far is that there are two spans of time when the waveform is at x, once going up and once going down, and we have only accounted for one of them. For the sine wave, on each span of time the absolute value of the derivative is the same, and so we have only to multiply by a factor of 2. Finally the correct formula for the PDF for a sine wave is

$$P(x)=\frac{1}{\pi\sqrt{A^2-x^2}} \quad \text{for } -A\leqslant x\leqslant A \tag{16.6}$$

and

$$P(x)=0 \quad \text{for } |x|>A.$$

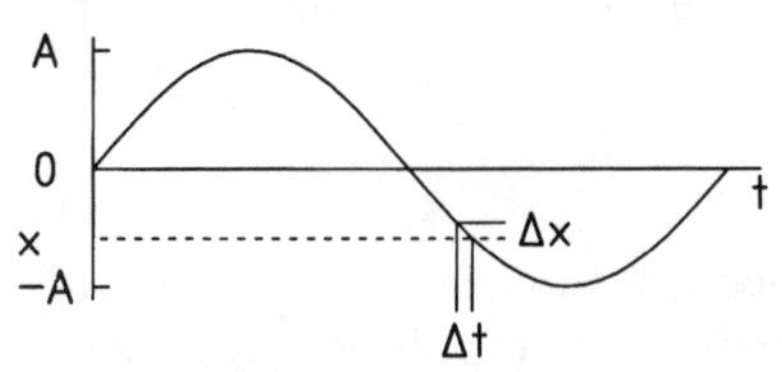

FIGURE 16.2. One cycle of a sine wave showing how a range of values Δx near value x implies a span of time Δt.

PDF for a General Waveform

In the sine wave example above there were only two places where the waveform had any given value x, and the absolute value of the derivative was the same at both of them. This kind of good luck is rare, and therefore the general expression for the PDF is more complicated,

$$P(x)=\frac{1}{T_D}\left[\frac{1}{|dx/dt|_{x_1=x}}+\frac{1}{|dx/dt|_{x_2=x}}+\ldots\right]. \tag{16.7}$$

The extended sum above means that $P(x)$ gets a contribution from the inverse of the derivative every time that the function takes on the value x. This equation was further generalized to cover the case where the waveform is not periodic. Therefore, the signal duration T_D replaces the period T. For a non-periodic signal the best we can do is to evaluate $P(x)$ over some long duration of time, hopefully long enough to give an accurate sample of the statistics of the signal. For a periodic signal, the best choice of duration is precisely one period.

The PDF depends upon the shape of the waveform. Therefore, if $x(t)$ is a complex periodic waveform, the PDF depends upon the relative phases of the harmonics. If the waveform is defined by a Fourier series, it becomes tedious to calculate each denominator in Eq. (16.7), and, what is worse, it is hard to invert the function $x(t)$ to find $t(x)$ so that one knows the values of time where the derivative should be evaluated. Therefore, with the general complex wave, the only reasonable approach to calculating the PDF is to let a computer evaluate the function at many points and to construct a histogram from the values.

PDF FOR THE SUM OF TWO FUNCTIONS

Suppose that waveform x_3 is the sum of two waveforms x_1 and x_2,

$$x_3(t)=x_1(t)+x_2(t). \tag{16.8}$$

Suppose also that the PDF for wave $x_1(t)$ is $P_1(x)$ and the PDF for wave $x_2(t)$ is $P_2(x)$. Then, if x_1 and x_2 are uncorrelated functions, the PDF for $x_3(t)$ is given by a convolution,

$$P_3(x)=P_1(x)*P_2(x). \tag{16.9}$$

The proof of this important formula is straightforward. $P_3(x)$ is the probability density that $x_3(t)$ has some value x. This density is calculated from the probability density that function $x_1(t)$ has some value x', and at the same time, function $x_2(t)$ has the value that is required so that $x'+x_2=x$. In other words, it involves the product

$$P_1(x')P_2(x-x'). \tag{16.10}$$

But the value of x' might be anything at all, so we integrate over all those possibilities, i.e.

$$P_3(x)=\int_{-\infty}^{\infty} dx' P_1(x')P_2(x-x'), \tag{16.11}$$

which is simply the definition of convolution.

Independence

The simplicity of the above proof belies the complexity of the requirement that x_1 and x_2 must be uncorrelated in order for the formula to apply. It is legitimate to

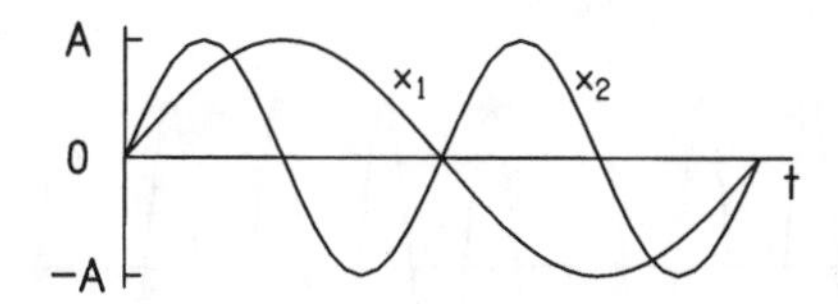

FIGURE 16.3. A sine function, $x_1(t)$, and its octave, $x_2(t)$.

convolve two densities, P_1 and P_2, as in Eq. (16.10), only if these densities are independent. This means that fixing the value of $x_1(t)$ must not restrict the value of $x_2(t)$ in any way.

There are some cases where lack of correlation is easy to show so that convolution of PDFs clearly applies. An exercise below considers the case where $x_1(t)$ is a constant. Then knowing the value of $x_1(t)$ does not restrict the values of $x_2(t)$ and the convolution formula applies.

In other cases, functions $x_1(t)$ and $x_2(t)$ are clearly correlated. Suppose $x_1(t)=\sin(\omega_1 t)$ and $x_2(t)$ is an octave higher, $x_2(t)=\sin(2\omega_1 t+\phi)$. These functions are not adequately independent for the convolution formula to apply. To show the lack of independence we need to choose some value of ϕ. Suppose we choose $\phi=0$. Then fixing a value of $x_1(t)$ does restrict the value of $x_2(t)$. For instance, whenever $x_1(t)$ is equal to zero, we know that $x_2(t)$ must also be zero, as shown in Fig. 16.3.

Another way to show that the convolution formula cannot apply in this case is simply to realize that P_3 must depend upon the phase ϕ. The maximum value attained by the sum which is $x_3(t)$, depends upon ϕ and so must the PDF P_3. However, neither $P_1(x)$ nor $P_2(x)$ has any dependence on ϕ, and so their convolution can't.

The problem with sines that are separated by an octave is that the two functions are locked together. In order for the convolution formula to apply to the sum of two periodic functions, we need those functions to slide with respect to one another. Put differently, we need the sum $x_3(t)$ to roll on an oscilloscope screen. But how much do they need to roll? Consider the case of a mistuned octave,

$$
\begin{aligned}
x_1(t) &= \sin(\omega_1 t) \\
x_2(t) &= \sin(2.1\omega_1 t) \qquad (16.12) \\
x_3(t) &= x_1(t)+x_2(t).
\end{aligned}
$$

These functions are less correlated than the functions separated by an octave. We expect that the convolution formula would work better for this case, but it does not work perfectly. We realize that after $x_1(t)$ has gone through 10 complete cycles, function $x_2(t)$ will have gone through precisely 21 complete cycles, and the sum x_3 will be ready to start over again. Therefore, x_1 and x_2 are not completely uncorrelated. For instance, if $x_1(t)$ happens to be at a maximum, then there are only 10

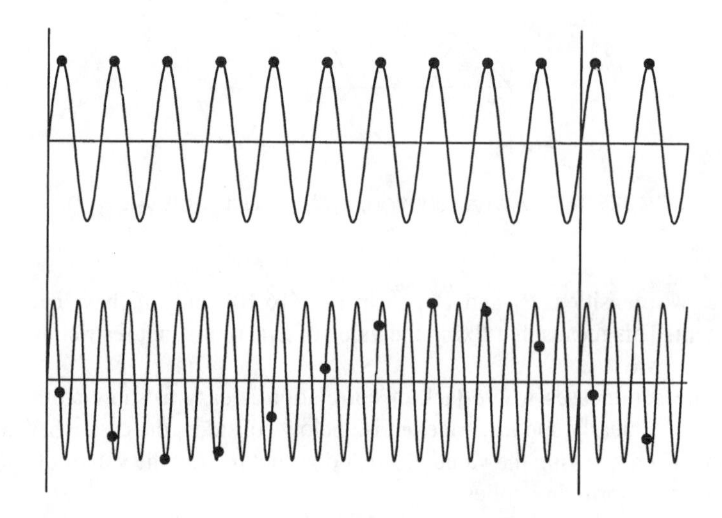

FIGURE 16.4. A reference sine wave and another sine wave with a frequency that is 2.1 times greater. Because 2.1 is a rational number, choosing a value for the reference sine (such as a peak shown here) limits the number of possible values for the second wave.

different possible values that function $x_2(t)$ can have, as shown in Fig. 16.4. Fixing $x_1(t)$ puts some restrictions on the allowed values of $x_2(t)$.

The above example suggests that the convolution formula for P_3 may apply if the waveform x_3 has a long period, much longer than the period of the lowest frequency component. A long period means that particular values of constituent waves x_1 and x_2 do not coincide frequently. However, a long period is not a sufficient condition for the convolution formula to apply. To see the problem, we have only to recall that a series of equally spaced Fourier components can result in either noise or a periodic pulse train, depending on the phases. For the noise, the phases are random variables and the convolution formula applies. Therefore, the convolution of the PDFs for the constituent sines should be the expected Gaussian PDF. For the pulse train, the PDF is a pair of delta functions, one at each of the two values that the pulse waveform has. This is quite different from the convolution of sine PDFs, no matter how long the period.

PDF FOR RANDOM EVENTS

Above, the probability density function has been applied to deterministic waveforms. It can similarly be applied to events that have a certain amount of randomness, with the additional advantage that probability densities may be combined with fewer concerns about the statistical independence of the distributions. We illustrate these ideas below with an example in which the events are completely random.

EXAMPLE 2: *Sum and difference of random variables*

A computer random number generator generates real numbers in the range from 0 to 1. Any number within that range is equally probable. We will use such a random number generator to create data for a numerical experiment in which we make successive observations. On each observation we find two random numbers x_1 and x_2. These two numbers are uncorrelated because choosing the first has no effect on the value of the second.

On each observation we form two new variables, the Sum (x_1+x_2) and the Difference (x_1-x_2). To display the results of this experiment we make a dot-raster plot in which the values of the variables are plotted as dots along the horizontal axis, and for each successive observation we increase the position along the vertical axis by one tiny unit. The dot-raster display is particularly appropriate if x_1 and x_2 represent the times of occurrences of two events during the observation interval.

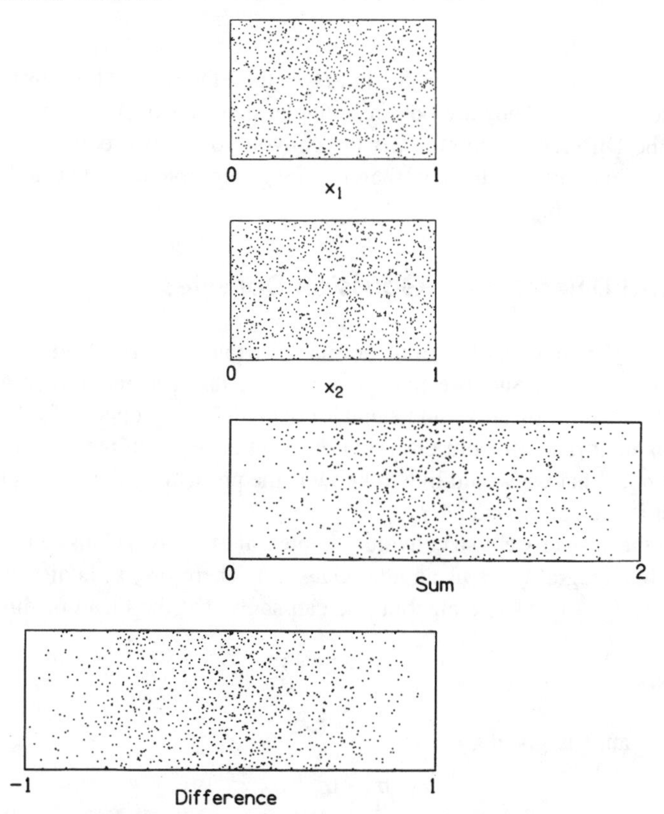

FIGURE 16.5. Dot-raster plots showing 999 observation intervals, each on a separate horizontal line. In each observation interval there are two random variables, x_1 and x_2, and their Sum and Difference.

The results of the experiment are shown in Fig. 16.5 for 999 observations. The plots for the original random variables x_1 and x_2 show a homogeneous distribution of events over the range from 0 to 1. Because the original variables are in the range 0 to 1, the Sum is in the range 0 to 2, and the Difference is in the range -1 to 1. Figure 16.5 shows that the distributions for Sum and Difference are not homogeneous. Instead, the dots tend to cluster towards the middle of the ranges.

To understand how the dots cluster for the Sum and Difference statistics in the example above, we find a PDF for these statistics by combining PDFs for the original variables. For definiteness we derive the PDF for the Difference.

Because the Difference is $x_1 - x_2$, we get a difference of x when the second variable x_2 has some value x_2 and the first variable x_1 has the value $x_2 + x$.

Therefore, the PDF for the Difference, P_D, is given by the convolution integral

$$P_D(x) = \int dx_2 P_1(x_2 + x) P_2(x_2), \tag{16.13}$$

where P_1 is the PDF for variable x_1 and P_2 is the PDF for x_2. For our example, P_1 and P_2 are the same. They are both rectangular densities over the range 0 to 1. The PDF for the Difference, therefore, is the convolution of two equal rectangles, and this convolution turns out to be a triangle. The generation of the triangular form can be seen in Fig. 16.6.

Sums and Differences—Gaussian Densities

The sum of two variables from normal distributions is a matter of particular practical importance. Suppose that x_1 is a value taken at random from Gaussian distribution P_1 with mean $\overline{x_1}$ and standard deviation σ_1. Suppose further that x_2 is a value taken at random from Gaussian distribution P_2 with mean $\overline{x_2}$ and standard deviation σ_2. What we would like to know is the probability density P_3 for variable x_3, which is the sum $x_3 = x_1 + x_2$.

To find the answer we do a convolution integral of the two Gaussian densities and find that this integral gives us another Gaussian. Therefore, x_3 is distributed with a normal density. After some algebra one can show that the mean of this density is

$$\overline{x_3} = \overline{x_1} + \overline{x_2}, \tag{16.14}$$

and the variance is given by

$$\sigma_3^2 = \sigma_1^2 + \sigma_2^2. \tag{16.15}$$

An example of the density for the sum of two normally distributed variables is given in Fig. 16.7.

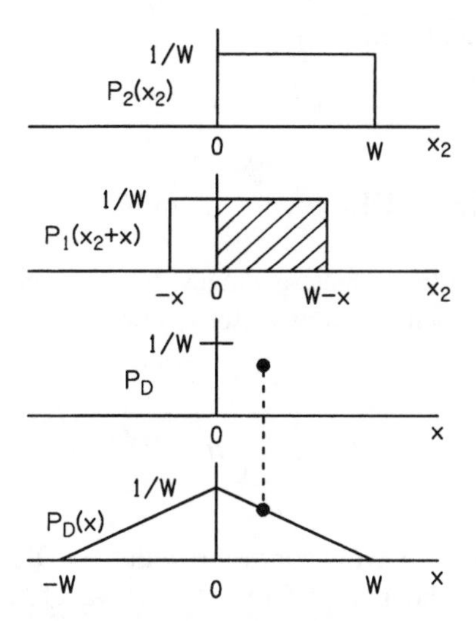

FIGURE 16.6. The densities P_1 and P_2 are both rectangles with range 0 to W (width). To convolve them we slide P_1 along the x_2 axis, as shown, and do the integral by finding the area of the overlap for each value of the displacement x. If the rectangular distributions are made by a computer random number generator, the width W is equal to 1.

The distribution of the difference of two random variables is similar. If $x_4 = x_2 - x_1$ where x_1 and x_2 are again drawn from densities P_1 and P_2, then x_4 is normally distributed with a mean given by the difference of the two original means,

$$\overline{x_4} = \overline{x_1} - \overline{x_2}, \tag{16.16}$$

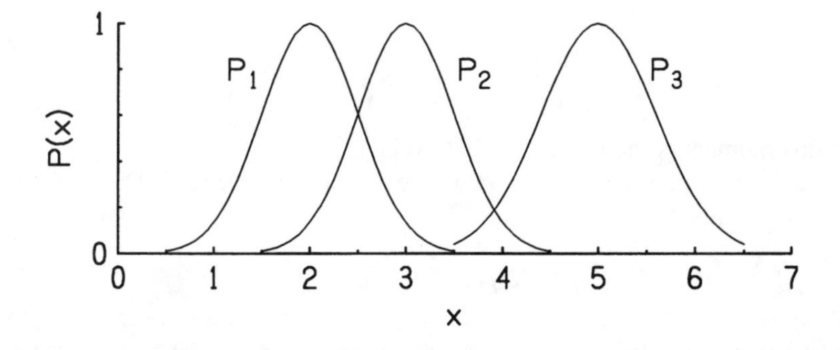

FIGURE 16.7. Densities for x_1 and x_2 have means of 2 and 3, respectively, and the same variance. The density for $x_1 + x_2$ has a mean of 5 and a standard deviation that is larger by a factor of $\sqrt{2}$.

and the variance is again given by

$$\sigma_4^2 = \sigma_1^2 + \sigma_2^2. \tag{16.17}$$

AVERAGES AND THE PDF

The PDF can be used to evaluate averages. For example, the average power is given by the second moment of the waveform. For a periodic signal with period T, the average power is

$$\bar{\mathcal{P}} = \frac{1}{T}\int_0^T dt\ x^2(t), \tag{16.18}$$

where $\mathcal{P}$ is used for power to avoid confusion with the PDF, P. This average power can also be calculated without any reference to the waveform as a function of time,

$$\bar{\mathcal{P}} = \int_{-\infty}^{\infty} dx\ P(x) x^2. \tag{16.19}$$

A similar equality must hold for any arbitrary moment of waveform x, or indeed, for any function F of waveform x, i.e.

$$\frac{1}{T}\int_0^T dt\ F[x(t)] = \int_{-\infty}^{\infty} dx\ P(x) F(x). \tag{16.20}$$

CENTRAL MOMENTS

The average values of powers of a waveform are known as "moments." With a bar to represent the average as before, the m-th moment is defined as

$$\overline{x^m} = \frac{1}{T}\int_0^T dt\ x^m(t) = \int_{-\infty}^{\infty} dx\ P(x) x^m. \tag{16.21}$$

The first moment is the average, or DC value, $\bar{x}$.

Variations around the average value are "central moments." The m-th central moment is defined as μ_m,

$$\mu_m = \overline{(x - \bar{x})^m}. \tag{16.22}$$

By definition μ_1 is zero. The second central moment is the variance about the mean, which is the average AC power in the signal,

$$\mu_2 = \overline{(x - \bar{x})^2}. \tag{16.23}$$

If $\bar{x}$ is zero then the AC power is the same as the power $\bar{\mathcal{P}}$ in Eq. (16.18) or Eq. (16.19). The RMS value is $\mu_2^{1/2}$. Higher-order central moments are usually normalized by an appropriate power of the RMS value (Hsueh and Hamernik, 1990). The purpose of doing so is that moments normalized in this way do not change if the waveform is multiplied by a simple constant factor.

The "skewness" is given by the third central moment,

$$\text{Skewness} = \mu_3 / \mu_2^{3/2}. \tag{16.24}$$

Probability densities that extend to one side of the mean more than to the other are skewed. Symmetrical densities, like the Gaussian, have a skewness of zero.

The "kurtosis" is given by the fourth central moment,

$$\text{Kurtosis} = \mu_4 / \mu_2^2. \tag{16.25}$$

The kurtosis is a way to measure the power fluctuation. We show that as follows:

At any time t the instantaneous AC power is $[x(t) - \bar{x}]^2$. The difference between this power and the average AC power is

$$\Delta\mathcal{P} = [x(t) - \bar{x}]^2 - \mu_2. \tag{16.26}$$

The average value of $\Delta\mathcal{P}$ is zero. To find an average measure of the power variation we square $\Delta\mathcal{P}$ to find a mean-square power fluctuation,

$$\overline{(\Delta\mathcal{P})^2} = \overline{[(x - \bar{x})^2 - \mu_2]^2}. \tag{16.27}$$

Expanding the square, we find

$$\overline{(\Delta\mathcal{P})^2} = \overline{(x - \bar{x})^4} - 2\overline{(x - \bar{x})^2}\mu_2 + \overline{\mu_2^2}, \tag{16.28}$$

which is

$$\overline{(\Delta\mathcal{P})^2} = \mu_4 - 2\mu_2^2 + \mu_2^2 \tag{16.29}$$

or

$$\overline{(\Delta\mathcal{P})^2} = \mu_4 - \mu_2^2. \tag{16.30}$$

We know that the difference on the right-hand side is positive because it is the square of a real number. Therefore, the power fluctuation is minimized when the fourth central moment μ_4 is minimized. This is the kurtosis, except for normalization.

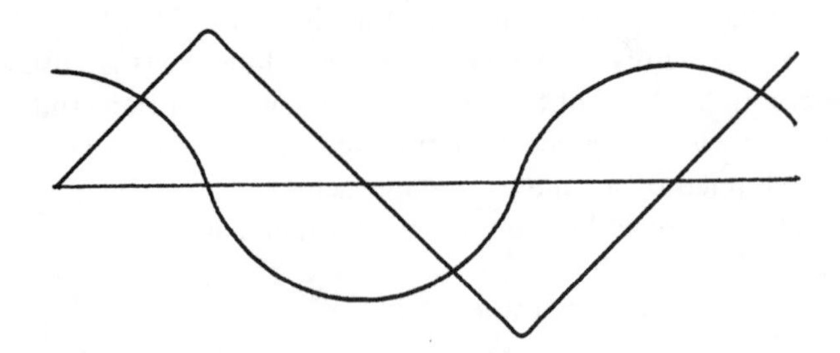

FIGURE 16.8. The triangle wave and the pseudotriangle. Both waveforms have the same amplitude spectrum, consisting of harmonics 1,3,...23. The pseudotriangle was constructed using Pumplin's algorithm and has the smallest possible power fluctuation for the given amplitude spectrum (Hartmann and Pumplin, 1991).

Minimum-Fluctuation Waveforms

The random waveforms known as "low-noise noise" are based upon minimum kurtosis (Pumplin, 1985). Pumplin's technique can be applied to any periodic waveform that can be written as a Fourier series,

$$x(t)=\sum_{n=1}^{N} C_n \cos(n\omega_0 t+\phi_n). \tag{16.31}$$

One begins with a desired amplitude spectrum having an arbitrary, but finite, number of components, N, and then adjusts the values of the phases, ϕ_n, to minimize the kurtosis. The result of applying the technique to the spectrum of a triangle wave is shown in Fig. 16.8.

Even-numbered central moments higher than the fourth are also measures of the power fluctuation. Higher-order moments μ_m give increased weight to those power fluctuations that differ considerably from the average power. In the limit of very large m the value of μ_m is determined entirely by the single value of $x^2(t)$ that differs the most from the average. Therefore, the normalized moment in the limit that m becomes infinite is the crest factor (see Chapter 3).

In principle, one could generate a minimum-fluctuation waveform by minimizing any of the even-numbered central moments. In practice, Hartmann and Pumplin (1988, 1991) found that minimizing the fourth moment leads to small values of the crest factor, and they conjectured that minimizing the fourth moment leads to small values of all the other moments too.

Schroeder (1970) developed a formula for selecting the phases, ϕ_n, that lead to a small crest factor. The n-th phase angle, in radians, is given by

$$\phi_n = \phi_1 - \frac{\pi}{\mu_2} \sum_{l=1}^{n-1} (n-l) C_l^2, \tag{16.32}$$

where μ_2 is the total power,

$$\mu_2 = \frac{1}{2} \sum_{n=1}^{N} C_n^2, \tag{16.33}$$

and ϕ_1 can be chosen at will. The formula is particularly easy to use for the case of a flat spectrum where C_n is a constant, independent of n. Then, using sum (C21) in Appendix C leads to

$$\phi_n = \phi_1 - \pi n(n-1)/N, \tag{16.34}$$

or, because the term that is linear in n is only a time shift,

$$\phi_n = \phi_1 - \pi n^2/N. \tag{16.35}$$

Schroeder's phase formula produces waveforms that have small power fluctuations as measured by both the kurtosis and the crest factor. Pumplin's algorithm does somewhat better for both measures, but the choice of Schroeder phase is a considerable improvement over random phases or alternating sine and cosine phases. Further, Schroeder's formula has the advantage that it is very simple to calculate.

The principal limitation of Schroeder's formula is that it actually generates a chirp. This is not necessarily a problem if the components are widely spaced, but if the components are closely spaced, as in the case of a noise signal, the signal is effectively a sine tone with a frequency that decreases from the top of the band to the bottom. The duration of the chirp is the reciprocal of the frequency spacing. At the end of the frequency sweep there is a brief interval with erratic fluctuations before the next chirp begins. Pumplin's algorithm does not generate such a special signal.

EXERCISES

Exercise 1: Normalization

A probability density $P(x)$ must be normalized. In other words, it must satisfy the equation

$$\int_{-\infty}^{\infty} dx\, P(x) = 1,$$

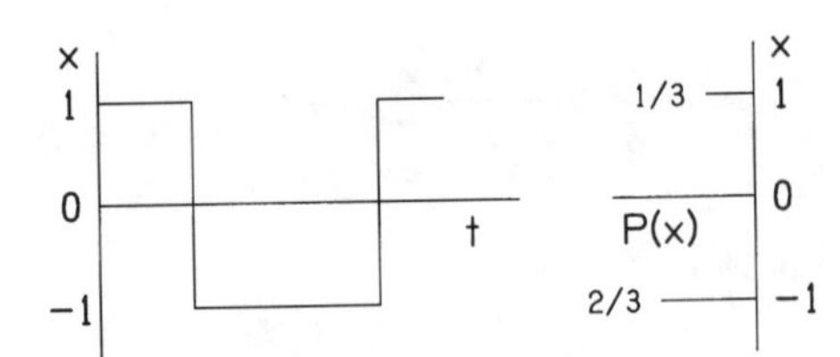

FIGURE E.1. A rectangular waveform and its PDF.

because the value of the waveform falls within the interval from $-\infty$ to $+\infty$ with absolute certainty.

(a) Show that Eq. (16.6) for the PDF for a sine wave is normalized.

(b) The PDF for a square wave, or any rectangular wave function that is constructed from horizontal line segments, poses a problem because the derivative is zero on the horizontal portions, and the formula for the PDF has the derivative in the denominator. Therefore, the PDF is a sum of delta functions. The problem of scaling the PDF is solved by the requirement for normalization and the fact that the integral of a delta function is unity. Use the normalization requirement to show that for a rectangular wave having a value of $+1$ one-third of the time and a value of -1 for two-thirds of the time, the PDF is

$$P(x)=\frac{1}{3}\,\delta(x-1)+\frac{2}{3}\,\delta(x+1).$$

A view of the relationship between waveform and PDF is given in Fig. E.1.

Exercise 2: The triangle and sawtooth

The triangle wave, with period T and amplitude A, has a derivative that always has the same absolute value, namely $4A/T$.

(a) Use the technique employed to calculate the PDF for the sine, given in Eq. (16.6), to show that the PDF for the triangle wave is

$$P(x)=1/(2A) \quad \text{for } |x|\leqslant A$$

$$P(x)=0 \quad \text{for } |x|>A.$$

(b) Show that this expression for $P(x)$ is normalized.

(c) The sawtooth wave is also made from straight line segments. Show that the PDF for the sawtooth wave is the same as for the triangle.

Exercise 3: The parabola wave

The parabola wave is a series of linked upward and downward parabolas, as shown in Exercise 6 of Chapter 5. For instance, on the half-period interval from $-T/4$ to $T/4$ the wave with unit amplitude is given by the downward parabola

$$x(t)=1-(4t/T)^2.$$

Show that the PDF is given by

$$P(x)=\frac{1}{2\sqrt{1-|x|}} \quad \text{for } |x|\leq 1$$

$$P(x)=0 \quad \text{for } |x|>1.$$

Exercise 4: Additive constant

(a) Suppose that function x_3 is a sum of a sine and a constant, $x_3(t)=\sin(\omega_1 t)+c_1$. Show that the PDF P_3 is given by the convolution of the PDF for the sine and the PDF for the constant. [Note: the PDF for the constant is $P_1(x)=\delta(x-c_1)$.]

(b) Suppose that $x_1(t)=c_1$ and $x_2(t)=c_2$, where c_1 and c_2 are both constants. Show that both common sense and the convolution formula agree that the PDF for the sum of these two functions is

$$P_3=\delta(x-c_1-c_2).$$

Exercise 5: Average power

Show that the average power in a sine can be calculated from the PDF. That is, show that

$$\frac{1}{T}\int_0^T dt\, A^2 \sin^2(2\pi t/T)=\int_{-A}^{A} dx\, \frac{1}{\pi\sqrt{A^2-x^2}}\, x^2.$$

Exercise 6: Uncertainty of differences

Levels $\overline{L_1}$ and $\overline{L_2}$ are average levels found in an experiment which also finds their standard deviations σ_1 and σ_2. If $\overline{L_1}=15.2(\pm 0.7)$ dB, and $\overline{L_2}=14.3$ (± 1.2) dB, show that the most probable difference is $\overline{L_2}-\overline{L_1}=-0.9(\pm 1.4)$ dB. Thus, while L_1 and L_2 have rather small relative errors, their difference is quite uncertain. It might not even be negative.

Exercise 7: Lorentzian densities

The Lorentzian density is given by the function

$$P_a(x) = \frac{1}{\pi}\frac{a}{x^2+a^2}.$$

The maximum value is $1/(a\pi)$ and the function is normalized so that its area is 1. The half-width at half-height (hwhh) is a.

Show that the convolution of two normalized Lorentzians, P_a and P_b, with hwhhs a and b respectively, is another normalized Lorentzian with hwhh equal to $a+b$.
Hint: The problem is much easier if you make use of the fact that the Fourier transform of the convolution of P_a and P_b is the product of the Fourier transforms of P_a and P_b. The Fourier transform of P_a is

$$\mathcal{F}[P_a] = \int_{-\infty}^{\infty} dx\ P_a(x)e^{-iqx} = e^{-a|q|}.$$

Exercise 8: Kurtosis

Show that the kurtosis of a Gaussian noise is equal to 3.

Exercise 9: Time machines

Evaluating a PDF is like a "time machine." So is shooting a movie at different locations. How far can you run with this analogy?

CHAPTER 17

Beats and Amplitude Modulation

Elected silence sing to me
And beat upon my whorlèd ear.
GERARD MANELY HOPKINS

The topic of beats and modulation can be perplexing at first. A variety of factors, technical, semantic and perceptual, conspire to cause confusion. We begin with the topic of beats.

BEATS OF EQUAL-AMPLITUDE SINES

Suppose that two sine tones, having the same amplitude and having nearly the same frequency, are added together. The general expression for the signal is then given by

$$x(t)=\sin(\omega_1 t)+\sin(\omega_2 t+\phi). \tag{17.1}$$

For example, the frequencies might be $f_1=1000$ Hz and $f_2=1001$ Hz, so that $\omega_1=2\pi\times 1000$ and $\omega_2=2\pi\times 1001$.

Equation (17.1) correctly describes the signal in terms of its Fourier components at 1000 and 1001 Hz, and from the spectral point of view there is nothing more to be said. However, a listener listening to the sum of 1000 and 1001 Hz sine tones does not have the impression of listening to two different frequencies. Instead the listener hears a single pure tone, with a pitch corresponding to about 1000.5 Hz and with a loudness that waxes and wanes with a period of one second. The loudness variation is known as beats.

The beat phenomenon is not the result of human hearing, nor is it the result of nonlinear processing. It is present in the linear response of a system. For example, if signal $x(t)$ is displayed on an oscilloscope, which is a linear device, the pattern shows the same amplitude variation as is heard by a listener.

It is possible to find a mathematical expression that captures the beating phenomenon by introducing the idea of a time-dependent amplitude. Within the formal

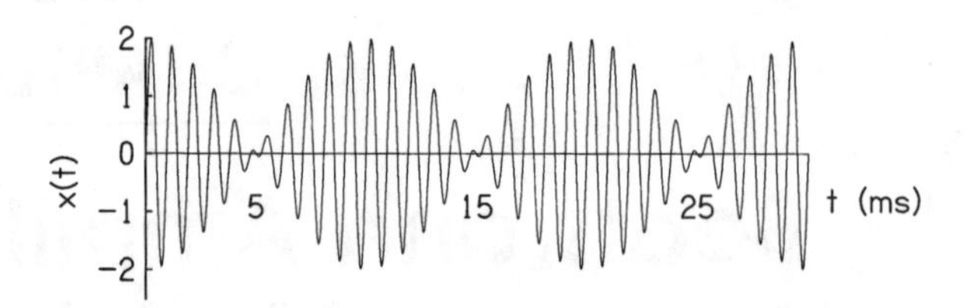

FIGURE 17.1. The sum of 1000 Hz and 1100 Hz sines with equal amplitude.

context of the Fourier transform, amplitudes cannot be time dependent. The mathematical treatment of beats requires a violation of the usual rules for the definition of amplitude, but the insight that is gained makes it worthwhile to bend the rules.

To obtain a form with a time-dependent amplitude $A(t)$, we try to represent $x(t)$ by something of the form $A(t)\exp(i\theta)$. A good way to begin is to rewrite Eq. (17.1) as

$$x(t)=\mathrm{Im}[e^{i\omega_1 t}+e^{i(\omega_2 t+\phi)}], \tag{17.2}$$

where the symbol ''Im'' means that we take the imaginary part of what follows.

Then, with the difference between the two angular frequencies defined as

$$\Delta\omega=\omega_2-\omega_1=2\pi(f_2-f_1), \tag{17.3}$$

and the average angular frequency defined as

$$\bar{\omega}=(\omega_1+\omega_2)/2, \tag{17.4}$$

Eq. (17.2) becomes

$$\begin{aligned} x(t)&=\mathrm{Im}[e^{-i[(\Delta\omega/2)t+\phi/2]}+e^{i[(\Delta\omega/2)t+\phi/2]}]e^{i(\bar{\omega}t+\phi/2)} \\ &=2\cos\left(\frac{\Delta\omega}{2}t+\frac{\phi}{2}\right)\mathrm{Im}[e^{i(\bar{\omega}t+\phi/2)}]=2\cos\left(\frac{\Delta\omega}{2}t+\frac{\phi}{2}\right)\sin(\bar{\omega}t+\phi/2). \end{aligned} \tag{17.5}$$

This expression shows the essence of the beating effect. There is rapid oscillation at the average frequency, and there is a slowly oscillating prefactor that represents the waxing and waning of the signal level. The prefactor oscillates at a rate given by $(f_2-f_1)/2$, but there are nevertheless (f_2-f_1) maxima per second, one when the cosine prefactor is $+1$, the other when it is -1. The ear cannot distinguish between these.

Figure 17.1 shows an oscillogram, 30 ms in duration, in which two sines of equal amplitude are beating. One sine has a frequency of 1000 Hz, the other a frequency of 1100 Hz. These two frequencies differ by 10 percent, one hundred times greater than our auditory example where the frequencies differed by 0.1 percent. Such a

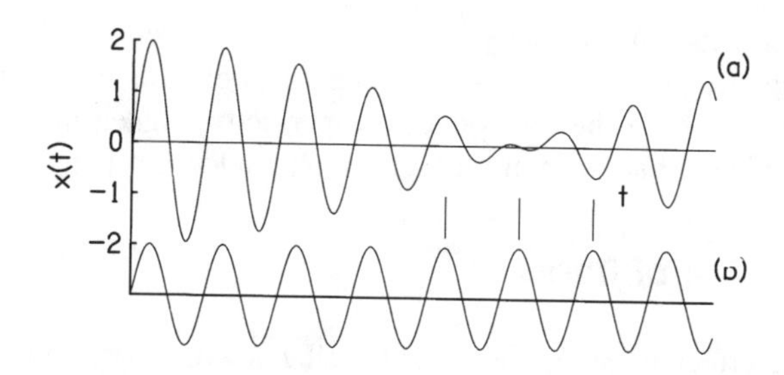

FIGURE 17.2. A beating pair (a) and a sine tone (b). Both have a frequency of 1050 Hz to illustrate the phase reversal at zero amplitude of the beating pair.

large frequency difference makes it possible to see the details within the beat cycle in a figure of the waveform such as Fig. 17.1. By contrast, the auditory example has 100 times more cycles of the waveform within any beat cycle. [Note: Except for the visual advantage, the case of 1000 Hz and 1100 Hz is a rather bad example of beating. When one listens to two tones with frequencies of 1000 and 1100 Hz one perceives two tones that sound very rough, like two musical instruments that are badly out of tune. One cannot follow the beats in time.]

Equation (17.5) is mathematically correct, but it is not quite what we were seeking when we began. We were looking for a time-dependent amplitude, but we found a slowly oscillating cosine prefactor instead. The problem is that the cosine function is negative half the time, whereas an amplitude can never be negative. To get it right, we can replace the cosine factor by using the trigonometric relation

$$2\cos(\Delta\omega\, t/2+\phi/2)=\pm\sqrt{2[1+\cos(\Delta\omega\, t+\phi)]}, \tag{17.6}$$

and then associate only the positive square root with the amplitude.

We still have to deal with the fact that the slowly varying cosine term in Eq. (17.5) creates an alternating sign. Our only recourse is to associate that alternating sign with the rapidly oscillating part. Therefore, the price that we pay for getting the amplitude formally correct is that we have introduced discontinuities into the function that was otherwise oscillating smoothly at the average frequency of $\bar{\omega}$. That means that we must write that function (which is the signal with the amplitude variation removed) as

$$=\text{Sign}\left[\cos\left(\frac{\Delta\omega}{2}t+\frac{\phi}{2}\right)\right]\sin(\bar{\omega}t+\phi/2). \tag{17.7}$$

This function changes its sign whenever the amplitude goes to zero. *A priori* it seems egregiously perverse to introduce such discontinuities, even though they take place where the amplitude is momentarily zero. As we shall see below, however, there is an advantage to doing so.

Figure 17.2 shows the sudden phase change. Part (a) is the beginning of the signal in Fig. 17.1, expanded horizontally by a factor of four. Part (b) shows a sine tone with the average frequency, 1050 Hz. A comparison of parts (a) and (b) shows that the peaks and valleys in the two waves are perfectly aligned until the amplitude goes to zero. At just this point in time the signal suddenly changes its sign.

Physical Nature of Beats

The beating effect is easily heard and well known. A common mistake is to imagine that the beat frequency $\Delta\omega$ must somehow appear in the spectrum of a tone. It does not. The spectrum consists only of the original two frequencies. Beating is a way to describe the behavior of a signal in time and is not directly related to the spectrum.

Further confusion in this direction is caused by the fact that if the signal is passed through a system that is nonlinear, then a component with a frequency of $\Delta\omega$ *does* appear in the output spectrum. This is the simple-difference tone, and it occurs in a system with even-order distortion, for example when there is a distortion product in the output that is proportional to the square of the input signal. None of this has anything to do with beats. Beats occur in a system that is perfectly linear.

BEATS OF SINES WITH UNEQUAL AMPLITUDES

Because beating is a form of wave interference, one would suppose that when the amplitudes of the two sines are not the same, the mutual cancellation will be less complete and the resulting time-dependent amplitude will not become as small as zero. The formal treatment below shows that this is indeed what happens. However, there is more. There is another effect, not anticipated by the above reasoning, namely periodic changes in the frequency of the beating tone.

To study beating with unequal amplitudes, we begin as before with the sum of two sines,

$$x(t)=a\ \sin(\omega_1 t)+b\ \sin(\omega_2 t+\phi). \tag{17.8}$$

With $\Delta\omega$ and $\overline{\omega}$ defined as above, we separate the signal into two factors, one with the average frequency, the other with a low frequency (half the difference frequency),

$$x(t)=\mathrm{Im}[ae^{-i(\Delta\omega\, t/2+\phi/2)}+be^{i(\Delta\omega\, t/2+\phi/2)}]e^{i(\overline{\omega}t+\phi/2)}, \tag{17.9}$$

or

$$x(t)=\mathrm{Im}\left[(b+a)\cos\left(\frac{\Delta\omega}{2}t+\frac{\phi}{2}\right)+i(b-a)\sin\left(\frac{\Delta\omega}{2}t+\frac{\phi}{2}\right)\right]e^{i(\overline{\omega}t+\phi/2)}. \tag{17.10}$$

Because our goal is to find a time-dependent amplitude, we write $x(t)$ in the form $A(t)\exp(i\theta)$, where $A(t)$ is the magnitude of the slowly varying factor in the square brackets,

$$A(t)=\sqrt{a^2+b^2+2ab\cos(\Delta\omega t+\phi)}. \tag{17.11}$$

The minimum amplitude is $|a-b|$. If sine amplitudes a and b are not equal, amplitude A never becomes zero. This was our anticipated result. If a and b are both equal to unity then amplitude A agrees with the amplitude in Eq. (17.6), as it must.

The instantaneous phase of the expression in Eq. (17.10) is complicated. It includes two parts that come from the average signal: a constant part $\phi/2$ and a part that is directly proportional to time, namely $\bar{\omega}t$. Then there is a part that is the argument of the factor in square brackets. We shall call the argument θ'. The tangent of θ' is given by the ratio of the real and imaginary parts within the square bracket,

$$\tan\theta'=\frac{b-a}{b+a}\tan(\Delta\omega\, t/2+\phi/2). \tag{17.12}$$

This equation can be solved for θ'

$$\theta'=\tan^{-1}\left[\frac{b-a}{b+a}\tan(\Delta\omega\, t/2+\phi/2)\right], \tag{17.13}$$

though we must bear in mind that the principal value of the arctangent may not be adequate when the real part in the square brackets of Eq. (17.10) becomes negative [see the treatment of this problem in Eq. (2.6) of Chapter 2].

Frequency Modulation in Beats

The instantaneous frequency is the time derivative of the phase,

$$\omega(t)=d/dt(\phi/2+\bar{\omega}t+\theta'). \tag{17.14}$$

After some algebra, the derivative of θ' simplifies considerably and the instantaneous frequency becomes

$$\omega(t)=\bar{\omega}+\frac{(\Delta\omega/2)(b^2-a^2)}{a^2+b^2+2ab\cos(\Delta\omega\, t+\phi)}. \tag{17.15}$$

People who are familiar with the sound of beats find the frequency variation shown in Eq. (17.15) to be surprising. When the cosine in the denominator becomes equal to -1, the excursion to higher frequencies can be considerable. However, Eq.

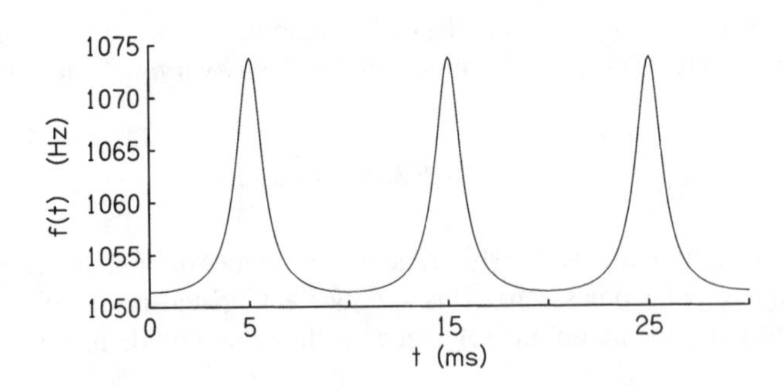

FIGURE 17.3. The instantaneous frequency of a beating pair of sines with frequencies of 1000 Hz and 1100 Hz and an amplitude ratio of 0.9 to 1. The phase ϕ is zero, so the features of this figure can be directly compared with Fig. 17.1. Feth and O'Malley (1977) showed that this signal can be distinguished from a signal with a ratio 1 to 0.9 with no error at all.

(17.11) shows that it is at just this point that the amplitude goes through a minimum, and this tends to obscure the frequency variation. Helmholtz (1885 Appendix XIV) claimed to be able to hear the associated pitch change, though his contemporary Stumpf did not. Békésy (1960, p. 575) sided with Helmholtz. Finally, Jeffress (1968) discovered that anyone can hear the pitch change by listening to only a portion of the beat cycle, a time slice that includes the minimum amplitude and maximum pitch change.

Figure 17.3 shows the frequency modulation for the 1000-Hz and 1100-Hz beating pair, as in Figs. 17.1 and 17.2, except that the amplitude of the lower frequency component a is 90% of the amplitude of the higher component b. As before, the frequency difference of 100 Hz must be reduced in order to get perceptually countable beats.

In the limit that $a=b$, the frequency excursion diverges, but it does so only when the amplitude is strictly zero [compare with Eq. (17.11)]. This is the behavior expected from our treatment of the beats of equal-amplitude sines. The discontinuity in the function in Eq. (17.7) appears as a momentarily infinite frequency because the phase changes instantaneously by 180 degrees, as shown in Fig. 17.2.

Pitch Shift

Whether or not one can hear the time-varying pitch in an extended beating tone, one can detect a change in the overall pitch of a beating tone when the relative amplitudes a and b are changed (Helmholtz knew about that too). The pitch increases when the amplitude of the higher frequency component b becomes larger than the amplitude of the lower frequency component a. This effect was studied by Feth (1974) who showed that a listener's ability to discriminate between conditions $a>b$ and $b>a$ can be predicted by calculating an envelope-weighted average of the instantaneous frequency in Eq. (17.15), and in Fig. 17.3 (the EWAIF model). Subsequently, Feth, O'Malley, and Ramsey (1982) studied the pitch difference between these two conditions and concluded that pitch was the most likely basis for discrimination.

In 1993, Anantharaman, Krishnamurthy, and Feth (1993) concluded that psychoacoustical data could be equally well fitted by a model in which instantaneous frequency was weighted by the instantaneous intensity (IWAIF model). Dai (1993) found that the IWAIF was actually more successful than the EWAIF. The IWAIF model is considerably easier to calculate because it corresponds to the centroid of the spectrum of the beating tones. This is an example of the situation described in Chapter 16, where the time order of events, here instantaneous frequency, is unimportant. The calculation can be done with a probability density function that is the power spectrum. The predicted pitch is the frequency given by Eq. (17.15) when the cosine factor in the denominator is set equal to zero.

The frequency variation caused by beating components can be an important ingredient in the perception of dense spectral complexes (for example, profile analysis experiments) when the separation of components becomes small. Weighting instantaneous frequency by the envelope, or by the intensity, is an idea that logically requires all the spectral components to be in the same critical band. The work of Versfeld and Houtsma (1992, 1995) on beating tone pairs suggests that the EWAIF model can account for the detection of amplitude changes in two-tone complexes only if the two tones are separated by a semitone or less.

AMPLITUDE MODULATION

The above treatment of beats began with the spectrum and derived an expression for a time-varying amplitude. The following treatment of modulation reverses the order of those steps.

The idea and terminology of amplitude modulation (AM) owes much to a venerable body of radio engineering. The AM radio signal begins with a sine oscillator with an angular frequency ω_c in the range of megahertz or hundreds of kilohertz. This is the radio-frequency (RF) signal and it is called the "carrier." The amplitude of this signal is varied (modulated) by an audio signal, $s(t)$, which is the program material to be broadcast. The audio material has a maximum frequency of 10 or 20 kHz. The changes caused by the modulation are, therefore, at least a factor of ten slower than the oscillations of the carrier. The final radio signal is then

$$x(t)=[1+s(t)]\sin(\omega_c t). \tag{17.16}$$

The quantity in the square brackets represents the amplitude. It changes as audio signal $s(t)$ changes, but in order to be a proper amplitude it must remain positive. Therefore, signal s must not become more negative than -1. If it does, the AM signal is said to be "overmodulated" and distortion may result. It is also generally assumed that $s(t)$ is entirely oscillatory, i.e. that $s(t)$ has no dc value. The only dc value in the amplitude comes from the offset, the number 1.

We now consider the special case where the audio signal is a pure tone with frequency ω_m and amplitude m,

$$x(t)=[1+m\cos(\omega_m t+\phi)]\sin(\omega_c t). \tag{17.17}$$

The quantity m is positive and it should not be greater than 1, as will be shown below. Expressed as a percent, m is known as the "modulation percentage."

When an AM signal is used in hearing research, there are no radio frequencies involved; all signals are at audible frequencies. The mathematics of the AM signal, however, remains the same, and the same terminology (carrier, modulating signal, modulation percentage) etc. is used.

The Spectrum of AM

It is revealing to calculate the spectrum of the AM signal. The spectrum becomes apparent when we rewrite the function $x(t)$ as a sum of sines or cosines. This is easy to do using the trigonometric identity,

$$\cos(B)\sin(A)=\frac{1}{2}[\sin(A+B)+\sin(A-B)]. \tag{17.18}$$

Therefore,

$$x(t)=\sin(\omega_c t)+(m/2)\sin[(\omega_c+\omega_m)t+\phi]+(m/2)\sin[(\omega_c-\omega_m)t-\phi]. \tag{17.19}$$

The spectrum includes a component at the carrier frequency as well as modulation components that are above and below the carrier. The modulation components are known as "sidebands." They are displaced from the carrier by the audio modulation frequency ω_m. In an AM signal the upper and lower sidebands always have the same amplitude. Their amplitudes are half that of the carrier if $m=1$, i.e. if the modulation percentage is 100%.

The negative frequencies in Fig. 17.4 come from the negative frequencies in its sine-tone factors, as described at the beginning of Chapter 8. The carrier itself has a positive and negative frequency. The modulator also has a positive and negative frequency. The latter can be seen to be responsible for the fact that there are two sidebands, by writing Eq. (17.17) in terms of complex exponentials using Euler's formula.

If the modulating signal $s(t)$ contains several different frequencies, the spectrum of signal $x(t)$ can be calculated by repeating the step above for each component frequency in the modulation. It is evident that each modulation Fourier component leads to an upper sideband and to a lower sideband. If the spectrum of the modu-

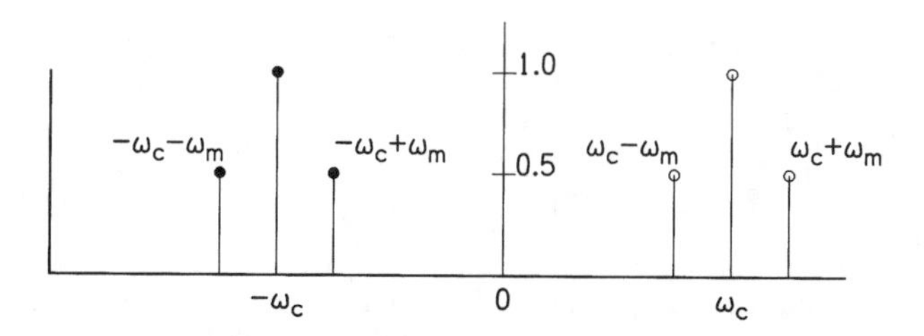

FIGURE 17.4. The amplitude spectrum of an amplitude modulated signal, $m=1$.

lating signal is a continuous band, as shown in Fig. 17.5(a), then the spectrum of the modulated signal is simply given by placing continuous sidebands on either side of the carrier, as shown in Fig. 17.5(b).

Power in an AM Signal

The average power of a signal is increased when the carrier is modulated. By the definition in Eq. (3.3) the average power is given by averaging the square of the signal over time. If the signal is the AM signal from Eq. (17.16), then the average power becomes

$$\bar{P}=\lim_{T_D\to\infty}\frac{1}{T_D}\int_0^{T_D}dt[1+s(t)]^2\sin^2(\omega_c t). \tag{17.20}$$

To do this integral it is convenient to rewrite the squared trigonometric functions in an expanded form, e.g.

$$\sin^2(\omega_c t)=\frac{1}{2}-\frac{1}{2}\cos(2\omega_c t). \tag{17.21}$$

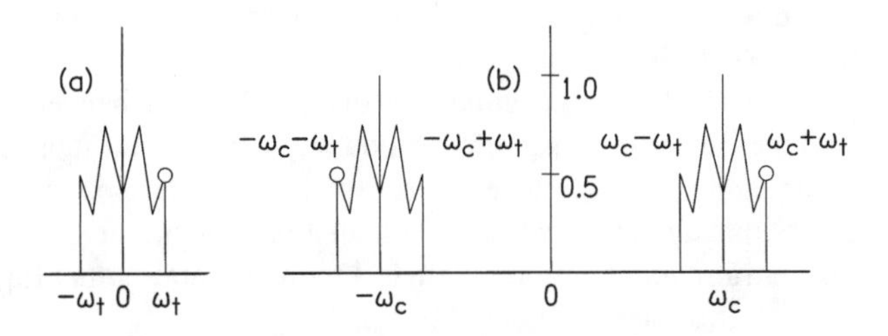

FIGURE 17.5. (a) Spectrum of a broad-band audio signal. (b) Spectrum of a sine with frequency ω_c modulated by the signal of (5a). The positive frequency of one particular sine component in the modulator is labelled with a circle and followed through the modulation process.

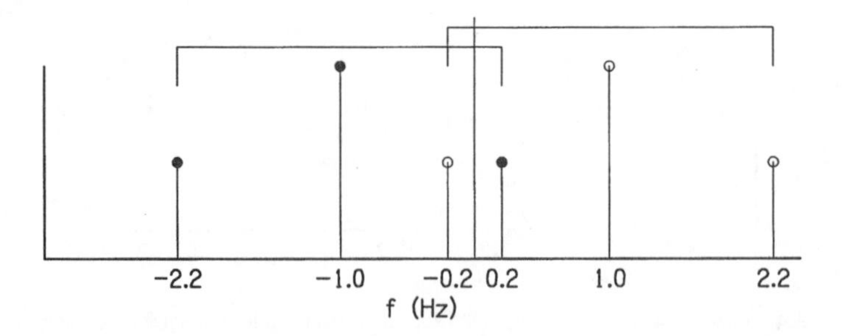

FIGURE 17.6. Amplitude spectrum of an AM signal where the carrier is 1000 Hz and the modulation is 1200 Hz. Because the modulation percentage is 100% the sideband amplitudes are half the carrier amplitude. Brackets show positive and negative carrier frequencies and their associated sidebands.

Whether or not further simplification is possible depends upon the circumstances. In the usual case, signal $s(t)$ has no average value and has no spectral components at the carrier frequency or at twice the carrier frequency. In this case the average power becomes

$$\overline{P}=\frac{1}{2}+\frac{1}{2}\frac{1}{T_D}\int_0^{T_D} dt\; s^2(t), \tag{17.22}$$

or simply

$$\overline{P}=\frac{1}{2}+\frac{1}{2}\overline{P_A}, \tag{17.23}$$

where $\overline{P_A}$ is the average power in the modulating signal itself.

Equation (17.23) says that the power in the modulated signal is equal to the power in the carrier (1/2) plus half the power in the modulating signal. In a modulation detection task it may be necessary to compensate for the increased total power as the modulation is increased, if one wants to avoid giving the listener a loudness cue (e.g. Yost, Sheft, and Opie, 1989).

In order for Eq. (17.23) to hold good, the overlap integral between the modulating signal and the carrier must be zero. (Modulation and carrier must be orthogonal.) If modulation and carrier are independent tones with line spectra, there is only a small probability that a component of the modulation coincides exactly with some frequency in the carrier, and therefore, there is a high probability that Eq. (17.23) is valid. Exceptions to this case are treated in Exercise 12.

There is, of course, nothing in the AM process itself that requires that the signal and the carrier be in different frequency ranges. It is not even necessary that the carrier frequency be higher than the modulation frequency. The mathematics that we have developed above still holds good. Figure 17.6 shows what happens when the carrier is 1000 Hz and the modulation is 1200 Hz. For positive frequencies there are sidebands at 2200 and 200 Hz, and it looks as though the sidebands are not symmetrically placed about the carrier. However, the symmetry becomes apparent when we consider both the positive and the negative frequency axis.

BALANCED MODULATION

Balanced modulation is just the same as amplitude modulation except that the modulating signal is not offset. Therefore, if signal $s_m(t)$ modulates signal $s_c(t)$ in a balanced modulator, the output signal is given by

$$x(t)=s_m(t)s_c(t). \tag{17.24}$$

The product is symmetrical in s_m and s_c, and there is no distinction between the modulating signal and the carrier. If s_m and s_c are cosine and sine waves with frequencies ω_m and ω_c, and relative phase ϕ, then

$$x(t)=\cos(\omega_m t)\sin(\omega_c t+\phi). \tag{17.25}$$

To make the spectrum obvious, we can use the trigonometric identity in Eq. (17.18) to write this as a sum of individual components,

$$x(t)=\frac{1}{2}\sin[(\omega_c+\omega_m)t+\phi]+\frac{1}{2}\sin[(\omega_c-\omega_m)t+\phi]. \tag{17.26}$$

Because the signal is the sum of two sines, the spectrum consists of only two components: they are the sum and difference frequencies. Unlike the case of AM, no trace remains of an original frequency, ω_c or ω_m, so long as neither carrier nor modulation contains a DC component.

If ω_c is a radio frequency carrier and ω_m is an audio modulating signal, the balanced modulation product is called ''double-sideband suppressed-carrier.'' With respect to AM, the modulation percentage is infinite.

Alternatively both ω_c and ω_m may be audio or subaudible frequencies. The above equations show that there is no real difference between balanced modulation and the beating of two sine tones with equal amplitudes. For example, the beating of a 1000-Hz sine and a 1001-Hz sine leads to the same signal as the balanced modulation of a 1000.5-Hz sine carrier by a 0.5-Hz modulating sine.

In practice, if we have the goal of creating a signal defined as the beating of equal-amplitude sines, then there are two important advantages to the balanced modulation technique: the frequency difference between the sidebands, or beat rate, can be more precisely controlled, and the amplitudes of the two components are automatically created equal.

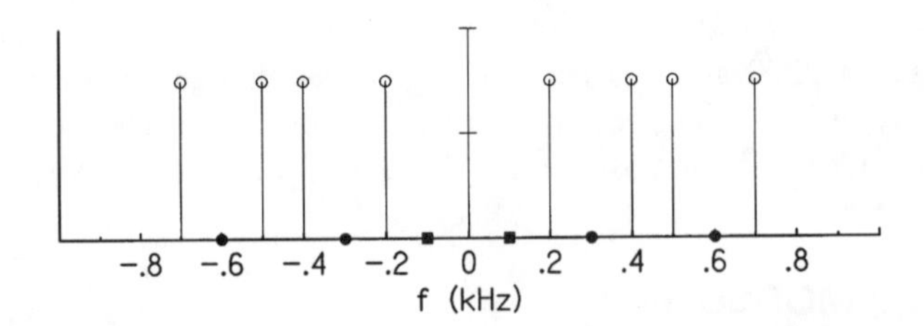

FIGURE 17.7. The spectrum of the product of a two-component tone and a sine tone. The components of the original tones (shown by solid circles and squares) do not appear in the output spectrum.

Balanced Modulation of Complex Sounds

In the case that signal and carrier are complex signals, the spectrum of the balanced modulation product consists of components with frequencies given by the sum of each carrier frequency with each modulating frequency and by the difference (absolute value) between each carrier frequency and each modulating frequency. To prove this requires that one simply repeat the trigonometric substitution of Eq. (17.18) for all possible pairings of carrier components with modulation components. For example, if $s_m(t)$ is the sum of 300 and 600 Hz, with equal amplitudes, and $s_c(t)$ is a sine of 100 Hz, the balanced modulation product has components at harmonics 2, 4, 5, and 7 of 100 Hz, as shown in Fig. 17.7.

BEATS AND THE FREQUENCY-DOMAIN GRATING

Chapter 13 introduced a spectrum called the ''rake,'' consisting of equally spaced sine components. There we considered the case where the spectral rake had an appreciable bandwidth. The frequencies of the highest and lowest components were far enough apart that one could distinguish pitches at both these edges. In this section we consider the case where the frequency separation between components, ω_0, is so small that a human listener cannot resolve the two edges. Then it makes sense to write the waveform $x(t)$ in an alternative way that exhibits beats.

We begin this alternative treatment by reproducing the rake from Eq. (13.37) except that we emphasize that ω_0 is small by writing it as $\delta\omega$. Further, the components will be labelled by subscript m instead of n because symbol n will ultimately be used for something else.

$$x(t) = \mathrm{Im}\, \frac{e^{i(m_B - 1/2)\delta\omega t} - e^{i(m_T + 1/2)\delta\omega t}}{e^{-i\delta\omega t/2} - e^{i\delta\omega t/2}}. \tag{17.27}$$

Including the bottom component number m_B and the top component number m_T, there are M components. The following steps are simpler to describe if M is an odd number, and we will assume it to be so. If M is odd, then there is a component in the center. We define its number to be $\bar{m}$. The number of components above $\bar{m}$ is $(M-1)/2$, and there is an equal number of components below. Therefore, the top and bottom component numbers can be written as $m_T = \bar{m} + (M-1)/2$ and $m_B = \bar{m} - (M-1)/2$.

The crucial step in our alternative treatment is to extract the component at the center as a common factor. Then

$$x(t) = \text{Im}\,\frac{e^{i\bar{m}\,\delta\omega\, t}[e^{-iM\,\delta\omega\, t/2} - e^{iM\,\delta\omega\, t/2}]}{e^{-i\,\delta\omega\, t/2} - e^{i\,\delta\omega\, t/2}}, \tag{17.28}$$

which reduces to

$$x(t) = \frac{\sin\, M\,\delta\omega\, t/2}{\sin\,\delta\omega\, t/2}\,\sin(\bar{m}\,\delta\omega t). \tag{17.29}$$

This expression contains several different oscillations; one has frequency $\bar{m}\,\delta\omega$, another has frequency $\delta\omega/2$, while a third is M times faster. The center component number $\bar{m}$ is presumed to be considerably larger than the number of components in the rake, M. Therefore, the signal in Eq. (17.29) is effectively a sine tone with frequency $\bar{m}\,\delta\omega$ and with an amplitude given by the complicated fraction in the prefactor. This prefactor describes the beats. Its periodic variation is given by $\delta\omega/2$. The expression for the amplitude can be written in terms of frequency separation δf,

$$B(t) = \frac{\sin\,\pi M\,\delta f\, t}{\sin\,\pi\,\delta f\, t}. \tag{17.30}$$

The idea of the frequency-domain grating is that the beating effect from Eq. (17.30) can be used to analyze a complex tone into its individual harmonics, by spreading them out in time. The object to be analyzed is a periodic complex tone with fixed harmonic amplitudes and phases,

$$x_1(t) = \sum_{n=1}^{N} A_n \cos(2\pi n f_0 t + \phi_n). \tag{17.31}$$

To make the grating, we add up M tones, just like $x_1(t)$ except that for each successive tone we increase the fundamental frequency by δf_0. For the m-th tone, then, the fundamental frequency is

$$f_0 + m\,\delta f_0. \tag{17.32}$$

The increment δf_0 is very small, e.g., 0.1 Hz. All the tones remain strictly periodic, and, because the fundamental shift is so small, the corresponding harmonics of all M tones are close together in frequency.

As given in Eq. (17.30), each harmonic of the complex tone goes through a beating pattern. The key to the frequency-domain grating is that each harmonic

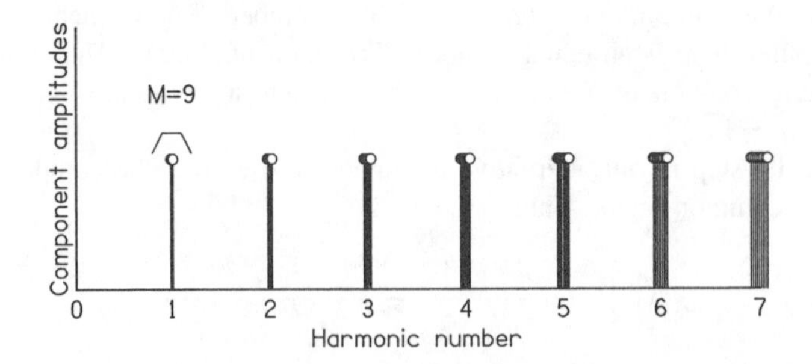

FIGURE 17.8. Adding up nine complex tones with slightly shifted fundamental frequencies creates a spectral rake for each harmonic.

beats at a different rate. The second harmonic beats twice as fast as the fundamental, the third harmonic beats three times as fast, etc. Thus, the amplitude of the n-th harmonic in the sum is

$$B_n(t)=A_n\,\frac{\sin(\pi Mn\,\delta f_0 t)}{\sin(\pi n\,\delta f_0 t)}. \tag{17.33}$$

The period of the entire pattern is $T_B=1/(\delta f_0)$, e.g. ten seconds for a separation of 0.1 Hz. The fundamental frequency appears once in that time period. The n-th harmonic appears on n occasions in that period, equally spaced. For example, the seventh harmonic appears at times $t=0,\ T_B/7,\ 2T_B/7,\ldots 6T_B/7$.

To illustrate the grating, we take a particular example, as used by composer Jean-Claude Risset. The periodic complex tone $x_1(t)$ is a low C, with fundamental frequency of 65.4 Hz, and it has the first seven harmonics ($N=7$) all of equal amplitude ($A_n=A_0$).

We analyze this tone by adding up nine versions ($M=9$) of $x_1(t)$ with fundamental frequencies 65.0, 65.1, 65.2, 65.3, 65.4, 65.5, 65.6, 65.7, and 65.8 Hz. The average fundamental is 65.4 Hz. The 63 components (9 complex tones$\times$7 harmonics) are shown in Fig. 17.8.

The "amplitudes" of the analyzed complex tone, as calculated from Eq. (17.33) are shown in Fig. 17.9. The true amplitudes are the magnitudes of the values that appear in the figure; both positive and negative values are retained in the figure to reduce the density of the spaghetti there. The individual amplitude peaks are heard as distinct musical tones, resulting in the melody shown in Fig. 17.9.

The heavy line in Fig. 17.9 shows the beat pattern for the first harmonic. This same pattern occurs for all the harmonics, but at different rates.

The melody in Fig. 17.10 is determined by the fact that the spectrum of the original complex tone $x_1(t)$ has the first seven harmonics. So long as the harmonic amplitudes are not zero, the order and timing of the notes in the melody do not depend upon the amplitudes or phases of the harmonics. Nor do they depend upon the number of complex tones added together to make the grating. Therefore, the

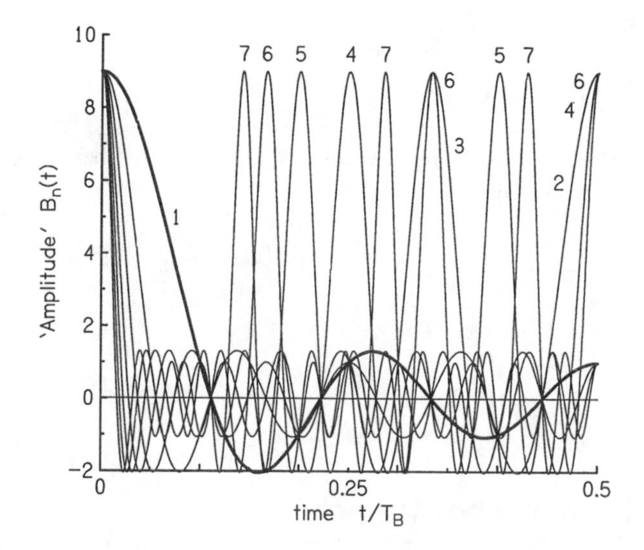

FIGURE 17.9. "Amplitudes" of the 7 harmonics in a complex tone, as revealed by the frequency-domain grating of 9 complex tones. The figure shows only half a beat cycle; the second half is just the first half played backward.

melody is appropriately called the "N=One to Seven" melody. The mathematical development of the grating shows that the frequency-domain grating analyzes tones in the same way that an optical diffraction grating analyzes white light into its constituent colors (Hartmann, 1985).

FIGURE 17.10. The tones of the "N=One to Seven" melody. As indicated in Fig. 17.9, the melody is a palindrome, where the time arrow shows the retrograde symmetry. The melody repeats indefinitely. The actual tones sound an octave lower than written here.

EXERCISES

Exercise 1: Plus ça change...

Replace the sines in Eq. (17.1) by cosines and derive the beating equation. In other words, show that if

$$x(t) = \cos(\omega_1 t) + \cos(\omega_2 t + \phi),$$

then

$$x(t) = 2\cos\left(\frac{\Delta\omega}{2}t + \frac{\phi}{2}\right)\cos(\overline{\omega}t + \phi/2).$$

Exercise 2: FM of a beating pair

Consider the case of two beating sines of unequal amplitude. The equation for the instantaneous frequency includes an average $\overline{\omega}$ and a deviation. Show that the maximum and minimum values of the deviation are respectively $(\Delta\omega/2)(b+a)/(b-a)$ and $(\Delta\omega/2)(b-a)/(b+a)$. Show that the overall frequency excursion is $(2\Delta\omega)ab/(b^2-a^2)$.

Exercise 3: IWAIF

Show that the IWAIF model predicts that the pitch for the signal in Fig. 17.3 is 1055 Hz.

Exercise 4: Spectrum via convolution

An AM signal is the product of a time-varying amplitude. $[1+s(t)]$, and a carrier $\sin(\omega_c t)$. Therefore the spectrum of the AM signal should be the convolution of the spectrum of $[1+s(t)]$ and the spectrum of the carrier. Show that this is so for $s(t) = \cos \omega_m t$.

Exercise 5: Power in SAMS

(a) Show that if a unit-amplitude sine carrier is amplitude modulated by a sine tone with modulation percentage m $(0 \leqslant m < 1)$, then the power in the sine-amplitude-modulated sine is $1/2 + (1/4)m^2$.

(b) Show that by adding the sine modulation, with a modulation percentage of 100, we increase the level of the signal by 1.76 dB.

Exercise 6: Four-quadrant devices

Electronic analog signal multipliers form the product of two input signals, $x(t)$ and $y(t)$. Possible values of the two signals are on the plot described by $[x,y]$ pairs.

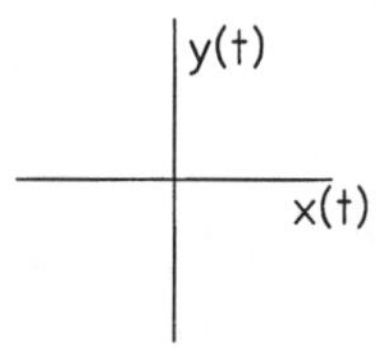

Explain why balanced modulation requires a four-quadrant multiplier whereas amplitude modulation requires only a two-quadrant multiplier.

Exercise 7: Overmodulation

For an overmodulated AM signal, the modulation percentage m is larger than 1. What happens in this case depends upon the kind of modulator.

(a) In a two-quadrant multiplier the output becomes zero when the modulating voltage becomes negative. Show that with a two-quadrant modulator, overmodulation leads to harmonic distortion in the sense that additional sidebands are added.

(b) Show that if modulation is done with a four-quadrant multiplier, then no new harmonics are introduced as m increases from a value less than 1 to a value greater than 1. Show also that if the overmodulated signal is detected (demodulated) using a half-wave rectifier and a low-pass filter, then distortion products, in the form of harmonics of the modulating signal, occur in the output of the detector. These distortion products do not occur if the signal is not overmodulated.

(c) Dye, Niemiec, and Stellmack (1994) studied the lateralization of a high-frequency modulated tone. Their signal was the sum of three sines of equal amplitude,

$$x(t)=\sin \omega_c t+\sin(\omega_c+\omega_m)t+\sin(\omega_c-\omega_m)t.$$

Show that this signal corresponds to 200 percent modulation, and show that the envelope of the signal is $|1+2\cos \omega_m t|$. This envelope has one major bump and one minor bump per cycle.

Exercise 8: Ring modulation

''Musique concrete'' is made by recording sounds, perhaps the sounds of musical instruments but not necessarily, and processing them electronically. A common technique is to modulate one sound by another. Balanced modulation generally creates a more dramatic processing effect than amplitude modulation. Explain why this is so with reference to the spectra for the modulated signals. [Note: Electronic musicians frequently refer to a balanced modulator as a "ring modulator" because multiplication of two signals can be done (but not well) with an electronic circuit that is a ring of four diodes.]

Exercise 9: Really equivalent?

The text says that balanced modulation of two sines is entirely equivalent to the beating of two sines. There is reason to doubt this statement because the two components in Eq. (17.1) have different starting phases, whereas the components in Eq. (17.26) have equal starting phases. Show that by choosing the values of phases and the origin of time you can make Eq. (17.26) agree with Eq. (17.1).

Exercise 10: Multiplied noise

Suppose that we need a very narrow band of noise, from 990 Hz to 1010 Hz, where the edges are so sharp that the noise level is down by 12 dB at 980 Hz and 1020 Hz. To create such a band with a filter would require passing the signal from a noise source through a bandpass filter with skirts of ± 840 dB/octave. That is serious filtering! By using a balanced modulator, however, it is possible to create this narrow noise band with a simple two-pole low-pass filter, with a skirt of only -12 dB/octave.

The technique is to create a narrow noise band from zero to 10 Hz, using a noise source and a two-pole low-pass filter, then to multiply this noise in a balanced modulator by a 1000-Hz sine tone. The result is called "multiplied noise." (a) Show that the multiplied noise band extends from 990 to 1010 Hz. (b) Show that the band is down by 12/dB at 980 and 1020 Hz.

Exercise 11: No DC allowed!

The spectrum for balanced modulation, as described after Eq. (17.26), is valid if neither signal has a DC component (i.e. a component at zero frequency.) Show why balanced modulation starts to resemble amplitude modulation if either signal acquires a DC component.

Exercise 12: Messy world

Equation (17.23) for the power in an AM signal is valid if the average value of the modulating signal is zero and if the modulating signal is orthogonal to the carrier and twice the carrier. Failing that, Eq. (17.23) must be corrected. Use Eq. (17.20) to show that the correction terms that must be added to the right-hand side of Eq. (17.23) are:

$$\Delta \bar{P} = \lim_{T_D \to \infty} \frac{1}{T_D} \int_0^{T_D} dt \left[s(t) - \cos(2\omega_c t) s(t) - \frac{1}{2} \cos(2\omega_c t) s^2(t) \right].$$

Exercise 13: FM zilch

Show that there is no FM in an AM signal as given by Eq. (17.19), so long as the modulation percentage is less than 100 percent.

Exercise 14: Even-numbered grating

Consider the frequency-domain grating when the number of tones that are added together is even.

(a) Show that $x(t)$ is again given by Eq. (17.29) except that $\bar{m}$ now is a half-integer number at the center of the band.

(b) If M is an odd number then $B(t)$ has period given by $1/(\delta f)$. Show that if M is an even number then $B(t)$ has period given by $1/(2\delta f)$ because $B(t)$ changes its sign after a time $1/(\delta f)$.

Note, however, for low beat rates, the beating sensation depends upon $B^2(t)$, and the sign change is not noticed. Therefore, for either odd or even values of M the perceived period is $1/(\delta f)$.

CHAPTER 18

The Envelope

And the noise was in the beast's belly.
SIR THOMAS MALORY

An audible sine tone may have thousands of cycles per second, but we perceive it as a steady sound. That is because its envelope is a constant. By contrast, signals that are heard as fluctuating have envelopes that change in time. For example, Fig. 18.1 shows a waveform constructed by adding five sine tones with different frequencies in a narrow band around 1000 Hz. It is evident that the waveform varies in time on two different time scales. One scale is the rapid variation at approximately 1000 Hz; the individual cycles are barely distinguishable in the figure. The other time scale is the slower variation, with a periodicity of one-tenth of a second, outlining the shape of the waveform. This slower variation is the envelope, shown by the dashed line, which *envelopes* (accent is on the second syllable of this verb) the signal. The envelope is a useful concept for a signal like a narrow band of noise where the high-frequency character of the signal is sinusoidal, but the signal appears to have an amplitude that fluctuates. There is little difference, if any, between the concept of a time-varying amplitude and an envelope.

FORMAL EVALUATION OF THE ENVELOPE

This section begins with a perfectly general signal $x(t)$ and derives its envelope $E(t)$. Any signal, periodic or not, with a finite number of components can be written as a sum of cosine waves:

$$x(t)=\sum_{n=1}^{N} C_n\cos(\omega_n t+\phi_n), \tag{18.1}$$

where amplitudes C_n are positive real numbers, and there are no restrictions on the frequencies that are included in the sum.

To find the envelope of $x(t)$ we begin by extracting a sinusoidal character by writing each frequency ω_n as the sum of a characteristic frequency plus a deviation, δ_n,

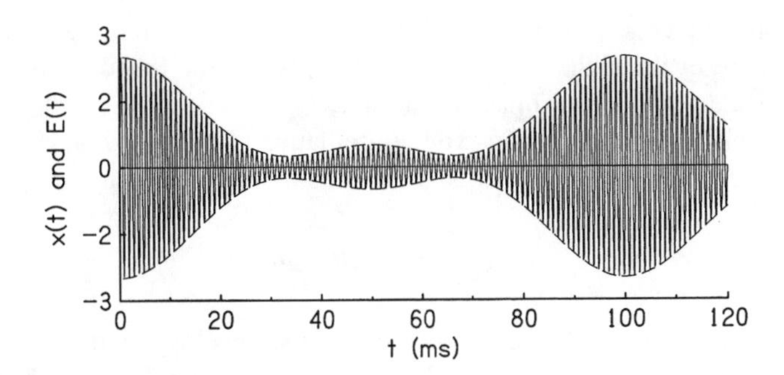

FIGURE 18.1. A waveform constructed by adding five sines with frequencies in a narrow band. The envelope is shown dashed.

$$\omega_n = \bar{\omega} + \delta_n . \tag{18.2}$$

Here $\bar{\omega}$ is the characteristic frequency. Its symbol makes it look like the average frequency, which it might be, but for the purpose of determining an envelope, the actual value of $\bar{\omega}$ is unimportant. This surprising fact will be the subject of one of the envelope rules below.

By using the trigonometric identity $\cos(A+B) = \cos A \cos B - \sin A \sin B$, we can write Eq. (18.1) as

$$x(t) = \sum_{n=1}^{N} C_n \cos(\delta_n t + \phi_n) \cos \bar{\omega} t - \sum_{n=1}^{N} C_n \sin(\delta_n t + \phi_n) \sin \bar{\omega} t. \tag{18.3}$$

This expression is of the form

$$x(t) = R(t) \cos \bar{\omega} t - I(t) \sin \bar{\omega} t, \tag{18.4}$$

where R is the sum of the cosine terms, and I is the sum of the sine terms:

$$R(t) = \sum_{n=1}^{N} C_n \cos(\delta_n t + \phi_n), \tag{18.5a}$$

and

$$I(t) = \sum_{n=1}^{N} C_n \sin(\delta_n t + \phi_n). \tag{18.5b}$$

Although the division of the time dependence into factors that oscillate at frequency $\bar{\omega}$ and the other factors, $R(t)$ and $I(t)$, is arbitrary, the point of making the division is that one supposes that $R(t)$ and $I(t)$ vary much more slowly than $\bar{\omega}$. Over an entire cycle of the oscillation at frequency $\bar{\omega}$ the functions $R(t)$ and $I(t)$ can be considered as approximately constant.

Equation (18.4) for $x(t)$ has an explicit oscillation at frequency $\bar{\omega}$, given by a sine term plus a cosine term. This can be written as a single cosine,

$$x(t)=E(t)\cos[\bar{\omega}t+\Phi(t)], \tag{18.6}$$

where $E(t)$ is a non-negative amplitude called the envelope, and $\Phi(t)$ is a phase.

The connection between $E(t)$ and functions $R(t)$ and $I(t)$ can be found by expanding the cosine in Eq. (18.6)

$$x(t)=E(t)\cos\bar{\omega}t\cos\Phi(t)-E(t)\sin\bar{\omega}t\sin\Phi(t), \tag{18.7}$$

whereupon by comparison with Eq. (18.4), we find that

$$E(t)\cos\Phi(t)=R(t), \tag{18.8a}$$

and

$$E(t)\sin\Phi(t)=I(t). \tag{18.8b}$$

Because of the fact that $\cos^2\Phi+\sin^2\Phi=1$, we know that

$$E^2(t)=R^2(t)+I^2(t), \tag{18.9}$$

and

$$E(t)=\sqrt{R^2(t)+I^2(t)}. \tag{18.10}$$

What we have done in developing Eqs. (18.6)–(18.8) is to show that any signal can be written as a cosine wave having a frequency of $\bar{\omega}$, so long as it also has a time-dependent phase $\Phi(t)$ and a time-dependent amplitude $E(t)$. The time-dependent amplitude is called the envelope. But the fact that we can always write a signal in the envelope form of Eq. (18.6) does not mean that it is always useful to do so. The envelope form explicitly extracts an oscillation at a characteristic frequency $\bar{\omega}$, and the envelope form is useful if this frequency is somehow representative of the signal as a whole. Usually, the envelope form is applied when the signal consists of a narrow band of frequencies, centered in the vicinity of $\bar{\omega}$. In the case of a narrow band, $E(t)$ and $\Phi(t)$ are slowly varying because they are derived from $R(t)$ and $I(t)$ which are slowly varying. That, in turn, is because the range of deviations δ_n is small.

The Complex Representation

The above derivation of the envelope was done using only real functions. There is a more economical way to describe the envelope using complex numbers. First, we note that by using Euler's formula, $\exp iA = \cos A + i \sin A$, we can combine Eqs. (18.5a) and (18.5b) to write down a complex number that combines both R and I:

$$R(t)+iI(t)=\sum_{n=1}^{N} C_n e^{i(\delta_n t+\phi_n)}. \tag{18.11}$$

Therefore, from Eq. (18.10), the envelope is given by the absolute value

$$E(t)=|R(t)+iI(t)|=\left|\sum_{n=1}^{N} C_n e^{i(\delta_n t+\phi_n)}\right|. \tag{18.12}$$

Below we will find several expressions that are even simpler.

ENVELOPE RULES

The absolute value in Eq. (18.12) puts the envelope in a form that makes a number of invariances evident. These can be described as envelope rules:

Envelope Rule 1, The frequency shift rule: "The envelope is unchanged by any rigid translation of the spectrum along the frequency axis." In other words, a constant frequency can be added to all the values of δ_n. In particular, we can add back the characteristic frequency $\bar{\omega}$. To be quite clear on this matter, we take a few lines to do it.

First, we know that $|e^{i\bar{\omega}t}|=1$. Next, because the product of absolute values is equal to the absolute value of the product, we can write Eq. (18.12) as

$$E(t)=\left|e^{i\bar{\omega}t}\sum_{n=1}^{N} C_n e^{i(\delta_n t+\phi_n)}\right|, \tag{18.13}$$

and therefore, from the definition in Eq. (18.2),

$$E(t)=\left|\sum_{n=1}^{N} C_n e^{i(\omega_n t+\phi_n)}\right|. \tag{18.14}$$

Equation (18.14) for the envelope looks a lot like Eq. (18.1) for the original signal $x(t)$. The only changes are to replace each cosine function, $\cos(\omega_n t+\phi_n)$, by an exponential function $\exp[i(\omega_n t+\phi_n)]$ and then to take the absolute value of the sum. This exponential form, obtained by the simple replacement process, is a practical way to calculate an envelope if the Fourier series for the waveform is known.

In general, the frequency shift rule says that the envelope of a signal is unchanged by a constant frequency shift. For instance, if a noise band is translated by balanced modulation (multiplication by a sine, as in Chapter 17) the envelope is unchanged.

The Analytic Signal

A direct connection between the signal $x(t)$ and the envelope is made by a function known as the *analytic signal*, given the symbol $\tilde{x}(t)$. The analytic signal is obtained by replacing each cosine in the signal of Eq. (18.1) by an exponential:

$$\tilde{x}(t)=\sum_{n=1}^{N} C_n e^{i(\omega_n t+\phi_n)}. \tag{18.15}$$

Unlike the real signal, the analytic signal is complex. From Euler's theorem, we know that each cosine in the real signal includes a term $\exp[i(\omega_n t+\phi_n]$ and a term $\exp[-i(\omega_n t+\phi_n)]$, i.e. both positive and negative frequencies. The analytic signal is what one gets by omitting all the negative frequency terms and multiplying by two.

It follows from Eq. (18.14) that the envelope is given by the absolute value of the analytic signal.

$$E(t)=|\tilde{x}(t)|. \tag{18.16}$$

Envelope Rule 2, The phase shift rule: "The envelope is unchanged if all phases of the signal components, ϕ_n are incremented by a common phase shift $\Delta\phi$." This rule comes out of Eq. (18.12) because a common phase shift simply multiplies the entire envelope by the factor $|\exp i\Delta\phi|$, and this absolute value is equal to one.

Envelope Rule 3, The time shift rule: "The envelope is unchanged if each component phase ϕ_n is incremented by a phase shift that is proportional to the component frequency ω_n." This transformation consists of the replacement

$$\phi_n \rightarrow \phi_n + t_0\omega_n, \tag{18.17}$$

where t_0 is a constant (independent of n) with dimensions of time.

The basis of this rule is that such shifts in phases are equivalent to a translation along the time axis,

$$t \rightarrow t+t_0. \tag{18.18}$$

A uniform translation like this does not change the shape of the envelope.

CALCULATION OF THE ENVELOPE

The frequency and phase rules above show that only *differences* in frequencies and phases matter so far as the envelope is concerned. These facts become even more evident from a general formula for the envelope.

To derive the general formula we use the rule that the absolute value of a complex number is obtained by multiplying the number by its complex conjugate and then taking the square root. Therefore, from Eq. (18.14),

$$E^2(t)=\sum_{n=1}^{N} C_n e^{i(\omega_n t+\phi_n)} \sum_{n'=1}^{N} C'_n e^{-i(\omega'_n t+\phi'_n)}. \tag{18.19}$$

The double sum over indices n and n' can be divided into terms where $n=n'$ and terms where n is always different from n',

$$E^2(t)=\left\{\sum_{n=1}^{N} C_n^2\right\}+\left\{\sum_{n=1}^{N}\sum_{\substack{n'=1\\ n'\neq n}}^{N} C_n C_{n'} e^{i(\omega_n t-\omega_{n'}t+\phi_n-\phi_{n'})}\right\}. \tag{18.20}$$

The equation can be further simplified by recognizing that positive and negative frequencies and phases appear symmetrically. For example frequencies $\omega_4-\omega_7$ and $\omega_7-\omega_4$ appear in the same way. The double sum can therefore be written with $n'<n$,

$$E^2(t)=\left\{\sum_{n=1}^{N} C_n^2\right\}+2\left\{\sum_{n=1}^{N}\sum_{n'=1}^{n-1} C_n C_{n'} \cos[(\omega_n-\omega_{n'})t+\phi_n-\phi_{n'}]\right\}. \tag{18.21}$$

To find $E(t)$ one takes the square root of the above. Only differences in frequencies and phases appear in the expression for the envelope.

EXAMPLE 1: *Five equally spaced and symmetrical components*

In this example we calculate the envelope for the signal shown in Fig. 18.1. The signal is the sum of sine components ($\phi_n = -\pi/2$) with an amplitude spectrum given by Fig. 18.2. The components are centered on 1000 Hz and separated by 10 Hz. We calculate the envelope by two methods, a simple method and a more general method.

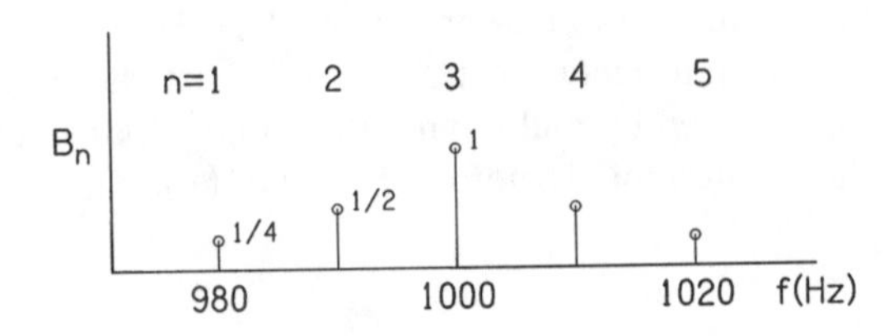

FIGURE 18.2 Spectrum of five equally spaced components in a narrow band at 1000 Hz.

To do this problem the easy way, we write E in terms of exponentials, per Eq. (18.11)

$$E(t) = \left| e^{i2\pi\cdot 1000t} \left[1 + \frac{1}{2} e^{i2\pi\cdot 10t} + \frac{1}{2} e^{-i2\pi\cdot 10t} + \frac{1}{4} e^{i2\pi\cdot 20t} + \frac{1}{4} e^{-i2\pi\cdot 20t} \right] \right|, \tag{18.22}$$

where we have explicitly extracted the oscillating factor at 1000 Hz. The absolute value of this factor is 1, and it won't trouble us further. Because of the symmetry of this spectrum, the exponentials conveniently sum to cosines, and the envelope becomes

$$E(t) = \left| 1 + \cos(2\pi\cdot 10t) + \frac{1}{2}\cos(2\pi\cdot 20t) \right|. \tag{18.23}$$

This envelope is shown dashed in Fig. 18.1.

EXAMPLE 2: *General envelope calculation*

An alternative calculation of the envelope uses the general function given in Eq. (18.21). To illustrate how it works we again calculate the envelope for the signal Example 1. First we calculate the sum of squared amplitudes,

$$\sum_{n=1}^{N} C_n^2 = 1 + \frac{1}{4} + \frac{1}{4} + \frac{1}{16} + \frac{1}{16} = \frac{13}{8}. \tag{18.24}$$

Next we evaluate the cross terms. For $N=5$ components in the spectrum there are $N(N-1)/2=5\times4/2=$ten cross terms. They are given by Table 18.1.

TABLE 18.1. Cross terms

n	n'	$A_nA_{n'}$	$f_n-f_{n'}$	$\phi_n-\phi_{n'}$
5	4	1/8	10	0
5	3	1/4	20	0
5	2	1/8	30	0
5	1	1/16	40	0
4	3	1/2	10	0
4	2	1/4	20	0
4	1	1/8	30	0
3	2	1/2	10	0
3	1	1/4	20	0
2	1	1/8	10	0

Adding together the terms from the table with the same frequency, we find

$$E^2(t)=\frac{13}{8}+2\left[\frac{5}{4}\cos(2\pi\cdot 10t)+\frac{3}{4}\cos(2\pi\cdot 20t)+\frac{1}{4}\cos(2\pi\cdot 30t)\right.$$
$$\left.+\frac{1}{16}\cos(2\pi\cdot 40t)\right]. \qquad (18.25)$$

To find the envelope we take the square root. It is not obvious that the expressions for $E(t)$ in Eqs. (18.23) and (18.25) are the same. It is left as an exercise in trigonometric identities to show that they really are.

The complexity of the general method in Example 2 suggests that whenever a spectrum offers a symmetry like that of Fig. 18.2 it is a good idea to take advantage of it. When there is no symmetry one must resort to the general formula of Eq. (18.21).

EXAMPLE 3: *Sum of first and tenth harmonics*

We consider the signal that is the sum of the first and tenth harmonics of 100 Hz:

$$x(t) = \sin(2\pi \cdot 100t) + \frac{1}{4}\sin(2\pi \cdot 1000t). \tag{18.26}$$

Using general formula Eq. (18.21), we find that the envelope is

$$E^2(t) = \frac{17}{16} + \frac{1}{2}\cos(2\pi \cdot 900t). \tag{18.27}$$

The envelope and the signal $x(t)$ are shown in Fig. 18.3.

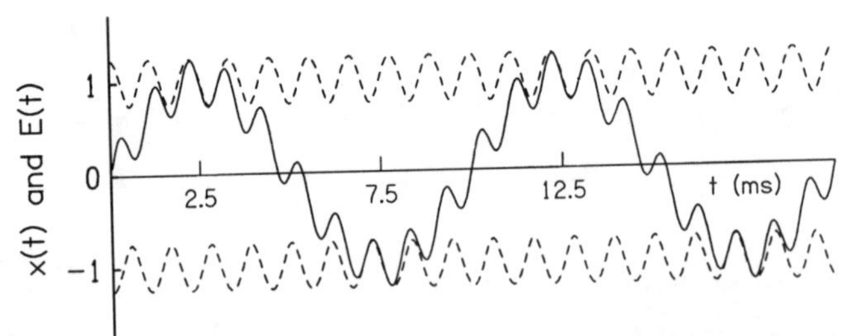

FIGURE 18.3. Illustration of a formally-correct but probably useless envelope. The signal is the sum of 100-Hz and 1000-Hz components. It is shown by a solid line. The envelope, calculated from Eq. (18.27) is shown by the dashed lines.

The examples above illustrate the important point that one can calculate an envelope for any signal whether or not it makes sense to do so. For Examples 1 and 2, the concept of an envelope makes sense. The time dependence of the signal can usefully be separated into a rapidly varying central component and much slower envelope variations. For Example 3, the envelope variations, which have the rate of the difference frequency $1000-100=900$ Hz, are not slow compared to the frequencies of the components, 100 and 1000 Hz.

The signals of these examples have perceptual characters that reflect the mathematics. The signal of Example 1 sounds like a tone with a pitch of about 1000 Hz and an amplitude that fluctuates, just as the mathematical analysis suggests. In the signal of Example 3, the 100-Hz and 1000-Hz components are analyzed by the auditory system into separate tones. Nothing useful is learned by calculating the envelope of the sum because there is no neural channel in which the two components coexist.

THE ENVELOPE AND THE HILBERT TRANSFORM

Above, the envelopes of signals were calculated from frequency representations. In each case the amplitudes and phases of the components were known, and this made a calculation of the envelope rather straightforward. Suppose, however, that we were given a signal as a function of time and were required to determine its envelope directly. What would we do then? In principle, the problem can be solved by finding the Hilbert transform of the signal, as we shall show below.

We recall that the analytic signal is similar to the real signal except that the negative frequency components have been excluded. This is the approach that we use to derive the new expression for the envelope. We begin with the real signal as the inverse Fourier transform of function $X(\omega)$,

$$x(t)=\frac{1}{2\pi}\int_{-\infty}^{\infty}d\omega\ e^{i\omega t}X(\omega). \tag{18.28}$$

The integral goes over all frequencies, positive and negative.

The analytic signal $\tilde{x}(t)$ is just the same, except (1) the negative frequencies are excluded and (2) the function is multiplied by two:

$$\tilde{x}(t)=\frac{2}{2\pi}\int_{0}^{\infty}d\omega\ e^{i\omega t}X(\omega). \tag{18.29}$$

We rewrite Eq. (18.29) by using the theta function,

$$\tilde{x}(t)=\frac{1}{\pi}\int_{-\infty}^{\infty}d\omega\ e^{i\omega t}X(\omega)\Theta(\omega), \tag{18.30}$$

where the theta function, $\Theta(\omega)$, is the unit step function,

$$\Theta(\omega)=0 \quad \omega<0$$

$$\Theta(\omega)=1 \quad \omega>0$$

$$\Theta(0)=1/2. \tag{18.31}$$

Equation (18.30) says that the analytic signal is given by twice the inverse Fourier transform of the product of $X(\omega)$ and $\Theta(\omega)$. However, we know that the inverse Fourier transform of a product is equal to the convolution of the inverse Fourier transforms. We already know that the inverse Fourier transform of $X(\omega)$ is $x(t)$. We define $\theta(t)$ as the inverse transform of $\Theta(\omega)$. Thus,

$$\tilde{x}(t)=2\int_{-\infty}^{\infty}dt'x(t')\,\theta(t-t'). \tag{18.32}$$

From Eq. (8.93) we know the Fourier transform for a theta function of time. To find the inverse Fourier transform for the theta function of frequency, we need to change the sign of i and divide by 2π so that

$$\theta(t)=\frac{1}{2}\,\delta(t)+\frac{i}{2\pi}\,\frac{1}{t}. \tag{18.33}$$

Therefore, the analytic signal is given by

$$\tilde{x}(t)=2\int_{-\infty}^{\infty}dt'x(t')\left[\frac{1}{2}\,\delta(t-t')+\frac{i}{2\pi}\,\frac{1}{t-t'}\right] \tag{18.34}$$

or

$$\tilde{x}(t)=x(t)+i\mathcal{H}[x(t)], \tag{18.35}$$

where, $\mathcal{H}[x(t)]$ is the Hilbert transform, defined by the integral

$$\mathcal{H}[x(t)]=\frac{1}{\pi}\int_{-\infty}^{\infty}dt'\,\frac{x(t')}{t-t'}. \tag{18.36}$$

Finally, the envelope is given by

$$E(t)=|x(t)+i\mathcal{H}[x(t)]|, \tag{18.37}$$

or

$$E(t)=\sqrt{x^2(t)+\{\mathcal{H}[x(t)]\}^2}. \tag{18.38}$$

We have succeeded in finding an expression for the envelope that is entirely in the time domain, without reference to frequency. In those cases where the Hilbert transform can be found from the integral in Eq. (18.36), the envelope can be found from Eq. (18.38). However, the integral in Eq. (18.36) converges slowly and there are many practical cases in which it cannot be used. In fact, given an arbitrary waveform, perhaps sampled from the real world and stored in computer memory, the most reliable way to compute the envelope is to find the Fourier transform, with both positive and negative frequency components, to reset the amplitudes of the negative frequency components to equal zero, to find the inverse Fourier transform, and finally to take twice the absolute value. Thus, in the end, we do not use the Hilbert transform to avoid the frequency representation. Instead, we do the reverse and use the frequency representation to calculate the Hilbert transform.

DAVID HILBERT (1862–1943)

David Hilbert was born at Königsberg on Jan 23, 1862. He studied mathematics at Königsberg and elsewhere in Germany, and in 1895 was appointed to a professorship at Göttingen where he remained until his retirement in 1930. His early work was in pure mathematics, algebraic invariants, number theory, and geometry. In 1909, he turned his attention to the area of mathematical physics, which, he said, ''was too difficult to be left to the physicists.'' His work on integral equations, including the Hilbert transform, is part of that effort (he invented the word ''spectrum''). Also part of his sojourn in physics is work on the theory of gasses and relativity. He, together with Minkowski and Klein, set standards of mathematical rigor for the rapid developments in theoretical physics in the early part of the century. Hilbert died in Göttingen on February 17, 1943.

MORE ENVELOPE RULES

Properties of the Hilbert transform are treated in Appendix F. Some of the formal properties are useful in proving facts about the envelope. For example, if signal $x(t)$ is an even function of time, then $\mathcal{H}[x(t)]$ is an odd function, and if signal $x(t)$ is an odd function of time, then $\mathcal{H}[x(t)]$ is even. Because the squares of both even and odd functions are even we have a further envelope rule:

Envelope Rule 4, The symmetry rule: ''If a signal is either an even function of time or an odd function of time, its envelope is an even function of time.''

A second example concerns envelope power. The envelope power is the long-term average value of $E^2(t)$, defined analogously to the familiar concept of signal power, which is the long-term average value of $x^2(t)$. Because the power in the Hilbert transform of a signal is equal to the power in the signal itself, Eq. (18.38) leads to a further envelope rule:

Envelope Rule 5, The power rule: ''The envelope power is precisely twice the signal power.''

Envelope Rule 6, The mirror rule: ''For a waveform having equally spaced components, the envelope is unchanged if the spectral amplitudes and phases are reflected about the center frequency, with all the phase angles replaced by their negatives.''

The meaning of this rather complicated rule should become clearer as it is proved. The proof uses a special case with only three components, but in the end it will be obvious that the proof can be extended to an arbitrary number of components.

Consider a signal with an envelope given by

$$E(t)=|C_1e^{i(\omega_1t+\phi_1)}+C_2e^{i(\omega_2t+\phi_2)}+C_3e^{i(\omega_3t+\phi_3)}|. \qquad (18.39)$$

It is evident that $E(t)$ is unchanged if we reverse the signs of all the frequencies and

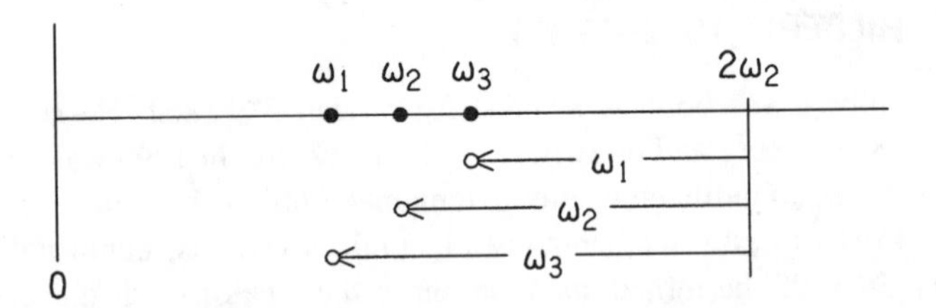

FIGURE 18.4. When the signs of frequencies and phases are reversed and then the frequencies are added to twice the center frequency, ω_2, the spectrum is replaced by its mirror image. Frequency ω_1 goes to where ω_3 used to be and vice versa. This manipulation does not change the envelope.

phases. This just reverses all the phases of the complex vectors that we are adding, and the final path length remains the same. Next, we shift the entire spectrum by adding $2\omega_2$ to all the frequencies. By the frequency shift rule, this too leaves the envelope unchanged. Figure 18.4 shows how this manipulation leaves frequency ω_2 the same while turning ω_1 into ω_3 and vice versa.

Then the expression for the envelope becomes

$$E(t)=|C_1e^{i(\omega_3t-\phi_1)}+C_2e^{i(\omega_2t-\phi_2)}+C_3e^{i(\omega_1t-\phi_3)}|. \tag{18.40}$$

Thus, as stated in Rule 6, the amplitude spectrum is reversed. Further, the phase spectrum has also been reversed, and replaced by its negative. For example, the amplitude and phase of frequency ω_3 are now C_1 and $-\phi_1$. But the envelope remains the same.

THE ENVELOPE AND PERCEPTION

The mirror rule was used by Richards (1988, 1990) to study the human ability to compare envelope fluctuations in two separated frequency bands. The interest in this topic arose from the phenomenon of comodulation masking release or CMR (Hall, Haggard, and Fernandes, 1984; Cohen and Schubert, 1987) wherein a signal in narrow-band noise becomes easier to detect if another noise band, having the same envelope, is added at a remote frequency. The work by Richards established that listeners could recognize envelope similarities between noises in different frequency bands, even if the spectra were reversed by the mirror operation. By contrast, listeners could not recognize spectral shape similarities across bands if the envelope fluctuations in the two bands were different.

Apart from the CMR, interest in the envelope has been sparked by the discovery that energy detection theory is not able to account for the detection of sine signals in narrow-band noise. In a roving-level experiment, Kidd, Mason, Brantley, and Owen (1989) found that the ability to detect a sine signal depended very little on whether reliable energy cues were or were not available. In the control condition of the experiment by Kidd *et al.* energy cues were reliable. In the test condition, the level of the noise band was randomly varied over a 32-dB range on each presentation, making energy cues unreliable. A comparison of the two conditions found that signal detection thresholds were the same. Only for brief duration stimuli did the listeners appear to be using energy cues. Similar results were found by Richards (1992), who also found that energy detection theory failed to account for the bandwidth dependence of threshold, whether or not the noise band level was randomized.

The observation that the detection of a sine signal in narrow-band noise cannot be predicted from only the increment in energy caused by the signal does not come as a surprise. Many listeners find that the best way to detect a sine signal in narrow-band noise is by the increased ''stability'' in the stimulus. With the signal added, the stimulus is smoother and seems to fluctuate less. The comparison of envelope fluctuations across conditions is an obvious candidate for the detection criterion. It is also reasonable that this criterion should begin to fail at short duration when it is not possible to track fluctuations well.

The envelope of a signal plays a particularly important role in binaural hearing. Because the neural channel that compares signal timing in the two ears is noisy, interaural time differences for steady-state tones are not registered for tones or tone components above 1.5 kHz. Even at 1 kHz the interaural timing is weak (Perrott and Musicant, 1977). However, listeners can make use of interaural differences between the *envelopes* of high-frequency signals in the two ears including timing differences (Henning, 1974; McFadden and Pasanen, 1976; Hafter, Dye, Wenzel, and Knecht 1990). The interaural time cues and interaural intensity cues present in the envelope appear to be used differently depending on the particular binaural task (Bernstein and Trahiotis, 1996). An interesting parallel is the fact that high-frequency vibrations (e.g. 1000 Hz) cannot be detected by the skin, but detection becomes possible if the vibrations are amplitude modulated at a lower frequency such as 200 Hz (Lamoré, Muijser, and Keemink, 1986).

The comparison between two ears that is involved in localization, or in the detection of binaural beats, or in the binaural detection of a signal in noise, is often thought to be based on cross-correlation between the two ears. For a mathematical treatment of the correlation between envelopes in narrow bands having different configurations of signal and noise, the reader should see the 1995 article by van de Par and Kohlrausch. Additional discussion of the detection of signal in narrow-band noise will be found in Chapter 20 on modulation and Chapter 23 on noise.

It is possible that envelope fluctuations also play a role in the segregation of tones that have different fundamental frequencies. For example, in the mistuned harmonic experiment, described in Chapter 6, the presence of a mistuned harmonic will result in fluctuations in the envelope in a critical band filter containing the mistuned harmonic (Lee and Green, 1994). In order to register the anomalous variation in envelope caused by the mistuning, the auditory system must only maintain synchrony with variations having a frequency of the order of the difference between successive harmonics in the complex tone. This contrasts with the conclusions of

Hartmann, McAdams, and Smith (1990). There, the rapid drop in the detectability of a mistuned harmonic with increasing frequency suggested a model in which detection was mediated by anomalies in the synchrony to the spectral components themselves.

ENSEMBLE-AVERAGE ENVELOPE

We suppose that there are two sine waves having the same frequency but different amplitudes and a random phase relationship. The sum is given by

$$x(t)=C_1\cos(\omega t)+C_2\cos(\omega t+\phi). \tag{18.41}$$

The two waves are defined such that $C_2 \leqslant C_1$. For example, the first wave might be a direct sound and the second might be a reflection from a near wall. From Eq. (18.21) we know that the square of the envelope is

$$E^2(\phi)=C_1^2+C_2^2+2C_1C_2\cos\phi. \tag{18.42}$$

It will simplify notation below to work in terms of C_1 and the ratio $r=C_2/C_1$. By definition $0 \leqslant r \leqslant 1$.

Because we do not know anything about the relative phase, we decide to average over all its possible values in an ensemble average. Averaging quantity E^2 is easy. It leads to the RMS value $\sqrt{\langle E^2\rangle}=\sqrt{C_1^2+C_2^2}$. Suppose, however, that we want to average the envelope itself to find $\langle E\rangle$. That calculation requires

$$\langle E\rangle=\frac{1}{2\pi}\int_0^{2\pi} d\phi\, E(\phi). \tag{18.43}$$

This integral can be related to an elliptic integral function. The first step is to use the trigonometric formula $\cos\phi=1-2\sin^2(\phi/2)$. Then a little algebra leads to

$$\langle E\rangle=\frac{2}{\pi}C_1(1+r)\mathcal{E}\left[\frac{4r}{(1+r)^2}\right], \tag{18.44}$$

where $\mathcal{E}$ denotes the complete elliptic integral of the second kind,

$$\mathcal{E}(\zeta)=\int_0^{\pi/2} d\phi\sqrt{1-\zeta\sin^2\phi}. \tag{18.45}$$

The elliptic integral is tabulated, and there are good series approximations for it [e.g. Abramowitz and Stegun (1964), p. 692] used by popular computer mathematics packages. An accurate series approximation was used to calculate the ensemble-averaged envelope in Fig. 18.5.

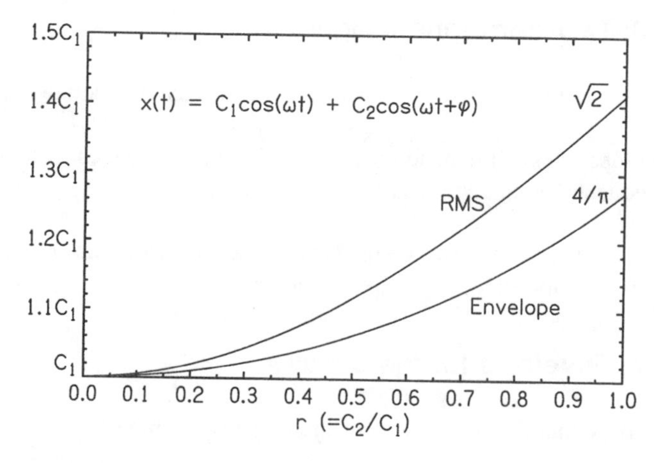

FIGURE 18.5. The ensemble-averaged envelope, or amplitude, for the sum of two pure tones having the same frequency but random relative phases as a function of the relative amplitudes of the tones. The RMS envelope is given for comparison.

The ensemble average calculation above applies equally well as a temporal average for beating sines.

EXERCISES

Exercise 1: Phase shift rule and the crest factor

Consider a signal that is a sum of cosines,

$$x(t)=\sum_{n=1}^{N} C_n \cos(\omega_n t).$$

Use the phase shift rule to show that the envelope is the same if all the cosine functions are replaced by sine functions.
[Note: Recall from Chapter 3 that adding up cosine functions corresponds to the choice of phases leading to the largest possible crest factor for a signal. Yet this exercise shows that we get the same envelope if we add up sine functions instead. We conclude that the sum of sine functions also leads to a rather large crest factor.]

Exercise 2: Constant envelope

Use the analytic signal to show that the envelope of $x(t)=A \sin \omega t$ is equal to A. Then use the Hilbert transform to show the same thing.

Exercise 3: Two-component signal

Use Eq. (18.21) to show that if a waveform has only two components ($N=2$) then:
(a) The envelope does not depend on whether the higher frequency component or the lower frequency component has the greater amplitude.

(b) The relative phase of the two components affects only the starting phase of the envelope and does not affect the shape of the envelope.

Exercise 4: Envelope for the octave

Consider the signal that is the sum of the first two harmonics of 100 Hz,

$$x(t)=\cos(2\pi\cdot 100t)+\cos(2\pi\cdot 200t).$$

Show that the envelope is given by

$$E(t)=\sqrt{2+2\cos(2\pi\cdot 100t)}.$$

Explain why the concept of the envelope is of doubtful use for this signal.

Exercise 5: The envelope envelops

Figures (like Fig. 18.1 or even Fig. 18.3) showing signals and their envelopes always have the signal enclosed by the envelope, but with positive and negative peaks of the signal in contact with the envelope. Explain why this is so, in terms of the definition,

$$x(t)=E(t)\cos[\bar{\omega}t+\Phi(t)].$$

Exercise 6: Bandwidth effects

For a band of noise, the envelope fluctuation is the sum of many frequencies. Show that in the calculation of $E^2(t)$, the largest of these frequencies is equal to the bandwidth of the noise, i.e. it is the difference between the largest and smallest frequencies in the band. [Note: This is why the envelope concept applies in cases of narrow bands.]

Exercise 7: Consistency check

Show that Eq. (18.23) and Eq. (18.25) are actually the same by squaring Eq. (18.23). [Hint: Recall that for any angle A, $\cos^2 A=(1+\cos 2A)/2$.]

Exercise 8: Case of the missing fundamental

Consider the complex periodic waveform consisting of harmonics 3, 4, and 5 of a missing fundamental of frequency 200 Hz,

$$x_1(t)=\sin(2\pi\cdot 600t)+\sin(2\pi\cdot 800t)+\sin(2\pi\cdot 1000t).$$

Consider now the "pitch-shift" stimulus

$$x_2(t)=\sin(2\pi\cdot 620t)+\sin(2\pi\cdot 820t)+\sin(2\pi\cdot 1020t).$$

Show that $x_1(t)$ and $x_2(t)$ have the same envelope.

The fact that the envelope is unchanged by a constant increment in the frequency of each component played an important role in the history of pitch perception. Early explanations for the pitch of missing-fundamental signals like $x_1(t)$ relied on the periodicity of the envelope to provide the cue for pitch. But if that were true, then a frequency-shifted signal like $x_2(t)$ should have the same pitch as $x_1(t)$ because the envelopes are the same. Experiments showed, however, that $x_2(t)$, with a shift of 20 Hz starting with the third harmonic, has a higher pitch than $x_1(t)$, by about 20/3=6 Hz. This result led to the abandonment of envelope theories of complex tone pitch perception (deBoer, 1956; Plomp, 1976).

Exercise 9: Envelope power

Consider a tone that is made by shaping a sine tone with an exponentially decaying envelope.

(a) Show that the envelope power is a decaying exponential of the form

$$P(t)=P_0e^{-t/\tau},$$

where P_0 is the initial power at time zero.

(b) Show that the energy in this pulse is $P_0\tau$.

(c) If the decay rate is 1 dB/ms, show that τ=4.343 ms (i.e. 10 $\log_{10}e$).

(d) If the energy in the decaying pulse (τ=4.343 ms) is the same as the energy in a rectangularly gated pulse 20 ms in duration, show that P_0 is greater than the power of the rectangular pulse by 6.6 dB.

CHAPTER 19

Frequency Modulation

Warble child; make passionate my sense of hearing.
W. SHAKESPEARE, *LOVE'S LABOUR'S LOST*

Frequency modulation (FM) is like amplitude modulation in that there is a carrier signal $x(t)$, some property of which is modulated. We do not lose much generality by assuming that the carrier is a sine wave,

$$x(t) = A \sin[\Theta(t)], \tag{19.1}$$

where $\Theta(t)$ is the instantaneous phase, in radians, as introduced in Chapter 1. What is known as the "angular frequency" is the time derivative of the instantaneous phase, $\omega = d\Theta/dt$. Therefore, the phase is the integral of the time-dependent angular frequency,

$$\Theta(t) = \int_{t_0}^{t} dt' \, \omega(t'), \tag{19.2}$$

where t_0 is an arbitrary reference time, and the frequency depends on time.

We consider the important case where the frequency is given by a constant carrier frequency, ω_c, plus a sinusoidal deviation,

$$\omega(t) = \omega_c + \Delta\omega \cos(\omega_m t + \phi), \tag{19.3}$$

where $\Delta\omega$ is the maximum excursion of the instantaneous frequency from ω_c, and ω_m is the modulation frequency. Therefore, from Eq. (19.2) the phase is given by

$$\Theta(t) = \omega_c t + \frac{\Delta\omega}{\omega_m} \sin(\omega_m t + \phi) + \Theta_0, \tag{19.4}$$

where Θ_0 is a constant of no significance, coming from the lower limit t_0. An example of the instantaneous phase is given in Fig. 19.1 ($\Theta_0 = 0$). The quantity $\Delta\omega/\omega_m$ appearing in Eq. (19.4) is the modulation index, β,

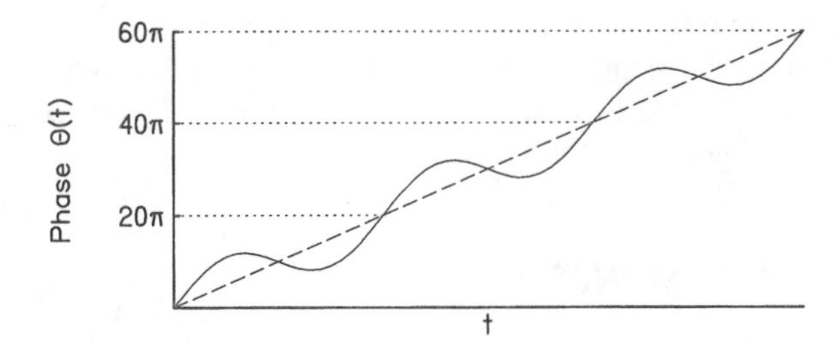

FIGURE 19.1. The instantaneous phase of a frequency modulated sine signal is shown by the solid line. The modulation frequency is one-tenth of the carrier frequency, and the modulation index is β=18.85. The dashed line is a reference for no modulation, β=0. See Exercise 19.1 for a discussion of the negative-slope regions.

$$\beta=\frac{\Delta\omega}{\omega_m}=\frac{\Delta f}{f_m}. \tag{19.5}$$

It is important to realize that β is not the percentage change in frequency. Instead, it is the change in frequency relative to the modulation rate. In this respect, β is quite different from the modulation percentage m that appears in the description of AM.

Finally, substituting Eq. (19.5) into (19.4) and (19.4) into (19.1) we find the FM signal to be given by

$$x(t)=A\ \sin[\omega_c t+\beta\ \sin(\omega_m t+\phi)]. \tag{19.6}$$

The amplitude of the signal is constant and therefore, unlike the case of AM, the power in the signal is not affected by modulation. Hereafter we set $A=1$.

We next find the spectrum of the FM signal in Eq. (19.6). The first step is to rearrange the expression somewhat by using the trigonometric identity

$$\sin(A+B)=\sin A\ \cos B+\cos A\ \sin B \tag{19.7}$$

to obtain

$$x(t)=\sin(\omega_c t)\cos[\beta\ \sin(\omega_m t+\phi)]+\cos(\omega_c t)\sin[\beta\ \sin(\omega_m t+\phi)]. \tag{19.8}$$

The advantage of this form is that the signal dependencies on ω_c and ω_m are separated into different factors. It then becomes possible to deal with the factors in ω_m by a series expansion. The form of the series expansion depends upon the modulation index. If modulation index β is small the modulation is called "narrow band," and we can proceed by a Taylor expansion in β. Successive terms in this expansion are successive powers of β. If the modulation index is not small (a frequent occurrence) the modulation is "wide band," and we use a Fourier series.

Successive terms in this expansion are successive harmonics of frequency ω_m. As we shall see below, the amplitudes of the harmonics are Bessel functions. First, we deal with the narrow-band case, when β is small. In many FM detection problems there is no need to go beyond this case.

NARROW-BAND FM (NBFM)

If modulation index β is small ($\beta \ll \pi/2$) then the function $\beta \sin(\omega_m t + \phi)$ is also small, because the sine function is never greater than 1. Therefore, we can simplify Eq. (19.8) by using two approximations: (1) the cosine of a small angle is equal to 1, and (2) the sine of a small angle is equal to the angle itself.

With these approximations, Eq. (19.8) becomes

$$x(t) = \sin(\omega_c t) + \beta \sin(\omega_m t + \phi)\cos(\omega_c t). \tag{19.9}$$

The product of sine and cosine in Eq. (19.9) can be expanded to obtain a sum of sine functions,

$$x(t) = \sin(\omega_c t) + \frac{\beta}{2} \sin[(\omega_c + \omega_m)t + \phi] - \frac{\beta}{2} \sin[(\omega_c - \omega_m)t - \phi]. \tag{19.10}$$

Equation (19.10) is in a form that makes the spectrum apparent. It shows that the NBFM spectrum has a carrier and two sidebands, separated from the carrier by $\pm\omega_m$. This spectrum is very similar to the spectrum of an AM signal. Recall from Eq. (17.19) that the AM signal is given by

$$x(t) = \sin(\omega_c t) + \frac{m}{2} \sin[(\omega_c + \omega_m)t + \phi] + \frac{m}{2} \sin[(\omega_c - \omega_m)t - \phi]. \tag{19.11a}$$

In fact, the power spectrum of FM in the narrow-band approximation is identical to the power spectrum of the AM signal. The sideband amplitudes, given by $m/2$ for AM, are given by $\beta/2$ for NBFM. Both spectra are shown in Fig. 19.2. The power spectrum considers only the magnitude of the Fourier components, paying no attention to their relative phase. It is only in the matter of relative phase that AM and NBFM differ.

Phasor Comparison of AM and NBFM

To appreciate the role that sidebands play in creating AM or NBFM we note that the only difference between Eq. (19.10) for NBFM and Eq. (19.11a) for AM is the sign of the lower sideband. The sign makes a considerable difference in the action of the sidebands on the signal. What is normally said in textbooks on the topic is

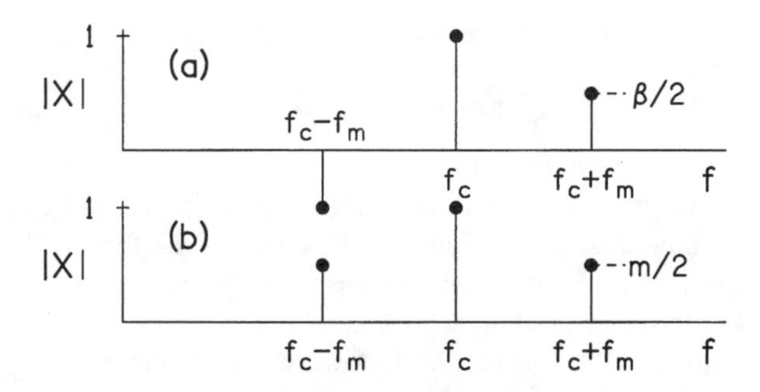

FIGURE 19.2. (a) Amplitude spectrum of the NBFM signal in Eq. (19.10). (b) Amplitude spectrum of the AM signal in Eq. (19.11a). The two spectra should be identical. However, part (a) employs a trick whereby some phase information sneaks into an amplitude plot by creating the fiction of a negative absolute value. A 180-degree phase shift is represented as a negative amplitude.

that for AM the sidebands are in phase with the carrier whereas in NBFM the sidebands are in quadrature with the carrier. This statement is true, but it is not obvious. In this section we spend some time showing that it is true. The reader who is not particularly interested in the relative phase of sidebands in AM and FM is advised to move on to the next section.

The basic difficulty with discussing sideband phase is that there are a lot of different ways that one can write the equation for a modulated signal in terms of its sidebands. For the case of AM for example, one can immediately write three other equations like Eq. (19.11a),

$$x(t)=\cos(\omega_c t)+\frac{m}{2}\cos[(\omega_c+\omega_m)t+\phi]+\frac{m}{2}\cos[(\omega_c-\omega_m)t-\phi] \tag{19.11b}$$

$$x(t)=\cos(\omega_c t)+\frac{m}{2}\sin[(\omega_c+\omega_m)t+\phi]-\frac{m}{2}\sin[(\omega_c-\omega_m)t-\phi] \tag{19.11c}$$

$$x(t)=\sin(\omega_c t)+\frac{m}{2}\cos[(\omega_c+\omega_m)t+\phi]-\frac{m}{2}\cos[(\omega_c-\omega_m)t-\phi]. \tag{19.11d}$$

These equations seem to involve different phase relationships between the sidebands and the carrier, but all of them are valid descriptions of AM. In fact, they come from the following products of amplitudes and carriers.

$$x(t)=[1+m\cos(\omega_m t+\phi)]\sin(\omega_c t) \tag{19.12a}$$

$$x(t)=[1+m\cos(\omega_m t+\phi)]\cos(\omega_c t) \tag{19.12b}$$

$$x(t)=[1+m\ \sin(\omega_m t+\phi)]\cos(\omega_c t) \tag{19.12c}$$

$$x(t)=[1+m\ \sin(\omega_m t+\phi)]\sin(\omega_c t). \tag{19.12d}$$

Equation (19.11a) comes from Eq. (19.12a), (19.11b) from (19.12b), etc. What we need to do is to find out what all the different descriptions in Eq. (19.11) have in common, and how that differs from an FM signal of the form of Eq. (19.10).

In general, the way to determine the nature of modulation, given the carrier and sidebands, is to use a phasor approach. A good way to begin is to express all trig functions as positive cosines, using the following trig identities:

$$\sin(\Theta)=\cos(\Theta-\pi/2) \tag{19.13}$$

$$-\cos(\Theta)=\cos(\Theta+\pi) \tag{19.14}$$

and

$$-\sin(\Theta)=\cos(\Theta+\pi/2). \tag{19.15}$$

After the reduction to a sum of positive cosine functions, the signature of an AM signal is that the average of the phases in the two modulation terms must equal the phase of the carrier term for every value of time, or else be equal except for a relative phase of π.

For Eq. (19.11b) the AM signature is evident because all terms are already positive cosines. The average phase for the modulation terms in the square brackets is $\cos(\omega_c t)$, in agreement with the carrier phase. A little more challenging is Eq. (19.11c), which reduces to

$$x(t)=\cos(\omega_c t)+\frac{m}{2}\cos[(\omega_c+\omega_m)t+\phi-\pi/2]+\frac{m}{2}\cos[(\omega_c-\omega_m)t-\phi+\pi/2], \tag{19.16c}$$

or Eq. (19.11d) which reduces to

$$x(t)=\cos(\omega_c t-\pi/2)+\frac{m}{2}\cos[(\omega_c+\omega_m)t+\phi]+\frac{m}{2}\cos[(\omega_c-\omega_m)t-\phi+\pi]. \tag{19.16d}$$

For Eq. (19.16c) the average of modulation phases is $\omega_c t$, equal to the phase of the carrier. For Eq. (19.16d) the average of modulation phases is $\omega_c t+\pi/2$, which differs by π from the carrier phase. According to the rule then, all of these are cases of AM.

The amplitude (or envelope) can be represented by phasors. A phasor is a two-dimensional vector in the complex plane, obtained by replacing every positive cosine ($\cos\Psi$) in the formula for the signal by an exponential ($e^{i\Psi}$). According to the usual rules of vector addition, a resultant phasor for the signal is given by adding phasors with lengths given by amplitudes and with angles relative to the carrier.

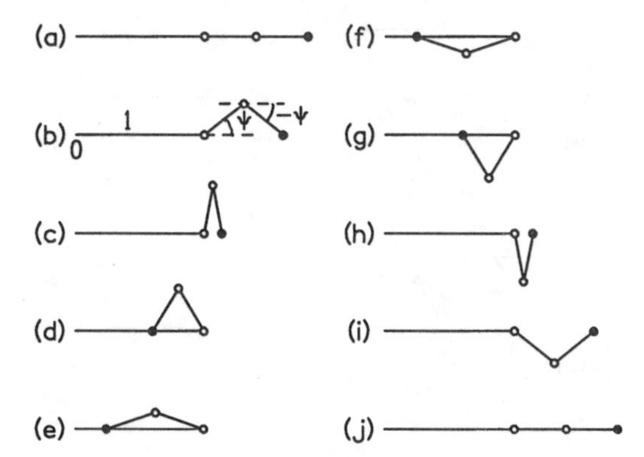

FIGURE 19.3. The phasor diagrams show the AM signal at successive instants in time. Successive modulation phases are separated by 40 degrees, so that 9 steps make a complete rotation. The long line (length 1) starting from the origin on the left is the carrier, the shorter lines have length $m/2$ and represent the sidebands. The upper sideband rotates counterclockwise, the lower clockwise. The final amplitude of the vector addition is the length from the origin to the filled circle. Because the amplitude is modulated, this length varies with time. Some details are explicitly shown in part (b). Angle Ψ is equal to $\omega_m t+\phi$.

Phasor diagrams for an AM signal at successive instants in time are shown in Fig. (19.3).

The NBFM signal can be analyzed in the same way. We first apply the trigonometric identities to write the waveform as a sum of positive cosines. Then the signature of NBFM is that the average of the two modulation phases should differ from the carrier phase by $\pm\pi/2$. This is what is meant by the statement that the sidebands are in quadrature with the carrier. For example, Eq. (19.10) becomes

$$x(t)=\cos(\omega_c t-\pi/2)+\frac{\beta}{2}\cos[(\omega_c+\omega_m)t+\phi-\pi/2]+\frac{\beta}{2}\cos[(\omega_c-\omega_m)t-\phi+\pi/2]. \tag{19.17}$$

The average phase for the sidebands is $\omega_c t$, to be compared with the carrier phase, $\omega_c t-\pi/2$. The signal therefore represents NBFM. The approximate constancy of the amplitude can be seen from the phasor diagram in Fig. 19.4.

Quasi FM (QFM)

The waveform described by Eq. (19.9) or in expanded form by Eq. (19.10) is an approximation to the FM waveform of Eq. (19.8). The approximation is valid in the narrow-band limit, the case of small modulation index. The waveform of Eq. (19.10) can also be created exactly. In this context Eq. (19.10) is not just an

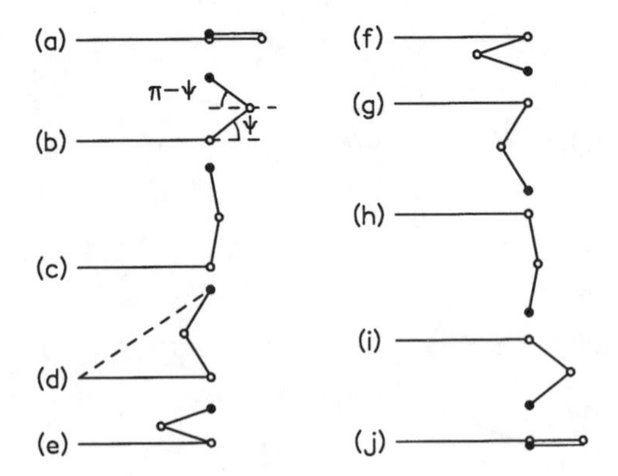

FIGURE 19.4. The phasor diagrams are just the same as in Fig. 19.3, except that here the sideband phases are those of NBFM. The amplitude is given by the distance between the origin on the left and the filled circle, as explicitly shown by the dashed line in part (d). If the sidebands are small enough the amplitude is almost constant. The sideband amplitudes in this figure are rather too large to qualify as NBFM.

approximation; it is an actual waveform, known as Quasi-FM (QFM). Its main feature of interest is that its power spectrum is identical to the spectrum of AM. Only the relative phases of the sidebands are different. A QFM signal can be generated by the circuit shown in Fig. 19.5.

AM in QFM

Because QFM is not true FM, there is some amplitude modulation in QFM. It can be calculated from the amplitude envelope of Eq. (19.10).

$$A=\left|1+\frac{\beta}{2}\,e^{i(\omega_m t+\phi)}-\frac{\beta}{2}\,e^{-i(\omega_m t+\phi)}\right| \tag{19.18}$$

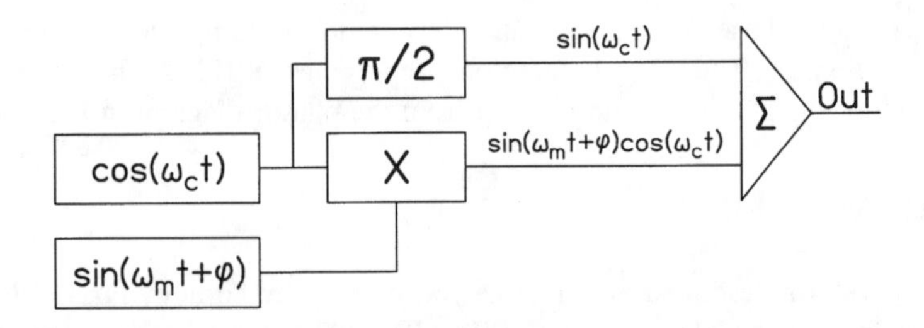

FIGURE 19.5. A circuit that creates QFM. It uses two oscillators, a 90-degree phase shifter, an analog multiplier (balanced modulator), and a summer (mixer).

$$=|1+i\beta\ \sin(\omega_m t+\phi)|=\sqrt{1+\beta^2\ \sin^2(\omega_m t+\phi)}. \tag{19.19}$$

If β is less than 1, a reasonable approximation for the amplitude modulation is given by the leading term in the series expansion of the square root (see Appendix C). The AM modulation frequency is $2\omega_m$. A few AM percentages are as follows: For β=0.1, 0.25%. For β=0.2, 1%. For β=0.5, 6%. For β=1, 20%.

WIDE-BAND FM

Our goal in this section is to calculate the spectrum of the general FM signal, where the modulation index β is not necessarily small. Because β is not much less than $\pi/2$, we cannot make the NBFM assumption.

We begin again with the general signal in Eq. (19.8). This equation is composed of two terms, $\sin(\omega_c t)\cos[\beta\sin(\omega_m t+\phi)]$ and $\cos(\omega_c t)\sin[\beta\sin(\omega_m t+\phi)]$. Our plan of attack is this: We first expand $\cos[\beta\sin(\omega_m t+\phi)]$ and $\sin[\beta\sin(\omega_m t+\phi)]$ in their Fourier series. These will both be sums over the harmonics of the modulation frequency ω_m. Because the first of these functions is an even function of time it will be a cosine series. The second one is an odd function of time and will be a sine series. Then the two terms in Eq. (19.8) will become two series in which each term is a product of a sine and a cosine. By expanding the products into sums using trigonometric formulas, we will get the spectrum. We now undertake the steps described above.

The Fourier series for both $\cos[\beta\sin(\omega_m t+\phi)]$ and $\sin[\beta\sin(\omega_m t+\phi)]$ may be found simultaneously by considering the periodic complex exponential function

$$v(t)=e^{i\beta\ \sin\ \omega_m t}, \tag{19.20}$$

where we have simplified the notation by choosing ϕ=0. The period of this function is given by $T=2\pi/\omega_m$. The real part of $v(t)$ gives the cosine function, the imaginary part, the sine function.

The Fourier series of Eq. (19.20) can be written in the form of Eq. (8.62)

$$v(t)=\frac{1}{2\pi}\sum_{n=-\infty}^{\infty} V_n e^{in\omega_m t}, \tag{19.21}$$

where, from Eq. (8.68), the Fourier coefficients are

$$V_n=\frac{2\pi}{T}\int_{-T/2}^{T/2} dt\ v(t)e^{-in\omega_m t}=\frac{2\pi}{T}\int_{-T/2}^{T/2} dt\ e^{i(\beta\ \sin\ \omega_m t-n\omega_m t)}.$$

Letting $x=\omega_m t$, we have

$$V_n = \int_{-\pi}^{\pi} dx \; e^{i(\beta \sin x - nx)}. \tag{19.22}$$

After integration with respect to x, the result of the integral will be a function of β and n. Parameter n is an integer; it plays the role of an index. The integral in Eq. (19.22) is almost exactly the function known as *the Bessel function of the first kind*, $J_n(\beta)$,

$$J_n(\beta) = \frac{1}{2\pi} \int_{-\pi}^{\pi} dx \; e^{i(\beta \sin x - nx)}. \tag{19.23}$$

Actually,

$$V_n = \frac{1}{2\pi} J_n(\beta). \tag{19.24}$$

Bessel functions for different values of n are frequently tabulated, for example Abramowitz and Stegun (1964). They can be calculated using algorithms given by Press *et al.* (1986). As Eq. (19.24) says, the coefficients of the Fourier series in Eq. (19.21) are given by these Bessel functions.

From Eq. (19.21) it is evident that we need to consider positive and negative values of n. Below we show that J_n is even or odd under reversal of the sign of index n as n itself is an even or odd integer.

Bessel Function Index Symmetry

The symmetry of the Bessel functions can be studied by considering $J_{-n}(\beta)$:

$$J_{-n}(\beta) = \frac{1}{2\pi} \int_{-\pi}^{\pi} e^{i(\beta \sin x + nx)} dx = (-1)^{2n} \frac{1}{2\pi} \int_{-\pi}^{\pi} e^{i(\beta \sin x + nx)} dx.$$

$$= (-1)^n e^{-in\pi} \frac{1}{2\pi} \int_{-\pi}^{\pi} e^{i(\beta \sin x + nx)} dx = \frac{(-1)^n}{2\pi} \int_{-\pi}^{\pi} e^{i[\beta \sin x - n(\pi - x)]} dx.$$

Let $y = \pi - x$, then

$$J_{-n}(\beta) = \frac{(-1)^n}{2\pi} \int_{2\pi}^{0} e^{i[\beta \sin(\pi - y) - ny]} (-dy) = \frac{(-1)^n}{2\pi} \int_{0}^{2\pi} e^{i(\beta \sin y - ny)} dy$$

$$= (-1)^n \frac{1}{2\pi} \int_{-\pi}^{\pi} e^{i(\beta \sin y - ny)} dy = (-1)^n J_n(\beta). \tag{19.25}$$

In the last step, we have used the fact that the integrand $e^{i(\beta \sin y - ny)}$ is a periodic function with its period $T=2\pi$.

Therefore, for even n we have

$$J_{-n}(\beta)=J_n(\beta), \tag{19.26}$$

and for odd n

$$J_{-n}(\beta)=-J_n(\beta). \tag{19.27}$$

Writing out the Fourier series term by term, and using Eqs. (19.26) and (19.27) to combine the positive and negative terms of equal magnitude of n, we get

$$v(t)=e^{i\beta \sin \omega_m t}=J_0(\beta)+2[J_2(\beta)\cos 2\omega_m t+J_4(\beta)\cos 4\omega_m t+\ldots]$$
$$+2i[J_1(\beta)\sin \omega_m t+J_3(\beta)\sin 3\omega_m t+\ldots]. \tag{19.28}$$

But from Euler's formula we know that

$$e^{i\beta \sin \omega_m t}=\cos(\beta \sin \omega_m t)+i \sin(\beta \sin \omega_m t). \tag{19.29}$$

Equating real and imaginary parts, we get

$$\cos(\beta \sin \omega_m t)=J_0(\beta)+2[J_2(\beta)\cos 2\omega_m t+J_4(\beta)\cos 4\omega_m t+\ldots], \tag{19.30}$$

and

$$\sin(\beta \sin \omega_m t)=2[J_1(\beta)\sin \omega_m t+J_3(\beta)\sin 3\omega_m t+\ldots]. \tag{19.31}$$

The spectrum of a frequency-modulated signal is now readily obtained. We recall the starting place, Eq. (19.8), which is

$$x(t)=\sin \omega_c t \cos(\beta \sin \omega_m t)+\cos \omega_c t \sin(\beta \sin \omega_m t). \tag{19.32}$$

Substituting Eq. (19.30) and (19.31) into (19.32), and then utilizing trigonometric sum and difference formulas, we finally get

$$x(t)=J_0(\beta)\sin \omega_c t+J_1(\beta)[\sin(\omega_c-\omega_m)t-\sin(\omega_c+\omega_m)t]+J_2(\beta)[\sin(\omega_c$$
$$-2\omega_m)t+\sin(\omega_c+2\omega_m)t]+J_3(\beta)[\sin(\omega_c-3\omega_m)t-\sin(\omega_c+3\omega_m)t]+\ldots \tag{19.33}$$

Equation (19.33) describes a time function consisting of a carrier and an infinite number of sidebands, separated from the carrier by frequency differences $\pm\omega_m$, $\pm 2\omega_m$ ..., etc.. By contrast, the AM signal had a carrier and only a single set of

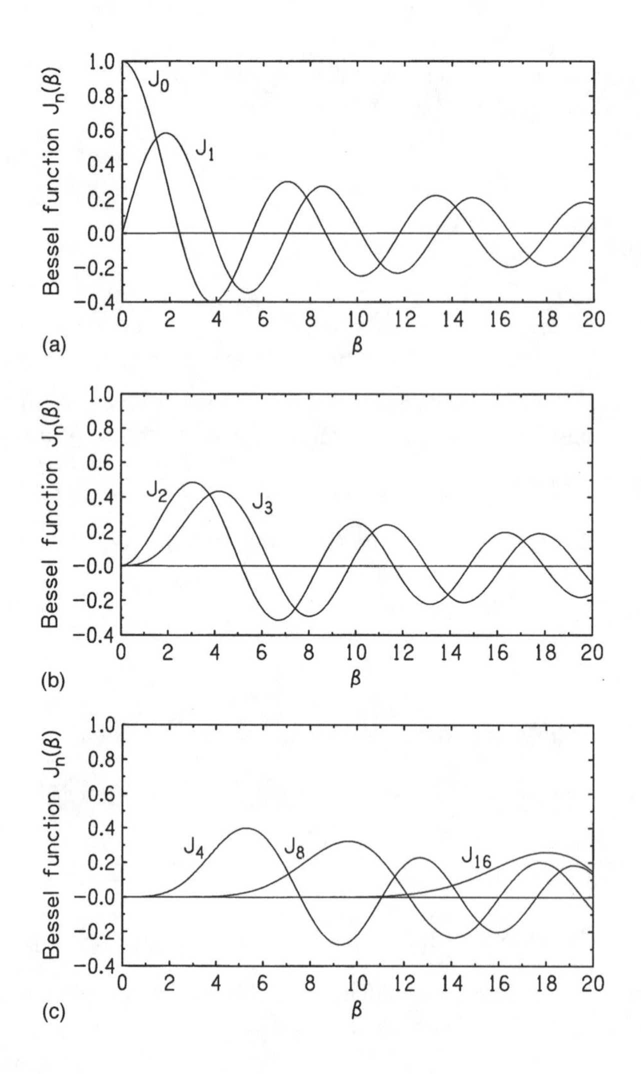

FIGURE 19.6. Bessel functions for several values of n. All Bessel functions except for J_0 are zero at the origin. For small values of β, function $J_n(\beta)$ is proportional to β^n.

sidebands. The amplitudes of the sidebands of the FM signal are given by Bessel functions, a few of which are plotted in Fig. 19.6.

EXAMPLE: ***Spectrum calculation***

The task in this example is to find the amplitude spectrum of a 1000-Hz sine that is frequency modulated at a rate of 100 Hz through a frequency excursion of ±200 Hz. We need to consider only the first four sidebands.

Solution:

The modulation index is β=200/100=2. Values of the Bessel functions J_0 through J_4 can be read off Fig. 19.6. More precise values from tables follow: The relative amplitude of the carrier is $J_0(2)=0.224$. The relative amplitude of the first sidebands is $J_1(2)=0.577$. The relative amplitudes of second, third, and fourth sidebands are respectively $J_2(2)=0.353$, $J_3(2)=0.129$, and $J_4(2)=0.034$. Therefore, the amplitude spectrum for positive frequencies looks like Fig. 19.7.

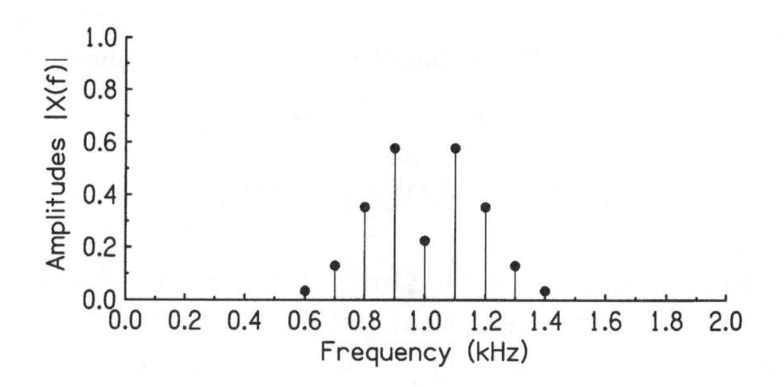

FIGURE 19.7. The amplitude spectrum of the example FM signal for positive frequencies. The amplitude spectrum for negative frequencies is the same because the amplitude spectrum is an even function of frequency.

As Fig. 19.7 shows, frequency modulation is capable of generating a wide variety of complicated spectra starting with only two sine signals, a carrier and a modulator. Further, the frequency modulation of one sine by another is particularly easy to do using a computer. If the carrier sine is stored as a table in memory, then an FM signal can be obtained by modulating the rate at which one reads through the table. This technique was exploited by John Chowning (1973) to create musical tones by digital synthesis. It was then adapted by the Yamaha Corporation to make the DX-series of synthesizers. The DX synthesizers were enormously popular. They provided many diverse timbres at low cost because they needed little memory. A major disadvantage of the FM technique comes from the fact that the Bessel functions oscillate. Therefore, it is difficult, if not impossible, for a musician to develop any intuition for the relationship between changes in modulation index and changes in tone timbre.

Bandwidth of an FM Signal

The magnitudes of the carrier and sideband terms depend on the modulation index, β, through the Bessel functions. To find the bandwidth of an FM signal, we use the fact that the higher-order Bessel functions, $J_n(\beta)$ with $n \gg 1$, are small until

β becomes approximately as large as n. They then increase to a peak, decrease again, and eventually oscillate like damped sinusoids. This behavior can be seen in Fig. 19.6, and it applies in general. Therefore, for $\beta \gg 1$ the number of significant sideband frequencies in the frequency-modulated wave is approximately equal to β. Since the sidebands are all separated by f_m Hz ($\omega_m/2\pi$), the sideband width on each side of the carrier is $B \approx \beta f_m$. Since there is a set of sidebands on either side of the carrier, the bandwidth, $2B$, of the FM signal for $\beta \gg 1$ is approximately

$$2B \approx 2\beta f_m = 2\frac{\Delta f}{f_m} f_m = 2\Delta f. \tag{19.34}$$

This result is intuitively sensible; the bandwidth covers the range of the frequency excursion, $\pm\Delta f$, itself.

DETECTION OF FREQUENCY MODULATION

An FM-detection experiment has the goal of finding the smallest modulation index β for which a frequency-modulated tone can be distinguished from a steady tone having no modulation. The seminal studies of FM detection have generally covered a wide range of modulation parameters, but the detection task is not homogeneous. In fact, the way that listeners detect frequency modulation depends upon the modulation rate, f_m.

If the modulation rate is high, approaching a critical bandwidth, it is no longer profitable to think of the signal as a tone, some property of which is modulated. Instead, the carrier and the sidebands all act like different tones. Therefore, the insightful approach is a spectral analysis, as described at the beginning of this chapter. Because the detection task involves only small values of β the narrow-band FM approximation is valid. There are only two sidebands, and the nature of the FM-detection task is the detection of one of the sidebands (usually the lower sideband) in the presence of the carrier as a masker (Hartmann and Hnath, 1982). In the end, FM detection becomes a question of tone-on-tone masking, not basically different from AM detection, as described in more detail in Chapter 20.

If the modulation rate is not high, the detection of frequency modulation becomes more interesting. The ability to detect low-rate frequency modulation has frequently been used to measure frequency difference limens. It is an alternative to a forced-choice task employing several tones of slightly different frequency (Chapter 12). An early FM-detection experiment, often cited, was done by Shower and Biddulph (1931). To measure frequency difference limens in this way requires some standards. First, FM detection depends upon the rate of modulation f_m. Normally, the most detectable rate is used; a value of $f_m = 4$ Hz is a popular choice, supported by Zwicker's thresholds (1952); Demany and Semal (1989) suggest 2 Hz instead. Second, FM detection depends upon the modulation waveform. Normally, a sine waveform is used, though the waveform used by Shower and Biddulph was roughly trapezoidal, a fact that has often been overlooked.

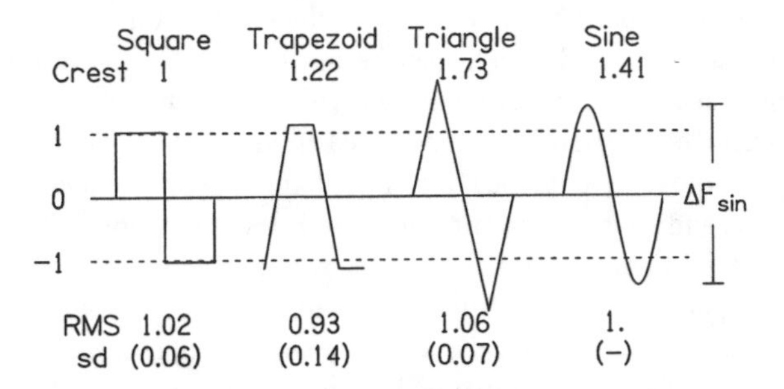

FIGURE 19.8. FM waveforms that are equally detectable. The trapezoid spends 50% of its time at the extremes. Under each waveform is the RMS value for barely detectable modulation, as measured by Hartmann and Klein (1980) for a modulation rate of 4 Hz and an 800-Hz sine-wave carrier. These are averaged over three listeners; the standard deviation (N-1 weight) is given in the bottom line. RMS values are expressed relative to the RMS for the sine waveform, which is shown by a dashed line.

Modulation Waveforms

The detection of FM depends upon the modulation waveform. Figure 19.8 shows four different waveforms that are found to be equally detectable. The RMS frequency deviations for just-detectable modulation, relative to the RMS deviation for a just-detectable sine-wave modulation, are shown below each waveform. These RMS values are all approximately equal. The maximum excursions are different in each case because the crest factors, shown above the waveforms, are different.

The fact that the RMS values are equal to within experimental error suggests that the auditory system performs an average of the FM excursions that is equivalent to a root-mean-square average. However, there are other models, quite different from RMS-averaging, that also make successful predictions. A model in which FM is detected on the basis of the fundamental component of the modulation waveform makes predictions that agree equally well with the data. As a practical matter, applying either the RMS model or the fundamental-only model leads to reasonable estimates of the magnitude of the FM sensation for different waveforms.

The waveform used by Shower and Biddulph spent 22% of the time at the highest frequency, 33% of the time at the lowest, and 45% in transition. Thus, it was similar to the trapezoid (50% at extremes) but it was asymmetrical.

Modulation Channels

Using frequency-modulation detection to measure frequency difference limens includes the tacit assumption that the role of the modulation is only to introduce a change in what would otherwise be a stationary percept. Essentially, it assumes that frequency modulation is detected as a modulation in pitch, or possibly in loudness.

An alternative idea is that FM is detected in channels that are specifically devoted to modulation detection, and which may be inactive in the absence of modulation. To the extent that such channels are involved, FM detection might actually be unrelated to frequency difference limens measured by other techniques.

There is physiological evidence for modulation-specific channels, both in the cochlear nucleus and in the inferior colliculus (Rees and Moller, 1983), and also in the cortex (Whitfield and Evans, 1965). But whether these cells play a significant role in human perception is an open question. There is some psychophysical evidence that they do, based upon the idea that sensory channels are indicated when selective adaptation can be demonstrated. Kay and Matthews (1972) found that the threshold for FM detection could be increased by a factor of 3 by previous stimulation with an FM tone. They also found that amplitude modulation adapts AM, but does not adapt FM. The FM-adaptation experiments suggested FM-specific channels, less than an octave in width, for modulation frequencies in the range 2 to 100 Hz. Tansley and Suffield (1983) measured the time course of adaptation. Their work suggested that asymptotic adaptation occurred only after a long exposure, more than 10 minutes, which is considerably longer than found by Kay and Matthews.

Evidence against the tuned-channels concept was found by Klein (Klein and Hartmann, 1979), using a complex FM waveform consisting of a fundamental (4 Hz) and its third harmonic (12 Hz). It was found that the ability to distinguish between the complex modulation waveform and a sine modulation waveform depended upon the relative phases of the two components of the complex waveform. Because 4 Hz and 12 Hz are expected to be in different modulation channels, the FM-channels model predicts that such phase dependence should not occur.

Modulation-Detection Strategies

If FM detection is regarded as just another way to measure frequency difference limens then one would expect it to agree with a forced-choice task using several tones with slightly different frequency, known as a ''pulsed-tones'' task. The two methods can be compared only if they are well defined. Then signal detection theory can be applied to them both, and the results can be compared.

The most common pulsed-tone method presents two tones in succession, and the listener's task is to say which one has the higher frequency (Wier, Jesteadt, and Green, 1977b). This two-interval, forced-choice task is a standard in signal detection theory. Each tone leads to a distribution on an internal pitch axis, and the presence of a frequency increment shifts the distribution. With this construction, it is possible to find the frequency difference limen Δf_{PT} corresponding to any value of d' (Chapter 24).

The most common FM-detection experiment also presents two intervals, one with modulation, the other with none, and the listener's task is to say which interval is modulated. A signal-detection model for this task was invented by Hartmann and Klein (1980, 1981). According to this model a listener detects modulation by sampling the signal at different times. Each sample leads to a pitch estimate, and the *difference* between successive pitch estimates is the decision variable on which detection is based. If the listener knows, or otherwise assumes, the modulation rate then it is to the listener's advantage to adopt a strategy in which the samples are spaced by half a period. Therefore, the model is called the *correlated-differencing model.* In no case, however, is it assumed that the listener knows the phase of the modulation. The power of the model comes from the fact that the pitch estimates are drawn from the same distribution that applies in the case of the pulsed-tone experiment. Therefore, the model can relate modulation-detection, ΔF_{sin}, and pulsed-tone, Δf_{PT}, measures of the frequency difference limen.

The correlated-differencing model predicts that ΔF_{sin} (the peak-to-peak excursion as defined by Fig. 19.8) is a factor of 1.75 greater than Δf_{PT}. This result agrees with experiments done in the intermediate frequency range. However, Fastl (1978) showed that as the carrier frequency varies from 300 Hz to 8 kHz, the ratio $\Delta F_{sin}/\Delta f_{PT}$ varies from 4 to 1, an observation that the model cannot explain. Demany and Semal (1989) gave the model a new life by pointing out that the ratio $\Delta F_{sin}/\Delta f_{PT}$ does not vary with carrier frequency if the pulsed-tone task is done with brief tones, comparable to the duration of instantaneous samples in the modulation detection model. It appears, therefore, that the distribution of instantaneous samples suffers degradation at a low carrier frequency where it becomes difficult to estimate a pitch in a brief duration.

The correlated-differencing model also predicts the form of the psychometric functions. It predicts that near threshold, the percentage of correct responses should grow quadratically with ΔF_{sin} for FM detection, whereas it grows only linearly with Δf_{PT} in the pulsed-tone task. (The quadratic behavior arises because the decision variable is a difference from two distributions.) Psychometric functions like this were seen by Jesteadt and Sims (1975). FM-detection psychometric functions measured by Moore and Sek (1995) also show quadratic dependence: d' depends quadratically on the modulation index β, and this behavior continues for d' as high as 2 or 3. By contrast, the quadratic dependence predicted by the correlated-differencing model gives way to a more gradual increase when d' approaches 1.

Not much is known psychophysically about the kind of rapid frequency-differencing operation that is postulated as the basis of the correlated-differencing model. It seems possible that it might be subject to adaptation, especially over the course of minutes of exposure as required by Tansley and Suffield.

It also seems possible that the correlated-differencing model might account for the phenomenon known as "modulation detection interference" (Yost and Sheft, 1990; Wilson, Hall, and Grose, 1990). The MDI effect is that the ability to detect frequency modulation of a target sine tone is decreased by modulating another sine tone having a carrier frequency that is remote from the carrier of the target signal. Because the target and interfering tone are in different critical bands, the MDI is a cross-channel effect. In the context of the correlated-differencing model of modulation detection, MDI would be a disruption of the correlated-differencing strategy.

FM-to-AM Conversion

Any experiment on FM detection must take cognizance of the FM-to-AM conversion effect. If a tone is passed through a filter then the frequency dependence of the transfer function of the filter leads to an amplitude modulation whenever the frequency of the tone is modulated. For instance, when an FM tone is heard in a room, the carrier has to be either at the peak, or in the valley, or on the skirt of some room resonance, leading to an uncontrolled amplitude modulation. To avoid an AM artifact, FM-detection experiments are done with headphones, and even then resonances of the headphone-outer-ear system lead to AM contamination when the carrier frequency is above several kHz (Henning, 1966).

Playing a violin with vibrato (FM) leads to amplitude modulation of the harmonics of the tone because of the resonances of the violin box (Mathews and Kohut, 1973). In fact, FM-to-AM conversion *within the auditory system* may ultimately be the mechanism by which listeners detect FM. Zwicker (1956) and Maiwald (1967a,b) have suggested that FM is detected when it produces a level change of 1 dB in an auditory filter. Because the upper skirt of an auditory filter is sharper than the lower (at least at high signal levels) one would expect FM to be detected on the upper skirt. This model of FM detection is further discussed in Chapter 20.

EXERCISES

Exercise 1: Instantaneous phase

Figure 19.1 shows a plot of instantaneous phase vs. time. This plot has a region of negative slope.

(a) Show that this corresponds to a negative frequency.

(b) Show that a negative slope occurs when $\Delta\omega$ is greater than ω_c.

(c) What range of β values is allowed in Fig. 19.1 if the frequency is required to be positive at all times?

Frequency excursions that cross zero in this way are not normal for analog frequency-modulated devices. They are common in digital FM, as used in FM synthesizers to generate complex musical tones.

Exercise 2: Vibrato

Vibrato is a slow ($\approx$6 Hz) FM used by musicians as an ornament. In violin playing and singing it is virtually obligatory. Studies by Seashore (1938/1967) and

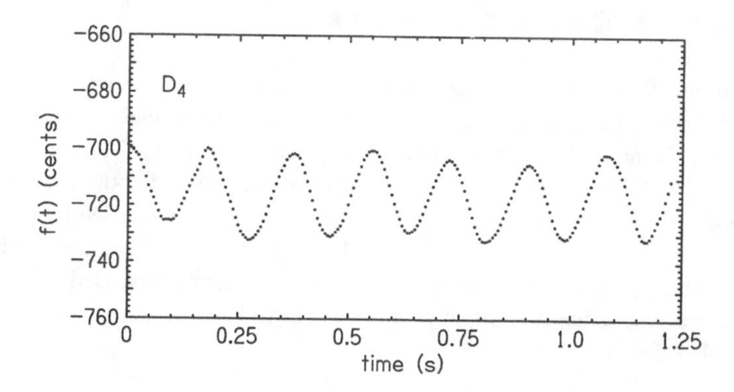

FIGURE E.2. Viola tone instantaneous frequency re 440 Hz.

others find that the modulating waveform is mainly sinusoidal and that the peak-to-peak frequency variation is about half a semitone (3% of the fundamental frequency).

(a) Show that for a tone with a fundamental frequency of 440 Hz the modulation index corresponding to the above parameters is $\beta=1.1$.

(b) Is this narrow-band FM?

(c) The waveform of a violin is complex with many high harmonics. The waveform retains its harmonic character throughout the vibrato cycle. Show that this implies that the modulation index for the n-th harmonic is proportional to n.

(d) Figure E.2 shows the instantaneous frequency of a viola tone (D_4) measured by Brown and Vaughn (1995) using the fundamental frequency tracker of Brown and Puckette (1993). The vertical axis is in cents with respect to A440 (see Chapter 11). First show that the tone is actually the note D_4. Next show that the modulation index is approximately $\beta=0.9$.

Exercise 3: Extending NBFM

The NBFM approximation to FM given in Eq. (19.10) is correct to first order in β. It is, of course, possible to extend the series to higher powers of β, starting with Eq. (19.8). Use the expansion of the cosine function given in Appendix C to include also the terms of order β^2. This leads to a signal with second sidebands, displaced from the carrier by $\pm 2\omega_m$. [Note: None of this has anything to do with QFM, which is not an approximation and does not involve β to higher power than the first.]

Exercise 4: Not as bad as it looks

Equation (19.19) calculates the amount of AM in QFM. It shows that by adding QFM (finite β), the amplitude is caused to oscillate above the value of amplitude when there is no QFM (β=0). This is one reason why QFM is not true FM. Equation (19.19) does not, however, give the correct result for the amount of AM in NBFM. In other words, it does not correctly give the error caused by the narrow-band approximation. The error in Eq. (19.19) appears to be of order β^2. Show that if one works consistently to order β^2, keeping the term introduced in Exercise 3 above, there is a cancellation that reduces the term in β^2 to zero. The error, therefore, is of order higher than β^2.

Exercise 5: Wide-band FM spectrum

A 1000-Hz carrier is frequency modulated at a rate of 100 Hz. Use Fig. 19.6 to find a frequency excursion Δf such that the spectrum of the signal has no amplitude at the carrier frequency of 1000 Hz. [Ans: Δf=240 Hz.]

Exercise 6: Generalized-trapezoid waveform

The triangle, the 50% trapezoid, and the square waveforms are all variations on the general trapezoid. The only difference is the fraction of the period spent at the extremes, respectively 0, 0.5, and 1. Let this fraction be parameter p.

(a) Show that the crest factor for the general trapezoid is given by

$$\text{Crest factor} = \sqrt{\frac{3}{1+2p}}.$$

(b) Show that if the RMS model of FM perceptual averaging is correct then the ratio of the peak frequency excursion for any trapezoid to the peak frequency excursion of a perceptually equivalent sine waveform is

$$\text{Ratio} = \sqrt{\frac{3/2}{1+2p}}.$$

CHAPTER 20

Modulation Detection and Perception

Sweet tones are remembered not.
PERCY BYSSHE SHELLEY

Because rooms and other sound-transmission media have transfer functions that are not flat, it is common for frequency modulation to induce amplitude modulation. Therefore, listeners are often exposed to both kinds of modulation at once. Because art follows nature, research on modulation perception has profited considerably from experiments in which listeners are presented with a combination of AM and FM. The combination is known as mixed modulation.

MIXED MODULATION

The mixed-modulation waveform is the same as the FM waveform given in Eq. (19.1) except that the amplitude, A, is a function of time,

$$x(t)=A(t)\sin[\Theta(t)]. \tag{20.1}$$

As in Eq. (19.2), the instantaneous phase is the integral of the time-dependent angular frequency,

$$\Theta(t)=\int_{t_0}^{t} dt'\,\omega(t'). \tag{20.2}$$

The frequency, $\omega(t)$, is given by a constant carrier frequency, ω_c, plus a sinusoidal deviation, as in Eq. (19.3),

$$\omega(t)=\omega_c+\Delta\omega\,\cos(\omega_m t+\phi). \tag{20.3}$$

The amplitude also has a sinusoidal variation, as given by the AM equation (17.17),

$$A(t)=A_0(1+m\,\cos\,\omega_m t), \tag{20.4}$$

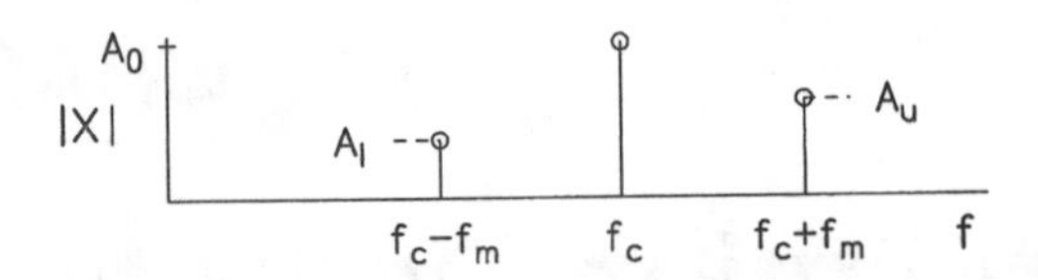

FIGURE 20.1. In mixed modulation the upper and lower sideband amplitudes may be different. Only two sidebands are shown because the FM modulation index is presumed to be small.

where A_0 is a constant, and in the classic mixed-modulation case the modulation frequency for AM is the same as the modulation frequency for FM. The relative phase of FM with respect to AM is given by ϕ. When $\phi=0$ the FM and AM are in phase, which means that a maximum in frequency coincides with a maximum in amplitude. The two parameters of modulation are the amplitude modulation percentage m and the frequency modulation index β, where $\beta=\Delta f/f_m$. Both parameters are non-negative numbers.

Detection of Mixed Modulation at High Modulation Frequency

In the high-modulation-frequency region, the modulation index is often small. This occurs because the denominator, f_m, is large and also because the numerator, Δf, tends to be small in a detection experiment. Therefore, β is considerably less than 1, and the narrow-band FM approximation is valid. There are only two sidebands of importance, and if f_m is large, they are well separated in frequency. As a result, the spectral approach to modulation detection becomes useful.

The amplitudes of the upper and lower sidebands are given by A_u and A_l,

$$A_u=\frac{A_0}{2}\sqrt{m^2+\beta^2+2m\beta\cos\phi}, \tag{20.5a}$$

and

$$A_l=\frac{A_0}{2}\sqrt{m^2+\beta^2-2m\beta\cos\phi}. \tag{20.5b}$$

Equations (20.5) show that, unlike AM or FM alone, the sideband levels are not necessarily the same, as shown in Fig. 20.1. The range of possible sideband levels depends on the ratio $r=\beta/m$. When this ratio is unity, the relative sideband levels can be anything at all, depending only on the relative phase ϕ. For other values of r the range is restricted, as shown in Fig. 20.2.

Modulation is detected by detecting at least one of the sidebands, which are small, in the presence of the carrier signal as a masker. The best strategy is shown

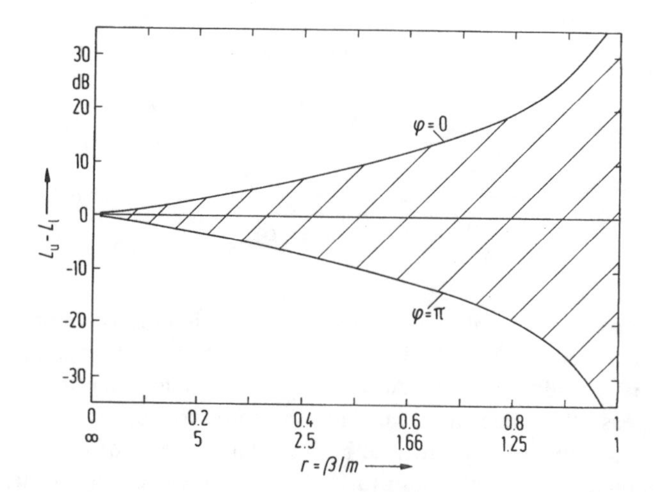

FIGURE 20.2. The difference between the upper-sideband level and the lower-sideband level is $L_u - L_l = 20 \log(A_u/A_l)$. The range of allowed differences is shown by the hatched region. This range is a function of the modulation ratio r, or equivalently its inverse $1/r$. (From Hartmann and Hnath, 1982.)

in Fig. 20.3(a). This strategy is to monitor an auditory filter that emphasizes a sideband while attenuating the carrier. Because the upper skirt of the filter is steeper than the lower, detection occurs in an auditory filter with a characteristic frequency CF below the carrier. Then the carrier is maximally attenuated. Ultimately the sideband (lower sideband) is detected by the presence of beats in the output of the filter.

To see how FM and AM interact, we consider the case when they are in phase (ϕ=0). Then Eqs. (20.5) become

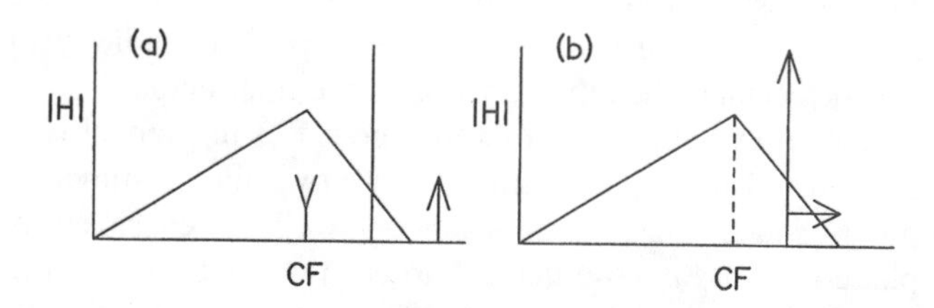

FIGURE 20.3. Two views of the detection of mixed modulation on the upper skirt of an auditory filter with characteristic frequency CF. The phase relationship between AM and FM is destructive, ϕ=0. (a) The spectral view (Hartmann and Hnath): Adding FM to AM decreases the lower sideband, making it harder to detect the modulation. (b) The temporal view (Zwicker, Maiwald, Goldstein): Because AM and FM are in phase, their effects on the amplitude of the signal are opposing, making it harder to detect the modulation. (Their effects would add constructively if the signal were on the lower filter skirt.)

$$A_u = \frac{A_0}{2}(m+\beta), \tag{20.6a}$$

and

$$A_l = \frac{A_0}{2}|m-\beta|. \tag{20.6b}$$

Suppose that we set β equal to zero (AM only) and find the value of m for which the modulation is just detectable. It is the lower sideband with amplitude A_l that is detected. Next we add some FM. According to Eq. (20.6b) the effect of increasing β is to decrease the value of A_l, making the lower sideband less detectable, as indicated by the downward arrowhead in Fig. 20.3(a). Thus for $\phi=0$, AM and FM are in a destructive phase relationship. The reader should remember this fact while reading the next section.

Detection of Mixed Modulation at Low Modulation Frequency

When the modulation frequency is low, it seems natural to think about the perception of the modulation in terms of the variations in a single tone. AM causes the loudness of the tone to go up and down; FM causes the pitch of the tone to go up and down. At the detection level it may not be clear to the listener which of these perceptual quantities is varying, and there are some listeners for whom the distinction is vague at any level. But although the percept may be ambiguous, the low-modulation frequency range is defined as the range in which the listener can track, or can imagine tracking, some signal variation as a function of time. It is a temporal region.

Within the class of temporal models are the FM-to-AM conversion models of Zwicker (1956), Maiwald (1967a,b), and Goldstein (1967a). This model says that FM is detected by converting it to AM on the skirt of an auditory filter. FM is presumed to be detectable when the variation in the filter output is 1 dB. The role of mixed modulation experiments within the context of this model is to allow the extents of AM and FM to be parametrically varied within the stimulus itself.

To see how FM and AM interact in this model, we again consider the case when they are in phase ($\phi=0$). Suppose that we do an AM detection experiment ($\beta=0$) and find the value of m for which the amplitude variation is just detectable. Next we add some FM, as shown in Fig. 20.3(b). Again we find that FM and AM are in a destructive phase relationship: at times when the AM causes the amplitude of the signal to increase, the FM causes the frequency to increase, which (because the signal is on the upper skirt of the filter) tends to cause the amplitude to decrease, counteracting the effect of the AM. This makes the modulation harder to detect. Thus, the approach to modulation detection from the low f_m limit leads to the same qualitative prediction as the approach from the high. Adding FM in phase to AM

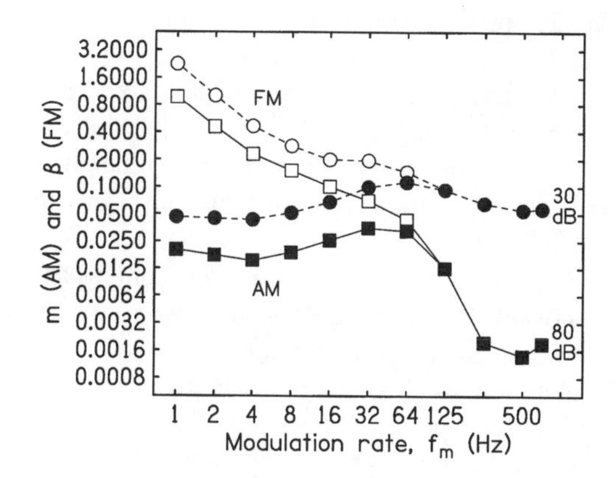

FIGURE 20.4. Just-detectable values of the modulation index β for FM and the modulation percentage m for AM as a function of modulation frequency for a 1000-Hz carrier. Results for two carrier levels are shown, 30 and 80 dB SPL. (After Zwicker (1952).)

leads to higher detection threshold. Further, in both approaches the effect depends quantitatively on the upper-frequency slope of the critical band filter.

MODULATION DETECTION AND THE CRITICAL BAND

Zwicker (1952) measured the threshold for the detection of FM and AM as a function of the modulation frequency. Some of his data are shown in Fig. 20.4. Plotted there are the values of β and m for modulation detection for a 1000-Hz carrier and modulation rates shown on the horizontal axis. The scales for β and m are logarithmic, which makes them proportional to the levels of the modulation sidebands [see Eqs. (19.10) and (19.11) giving the sideband amplitudes relative to the carrier as $\beta/2$ and $m/2$].

It is evident that at low modulation frequencies the threshold values of β and m are very different. In fact β becomes so large at low values of f_m that the narrow-band approximation is no longer valid. As the modulation frequency increases, the threshold values of m and β become the same. The frequency for which this occurs is known as the critical modulation frequency, CMF. In Fig. 20.4 it appears to be about 80 Hz.

Zwicker interpreted the CMF as a measure of the critical band width. His analysis took the spectral (high-f_m) view of modulation detection and noted that the only difference between AM and narrow-band FM is in the phases of the sidebands. Zwicker argued that when the modulation frequency is small, the carrier and both sidebands all appear in the same critical band. Then the auditory system is sensitive to the relative phases of the sidebands, and thresholds for AM and FM detection can be different. When the modulation frequency becomes as high as the CMF, however, the sidebands appear in different critical bands. The ear no longer can keep track of their relative phases and therefore the detection thresholds must be the same. According to this model the critical bandwidth should be just two times the CMF. For example, the CMF in Fig. 20.4 suggests that the critical bandwidth at 1000 Hz should be 160 Hz. Comparison between critical bandwidths determined in this way and the Bark scale determined by masking techniques showed good agreement (Zwicker and Fastl, 1990).

This view of modulation detection was amended somewhat by Hartmann and Hnath (1982) who showed that at the CMF and above, only one of the two sidebands, usually the lower sideband, contributes to detection of the modulation. They showed this by eliminating what was thought to be the unused sideband and finding that the threshold did not change. Therefore, the CMF measure is subject to many of the same uncertainties due to off-frequency listening as appear in tone-on-tone masking. Nevertheless, the logic of the critical-band analysis requires that the critical bandwidth be at least twice the CMF. Unfortunately, the Munich critical bands, determined in part by the CMF, are already wider than the Cambands.

FM AND AM DETECTION UNIFIED?

If frequency changes are detected according to place principles, then both frequency modulation and amplitude modulation ought to be detected on the basis of the modulated output of an auditory filter. This idea led Zwicker (1956) and Maiwald (1967a,b) to postulate that AM detection and FM detection are the same sort of process. In this model, FM is converted to AM on the upper skirt of an auditory filter, as shown in Fig. 20.3(b), and then is detected in the same way as AM. The steeper the slope, the more effectively a change in frequency can be converted into a change in amplitude. Comparing FM and AM detection led to a conversion rate, or slope, of -27 dB/Bark. This value of the upper slope was in reasonable agreement with other measures of the skirt of the auditory filter based on masking. Thus AM and FM detection were unified.

The mixed modulation MM experiments of Hartmann and Hnath (1982) tested the unified FM-AM model and found it to be inadequate. When the parameters of the unified envelope variation model were adjusted to fit the detection data for modulation that was mostly AM, the model failed to fit the data for mostly FM. The model predicted a rapid rise in threshold as the relative percentage of FM increases, much more rapid than observed experimentally. Therefore, these authors concluded that FM detection is not exclusively based on conversion to AM. It appears that there is an additional mechanism for detecting FM, presumably based on neural timing.

Demany and Semal (1986) compared AM and FM detection at different modulation frequencies. Because the AM and FM thresholds were found to have a different functional dependence on modulation rate, Demany and Semal agreed that AM and FM are detected by different processes. They also showed that AM and FM were distinguishable at the detection level when the modulation rate was low (circa

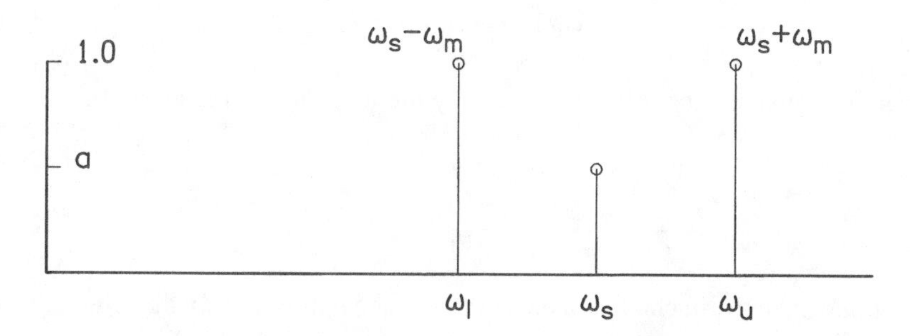

FIGURE 20.5. The two-tone masking stimulus, with lower (*l*) and upper (*u*) masking tones, and a weak signal (*s*) in the center.

2 Hz), but not when it was as high as 15 Hz. Moore and Sek (1992, 1995) measured AM and FM detection thresholds as a function of carrier frequency and modulation rate. To account for their data, they proposed a unified model in which AM and FM are both detected on the basis of excitation pattern changes for carrier frequencies above 6 kHz. For lower carriers they required an additional mechanism dependent on neural synchrony. However, this mechanism was only found to be operative for modulation rates below 10 Hz.

To summarize the above, three pairs of investigators who have studied AM, FM, and MM since the work of Maiwald appear to agree that AM and FM are not detected by a unified model for low carrier frequencies (where neural synchrony exists) and low modulation frequencies (less than 10 Hz... maybe lower). A dissenting note was raised by Edwards and Viemeister (1994) who pointed out that the detection of MM can be predicted by adding the d' values for detection of AM alone and FM alone, as expected for a single detection mechanism or, at least, a single decision statistic. By contrast, the square root of the sum of squares of d' values, expected for independent AM and FM detection, underestimated performance on MM. It is possible, however, that the conclusions of Edwards and Viemeister were heavily influenced by results obtained at high modulation rates, 10 Hz and above. There may yet be convergence on this question.

MASKING BY TWO TONES

In a modulation detection experiment there is a strong carrier, and there are weak sidebands above and below it. The task is to detect the sidebands. The reverse situation is the two-tone masking experiment. Here there are two strong sine-tone maskers of equal level and a weak sine tone at a frequency that is midway between them, as shown in Fig. 20.5. The task is to detect the weak tone at the center (Green, 1965; Phipps and Henning, 1976).

The stimulus is

$$x(t)=\sin(\omega_l t)+a\ \sin(\omega_s t+\phi)+\sin(\omega_u t), \tag{20.7}$$

where the three components are spaced by the modulation frequency ω_m,

$$\omega_l=\omega_s-\omega_m \tag{20.8a}$$

$$\omega_u=\omega_s+\omega_m . \tag{20.8b}$$

The principal experimental variable is the signal amplitude, a. In the absence of the signal, the two masking components beat at a rate of $2\omega_m$.

It is particularly interesting to study $x(t)$ for small values of $2\omega_m$ (less than a critical band) where Phipps and Henning found a strong dependence of detection threshold on phase ϕ. The small ω_m analysis given in 1982 by Hartmann follows:

The stimulus can be written as a sine tone with frequency ω_s multiplied by a modulation prefactor,

$$Ee^{i\Phi}=e^{-i\omega_m t}+ae^{i\phi}+e^{i\omega_m t}. \tag{20.9}$$

After some algebra the envelope, E, and the phase, Φ, can be written as

$$E(t)=[2+a^2+2a\ \cos(2\omega_m t)+4a\ \cos\ \phi\ \cos(\omega_m t)]^{1/2} \tag{20.10}$$

and

$$\Phi(t)=\tan^{-1}\left[\frac{a\ \sin\ \phi}{2\ \cos\ \omega_m t+a\ \cos\ \phi}\right], \tag{20.11}$$

where it is understood that the inverse tangent function is continued outside the principal-value range $-\pi/2$ to $\pi/2$ when the denominator becomes negative.

The relative phase, ϕ, determines the character of this stimulus. When $\phi=0$, angle Φ is zero, and adding the signal does not change the frequency of the stimulus. The effect of adding the signal is only to change the amplitude of the stimulus. Adding the signal changes the periodicity of the envelope; the period changes from $1/(2\omega_m)$ to $1/(\omega_m)$, as shown in Fig. 20.6(a).

When $\phi=\pi/2$, adding the signal does not change the envelope period. It remains $1/(2\omega_m)$, but the depth of the envelope is reduced, as shown in Fig. 20.6(b). The dramatic effect of adding the signal with $\phi=\pi/2$ is that the signal introduces frequency modulation. The frequency of the stimulus as a whole is given by ω_s plus a modulation increment given by

$$\Delta\omega=\frac{d\Phi}{dt}=\frac{2a\omega_m\ \sin(\omega_m t)}{4\ \cos^2(\omega_m t)+a^2}. \tag{20.12}$$

This frequency modulation is also shown in Fig. 20.6(a).

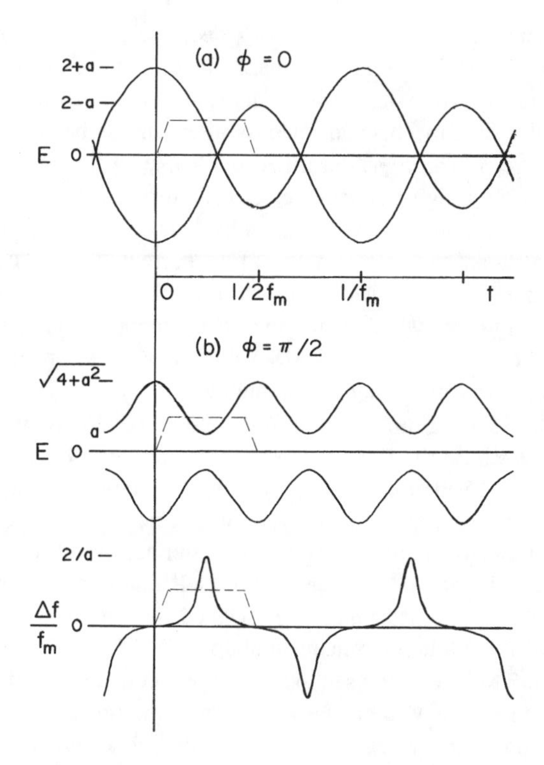

FIGURE 20.6. The figure shows envelopes for the two-masker stimulus with signal amplitude a=1/2. Part (a) is for ϕ=0; part (b) is for $\phi=\pi/2$. Part (b) also shows the modulation in instantaneous frequency, relative to modulation frequency f_m. The dashed lines indicate a stimulus segment that leads to much smaller threshold for $\phi=\pi/2$ (Phipps and Henning, 1976).

The fact that Phipps and Henning observed much lower thresholds for $\phi=\pi/2$ than for $\phi=0$ is probably due to the FM character that is added by the signal in the $\phi=\pi/2$ condition for the stimulus segments that they chose.

MODULATION TRANSFER FUNCTION

The modulation transfer function is a form of lowpass filtering that can be illustrated by a (somewhat lengthy) electrical analogy. We imagine an analog DC ammeter consisting of a pointer that is subject to a magnetic force due to electrical current. The pointer moves to the right or left as the current is positive or negative. The magnetic force operates against a spring that tends to restore the pointer to a zero reading at the center. If this meter is connected to an electronic oscillator that produces a signal with a frequency of 1 Hz, the pointer oscillates with wide excursions, one cycle every second. As the frequency of the oscillator increases it is found that the amplitude of the oscillations decreases. Because of its inertia the pointer cannot keep up with the rapid signal. As the frequency increases further, the pointer hardly budges from its position at dead center; it does not register the oscillations at all. It is a mechanical lowpass filter, and its behavior can be characterized by a transfer function, as described in Chapter 9.

In order to register a high-frequency electrical current with an electromechanical meter we need an AC ammeter containing a half-wave rectifier that produces a DC component proportional to the RMS value of the current. The DC component is just the first term in the Fourier series given in Eq. (5.47). The rectifier is followed by an ordinary DC ammeter which reads the DC component, and, assuming a sine signal, this can be interpreted as a RMS value, or a peak, or an envelope value, because these are all proportional to one another.

If now the electrical current is amplitude modulated slowly, the reading on the AC meter varies because the RMS value is changing. For a large, slow modulation of the amplitude the excursion of the pointer will be large. As the modulation frequency increases, the amplitude of the variation decreases because of the inertia in the movement of the AC ammeter. Again the meter acts as a lowpass filter, but this time it is the modulation and not the signal itself that is suppressed in the meter reading. The filtering character can be described by a transfer function, and it is called the temporal modulation transfer function, TMTF.

The idea of the TMTF can be used to model the fact that the auditory system does not respond instantaneously to changes in signals, and there is a limit to the speed of the modulation that it can register. Experimentally one studies this limitation by rapidly modulating a signal and observing the decrease in some property of the modulation perception as the modulation rate increases. An obvious problem with this technique is that as the modulation rate changes, the spectrum of the signal changes as well, and it becomes hard to say whether any result that is observed is caused by a TMTF or by some spectral effect. The way to avoid this problem is to use a signal that is already broadband and spectrally dense such as white noise.

Viemeister (1979) studied the ability of listeners to detect amplitude modulation of a broadband noise as a function of modulation rate using a two-interval forced-choice test; the modulation was present on one interval and absent on the other. He found a TMTF that could be well fitted by a one-pole lowpass filter, as described in Chapter 9. Significantly, the function was independent of signal level, except for very low levels. Level insensitivity makes the filter model implied in the TMTF concept more plausible. Viemeister repeated the experiment with noise bands about one octave wide by filtering the broadband noise after modulating it. His data, showing the TMTFs for different center frequencies, are given in Fig. 20.7.

The data in Fig. 20.7 can be fitted with a one-pole lowpass filter having a time constant between 2 and 3 ms (see Exercise 7 below). Time constants of this order appear frequently in data from experiments in which it is to the listener's advantage to respond rapidly to changes in a stimulus. Although 3 ms seems well established as a typical time constant for the TMTF, there is a striking difference between this integration time and integration times of 100 ms or longer found for signal detection

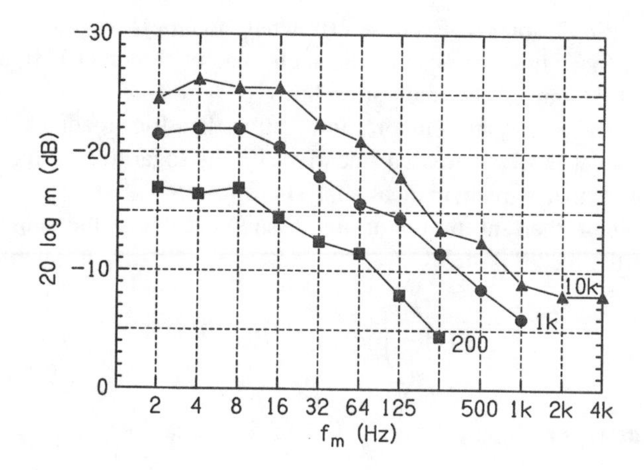

FIGURE 20.7. Threshold for detecting amplitude modulation on a wide band of noise as a function of the modulation rate, with the noise center frequency as a parameter. Shown on the vertical axis is the sum of the power in the sidebands relative to the carrier power. The plot is upside down, with the higher thresholds below the lower thresholds, so that it represents the power transfer of a filter, the TMTF. After Viemeister (1979).

and loudness integration, as described at the end of Chapter 4.

A possible response to the discrepancy in integration times that enjoys a widespread following is that there are both maximum and minimum integration times and that the auditory system can optimize the integration to cope with the particular task at hand (Eddins and Green, 1995). The flexibility of this model makes it attractive. There is, however, at least one case where it fails. Given the task of distinguishing between noise bands with different spectral density it turns out to be to the listener's advantage to integrate for as long as possible. Doing so maximizes the difference in the RMS power fluctuation of different densities. However, data on spectral density discrimination suggest an integration time of only 3 ms (Hartmann *et al.*, 1986), consistent with the minimum integration time seen in other broadband experiments.

An alternative approach to the discrepancy between integration times is that there is one true integration time, the short one of about 3 ms, and that the appearance of long integration times must be explained in some other way. One possible way is the multiple-looks model (Viemeister and Wakefield, 1991) noted in Chapter 4. Another possibility is the compressive transformation suggested by Penner (1978).

Penner's Compressive Model

The idea that temporal integration models could be brought into better agreement with experiment by allowing for the compressive nature of auditory transduction was introduced in Chapter 4 (Exercise 10), where loudness was integrated with an exponential memory function. Penner did such a calculation in (1978), but she went further. She also changed the integration law.

Penner's work began with temporal integration data for broadband noise where the threshold level in dB decreased linearly with the logarithm of the duration. A paraphrase of Penner's mathematical analysis follows:

As in Chapter 4, the sensation Ψ at time t is represented as the integral of neural firing R over the recent past,

$$\Psi(t)=\int_0^t dt'h(t')R(t-t'). \tag{20.13}$$

The firing rate is some compressive power of the signal intensity,

$$R(t)=R_0[I(t)/I_0]^p, \tag{20.14}$$

where I_0 is the threshold of hearing, and p is less than 1. Perhaps p equals 0.3 as suggested by the sone scale.

In a temporal integration experiment, intensity $I(t)$ is a constant, given by the threshold intensity, I_{Th}, for a duration T_D and then is shut off. The sensation therefore integrates to

$$\Psi(T_D)=R_0(I_{Th}/I_0)^p\int_0^{T_D} dt'h(t'). \tag{20.15}$$

Whereas temporal integration of the Plomp and Bouman (1959) type takes $h(t)$ to be an exponential memory function, Penner used a power-law memory function,

$$h(t)=0 \quad \text{for } t<0$$

$$h(t)=c_1t^{-\gamma} \quad \text{for } t>0, \tag{20.16}$$

where c_1 is a constant. In order for the memory function to vanish in the infinite past, γ must be positive. In order for the memory function to be integrable, γ must be less than 1.

Doing the integral, and dividing both sides of the equation by constants to expose the relevant variables on the right gives

$$\frac{(1-\gamma)\Psi(T_D)}{c_1R_0}=\left(\frac{I_{Th}}{I_0}\right)^p T_D^{(1-\gamma)}. \tag{20.17}$$

To compare with temporal integration data, we take ten times the log of both sides. The left side leads to a constant c_2, and

$$c_2=pL_{Th}+10(1-\gamma)\log T_D, \tag{20.18}$$

where L_{Th} is the predicted threshold in dB. Dividing by p gives the trading relationship between the threshold level and the duration of the noise burst, which is logarithmic in agreement with Penner's experiment.

$$c_2/p=L_{Th}+10[(1-\gamma)/p]\log T_D. \tag{20.19}$$

The experiment found a slope of -7.6 dB per decade of duration, so $7.6=10[(1-\gamma)/p]$. If $p=0.3$, as suggested by Stevens and others, then $\gamma=0.77$. Although it is not possible to make a direct comparison of integration times when the integration law is changed, Penner argued that the memory function with γ chosen to agree with the expected auditory compression, resembles exponential memory functions with time constants of the order of 3 ms, as found in minimum integration time experiments. The essence of this model is that one can accept a much shorter integration time if the low threshold intensities that are found for long durations are compressed by an auditory nonlinearity. To fit any given temporal integration slope, the more the compression (the smaller is the value of p) the quicker the memory function can decay (the larger is the value of γ).

The Binaural TMTF

Whereas the minimum integration times observed in monaural or diotic experiments are about 3 ms, the corresponding response times for binaural listening are much longer. The effect appears in a variety of binaural conditions and is generally known as "binaural sluggishness." For example, Grantham and Wightman (1978, 1979) found that a binaural masking level difference of 20 dB, seen as usual with fixed interaural parameters, could be reduced to zero if the interaural parameters varied in time. The rate of variation was only 4 Hz.

Binaural sluggishness also appears in sound localization (Blauert, 1972), particularly in the precedence effect. The precedence effect suppresses reflected sounds that come from the surfaces of a room so that a listener perceives a single sound, localized in the direction of the direct (unreflected) wave from the source. The precedence effect is usually demonstrated using impulsive sources, where the reflections from room surfaces would otherwise lead to multiple sound images, all coming from different directions. The effect can be studied experimentally with loudspeakers to the left and right of the listener, one with a leading impulse simulating the direct sound and the other with a lagging impulse, simulating a reflection. When the precedence effect operates, the listener hears only a single impulsive sound, localized at the leading speaker. Clifton (1987) showed that if the conditions of lagging click delay and lagging click level are set so that the precedence effect is barely established, then when the leading and lagging speakers are suddenly interchanged, the listener hears a double click, with ambiguous location. The double click continues until the listener has heard a number (often about eight) of clicks in the new speaker configuration. Then the precedence effect returns as normal (Freyman, Clifton, and Litovsky, 1991).

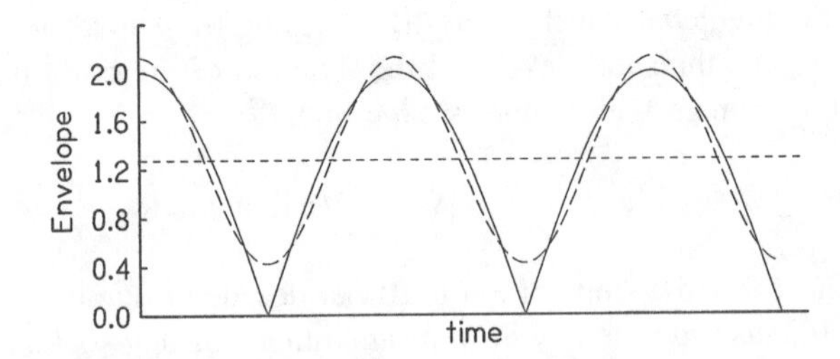

FIGURE 20.8. The solid curve shows the envelope for a beating pair of tones of equal amplitude. The dashed lines show the first two terms in the Fourier series, the constant component and the sum of the first two terms, as given in Eq. (20.24).

The phenomenon of binaural sluggishness seems paradoxical. On the one hand, the binaural system is known to be remarkably fast. It is able to encode and use interaural time differences as small as 5 μs in localizing sound sources. On the other hand, the binaural system is sluggish in responding to changes in the interaural differences. There are essentially two points of view on this problem. Possibly binaural sluggishness can be incorporated into standard binaural models as the result of a low-level process that integrates binaural differences (Stern and Trahiotis, 1995). However, some of the time constants that appear in versions of Clifton's effect are so long that it is difficult to imagine them as the result of a low-level dynamical process (Hartmann, 1996). Therefore, a second point of view regards sluggishness as representing a preparation time by the central nervous system.

ROUGHNESS

Comparison of Beats and AM

As noted in Chapter 17, both beating tones and amplitude modulation lead to fluctuation. The envelope of the signal for beating tones is given by Eq. (17.11),

$$A(t)=\sqrt{A_1^2+A_2^2+2A_1A_2\cos(\Delta\omega t+\phi)}, \tag{20.20}$$

where $\Delta\omega$ is the frequency difference between the tones, and A_1 and A_2 are the amplitudes of the lower and higher frequency tones, respectively. If these amplitudes are both equal to 1, then Eq. (20.20) simplifies to

$$A(t)=2|\cos(\Delta\omega t/2+\phi/2)|. \tag{20.21}$$

This means that the envelope is a full-wave-rectified cosine wave, as shown by the solid line in Fig. 20.8. The lowest finite-frequency Fourier component of the envelope in Eq. (20.20) or (20.21) has frequency $\Delta\omega$.

By contrast, the amplitude of an AM signal, from Eq. (17.17), is

$$A(t) = 1 + m\cos(\omega_m t + \phi). \tag{20.22}$$

It is of interest to compare the fluctuation caused by beats with the fluctuation caused by AM. The first point of comparison is the rate. Equations (20.20) and (20.22) show that the fluctuation due to beats has the same rate as the fluctuation due to AM if the frequency difference between the two beating tones is the same as the modulation frequency for AM.

Equations (20.21) and (20.22) show that the amplitude variations for AM (with $m=1$) and for beats cover the same range; both go from 0 to 2. Therefore, one might conjecture that both kinds of fluctuation would lead to the same amount of roughness, if the fluctuation rates were chosen to be the same.

However, the envelope of the AM tone is a smooth cosine, whereas the envelope of the beating pair has a sharper structure. Therefore, an alternative conjecture is that the amplitude variation in the beating pair would be less effective because the auditory system may not be able to follow all the details of the envelope structure for the beating tones. This prediction seems especially plausible when the fluctuation rate is high. Experiments by Terhardt (1974a) suggest that something rather like this happens. Working at a fluctuation rate of 40 Hz, he asked listeners to compare the roughness for AM tones having a variable modulation percentage m, with the roughness of beating pairs having equal amplitudes. He found that the beating pairs did have a reduced effectiveness.

It is not difficult to use the ideas of Fourier series to provide an analysis of this experiment. As a first approximation we suppose that the auditory system follows the beating pair with a modulation transfer function that only passes the lowest finite frequency component. That allows a direct comparison with AM. To find out what fraction of the envelope variation is due to the lowest frequency we need to look up the Fourier series for the full-wave-rectified cosine. From Eq. (5.48) we find that the Fourier series begins

$$A(t) = \frac{2}{\pi} - \frac{4}{\pi}(-1)^1 \frac{1}{(4-1)} \cos \Delta\omega t \ldots \tag{20.23}$$

Therefore, for the envelope of Eq. (20.21), with maximum value of 2,

$$A(t) \approx \frac{4}{\pi}\left(1 + \frac{2}{3}\cos \Delta\omega t\right). \tag{20.24}$$

Figure 20.8 shows the envelope of the beating pair together with these first two terms in its Fourier series.

Comparison with Eq. (20.22) for AM shows that beating tones with equal amplitude have a low-frequency envelope variation that looks like AM with a modulation percentage m equal to 2/3. Therefore, one would expect to match the roughness of a beating pair of tones having equal amplitudes by an AM signal with $m=2/3$. That is precisely what Terhardt found.

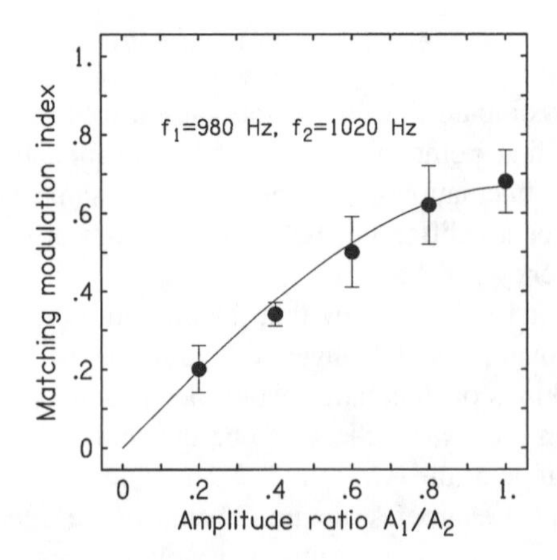

FIGURE 20.9. Symbols show the modulation percentage for AM that leads to the same roughness as beating pairs of sine tones with frequencies of 980 and 1020 Hz, and amplitudes A_1 and A_2 respectively, as measured by Aures. The AM tone had a carrier frequency of 1000 Hz and a modulation frequency of 40 Hz to resemble the beating pairs. The solid line shows the predicted modulation index obtained by assuming that the auditory system passes only the lowest frequency component of the envelope variations for the beating pair. In the limit that $A_1 = A_2$ the ratio is 2/3, as expected from the Fourier series for the full-wave-rectified cosine envelope.

Aures (1985) extended the measurements of Terhardt by letting the relative amplitudes of the beating pair vary. Therefore, the envelope was given by the more complicated formula in Eq. (20.20). Data from this experiment are shown by the solid symbols in Fig. 20.9. To compare this with the model, we need to find the first two Fourier components (DC term and lowest-frequency component) for the envelope given by Eq. (20.20). Unlike the simpler case of the full-wave-rectified envelope, the Fourier components cannot be found analytically. However, the Fourier transform can be done numerically. The ratio of the amplitude of the lowest-frequency component to the size of the DC component is given by the solid line in Fig. 20.9 and this line should agree with the psychophysical data. In fact, the agreement is excellent, supporting the idea that the amplitude averaging done by the auditory system resembles the extraction of the lowest component of the envelope variation for rapidly beating tones.

EXERCISES

Exercise 1: Mixed modulation sideband levels

Derive Eqs. (20.6) for sideband levels for modulations in phase in the narrow-band limit. To do this, expand Eq. (20.1) for the mixed modulation signal using Eq.

(20.4) for the modulated amplitude $A(t)$ and Eq. (19.9) for the frequency modulated $\sin\Theta(t)$. Because of the narrow-band limit you may neglect terms proportional to the product $m\beta$. Then expand the products of sine and cosine functions using the trigonometric identity in Eq. (B.21).

Exercise 2: Munich bands and the critical modulation frequency

Explain why Zwicker regarded Fig. 20.4 for a 1000-Hz carrier as indicating that the critical bandwidth at 1000 Hz is 160 Hz.

Exercise 3: m and β and sideband levels

Figure 20.4 shows that modulation parameters m and β are not particularly sensitive to signal level when the modulation rate, f_m, is low. Increasing the level from 30 to 80 dB SPL decreases the parameters by about a factor of 2. But when f_m is high (e.g. 500 Hz) the parameters, m or β, depend sensitively on level. There is no obvious reason why this should be so.

Show that the level dependence at high f_m can be understood from a spectral view because if the modulation parameters continued to decrease for 30 dB as they do for 80 dB then the sideband levels would fall below absolute threshold.

Exercise 4: Low frequency sideband switching

For a high carrier frequency, modulation is detected via the lower sideband, as shown in Fig. 20.3, but as the carrier frequency decreases the lower sideband becomes increasingly difficult to hear because of the rapidly rising threshold of hearing. Near a carrier frequency of 250 Hz, detection switches to the upper sideband (Sek and Moore, 1994). Possibly this switching effect is responsible for the fact that Zwicker and his colleagues, who used the critical modulation frequency to determine the critical bandwidth at low frequency, found critical bandwidths that are larger than the Cambridge bandwidths.

(a) Assume that the Cambridge critical bands from Chapter 10 are correct and show that the sideband switching effect is in the right direction to flatten the critical-band-width function at low frequency in the direction of the Munich bandwidths.

(b) Show that the sideband switching effect is not nearly large enough to account for the difference between Cambridge and Munich critical bandwidths.

[Hint: Compute Cambridge bandwidths, $B_C(f)$, from Eq. (10.22) and plot them against frequency $f-B_C(f)/2$ to simulate lower sideband modulation detection. Then plot $B_C(f)$ against frequency $f+B_C(f)/2$ to simulate upper sideband detection. Sideband switching corresponds to switching between these two plots of the critical bandwidth.]

Exercise 5: Second sidebands in FM detection

Consider FM in the high-f_m case where detection is mediated by individual sidebands. Because the sidebands tend to be masked by the carrier, the second sidebands should be easier to detect than the first because they are further away from the carrier. This advantage of the second sidebands should increase as f_m increases. However, the data show that β decreases rapidly as f_m increases, and the second sidebands decrease as β^2 whereas the first only decrease as β. Use this β dependence as well as typical critical band slopes from Chapter 10 to show that in actuality second sidebands are not important in detecting FM.

Exercise 6: Super-slick stimulus synthesis

An effective way to do the two-tone masking experiment is to create the masker as the balanced modulation product of a carrier (frequency ω_s) and a modulator (frequency ω_m). To make the signal one adds in an attenuated (a) and phase-shifted (ϕ) carrier. Show that this technique produces a stimulus with the following properties:

(a) The amplitudes of the two maskers are identical, and the amplitude of the signal is a.

(b) The signal frequency is exactly midway between the two maskers.

(c) The signal phase is fixed with respect to the maskers.

(d) There is no difference between advanced ($\phi=\pi/2$) and retarded ($\phi=-\pi/2$) conditions.

Exercise 7: Fitting the TMTF

The power transfer of a one-pole lowpass filter is given by Eq. (9.26). Therefore, to test this filter model of the TMTF you should try to fit the function $-10\log(1+\omega^2\tau^2)$ to the data in Fig. 20.7. Try fitting the 1-kHz data with a time constant $\tau=3$ ms.

Exercise 8: Origin of the TMTF

It is possible that the inability of the auditory system to track rapid changes in signal level is the result of peripheral filtering. Sharp peripheral filters would lead to long time constants and slow response. To estimate the effect, consider the time constant as it appears in Eq. (9.61) for the impulse response of a two-pole resonator with characteristic frequency ω_0. The time constant is $2Q/\omega_0$, which is equivalent to $1/(\pi\Delta f)$, where Δf is the bandwidth. Suppose that Δf is given by the Cambridge critical bandwidth in Eq. (10.22).

(a) Show that for 1 kHz, the time constant is between 2 and 3 ms, in agreement with Viemeister's data.

(b) Although Viemeister found shorter time constants for higher frequencies than for lower, the dependence of time constant on center frequency was not as strong as would be suggested by the critical band width. Use Eq. (10.22) and Fig. 20.7 to show that this is true.

Most models of the TMTF do not ascribe it to peripheral filtering. Instead, they treat it rather like the AC ammeter of the analogy in the text, a lowpass filter that follows a half-wave rectifier. Closely related is the measurement of the human ability to detect a temporal gap in broadband noise. Here too it appears that the ammeter analogy, with a time constant of 3 ms, is supported by the data (Forrest and Green, 1987).

CHAPTER 21

Sampled Signals

The sound is forced, the notes are few!
WILLIAM BLAKE

Modern acoustical research makes frequent use of digitized signals. There are three basic digital signal applications: (1) digital signal analysis-display, (2) digital signal recording-processing-reproduction, and (3) digital signal synthesis. These applications are discussed in turn.

In the case of analysis-display, an acoustical signal, called $x(t)$, is transduced by a microphone into an electrical signal and is then converted to a digital format by an analog-to-digital converter (ADC). In what follows, the digitized signal will be called $y(t)$. In a digital oscilloscope the digitized signal is displayed on a CRT. In a spectrum analyzer the digitized signal is processed by a discrete Fourier transformation or by digital filtering, and is then displayed. In such applications the signal has no future existence as an analog waveform. The signal goes in, and what comes out is a measurement of some kind.

Digital recording-processing-reproduction begins with a conversion from an analog signal, $x(t)$, to a digital format, $y(t)$, using an ADC. It ends with the conversion from a digital format to an analog signal, $w(t)$, by means of a digital-to-analog converter (DAC). Between these two conversions, the digitized signal may be stored (in computer memory, on digital tape, or on optical disc), or the signal may be processed, for example by digital filters or delay lines.

Digital signal synthesis is the computation of a digital signal, $y(t)$, by means of an algorithm and subsequent conversion to an analog signal, $w(t)$, by a DAC. The algorithm may be written in a high-level language for a general purpose computer or it may be a program on a dedicated signal processor constructed on a single piece of silicon. In this application the output signal $w(t)$ has had no previous existence in any analog form prior to conversion by the DAC.

These three applications have different purposes and are implemented differently. The first involves an ADC, the last involves a DAC, and the other one involves them both. However, the applications are similar in that they all deal with signals in the form of sampled data. One of the merits of the mathematical techniques that have been developed in preceding chapters is that they enable one to understand the sometimes unusual characteristics of sampled signals. The development continues below, maintaining the above notation for signals:

> Signal $x(t)$ is an analog input signal.
> Signal $y(t)$ is a sampled signal, perhaps in computer memory. The information is in the form of discrete samples, but we regard it as a function of a continuous time variable t. For times in between sample points $y(t)$ is zero.
> Signal $w(t)$ is an analog output signal. It might be sent to a listener's headphones for an experiment or to a loudspeaker for a concert.

THE DIGITIZED SIGNAL

The digitized signal is discrete, as shown in Fig. 21.1. It is actually discrete in both of its dimensions. First, it is a discrete function of time, which is to say that it is a sampled signal. Second, it is discrete in the vertical dimension because the numerical values of the digitized signal are in discrete integer steps.

In this section we deal with discreteness in the vertical dimension. It doesn't take long to dispense with this matter. The numerical values of the digitized function $y(t)$ are integers because an analog-to-digital converter has a digital output that is an integer, and a digital-to-analog converter has a digital input that is an integer.

The fact that integer values y_m cannot exactly represent all possible values in the continuum of an analog variable leads to the idea of "quantization error." The number of distinct integer values that a given sample can have is specified by a number of bits. The number of integer values is given by 2^b, where b is the number of bits. For example, an audio compact disc stores a sample as a 16-bit word; the word has one of $2^{16}=65536$ possible values. To illustrate quantization error, Fig. 21.1 shows a 3-bit system. This system is so crude that the quantization errors can easily be seen on a graph.

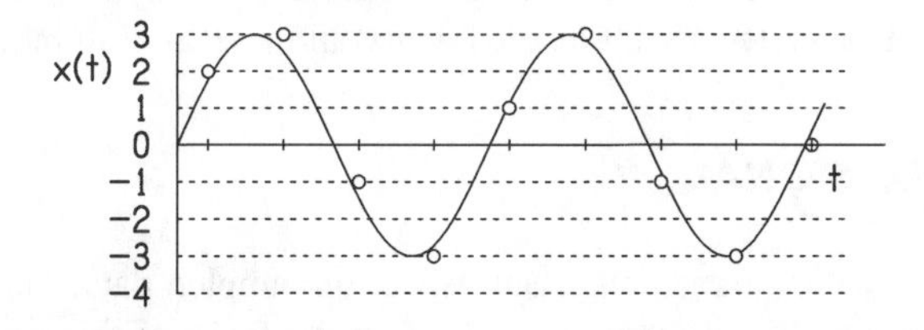

FIGURE 21.1. A signal $x(t)$, with amplitude 3, is shown by the continuous line. Tic marks on the horizontal axis show the regularly spaced sampling times. The sampled version of the signal, $y(t)$, is shown by the circles. The conversion of the signal to digital form is only a 3-bit conversion, so that only eight different values are possible for the circles, namely −4, −3, −2, −1, 0, 1, 2, and 3. The resulting quantization error can easily be seen. The digitized points do not agree with the signal.

Quantization Error Quantified

A simple and useful way to describe the quantization error is in terms of the resolution of the digitizing system. The system is able to resolve one integer unit, and the maximum signal size is 2^b units, where b is the number of bits that characterize the system. The ratio of signal to error is therefore 2^b, which is usually put in the form of a signal-to-noise ratio in decibels,

$$L_S - L_N(\mathrm{dB}) = 20 \log 2^b = 20b \log 2 = 6b. \tag{21.1}$$

The ratio of signal to quantization noise for a 16-bit system is therefore 6×16, or 96 dB. Each additional bit of resolution adds 6 dB to the ratio. Like other standard measures of dynamic range, this signal-to-noise ratio is a best case because the signal is assumed to be at its maximum possible value (see Exercise 8 for a minor improvement). Treating quantization error as a noise, as we have done, makes the further assumption that the signal and sampling are not synchronized. In the synchronized case the error appears as distortion, as discussed below in the section on oversampling.

When digital signal processing was an infant technology, DACs and ADCs had eight or twelve bits and quantizing noise was an important limitation. As digital technology has advanced, the role of quantization noise has become less important. DACs and ADCs with 16 bits are now standard and, if used properly, they are very good indeed. Converters with 18 or 20 bits are also available. Interpolating output devices, commonly used in compact disk players, can blur the rigid limits of finite word size. It is probable that in most digital signal processing applications the quantization noise is now smaller than the noise generated by other electronic means. (It is worth remembering that even nominally analog signals in the electrical domain are not free of quantization error because of the finite charge of the electron.) In a curious twist of fate, further advances in digital technology, pressured by the commercial interest in low-bit-rate encoding, have mainly been concerned with the development of systems with lesser fidelity and a smaller number of bits.

THE SAMPLED SIGNAL

We now deal with a more important aspect of sampled data, the fact that it is discrete along the temporal dimension. Because the focus of this section is on the sampling operation in time, we neglect quantization error and assume that when a sample is taken its numerical value is precisely equal to the value of the input. Therefore, the sampled signal $y(t)$ is the product of the input $x(t)$ and a sampling train, where the sampling train is a train of equally spaced delta functions, spaced by the sampling interval T_s. This product operation, shown in Fig. 21.2, represents the action of an ADC. Therefore,

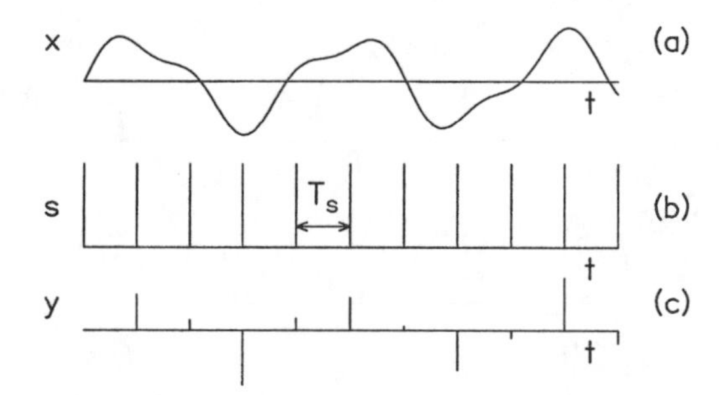

FIGURE 21.2. (a) The input signal as a function of time. (b) The sampling train. (c) The product of the input signal and the sampling train which is the sampled signal $y(t)$, also regarded as a function of a continuous time variable.

$$y(t)=x(t)s(t), \tag{21.2}$$

where $s(t)$ is a lattice sum on time, as described in Fig. 8.9,

$$s(t)=T_s \sum_{m=-\infty}^{\infty} \delta(t-mT_s)$$

or

$$s(t)=T_s \sum_{m=-\infty}^{\infty} \delta(t-t_m), \tag{21.3}$$

where time point t_m is the m-th sample time mT_s. The prefactor T_s gives $s(t)$ dimensions of unity and serves to make the Fourier transform of $s(t)$ a simpler function.

The data given by y are in the form of discrete numbers, a set $\{y_m\}$. The function $y(t)$ has no real existence. However, it is important because it illustrates limitations that affect the digital processing, as will be shown below. Function $y(t)$ is also what the output signal would be if the sampled data were simply converted by a DAC without further processing. Between the sample points it is zero.

We are particularly interested in the Fourier transform of the sampled signal, i.e. in $Y(\omega)$. Because $y(t)$ is the product of the input signal and the sampling train, $Y(\omega)$ is related to the Fourier transform of the input, $X(\omega)$, by a convolution with the Fourier transform of the sampling train, $S(\omega)$,

$$Y(\omega)=\frac{1}{2\pi}\int_{-\infty}^{\infty} d\omega' X(\omega')S(\omega-\omega'). \tag{21.4}$$

The function $S(\omega)$ is given by the Fourier transform of Eq. (21.3),

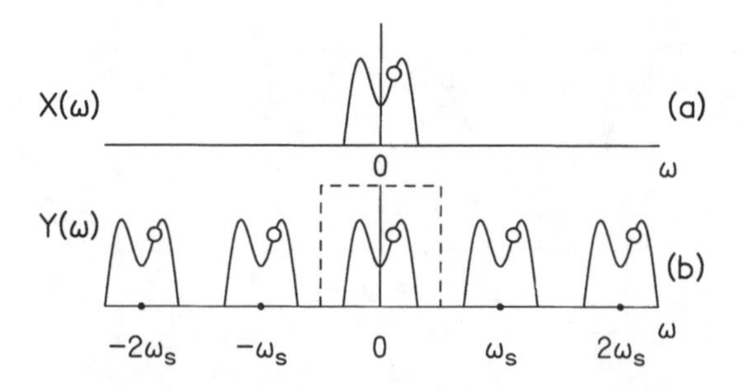

FIGURE 21.3. The input spectrum is shown in (a). The spectrum of the sampled signal is shown in (b). An arbitrarily chosen frequency is indicated by a circle to show how a particular frequency is displaced up and down the frequency axis by the sampling process. The dashed rectangle shows the window that contains all the correct information about the input signal. The upper frequency of the window is equal to half the sample rate.

$$S(\omega)=T_s\int_{-\infty}^{\infty} dt\; e^{-i\omega t}\sum_{m=-\infty}^{\infty}\delta(t-mT_s) \tag{21.5}$$

$$=T_s\sum_{m=-\infty}^{\infty} e^{-i\omega m T_s}. \tag{21.6}$$

From Eq. (8.84) this becomes

$$S(\omega)=2\pi\sum_{n=-\infty}^{\infty}\delta(\omega-n\omega_s), \tag{21.7}$$

where $\omega_s=2\pi/T_s$.

Therefore, the convolution integral for Y in Eq. (21.4) is

$$Y(\omega)=\int_{-\infty}^{\infty} d\omega' X(\omega')\sum_{n=-\infty}^{\infty}\delta(\omega-\omega'-n\omega_s)=\sum_{n=-\infty}^{\infty}X(\omega-n\omega_s). \tag{21.8}$$

This equation says that the Fourier transform of the sampled signal is given by repeated versions of the Fourier transform of the input signal. The term in the sum with $n=0$ is $X(\omega)$, precisely the Fourier transform of the input signal. Other values of n lead to images of the input signal displaced up and down the frequency axis by integral multiples of the sampling frequency, namely by $n\omega_s$. This is shown in Fig. 21.3.

Figure 21.3(a) shows the input $X(\omega)$, and Fig. 21.3(b) shows the sampled signal $Y(\omega)$. In drawing Fig. 21.3, several assumptions were made. First, it was assumed that $X(\omega)$ is real so that it can be described by a simple two-dimensional graph.

[There is no problem in principle if X is complex, only a more complicated figure. Alternatively if X is complex, we could regard the figure as a picture of the amplitude spectrum $|X|$.] Second, it was assumed that the repeated versions of $X(\omega)$ do not overlap. As can be seen from Fig. 21.3(b) overlapping does not occur if the highest frequency in the spectrum $X(\omega)$, is less than ω_s minus this highest frequency. In other words, if ω_{top} is the highest frequency in the signal, then ω_{top} must be less than $\omega_s - \omega_{top}$. This is equivalent to the statement that ω_{top} must be less than $\omega_s/2$, and this is a very important requirement. It is known as the Nyquist criterion.

> The Nyquist criterion: When a signal is sampled, the highest frequency in the signal must be less than one half of the sample rate.

By looking at the spectrum of the sampled signal, we realize that, although the sampled signal $y(t)$ looks very different from the input $x(t)$, all the essential information contained in $x(t)$ is actually present in the sampled signal in an uncorrupted form. We find it in the spectrum of the sampled signal between the frequencies $-\omega_s/2$ and $+\omega_s/2$.

Because the Nyquist criterion has been satisfied in the sampling procedure, no information has been lost. A frequency analysis of $y(t)$ will give a complete and correct analysis of $x(t)$ so long as the analysis is confined to frequencies less than $\omega_s/2$. However, as shown by Fig. 21.3, at frequencies above $\omega_s/2$ the sampled function $y(t)$ has spectral power whereas the original signal $x(t)$ has none.

EXAMPLE 1: *A sampled cosine tone*

To illustrate the spectrum of a sampled signal we consider a 1000-Hz pure tone, sampled at a rate of 5000 Hz. We choose to look at the signal in cosine phase so that $X(\omega)$ is real. The Fourier transform of the cosine is

$$X(\omega) = \pi[\delta(\omega - \omega_0) + \delta(\omega + \omega_0)], \tag{21.9}$$

where ω_0 is $2\pi \times 1000$. The Fourier transform of the sampled signal is, from Eq. (21.8),

$$Y(\omega) = \pi \sum_{n=-\infty}^{\infty} \{\delta[\omega - 2\pi(1000 - 5000n)] + \delta[\omega + 2\pi(1000 + 5000n)]\}. \tag{21.10}$$

The equation is perhaps more easily understood if it is expanded somewhat:

$$Y(\omega)=\pi\delta[\omega-2\pi(1000)]+\pi\delta[\omega+2\pi(1000)]+\pi\left\{\sum_{n=1}^{\infty}\delta[\omega-2\pi(1000-5000n)]+\delta[\omega+2\pi(1000-5000n)]+\delta[\omega-2\pi(-1000+5000n)]+\delta[\omega+2\pi(-1000+5000n)]\right\}. \tag{21.11}$$

Equation (21.11) shows how the spectrum of the original cosine (delta functions at ± 1000 Hz) appears again and again, centered about all the harmonics of the sampling frequency of 5000 Hz. The result is illustrated in Fig. 21.4.

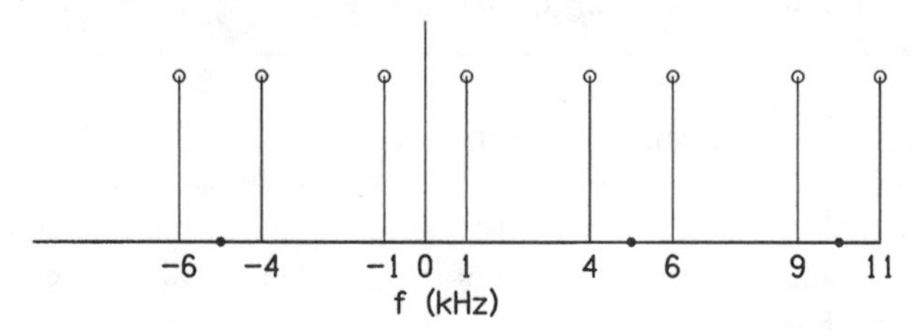

FIGURE 21.4. Spectrum of a 1000-Hz tone with 5 samples per cycle.

Undersampling

The importance of the Nyquist sampling criterion becomes apparent when it is violated, i.e. when there is a spectral component in the input signal that has a frequency greater than half the sample rate. This situation is known as ''undersampling.'' For a specific example, we consider a 1000-Hz cosine tone that is sampled at a rate of $4/3\times1000$ Hz, $\approx$1333 Hz. According to Eq. (21.8) the spectrum $Y(\omega)$ consists of satellites around harmonics of 1333 Hz. The satellites are displaced by the signal frequency, ±1000 Hz. Figure 21.5 shows the spectrum.

Having undersampled the input, we are now in an unfortunate situation. The spectrum contains a component at 333 Hz whereas the input contained no such component. The component at 333 Hz is known as an ''alias.'' It was originally a component at 1000 Hz, but the undersampling caused it to masquerade as a 333-Hz

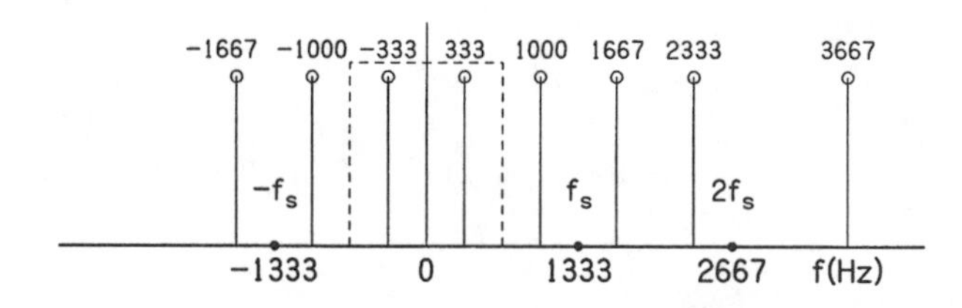

FIGURE 21.5. The spectrum of a 1000-Hz cosine sampled at a rate of 1333 Hz.

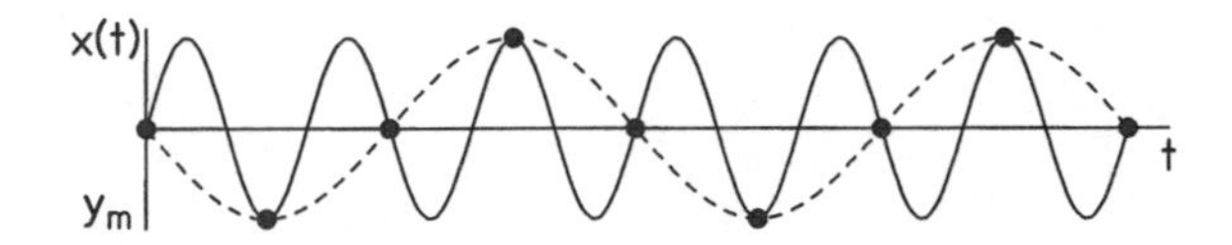

FIGURE 21.6. The solid curve shows a 1000-Hz tone that is sampled at 1333 Hz. The dashed curve shows the interpretation of the sampled signal, subject to the legitimate constraint that all frequencies in the sampled signal shall be less than half the sample rate.

component. Of course, the appearance of extraneous frequency components in a sampled signal is nothing unusual. It occurs whenever a signal is sampled, and we have seen it before. But in this example, there is an extraneous component with a frequency that is less than half the sample rate. It falls within the dashed rectangle, and that is something that does not occur when a signal is correctly sampled.

If we make an analysis of the samples $\{y_m\}$ obeying the standard rule that the analysis frequency shall be less than half the sample rate, we will find the component at 333 Hz, and there is no way that we can tell that this component is an alias and was not in the original signal. The image that the analysis procedure has of the signal is shown by the dashed line in Fig. 21.6. We observe that the dashed line has a frequency that is just one-third of the original frequency, namely 333 Hz, as expected from the spectral argument above.

If we reproduce the signal with a DAC, and lowpass filter the output at half the sample rate, the 333-Hz component will appear in the output as a large distortion component. It is a distortion of a particularly objectionable kind, with a frequency that is not a harmonic of the original component at 1000 Hz. We notice too that if the frequency of the original tone increases, say to 1100 Hz, the frequency of the alias decreases, in this case to 233 Hz. When a signal is undersampled there is nothing that can be done to repair the error. No operation, digital or analog, can disentangle the aliased components from components present in the original signal.

Anti-Aliasing Filters

Distortion due to aliasing can produce unpleasant surprises. Suppose that we wish to digitize some speech tokens from an FM broadcast. Because the material is broadcast speech, we believe that there is no significant power above 10 kHz. Therefore we choose to sample at a rate of 22 kHz, expecting to satisfy the Nyquist criterion. However, the audio signal that we have to deal with happens to include a 19-kHz tone. It is actually the 19-kHz pilot tone for the stereo subcarrier in the FM signal. Normally we would be unaware of this tone because the human auditory system is quite insensitive at frequencies this high. Unfortunately, the 19-kHz tone is too high to satisfy the Nyquist criterion, and it is aliased to a frequency of $22-19=3$ kHz. Now the sampled signal includes an extraneous tone just where the

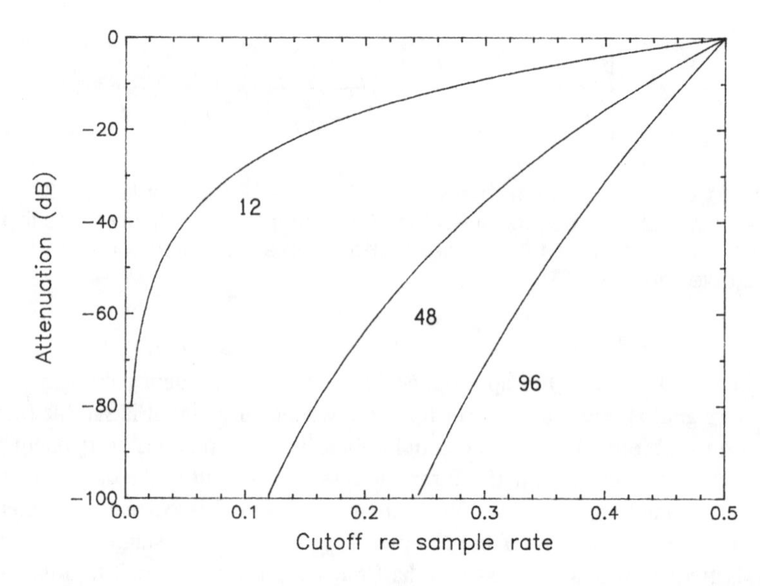

FIGURE 21.7. The attenuation at a frequency of half the sample rate produced by a lowpass filter with a cutoff frequency given on the horizontal axis as a fraction of the sample rate. Attenuations for lowpass filters with slopes of −12, −48, and −96 dB/octave are shown.

hearing system is most sensitive. Our speech tokens are irrevocably contaminated with a 3-kHz tone. Although our example has considered the aliasing of a sine tone, it should be understood that if a complex tone has any Fourier component with a frequency above half the sample rate then that component will be aliased.

In order to eliminate the effects of aliasing distortion, digitizing systems include a lowpass filter, called an ''anti-aliasing'' filter, in the electrical circuit that immediately precedes the ADC. The filter eliminates all components with frequencies above half the sample rate. Principles of economy usually make it attractive to get the highest possible bandwidth with the lowest possible sample rate consistent with the Nyquist criterion. Therefore, an anti-aliasing filter normally has a cutoff frequency that is quite close to half the sample rate, which means that it must have very steep skirts.

The requirements on anti-aliasing filters can be rather difficult. Figure 21.7 shows what is involved. The ratio of the cutoff frequency of the lowpass filter to the sample rate is shown on the horizontal axis. To satisfy the Nyquist criterion this ratio must be less than 0.5. The filter is expected to eliminate all components of the input analog signal that are above half the sample rate, even if they are only slightly above half the sample rate. The attenuation of a spectral component with a frequency equal to half the sample rate, the most difficult component for this filter to suppress, is plotted on the vertical axis. The parameter is the filter skirt, given in dB per octave. The figure shows that it is not easy to get a large attenuation of the

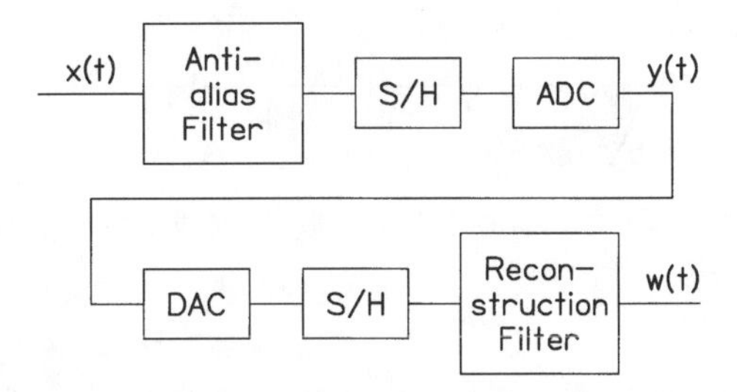

FIGURE 21.8. A digital record and reproduce system with sample-hold operations (S/H) external to the converters.

unwanted component. A filter with a skirt of −12 dB/octave (a two-pole filter) is essentially useless. Even if the sample rate is 20 times the filter cutoff frequency, there is only 40-dB attenuation of an unwanted component at half the sample rate. A steeper filter skirt is needed. Even with a steeper skirt one still needs to be conservative in choosing the cutoff frequency. With a skirt of 96 dB per octave, the filter cutoff frequency can only be about 0.3 of the sample rate if the unwanted component is to be attenuated by 70 dB.

THE OUTPUT SIGNAL

To begin the discussion of the output signal in a sampled data system, we consider an example where our goal is simply to create an accurate reproduction of the input signal. In other words, our goal is to make $w(t)$ identical to $x(t)$. Our problem is that the data have been stored as a sampled signal $y(t)$. From the spectral arguments above it is clear that provided that the Nyquist criterion was satisfied on input, we can make the output w identical to the input x by passing the sampled signal $y(t)$ through a lowpass filter. For example, the ideal lowpass filter shown by the dashed lines in Fig. 21.3 removes all the extra versions of the input spectrum. The result is that the output is equal to the input. The filter in this application is called a "reconstruction filter," or "smoothing filter." It should be evident that in a digital recording/reproducing system one needs both an anti-aliasing filter and a reconstruction filter. The requirements for these two filters are the same; they both need a cutoff frequency below half the sample rate and they both need steep skirts. In a practical digital recorder the filters may be of identical design, but they serve different purposes. A block diagram of a digital record and reproduce system is shown in Fig. 21.8.

Digital Synthesis

The output filtering requirements described above are not unique to the problem of reproducing a sampled analog input. The same considerations apply in digital synthesis, where signals are computed.

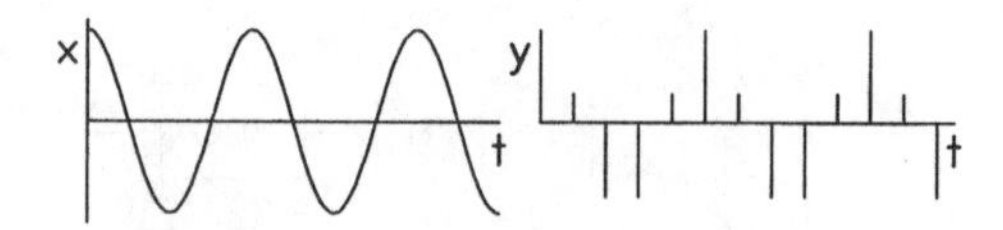

FIGURE 21.9. (a) A 1000-Hz cosine. (b) A 1000-Hz cosine computed by calculating 5 points per cycle.

As a synthesis example, suppose that we want to compute a 1000-Hz signal. The first step is to choose a sample rate; we choose 5000 Hz, a convenient number that satisfies the Nyquist criterion. We compute successive values of the function

$$y_m = A\ \cos(2\pi 1000 t_m), \tag{21.12}$$

where t_m is the m-th sample time. Thus

$$y_m = A\ \cos(2\pi 1000 m/5000). \tag{21.13}$$

There are, therefore, five sample points per cycle of the sine signal, as shown in Fig. 21.9. The values of y_m give the function $y(t)$ for the specific time points $t = t_m$. For all other values of t, $y(t) = 0$.

The spectrum of this signal is the same as the spectrum of the sampled signal shown in Fig. 21.4. In order for the output waveform, $w(t)$, to be a 1000-Hz cosine as intended, we need to eliminate all the high-frequency repetitions of the spectrum with a lowpass filter. This shows that there is a need for a reconstruction filter on the output of sampled data, whatever the origin of the samples.

Sample and Hold

The description of the output signal in the previous section assumed that the digital-to-analog conversion process initially produces an analog signal that is equal to the sampled data signal $y(t)$. This signal is a series of spikes with heights that vary according to the signal to be reproduced. Real DACs do not behave in quite this way. Instead of producing spikes, they hold each value for a brief span of time. Sometimes the "hold" operation is distinct from the conversion process of the DAC; it is assigned to an analog sample and hold circuit known as a "deglitcher." Sometimes the hold operation is part of an analog multiplexor that follows the conversion process. Whatever the circuitry, the output $w(t)$ becomes a series of flat-top pulses. The duration of the pulses cannot be greater than the time between successive samples, T_s. We define the duration of the pulses as pT_s, where $0 < p \leq 1$. An example is given in Fig. 21.10 with two different values of p.

The process of turning a series of delta functions, $y(t)$, into a series of pulses, $w(t)$, can be represented mathematically by convolving $y(t)$ with the box function $h(t)$ shown in Fig. 21.11.

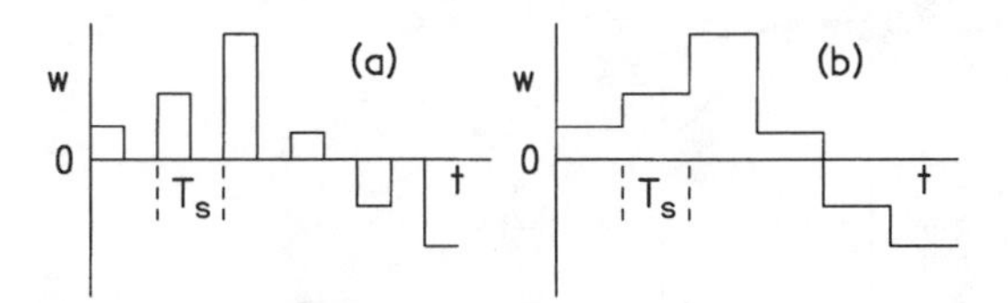

FIGURE 21.10. The output of a sample/hold for two different pulse widths pT_s. (a) p=0.5; (b) p=1.

We prove that as follows: First, from Eqs. (21.2) and (21.3) we have

$$y(t)=T_s \sum_{m=-\infty}^{\infty} x(t)\,\delta(t-t_m). \tag{21.14}$$

If $w(t)$ is the convolution of y with h then

$$w(t)=\int_{-\infty}^{\infty} dt' h(t-t')y(t'), \tag{21.15}$$

$$=T_s\int_{-\infty}^{\infty} dt' h(t-t') \sum_{m=-\infty}^{\infty} x(t')\,\delta(t'-t_m). \tag{21.16}$$

Using the selection property of the delta function to do the integral, we find:

$$w(t)=T_s \sum_{m=-\infty}^{\infty} x(t_m)h(t-t_m). \tag{21.17}$$

Equation (21.17) is just what we expected. It is a periodic series of boxes with heights given by the input values at the time points t_m and with widths equal to pT_s.

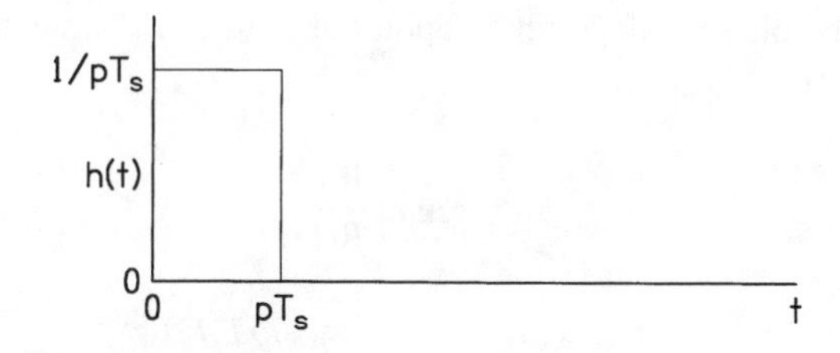

FIGURE 21.11. The sample/hold box function $h(t)$. This function is not periodic. What is explicitly shown is all there is.

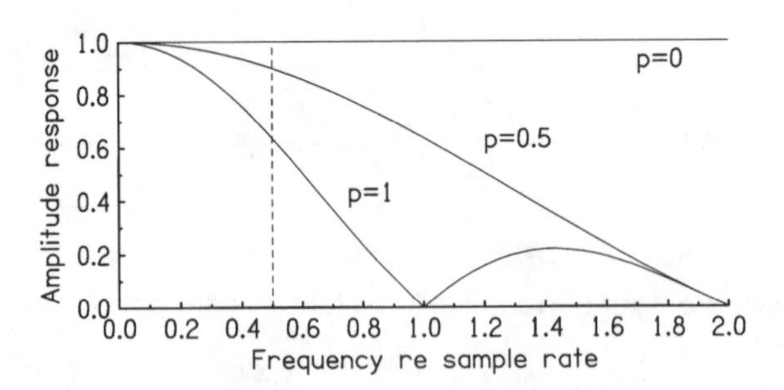

FIGURE 21.12. The amplitude response of the sample/hold filter function $|H(\omega)|$. Because of the Nyquist criterion the function is of practical interest only for abscissa values less than 0.5 (dashed line).

This is the output, $w(t)$, of a typical DAC sample/hold. In many cases, fraction p is equal to 1, and the output is a staircase function. After this output is filtered by the reconstruction filter the signal is ready to be heard.

We next study the spectral consequences of the sample/hold operation. Because the output $w(t)$ is the convolution of $y(t)$ and $h(t)$, the Fourier transform of the output $W(\omega)$ is the product of $Y(\omega)$ and $H(\omega)$, where $H(\omega)$ is the Fourier transform of the single rectangle, from Chapter 8,

$$H(\omega)=e^{i\omega pT_s/2}\frac{\sin(\omega pT_s/2)}{\omega pT_s/2}. \tag{21.18}$$

Because multiplication in frequency space simply means filtering, it is clear that output W is a filtered version of sampled signal Y where the transfer function of the filter is H. This transfer function has two parts: the absolute value gives the amplitude response, the complex factor gives the phase response. Because the exponential function in Eq. (21.18) has an absolute value of 1, the amplitude response is simply the sinc function. This function is 1 for $\omega=0$ and decreases to zero when $\omega pT_s/2=\pi$. Thus it is clear that the filter described by H is a lowpass filter. This response is shown in Fig. 21.12 for several values of the width fraction p. The phase response of the filter is characterized by the shift of the phase of a component with frequency ω. The phase shift depends upon the ratio of the imaginary part of H to the real part of H. It is

$$\phi_s=\arctan\left(\frac{\text{Im } H}{\text{Re } H}\right) \tag{21.19}$$

$$=\arctan\left[\frac{\sin(\omega pT_s/2)}{\cos(\omega pT_s/2)}\right]=\omega pT_s/2=\pi pfT_s. \tag{21.20}$$

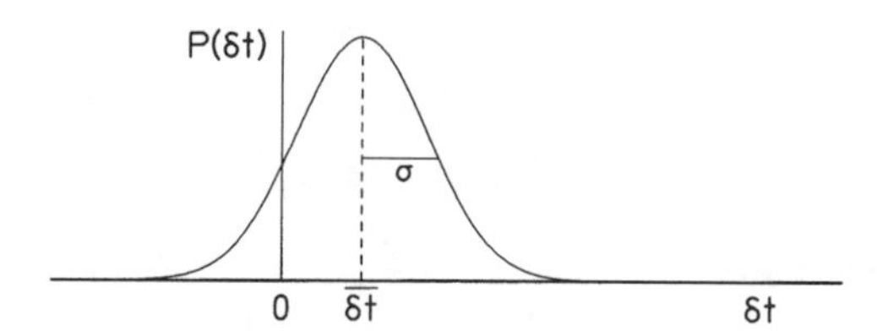

FIGURE 21.13. The probability density for the jitter, namely the discrepancy between the actual sample times and the correct sample times.

The phase shift is directly proportional to frequency, which means that it is simply a time delay. The constant of proportionality, $pT_s/2$, is the delay. Not surprisingly, it is half of the pulse width. Thus, the sample and hold process does not introduce any phase distortion that changes the shape of the output signal, it only shifts the signal along the time axis.

SAMPLING-JITTER NOISE

For accurate digital encoding and decoding, the operation of sampling must occur with precise regularity. If there is some random error in the timing of the clock that determines when the samples are taken, the result is an additional noise, as will be shown below. If there is a periodic error in timing, perhaps due to hum on the clock line, the result is a modulated signal. Below we deal with the case of random error in clock timing.

For definiteness we illustrate the noise that occurs when a digital signal is converted to analog form by a jittery DAC. We suppose that an accurate sine signal exists in a memory buffer. The buffer and the DAC have so many bits that they are arbitrarily precise. All errors, therefore, result from the fact that the clock that reads out the signal is not absolutely precise. As we watch this clock, we find that the discrepancy between the actual sample times and the intended sample times (the jitter) has a distribution shown in Fig. 21.13.

The density looks like a Gaussian function, which is certainly a likely form for the distribution, but what we have to say about the noise that is generated does not depend upon the form of the distribution. The average value of the discrepancy is $\overline{\delta t}$. It does not lead to noise; it simply leads to the wrong frequency in the output of the DAC. It might be zero. We are more interested in the standard deviation of the discrepancy, σ.

To calculate the noise power caused by the mistiming, we consider a particular sample, the j-th. Because the sample time is wrong by an amount δt_j there is an error in the reproduced signal, δx,

$$\delta x = A \sin \omega_0(t_j + \delta t_j) - A \sin \omega_0 t_j, \tag{21.21}$$

where ω_0 is the angular frequency of the sine, and t_j is the correct sample time.

Assuming that $\omega_0 \delta t_j$ is small we can expand the first term above in a Taylor series and find that

$$\delta x = A\,\omega_0 \delta t_j \cos \omega_0 t_j . \tag{21.22}$$

The average noise power is given by averaging the square of δx. Because the timing errors are random, they are independent of the actual times t_j. Therefore, we can do separate averages over the thousands of sample times needed to convert a signal and over the distribution of errors. The average of the squared cosine is 1/2. The average of the square of δt_j is simply σ^2. Therefore,

$$\overline{\delta x^2} = \frac{1}{2} A^2 \sigma^2 \omega_0{}^2 . \tag{21.23}$$

Primarily we are interested in the ratio of this noise power to the signal power, which is $A^2/2$. Therefore, the noise-to-signal ratio is given by the simple expression

$$N/S = \sigma^2 \omega_0^2 . \tag{21.24}$$

EXAMPLE 2: ***Calculate the N/S ratio for a 20-kHz sine reproduced by a DAC with an RMS timing jitter of 10 ns***

Solution: The level is given by

$$L_{\text{jitter noise}} = 20 \log \sigma \omega_0$$

$$L_{\text{jitter noise}} = 20 \log(4\pi \times 10^4 \times 10^{-8}) = -58 \text{ dB}. \tag{21.25}$$

DISCRETE FOURIER TRANSFORM

The discrete Fourier transform (DFT) is a way of calculating the Fourier transform of a signal with a digital computer. A DFT is used in theoretical work when the Fourier integral cannot be done by analytic means. A DFT must be used when Fourier transforming data extracted from the real world.

Machine computation places some restrictions on the mathematics of the Fourier transform. First, a signal $x(t)$ must have a finite duration T_D. Second, signal $x(t)$ and its Fourier transform $X(\omega)$ are no longer functions of continuous variables. The variables of time t and frequency ω can only have discrete values. As shown above, we obtain a signal as a function of a discrete time variable by sampling with the lattice function. The steps necessary to obtain a Fourier transform as a function of a discrete frequency are the mathematics of the DFT.

The key to the DFT is to remember that we automatically obtain a function of a discrete frequency variable in the Fourier series. That, in turn, depends upon the signal being periodic. What we do, therefore, is to force the signal to be periodic as shown in Fig. 21.14.

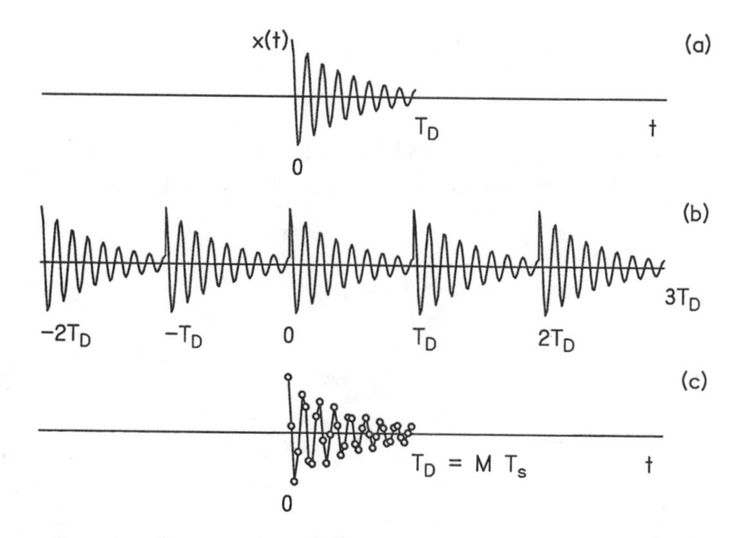

FIGURE 21.14. To do a discrete Fourier transform, we imagine that the signal of interest $x(t)$, shown in (a), is artificially extended periodically to make a function that is infinitely long, as shown in (b). The Fourier transform only requires that we sample one period, as shown in (c).

The Fourier transform of a periodically extended function was described in Chapter 8. From Eq. (8.81) we have the Fourier transform as a sum of harmonics,

$$X(\omega)=\sum_{n=-\infty}^{\infty}\delta(\omega-n\omega_0)X_n, \tag{21.26}$$

where X_n is given by Eq. (8.68), namely

$$X_n=\frac{2\pi}{T_D}\int_{-T_D/2}^{T_D/2}dt\ x(t)e^{-in\omega_0 t}, \tag{21.27}$$

and where $\omega_0=2\pi/T_D$.

Finding the harmonics of a periodic signal is a familiar subject from previous chapters of this book. What is new here is that the period is the entire duration of the signal of interest, and the fundamental frequency may be very low. If the signal is 5 seconds long then the fundamental frequency is 1/5 Hz, the second harmonic is 2/5 Hz, etc.

We next find the Fourier coefficients X_n for a sampled signal. If $x(t)$ is sampled with the lattice function, it becomes

$$x(t) \rightarrow T_s \sum_{m=1}^{M} x(t)\, \delta(t-mT_s), \tag{21.28}$$

where T_s is the sampling period, and M is the number of points in the signal, $M = T_D/T_s$. The delta functions make it easy to do the integral in Eq. (21.27), so that

$$X_n = \frac{2\pi T_s}{T_D} \sum_{m=1}^{M} x(mT_s) e^{-2\pi i n m T_s / T_D}, \tag{21.29}$$

or

$$X_n = \frac{2\pi}{M} \sum_{m=1}^{M} x_m e^{-2\pi i n m / M}, \tag{21.30}$$

where we have defined x_m as the value of $x(t)$ when t is at the m-th time point $t = mT_s$. Alternatively, we may calculate the A and B coefficients in a Fourier series,

$$A_n + iB_n = \frac{2}{M} \sum_{m=1}^{M} x_m e^{2\pi i n m / M}. \tag{21.31}$$

We are accustomed to seeing a harmonic series go on indefinitely. Not so for the sampled signal. The number of independent frequency components is limited by the Nyquist theorem. The maximum frequency is $f_{max} = 1/(2T_s) [= M/(2T_D)]$, so that harmonic number n is limited to the range $-M/2 \leqslant n \leqslant M/2$. Within this range there are $M+1$ values of n including $n=0$. For each value of n one expects two Fourier coefficients, the real and imaginary parts of X_n. However, because $x(t)$ is real, we have $X_n = X^*_{-n}$. The coefficients for $n<0$ are completely determined by the coefficients for $n>0$, and the number of independent parameters is reduced to the real and imaginary parts for $M/2$ positive frequencies plus a component at zero, a total of $M+2$ values. The actual number of independent values can be further reduced to M because the imaginary parts of both X_0 and $X_{M/2}$ must be zero. For $n=0$ the imaginary part of $\exp(2\pi i n m/M)$ is $\sin(0)$, and for $n = M/2$ the imaginary part is $\sin(\pi m)$, which is also zero. Thus M values of the signal x_m lead to M independent values in the DFT coefficients X_n.

FAST FOURIER TRANSFORM

The fast Fourier transform (FFT) is a mathematical technique of enormous technological importance. It has allowed inexpensive microcomputers (and even less expensive special chips) to do powerful signal analysis and synthesis on a scale that would have been otherwise impossible. The FFT is at the heart of modern digital spectrum analyzers. Auditory models, either real time or merely very fast, commonly begin with an FFT. Nevertheless, the FFT is basically just a discrete Fourier transform, and everything that has been said above about the DFT applies to the FFT also. What is different is that the FFT employs a particularly efficient means of evaluating the X_n. A brief description is given by Press *et al.* (1986).

If one asks about the number of multiplication operations that are required to do a complete DFT, as in Eq. (21.30), the answer appears to be of the order of M^2. There are M coefficients to be determined and each of them involves a sum of M terms. If one uses the FFT algorithm, however, the answer is of the order of $M \log_2 M$. The ratio of these is $(\log_2 M)/M$, which can be an enormous reduction in the number of operations. For a signal that is 65536 points long, the ratio is 16/65536, an improvement by a factor of more than 4000.

The efficiency of the FFT comes with a price to pay; it is necessary that the number of signal samples be a power of 2. If the signal is not long enough to fill an array having a length that is a power of 2, then one fills the empty slots with zeros, called "zero padding." By contrast, a simple DFT can work with arbitrary M. Engineers have generally found that the speed of the FFT is well worth the price.

OVERSAMPLING

The principles in this chapter are fundamental to digital signal synthesis, recording, and reproduction. However, the field of digital audio is rapidly moving, to say the least. It is worth considering how the current state of the art has altered the appearance of some of these principles. Among recent developments, the technique of oversampling is perhaps the most important. In practice, the term "oversampling" comprises a number of formally independent processes, including digital filtering, dithering, and noise shaping (Hauser, 1991).

In the first instance, oversampling means that the interface between the digital world and the analog world is done at a higher sampling rate than the digital data storage rate. For example, in a compact disc player with four-times oversampling, three output samples are interpolated between successive stored samples read off the disc itself. Whereas the disc delivers 44,100 input samples (per channel) every second, the interpolator sends 176,400 samples per second to the DAC. The interpolation is accomplished by adding three samples with zero value for every input value and then passing the result through a digital filter like the transversal filter described in Chapter 9. The coefficients in the digital filter actually weight four input samples in a running average to generate one interpolated output sample. Digital filters can be very accurate. They add no noise, and they can be without phase distortion, as shown in Exercise 8 of Chapter 9. The digital filter handles the difficult task of lowpass filtering the 44.1-kHz sample train at a cutoff frequency near 20 kHz (expected to be the upper limit of human hearing) as required by the Nyquist theorem. In the following stage, the DAC sampling rate is four times as large and the requirements on the analog reconstruction filter are only that it cut off at 176,400/2 Hz, a task that is easily done by inexpensive filters with little distortion of any kind. Thus the requirements of the Nyquist theorem, as presented in this chapter, are not obviated by the technique of oversampling; they are just moved to the digital domain.

A similar principle applies in recording. The ADC operates at a sample rate much higher than the rate at which the data will eventually be stored. Thus, a simple gentle analog lowpass filter can be used as an anti-aliasing filter at first. A decimator, which discards samples prior to storage, includes a digital filter that is ultimately responsible for removing the input components that have frequencies greater than half the storage rate and would lead to aliases in the stored signal. This digital filter can do other useful tasks such as adding noise!

Just why one would want to add noise to the input of a digital recording is not immediately obvious. The technique is known as ''dithering'' and its purpose is to reduce the distortion introduced by quantizing. It is evident that errors due to quantizing are most apparent when the signal is small. Then it may happen that the input signal varies for awhile in the vicinity of a quantized level without ever causing a jump to the levels immediately above or below. The output of the sampler is constant and the actual variation of the signal is lost in the sampling. By adding random noise, with zero average value and magnitude of the order of a quantized level, transitions up or down will occur. On the average, upward transitions will occur when the signal is at a relatively high value, and transitions to the lower level will occur when the signal is at a relatively low value. Thus adding the random noise (dither) has linearized the system in a statistical sense.

Consider too, the unfortunate consequences of quantizing a signal that has a frequency that is related in some simple way to the sample rate. Then the errors made by quantization do not appear randomly but are correlated with the signal. Such errors then appear as distortion and not as noise. For example, suppose that a sine signal with a frequency of 18.9 kHz is sampled at a rate of 44.1 kHz. These two frequencies are simply related because their ratio is precisely 3/7. That means that for every three cycles of the signal there will be exactly seven samples. Because the signal and sampling are synchronized in this way, the same errors will be made on every set of three signal cycles. The periodicity of the error signal is 18.9/3=6.3 kHz, and the second harmonic is 12.6 kHz. These two error components will be added as distortion products to the signal. It should be noted that a digital function generator that creates a periodic waveform by cycling a small buffer continuously necessarily correlates the fundamental frequency and the sampling rate. The waveform is subject to distortion of this kind.

If dither is added to the signal then the periodicity of the error signal is destroyed. In that way distortion components are converted into a random noise background which is much less objectionable. The optimum dither to use is a noise with a triangular probability density function covering two quantization levels. See Pohlmann (1985) or Watkinson (1994) for a simple explanation of this fact and of other topics in digital audio.

EXERCISES

Exercise 1: Sample rates

The sample rate of compact disc recording is 44.1 kHz per channel. What is the maximum frequency that can be recorded on this medium according to the Nyquist criterion? What is the information rate, measured in bits per second, in a compact disc recording given the sample rate of 44.1 kHz, two channels for a stereo signal, and a sample word length of 16 bits?

Exercise 2: Redundancy of a periodic sampled signal

Figure 21.9 shows $y(t)$, a 1000-Hz cosine function that is sampled at a rate of 5000 Hz. There are, therefore, exactly five samples for every cycle of the sine. Because of the simple frequency ratio, the function $y(t)$ is periodic (period=5 samples) and does not trace out the cosine function very well. Do you think that a more accurate representation of the 1000-Hz cosine could be obtained by sampling at a less commensurate sample rate, e.g. 4913 Hz?

Exercise 3: Sampling and balanced modulation of sines

Sampling and balanced modulation (Chapter 17) both involve the multiplication of an information-carrying signal by another function. In balanced modulation the other function is a sine carrier; in sampling it is a pulse train. Show how your knowledge of the spectrum of a balanced modulated signal enables you to understand the spectrum of the sampled signal.

Exercise 4: Narrow limit of the sample/hold function

(a) Show that in the limit that $p \to 0$ the sample/hold box function becomes a delta function. (The spike does not occur at exactly $t=0$.)

(b) Using a Fourier transform show that in this limit $H(\omega)=1$.

(c) Show from the convolution integral for $w(t)$ that in this limit, $w(t)$ must be equal to $y(t)$.

Exercise 5: The sample/hold filter function

The sample/hold filter function has consequences for the fidelity of a digital recording system.

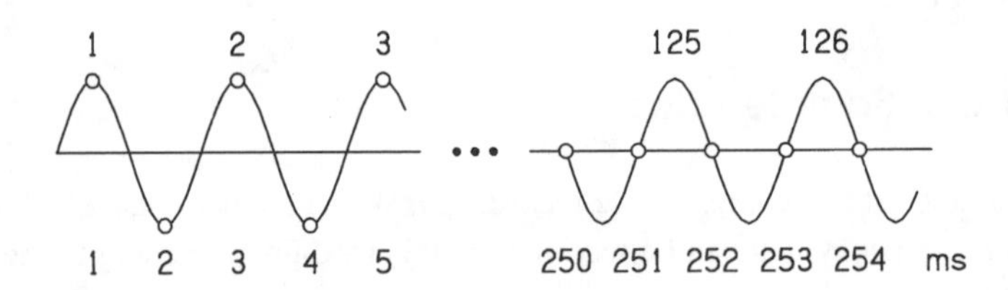

FIGURE E.6. The solid line shows a 499-Hz sine tone $x(t)$. The dots show the sampled signal, where samples are taken every millisecond. Numbers above count the peaks of the sine tone; numbers below count the samples.

(a) Show that in the worst case the lowpass-filter character of $H(\omega)$ leads to an attenuation of 3.9 dB. (Hint: The worst case occurs for the highest possible frequency.)

(b) In high-quality digital reproduction systems the linear distortion caused by the sample/hold filter is removed by an inverse filtering that is called, appropriately, ''sine-x-over-x correction.'' Because the output of the sample/hold must be passed through the reconstruction filter anyway, it is common to build this correction function into that filter circuit. Consider the case of direct digital synthesis. Is it possible to build the $(\sin x)/x$ correction into the algorithm that computes the signal in the first place?

Exercise 6: The sampling theorem, life at the edge of the Nyquist criterion

According to the Nyquist criterion, a sine tone can be reproduced perfectly if there are at least two samples per cycle, and if no frequency greater than half the sample rate is allowed to contribute to the reproduction.

The Nyquist criterion is more than just a rule of thumb. It is a theorem, often called ''The Sampling Theorem.'' Having the exalted status of a theorem makes the Nyquist criterion fair game for attacks without limit. The following is an attempt to disprove the theorem by showing that two samples per cycle are not enough to reproduce a sine tone. It will become the reader's task to defend the theorem.

Suppose we have a sine tone $x(t)$ with a frequency of 499 Hz, and we sample it at a sample rate of 1000 Hz. Thus, the Nyquist criterion is satisfied, not by much, but we are talking about a theorem here, so it should be enough. Figure E.6 shows a few samples of several cycles. For definiteness we chose the first sample to be at a maximum of the tone. Because the sample rate is slightly higher than twice the tone frequency, the third sample is taken a little in advance of the second maximum, though the displacement is so small that it's hard to see on the figure. On sample number 251, a time of 250 ms has elapsed, which means that the phase of the sine

tone has advanced by 0.250×499=124.75 cycles. The fraction of 3/4*ths* of a cycle now puts the samples at the zero crossings of the tone. It is evident that the function $y(t)$ has some form of slow amplitude variation which was not present in the original signal $x(t)$. Therefore, when $y(t)$ is reproduced the output will contain this completely spurious variation. The output is not a faithful reproduction of the input, and therefore the sampling theorem is disproved.
[Hint: Show that the amplitude variation is actually caused by a component at 501 Hz.]

Exercise 7: Periodicity of the discrete Fourier transform

For a signal with M samples there are only M independent values of Fourier coefficients.

(a) Show that this result is built into Eq. (21.30) in that X_{n+M} is equal to X_n.

(b) Show that this result corresponds to the fact that the spectrum of a sampled signal is a periodic function of frequency with a period equal to the sample rate, f_s.

Exercise 8: Quantization error requantified

The theoretical limit on signal-to-noise ratio due to quantization error is sometimes given as

$$L_S - L_N = 6.02b + 1.76 \text{ dB},$$

instead of Eq. (21.1). Show that this result follows from assuming that the signal is a sine wave with maximum possible amplitude ($2^{(N-1)}$) and that the quantization error is distributed with a rectangular PDF from $-1/2$ to $+1/2$.

Exercise 9: Cinematic sampling

Figure E.9 shows successive frames of a rotating wheel with a dot on its

FIGURE E.9. Successive pictures of a rotating wheel.

periphery. The sample rate (in frames per second) is about four times faster than the rotation rate. When the frames are projected in sequence the wheel appears to rotate clockwise.

(a) Show that the wheel will appear to rotate counterclockwise if the actual clockwise rotation is somewhat greater than half the sample rate.

(b) Compare the apparent counterclockwise rotation rate with the frequency of an alias, both with respect to the sample rate.

(c) Imagine that the wheel with a dot is replaced by a wheel with N_s spokes. Explain why the effective rotation frequency is increased by a factor of N_s. What are the implications for western movies?

Exercise 10: Generalized jitter

Extend the derivation of Eq.(21.24) to show that the noise to signal ratio is given by

$$N/S = \overline{\omega^2}\sigma^2,$$

Where $\overline{\omega^2}$ is the second moment of the spectrum (see Chapter 16).

CHAPTER 22

Nonlinear Distortion

Life is very nonlinear.
P. DALLOS

Ben Franklin once said that only two things in life are certain, death and taxes. He might have added nonlinear distortion. It is natural to regard nonlinear distortion as an unmitigated evil. Distortion is bad in an acoustical or audio system; it is responsible for some of the most unpleasant sounds that these systems can make. Distortion is also bad mathematically; much of the linear mathematics of preceding chapters, including Fourier transforms, is seriously compromised by nonlinearities.

But nonlinearities are an essential characteristic of life in all its forms, including the action of the inner ear. Physiologists observe that the only linear cochlea is a dead cochlea. And nonlinear distortion can be put to good use in the studio or laboratory for signal shaping (Beauchamp, 1979, 1982), or put to bad use in guitar fuzz boxes. Like it or not, we have to deal with it. There are two representations for systems that exhibit nonlinear distortion. One is the *memoryless* representation, the other is *dynamical*.

MEMORYLESS REPRESENTATION

The memoryless representation consists of an input-output characteristic, which is an equation for the output as a nonlinear function of the input. Output and input are functions of time, and the output at time t depends only on the input at time t. It does not depend on the input at previous times, and that is why the system is called "memoryless." There are different degrees of nonlinearity too. Nonlinearity may be gentle, in which case the output $y(t)$ can be written as a power series of the input signal, $x(t)$:

$$y(t)=c_1x(t)+c_2x^2(t)+c_3x^3(t)+\dots \tag{22.1}$$

The linear term is $c_1x(t)$, and for a linear system this is the only term there is. The coefficient c_1 is a positive or negative constant which is just the linear gain of the system. Nonlinearities are represented by higher-order terms. The terms in x^2 and x^4 are even-order terms; those in x^3 and x^5 are odd-order terms.

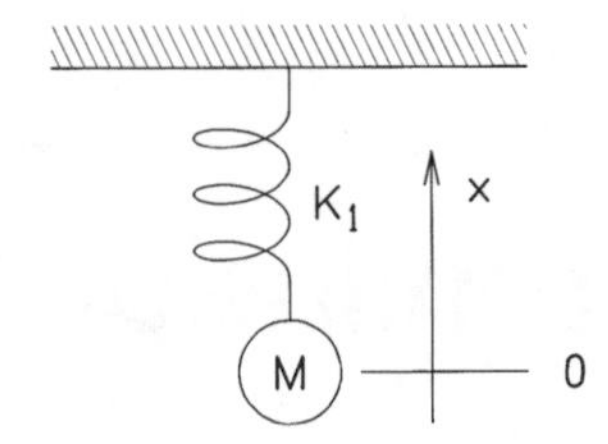

FIGURE 22.1. A mass and spring system.

The power series in Eq. (22.1) does not describe gross nonlinearities such as clipping or half-wave rectification. It has other limitations too. First, it assumes that the system is stationary, namely the coefficients c_n that represent the characteristics of the nonlinear system do not depend on time. Second, the memoryless assumption excludes many important systems. It is not even possible to represent a linear filter in this way. For a linear filter, the output depends on the input at previous times, as described in Chapter 9 with the convolution integral and the impulse response. It is not hard to imagine nonlinear systems that similarly depend on the input at previous values of time (e.g. Strube, 1986). Such systems are described by a dynamical representation.

DYNAMICAL REPRESENTATION

The dynamical representation can be illustrated by a mechanical system described by an *equation of motion.* We deal with a linear system first.

We imagine a mass M connected to a spring with spring constant K_1 as shown in Fig. 22.1. The mass is able to vibrate up and down about an equilibrium position at $x=0$. The equilibrium position is the result of a competition between the stiffness of the spring and gravity. Once the equilibrium position is established, gravity plays no further role in this problem because it is a constant force that is cancelled by the equilibrium stretch of the spring.

The mass provides inertia, and the spring provides a restoring force when the mass is displaced from its equilibrium position. If the displacement is given by variable x, the restoring force is given by $-K_1x$ (K_1 positive), by which we understand that if the displacement is upward then the restoring force is proportional to the displacement and is directed downward. It attempts to return the mass to the equilibrium position.

The equation of motion for such a system is given by Newton's third law, which says that the force is equal to the product of the mass and the acceleration

$$M d^2x/dt^2 = F, \tag{22.2}$$

or

$$Md^2x/dt^2 = -K_1 x. \tag{22.3}$$

Such a system is linear because the restoring force is a linear function of the displacement. The solution to the differential equation is the familiar sine function,

$$x(t) = A\ \sin(\omega_0 t + \phi), \tag{22.4}$$

with a frequency given by

$$\omega_0 = \sqrt{K_1/M}, \tag{22.5}$$

and with an amplitude and a starting phase that depend on the starting conditions of the motion.

An important fact is that the amplitude of the sine oscillation can be anything that we want. Because both sides of the equation of motion involve x to the first power, the amplitude of oscillation plays no role in the solution to this problem. It is only for a linear system that one can make such a statement.

It is worth noting that one can add dissipation in the form of viscous damping, to the equation without changing the linear status. Viscous damping adds another force term to the equation, a term that is proportional to the instantaneous velocity of the mass. The equation of motion becomes

$$Md^2x/dt^2 = -K_1 x - \gamma dx/dt. \tag{22.6}$$

This equation is also linear in x. The solution to the equation is a damped sine tone, and, under the usual conditions of slow damping, there is no nonlinear distortion.

The most commonly studied nonlinear dynamical system is one in which the force is a nonlinear function of the displacement. Possibly such a function is grossly nonlinear. If the oscillating mass bumps against a rigid wall when its displacement becomes too large, the nonlinearity is gross. Alternatively, the nonlinearity may be gentle, and we can represent the restoring force as a power series in the displacement. Then Eq. (22.2) becomes

$$Md^2x/dt^2 = -K_1 x - K_2 x^2 - K_3 x^3 + \ldots \tag{22.7}$$

Again it is assumed that coefficients K_n, representing the system, are not functions of time, i.e. the system is assumed to be stationary. Nonlinear equations in which the coefficients are not constant are ways to represent the action of systems that are called "self organizing."

Equation (22.7) is a classic form. When the displacement x is small, terms to the second and higher powers of x are relatively unimportant and the restoring force resembles that of the linear system in Eq. (22.3). As the displacement becomes larger the nonlinearity becomes an increasingly important consideration. A pendulum is a system of this type. Figure 22.2 shows the restoring force as a function of displacement for four different systems. All of the systems are linear for

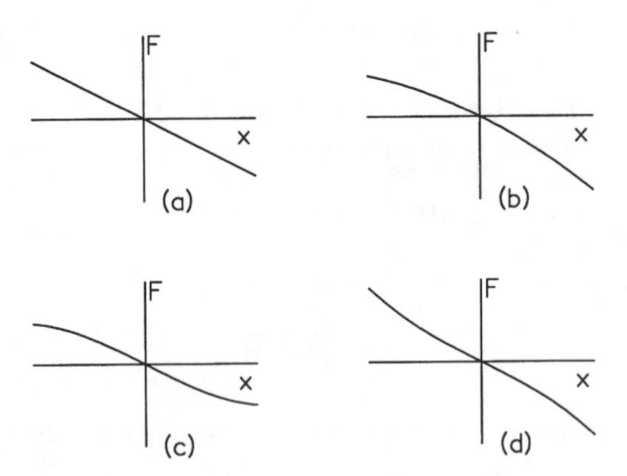

FIGURE 22.2. Restoring force as a function of displacement for a linear system (a) and for three nonlinear systems (b), (c), and (d).

small x, but systems (b), (c), and (d) depart from linearity for large displacement. Equation (22.7) was used by Helmholtz (1877/1954, his Appendix XII) to model distortion products of the ear. Helmholtz believed that quadratic nonlinearity, as shown in panel (b) of Fig. 22.2, would be a particularly important contributor to distortion in the middle ear because of the asymmetry of the eardrum mounting.

In panel (a) the system is linear. The slope of the straight line is $-K_1$. Panel (b) shows a system with a linear term and quadratic distortion,

$$M d^2x/dt^2 = -K_1 x - K_2 x^2. \tag{22.8}$$

Because x is an odd function and x^2 is even, the restoring force has no particular symmetry. Mechanically it represents a spring with elastic properties that are different depending on whether the spring is stretched or compressed. Real steel springs are like that only when they have been excessively distorted; rubber bands are always like that.

Panels (c) and (d) show systems with a linear term and cubic distortion,

$$M d^2x/dt^2 = -K_1 x - K_3 x^3. \tag{22.9}$$

Because x and x^3 are both odd, the restoring force maintains odd symmetry. Cubic distortion occurs very often. For instance, it is typically the largest contributor to distortion in loudspeakers. Panel (c) shows a ''softening'' nonlinearity, where the restoring force becomes relatively smaller for larger displacements because K_3 is negative. Panel (d) shows a ''hardening'' nonlinearity, where the restoring force becomes relatively larger for larger displacements because K_3 is positive.

With few exceptions there is no exact solution for a nonlinear differential equation of the form of Eq. (22.7). Approximate solutions are the subject of a vast literature.

The dynamics, both linear and nonlinear, in Eqs. (22.2) through (22.9) describe the free vibration of a tuned system. Quantity x represents the response (the output) given an impulsive input. To describe the response to an input signal one needs to include an input term somewhere in the equation of motion. Frequently the input is represented as a driving force and is added to the restoring force F.

Nonlinear equations of motion are not restricted to nonlinearity in the restoring force. Other forms of dynamical nonlinearity involve the damping of the system, which enters the dynamical equation as an amplitude factor multiplying the velocity. An example is Hall's (1974) nonlinear cochlea model. Models of the active cochlea reverse the sign of a damping term and feed energy back into the system in a way that depends on amplitude. An example is the Van der Pol oscillator model of the cochlea (Bialek and Wit, 1984) wherein an equation for the incremental pressure in the ear canal caused by otoacoustic emissions is given by x, where

$$d^2x/dt^2 = -\omega_0^2 x + (\gamma_1 - \gamma_2 x^2) dx/dt + E \cos \omega t. \tag{22.10}$$

Here ω_0 is the frequency of this tuned system without the nonlinearity, and $E \cos \omega t$ is the driving signal from the stimulus. The term in γ_1 is a force in the same direction as the velocity (dx/dt), that continually puts energy into the system. By itself, this function would cause oscillations to grow indefinitely. Catastrophe is avoided by the nonlinear term (wherein x^2 multiplies the velocity) which reduces this positive feedback or turns it negative when the pressure becomes large. It is important to note that the Van der Pol equation does not have stable solutions of a linear kind when the displacement is small because γ_1 is positive. The dynamics of this equation, as applied to otoacoustic emissions generated in the cochlea, have been discussed by Long, Tubis, and Jones (1991) among others.

The topic of cochlear mechanics is outside the scope of this book. The remaining treatment of nonlinear distortion will be in terms of the memoryless models. Such models have the advantage that they can be represented electronically without reactive elements (capacitors and inductors). One only needs resistors and semiconductors.

HARMONIC DISTORTION

Distortion can be observed to operate in several different ways. If we put a pure tone into a nonlinear system, we can observe the generation of harmonics. The effect is known as harmonic distortion.

Harmonic distortion is most easily studied in the power series representation, described by Eq. (22.1). Here there is an input signal that we assume to be a cosine with period T. The system distorts the signal leading to an output waveform that does not have a cosine shape as shown in Fig. 22.3. But, because the parameters are stationary, the system distorts each cycle of the signal in the same way. Therefore,

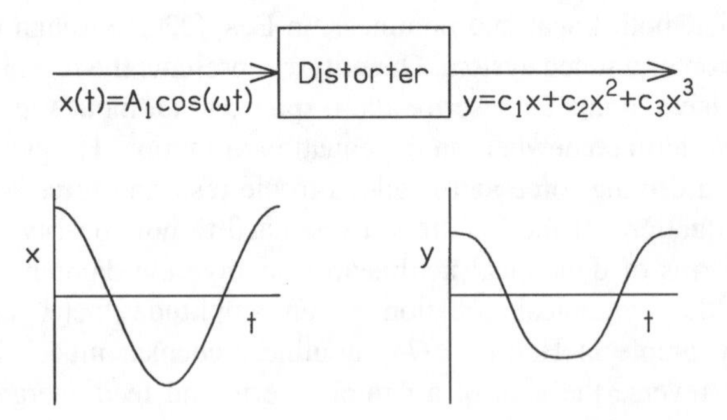

FIGURE 22.3. With a pure tone input $x(t)$, the distorter creates harmonic distortion, described by a power series.

the output is periodic with the same period, T. The output differs from the input only in that it has acquired harmonics; they are harmonics of the fundamental frequency $1/T$.

For definiteness, suppose that all the coefficients in the power series are zero except for c_2. Then the system becomes a "square-law distorter." We study it by putting in a cosine with amplitude A_1 and angular frequency ω_1. We are not concerned with the phase of the cosine. Because there is no other time reference, phase never plays a role in harmonic distortion described by a power series. Then, the output of the system is given by

$$y(t)=c_2A_1^2\cos^2(\omega_1 t). \tag{22.11}$$

It is much more useful to write the square of the cosine as a Fourier series. (Because the output is periodic we know that we can always describe the output of the distorter that way.) From a trigonometric identity, Eq. (22.11) can be written as

$$y(t)=c_2A_1^2\left[\frac{1}{2}+\frac{1}{2}\cos(2\omega_1 t)\right]. \tag{22.12}$$

The output $y(t)$ contains a DC term, $c_2A_1^2/2$. Therefore, this system is sometimes called a "square-law detector." It converts an oscillating signal into something that has a finite average value which could be detected on a DC-reading device. The output also includes a second harmonic. The useful observation here is that the highest harmonic number is the second, a direct result of the fact that the highest power in the power series is the second.

We next consider a nonlinear system where the only nonzero coefficient in the power series is c_3, a case of cubic distortion. With cosine input, the output becomes

$$y(t)=c_3A_1^3\cos^3(\omega_1 t). \tag{22.13}$$

Again, we can write this output as a Fourier series using a trigonometric identity:

$$y(t)=c_3A_1^3\left[\frac{3}{4}\cos(\omega_1 t)+\frac{1}{4}\cos(3\omega_1 t)\right]. \tag{22.14}$$

The highest frequency in the distorted output is the third harmonic, a direct result of the fact that the highest power in the power series is the third. There is also a Fourier component at the fundamental frequency, even though there is no linear term in the power series. The fundamental is created by the cubic distortion. The dependence on A_1^3 is important; it means that if the input signal increases by 1 dB, the cubic terms increase by 3 dB.

Symmetry

There is no DC term in the output of the cubic distorter. In fact, there is no even-numbered harmonic of any order, 0,2,4,.... This is what we expect for a distorter that has only odd powers in the power series. The reasoning is as follows: If the power series has only odd powers then when $\cos \omega_1 t$ changes sign, the output $y(t)$ changes its sign as well. The output is an odd function of the input. If there were an even-numbered harmonic, then with a trigonometric identity one could write that term as an even power of the cosine, for example as $\cos^2(\omega_1 t)$. But such a term would violate the requirement that the output must merely change its sign when the input changes its sign. Therefore, no even-numbered harmonic can exist. Similarly, if the power series includes only even powers of the input [e.g. Eq. (22.11)] there is no odd-numbered harmonic in the output.

If the power series contains both even and odd powers of the input it has no particular symmetry, and all harmonics are possible in the output. In previous examples, if coefficients c_2 and c_3 are not zero, the output contains all the harmonics from DC through the third.

Harmonics for the General Power Series

For the general case of memoryless distortion, the power series contains all powers of n, and so the output contains all harmonics. If the magnitudes of the coefficients c_n decrease rapidly with increasing n then it becomes possible to calculate the amplitudes of all the important harmonic distortion components, i.e. to find a Fourier series for the output $y(t)$,

$$y(t)=\sum_{k=0}^{\infty} b_k \cos(k\omega_1 t). \tag{22.15}$$

Here index k indicates the k-th harmonic.

As will be proved below, the Fourier coefficients b_k can be calculated from the binomial coefficients $\binom{n}{m}$,

$$b_k = 2\sum_{n=k}^{\infty}\left(\frac{1}{2}\right)^n \binom{n}{(n-k)/2} c_n \quad (k \neq 0), \tag{22.16}$$

and for $k=0$,

$$b_0 = \sum_{n=0}^{\infty}\left(\frac{1}{2}\right)^n \binom{n}{n/2} c_n \quad (k=0).$$

The binomial coefficients are given in terms of factorials of integers, in general,

$$\binom{n}{m} = \frac{n!}{m!(n-m)!}. \tag{22.17}$$

Equation (22.16) makes use of the fact that binomial coefficient $\binom{n}{m}$ is zero when index m is not a positive integer or zero. This means that only even powers of n can contribute to the DC term b_0. It also means that only odd powers of n can contribute to b_k for odd-numbered harmonics, i.e. for k odd, and that only even powers of n can contribute to b_k for even-numbered harmonics. The formula in Eq. (22.16) is derived in the next section, but first we illustrate its operation with an example:

EXAMPLE 1: *Cubic distortion*

Problem: Use Eq. (22.16) to find the Fourier series of $y=\cos^3 \omega_1 t$.

Solution: For this cubic distorter all the c_n are zero except for $n=3$, and the coefficient c_3 is equal to 1. Therefore, the sum in Eq. (22.16) reduces to a single term, the $n=3$ term. Further, because n is odd, b_k is zero for all even k.

For b_1 the equation gives

$$b_1 = 2\left(\frac{1}{2}\right)^3 \binom{3}{1} = \frac{1}{4}\frac{3!}{1!2!} = \frac{3}{4}, \tag{22.18}$$

and for b_3,

$$b_3 = 2\left(\frac{1}{2}\right)^3 \binom{3}{0} = \frac{1}{4}\frac{3!}{0!3!} = \frac{1}{4}. \tag{22.19}$$

For $k=5$ and higher, b_k must be zero because binomial coefficient $\binom{n}{(n-k)/2}$ is zero because $n-k$ is negative. Therefore, using b_1 and b_3 above, Eq. (22.15) gives

$$y=\frac{3}{4}\cos(\omega_1 t)+\frac{1}{4}\cos(3\omega_1 t), \tag{22.20}$$

in agreement with Eq. (22.14).

Derivation of the Harmonic Series for *n*-th Order Distortion

The derivation of the harmonic series formula in Eqs. (22.15) and (22.16) begins with the input expressed as a binomial,

$$\cos \omega_1 t=\frac{1}{2}(e^{i\omega_1 t}+e^{-i\omega_1 t}). \tag{22.21}$$

The n-th term in the distorted output can then be written as a binomial series, as in Appendix C, Eq. (C9):

$$[\cos \omega_1 t]^n=\left(\frac{1}{2}\right)^n \sum_{m=0}^{n} \binom{n}{m} e^{i(n-m)\omega_1 t}\times e^{-im\omega_1 t} \tag{22.22}$$

$$=\left(\frac{1}{2}\right)^n \sum_{m=0}^{n} \binom{n}{m} e^{i(n-2m)\omega_1 t}. \tag{22.23}$$

We let $k=n-2m$ in order to have a simple label, k, for harmonic number,

$$[\cos \omega_1 t]^n=\left(\frac{1}{2}\right)^n \sum_{k=-n}^{n} \binom{n}{(n-k)/2} e^{ik\omega_1 t}. \tag{22.24}$$

Then, explicitly extracting the $k=0$ term and exploiting the symmetry of the binomial coefficients, we find

$$[\cos \omega_1 t]^n=\left(\frac{1}{2}\right)^n\left[\binom{n}{n/2}+2\sum_{k=1}^{n}\binom{n}{(n-k)/2}\cos(k\omega_1 t)\right]. \tag{22.25}$$

The first term in the square bracket is the $k=0$ term. It gives the DC component of distortion.

This derivation shows how to calculate the amplitudes of all the harmonics that are generated by the n-th power of the cosine. To obtain the complete expression for the output, one multiplies all such terms by the coefficients c_n and sums over all values of n. The result is then Eqs. (22.15) and (22.16), which is what we set out to prove.

The formula for the harmonic series is particularly easy because the input is a cosine. If the input includes a phase shift, for example if the input is a sine, then the best approach is to redefine the origin of time to obtain an input that is only a cosine.

If the situation at hand does not permit a shift of time origin then the harmonic number k multiplies the phase angle in the output. In other words, if the input is $\cos(\omega_1 t+\phi)$ then the output is given by Eq. (22.25) with $\cos(k\omega_1 t)$ replaced by $\cos(k\omega_1 t+k\phi)$.

To illustrate the operation of the formula we calculate the amplitudes of a few harmonics in terms of the parameters of the memoryless power series, assuming that the power series contains all powers up through the 12th power and no others. Specifically, if

$$y(t)=\sum_{n=1}^{12} c_n x^n, \tag{22.26}$$

where

$$x=\cos \omega_1 t. \tag{22.27}$$

Then the Fourier series is

$$\begin{aligned}
y(t)=(1)&\left\{\frac{1}{2}c_2+\frac{3}{8}c_4+\frac{5}{16}c_6+\frac{35}{128}c_8+\frac{63}{256}c_{10}+\frac{231}{1024}c_{12}\right\}+\cos(\omega_1 t)\left\{c_1\right.\\
&+\frac{3}{4}c_3+\frac{5}{8}c_5+\frac{35}{64}c_7+\frac{63}{128}c_9+\left.\frac{231}{512}c_{11}\right\}+\cos(2\omega_1 t)\left\{\frac{1}{2}c_2+\frac{1}{2}c_4\right.\\
&+\frac{15}{32}c_6+\frac{7}{16}c_8+\frac{105}{256}c_{10}+\left.\frac{99}{256}c_{12}\right\}+\cos(3\omega_1 t)\left\{\frac{1}{4}c_3+\frac{5}{16}c_5++\frac{21}{64}c_7\right.\\
&+\frac{21}{64}c_9+\left.\frac{165}{512}c_{11}\right\}+\cos(4\omega_1 t)\left\{\frac{1}{8}c_4+\frac{3}{16}c_6+\frac{7}{32}c_8+\frac{15}{64}c_{10}\right.\\
&+\left.\frac{495}{2048}c_{12}\right\}+\cos(5\omega_1 t)\left\{\frac{1}{16}c_5+\frac{7}{64}c_7+\frac{9}{64}c_9+\frac{165}{1024}c_{11}\right\}+\cos(6\omega_1 t)\\
&\times\left\{\frac{1}{32}c_6+\frac{1}{16}c_8+\frac{45}{512}c_{10}+\frac{55}{512}c_{12}\right\}+\cos(7\omega_1 t)\left\{\frac{1}{64}c_7+\frac{9}{256}c_9\right.\\
&+\left.\frac{55}{1024}c_{11}\right\}+\cos(8\omega_1 t)\left\{\frac{1}{128}c_8+\frac{5}{256}c_{10}+\frac{33}{1024}c_{12}\right\}\\
&+\cos(9\omega_1 t)\left\{\frac{1}{256}c_9+\frac{11}{1024}c_{11}\right\}+\cos(10\omega_1 t)\left\{\frac{1}{512}c_{10}+\frac{3}{512}c_{12}\right\}\\
&+\cos(11\omega_1 t)\left\{\frac{1}{1024}c_{11}\right\}+\cos(12\omega_1 t)\left\{\frac{1}{2048}c_{12}\right\}.
\end{aligned} \tag{22.28}$$

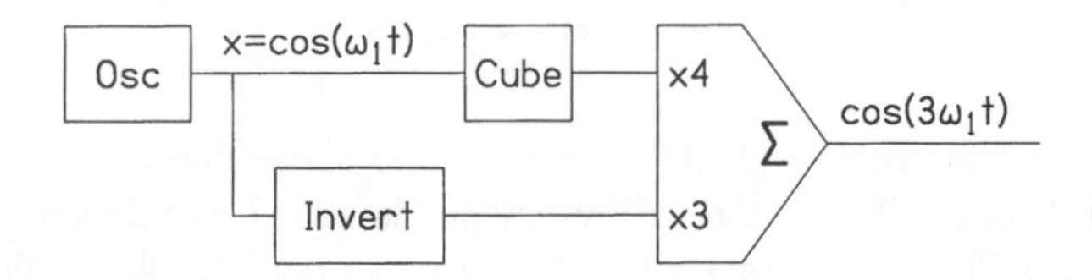

FIGURE 22.4. Block diagram of a third harmonic generator.

Because the highest power in the series is 12, there are no harmonics beyond the 12th. There was, however, no particular reason for choosing a maximum n of 12; the calculations of the above equation could have been continued indefinitely. The long equation above can be used when the power series is a single power. For example, if $y=x^5$ then the Fourier series for y is given by Eq. (22.28) with $c_5=1$ and all the other values of c_n equal to zero,

$$y(t)=\frac{5}{8}\cos(\omega_1 t)+\frac{5}{16}\cos(3\omega_1 t)+\frac{1}{16}\cos(5\omega_1 t). \tag{22.29}$$

The output consists of first, third, and fifth harmonics.

Summing Powers for a Given Harmonic

The previous section expanded a power of a cosine as a series of harmonics. The present section does the reverse. Suppose that we have a 1000-Hz cosine tone and we would like to create a 3000-Hz cosine tone. From our previous section we know that cubing the 1000-Hz tone will lead to a signal that has a 3000-Hz component. The only problem is that the cubed signal also contains the original component at 1000 Hz. Using a mixer, however, we could subtract out the 1000-Hz component in the cubed signal. A block diagram for doing this is shown in Fig. 22.4.

The figure shows that the weights at the mixer input are 4 and 3. These values were found by solving Eq. (22.28) for the cubed signal x^3 when $x=\cos(\omega_1 t)$,

$$y=\cos^3 \omega_1 t. \tag{22.30}$$

All coefficients c_n are zero except that $c_3=1$. Therefore, from Eq. (22.28)

$$\cos^3(\omega_1 t)=\frac{3}{4}\cos(\omega_1 t)+\frac{1}{4}\cos(3\omega_1 t). \tag{22.31}$$

The desired third harmonic signal is therefore

$$\cos(3\omega_1 t)=4\cos^3(\omega_1 t)-3\cos(\omega_1 t), \tag{22.32}$$

or, in terms of the input signal x,

$$\cos(3\omega_1 t) = 4x^3 - 3x. \tag{22.33}$$

The right-hand side of Eq. (22.33) is a particular series known as a Chebyshev polynomial. Specifically it is the Chebyshev polynomial $T_3(x)$. Because the mathematics of the Chebyshev polynomials has been worked out thoroughly, it is an easy matter to describe the generation of any harmonic as a sum of powers of the fundamental signal. The equation of special interest in our case is

$$\cos n\theta = T_n(\cos\theta). \tag{22.34}$$

As expected, the polynomial $T_n(x)$ includes odd or even powers of x as n is odd or even. The largest power in the polynomial is n. Table 22.1 gives the first seven polynomials.

TABLE 22.1. Chebyshev Polynomials

$$T_0(x) = 1$$
$$T_1(x) = x$$
$$T_2(x) = 2x^2 - 1$$
$$T_3(x) = 4x^3 - 3x$$
$$T_4(x) = 8x^4 - 8x^2 + 1$$
$$T_5(x) = 16x^5 - 20x^3 + 5x$$
$$T_6(x) = 32x^6 - 48x^4 + 18x^2 - 1$$

The polynomials in the table all come from the recurrence relation for the Chebyshev polynomials:

$$T_n(x) = 2xT_{n-1}(x) - T_{n-2}(x). \tag{22.35}$$

EXAMPLE 2: *Recurrence relation*

Problem: Use the recurrence relation to find Chebyshev polynomial $T_4(x)$.

Solution: The recurrence relation applied to T_4 says that

$$T_4 = 2xT_3 - T_2. \tag{22.36}$$

Substituting expressions from the above table for T_3 and T_2, we find

$$T_4 = 2x(4x^3 - 3x) - (2x^2 - 1), \tag{22.37}$$

and simplifying, we have

$$T_4 = 8x^4 - 8x^2 + 1, \tag{22.38}$$

in agreement with Table 22.1.

MEASUREMENT OF HARMONIC DISTORTION

The harmonic distortion caused by an acoustical, audio, or biological system is measured by putting a high-quality (low-distortion) sine tone into the system and examining what comes out. The output includes a linear response, which leads to a sine tone of the original frequency, possibly phase shifted and changed in amplitude. The output will also include harmonic distortion products.

The distortion can be specified in terms of the spectrum of the output, giving the levels of all the harmonics, normally with respect to the fundamental, which is the linear response. For example, the spectrum of the output of an amplifier might include a fundamental with a level of −4 dB (arbitrary reference level), a second harmonic with a level of −83 dB, and a third harmonic with a level of −56 dB. All other harmonics are too small to measure. One characterizes the distortion then by saying that the second harmonic is 79 dB down and the third harmonic is 52 dB down.

Alternatively the distortion can be specified by a single number that combines all the distortion products and expresses the combination as a percentage of the total RMS value of the output (Beranek, 1988, p. 676). Such a measurement can be done with the distortion analyzer shown in Fig. 22.5.

The operation of the analyzer is as follows: When the switch is open the RMS voltmeter reads the output of the system under test, as the system responds to the pure sine tone from the sine generator. The resulting voltage is called V_{TOTAL}. When the switch is closed the original sine is added to the output of the system under test. By adjusting the phase and amplitude of the admixed original sine, the linear response of the system under test can be cancelled and the RMS voltmeter

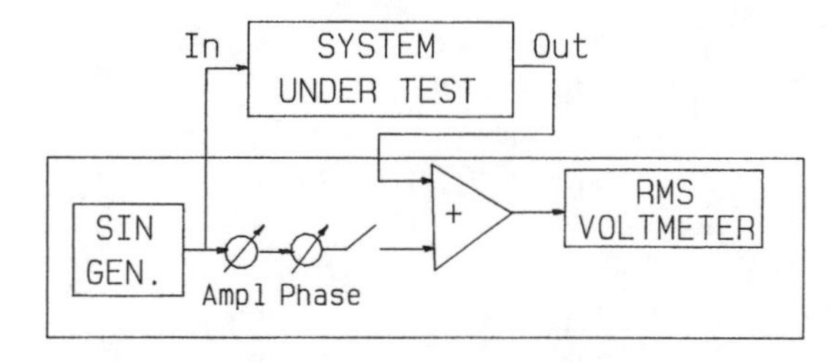

FIGURE 22.5. An analyzer that measures total harmonic distortion plus noise.

reads only the harmonic distortion, V_{DIST}. In fact, this voltmeter reading also depends on the noise added by the system under test, and the measurement made by this analyzer is often called "total harmonic distortion plus noise," (THD). We neglect the noise in the present treatment.

The THD is then given as a percent by

$$THD = 100 \frac{V_{DIST}}{V_{TOTAL}}. \tag{22.39}$$

It is easy to relate the measurement of the distortion percentage to the measurement of the individual distortion products. If the amplitude of the fundamental is C_1, and the amplitude of the n-th harmonic is C_n, then the distortion percentage is given by

$$THD = 100 \times \sqrt{\frac{C_2^2 + C_3^2 + C_4^2 + \ldots}{C_1^2 + C_2^2 + C_3^2 + C_4^2 + \ldots}}. \tag{22.40}$$

Essentially this formula adds up the power of the distortion products, divides it by the total power to get a relative measure, and then takes the square root to find an RMS value.

EXAMPLE 3: *Distortion percentage*

Problem: Find the percentage of harmonic distortion for the amplifier with second and third harmonic distortion described above.

Solution: If every term in the numerator and the denominator in Eq. (22.40) is divided by C_1^2, then all terms are referenced to the power in the fundamental. Therefore,

$$THD = 100 \times \sqrt{\frac{C_2^2/C_1^2 + C_3^2/C_1^2 + C_4^2/C_1^2 + \ldots}{1 + C_2^2/C_1^2 + C_3^2/C_1^2 + C_4^2/C_1^2 + \ldots}}. \tag{22.41}$$

From the spectral measurements we know that

$$10 \log(C_2^2/C_1^2) = -79 \text{ dB}, \tag{22.42}$$

and

$$10 \log(C_3^2/C_1^2) = -52 \text{ dB}. \tag{22.43}$$

Thus, $(C_2^2/C_1^2)=1.26\times10^{-8}$, and $(C_3^2/C_1^2)=6.31\times10^{-6}$. Therefore,

$$THD=100\times\sqrt{\frac{1.26\times10^{-8}+6.31\times10^{-6}}{1+1.26\times10^{-8}+6.31\times10^{-6}}}, \tag{22.44}$$

$$THD=0.25\%$$

It is typical for the distortion to be dominated by one or two terms; here it is dominated by the third harmonic. It is also typical for all the terms in the denominator to be negligible compared to the number 1. This simplifies the numerical calculation of the final answer above. When the distortion is small and one harmonic dominates entirely, it is easy to calculate the percentage of distortion. If the dominant harmonic is down by 40 dB the distortion percentage is 1%, if down by 60 dB the distortion percentage is 0.1%, etc.

INTERMODULATION DISTORTION

Intermodulation distortion occurs when a signal having more than one spectral component is put into a nonlinear device. To test a system for intermodulation distortion it is usual to put in a signal with two components, i.e. two different frequencies. A signal with two frequencies is often sufficient to generate a bewildering array of distortion products. With three input components there are even more, and with the complex sounds of everyday life it is hardly possible to speak of individual distortion products. Both for testing audio equipment and for studying the nonlinearity of the human ear, two components are normally quite enough.

This section introduces the mathematics of intermodulation distortion for the case of two input components only. The two input components are called *primaries*. They have frequencies f_1 and f_2 and by definition $f_2>f_1$. The input is therefore,

$$x(t)=A_1\cos(\omega_1 t)+A_2\cos(\omega_2 t+\phi), \tag{22.45}$$

where $\omega_1=2\pi f_1$ and $\omega_2=2\pi f_2$.

We consider the memoryless distorter, where the output is related to the input by the power series in Eq. (22.1). The general term in that series is $c_n x^n$, which we can write as

$$c_n x^n=c_n\left(\frac{1}{2}\right)^n[A_1(e^{i\theta_1}+e^{-i\theta_1})+A_2(e^{i\theta_2}+e^{-i\theta_2})]^n, \tag{22.46}$$

where we have adopted the shortened notation

$$\theta_1=\omega_1 t, \tag{22.47}$$

and

$$\theta_2 = \omega_2 t + \phi. \tag{22.48}$$

The expansion of Eq. (22.46) leads to all possible sums and differences of n values of θ_1 and θ_2. For example, we take the special case $n=5$: There will be a term with the exponent $i[\theta_1 - \theta_1 + \theta_2 + \theta_2 + \theta_2] = i3\theta_2$. This term represents simple harmonic distortion; it is the third harmonic of frequency f_2. The term $i[\theta_1 - \theta_1 + \theta_1 + \theta_2 - \theta_2] = i\theta_1$ contributes to the distortion component at the fundamental of frequency f_1. None of this is new; it simply points out that the general treatment of intermodulation distortion includes the harmonic distortion as well.

In addition to the harmonic distortion terms, there are terms with exponents of the form $i[\theta_1 + \theta_1 + \theta_1 - \theta_2 - \theta_2] = i[3\theta_1 - 2\theta_2]$. Such a term represents intermodulation distortion proper, leading to a spectral component with frequency $3f_1 - 2f_2$. The sum of the absolute values of the integers (here 3 and 2) is 5: It cannot be any more than 5 because we are considering the case $n=5$. But this sum can be less than 5. For instance, the term $i[\theta_1 + \theta_1 + \theta_2 - \theta_2 - \theta_2] = i[2\theta_1 - \theta_2]$, leads to the intermodulation distortion component with frequency $2f_1 - f_2$, and $2+1=3$.

Distortion products $3f_1 - 2f_2$ and $2f_1 - f_2$ are called difference tones because of the negative sign. Whenever the calculation of a difference tone frequency leads to a negative number, one has only to change the sign to find the frequency of a legitimate difference tone. There are also summation tones, for example $3f_1 + 2f_2$ and $2f_1 + f_2$. The summation tones have a plus sign.

EXAMPLE 4: *Cubic intermodulation distortion*

This example finds a Fourier series for the cube of a signal that is the sum of two cosine waves:

$$y(t) = (A_1 \cos \omega_1 t + A_2 \cos \omega_2 t)^3. \tag{22.49}$$

From trigonometric identities in Appendix B, the powers can be expanded in a series of harmonic terms, where the cosine function appears only to the first power (i.e. a Fourier series):

$$y(t)=A_1^3\left[\frac{3}{4}\cos\omega_1 t+\frac{1}{4}\cos 3\omega_1 t\right]+A_2^3\left[\frac{3}{4}\cos\omega_2 t+\frac{1}{4}\cos 3\omega_2 t\right]$$

$$+A_1^2A_2\left[\frac{3}{2}\cos\omega_2 t+\frac{3}{4}\cos(2\omega_1-\omega_2)t+\frac{3}{4}\cos(2\omega_1+\omega_2)t\right]$$

$$+A_2^2A_1\left[\frac{3}{2}\cos\omega_1 t+\frac{3}{4}\cos(2\omega_2-\omega_1)t+\frac{3}{4}\cos(2\omega_2+\omega_1)t\right].$$

(22.50)

There is no DC term, but there are first and third harmonics of the two individual primary frequencies. The intermodulation products include difference tones and summation tones. These are the cubic difference tones, so named because the lowest order of distortion in which they appear is the cube. They are present for fifth and higher order distortions too, however, as shown in the previous section.

General Intermodulation Amplitude

It is possible to calculate the amplitude of any intermodulation distortion product given a two-component input signal and a distorter that is described by a power series. The technique is similar to the technique used for harmonic distortion. The input signal is the sum of two cosines. This sum is a binomial, and the general n-th term in the output is given by the n-th power of this binomial. The expansion can therefore be written in terms of a binomial coefficient and powers of the individual cosines. Then, each cosine can itself be written as a binomial (as was done in the previous treatment of harmonic distortion). Both kinds of binomial appear in Eq. (22.46). In the end then, the general n-th term can be written as a series involving the product of three binomial coefficients. The description of the general formula can be divided into four kinds of terms:

$$x^n=DC+H1+H2+IM. \tag{22.51}$$

The terms DC are zero-frequency terms. Terms $H1$ and $H2$ are harmonics of f_1 and f_2, respectively. Terms IM are cross terms involving sums and differences of f_1 and f_2. These terms are calculated in Appendix K.

A distortion product amplitude, for harmonic or intermodulation distortion, is most easily calculated when there is only one way to combine frequencies f_1 and f_2 to get the frequency of the distortion product. For example, if $f_1=1000$ Hz, $f_2=1200$ Hz, and $n=3$ then there is only one way to get a distortion frequency of 800 Hz, namely $2\times1000-1200$. All the other distortion frequencies are higher than this. However, if $f_1=1000$ Hz and $f_2=2000$ Hz, still with $n=3$, then there are two ways to get a distortion component of 3000 Hz, namely $2\times2000-1000$ and the third harmonic of 1000 Hz. This is an example of *degeneracy*, a situation in which the amplitude of a particular distortion product comes from two or more different terms. In the general case degeneracy is common. For instance, if we allow terms with $n=8$ in addition to $n=3$, then with $f_1=1000$ Hz and $f_2=1200$ Hz a distortion component with frequency 800 Hz can be obtained as $4\times1200-4\times1000$. This component must be added, with phase information included, to the $n=3$ term $2\times1000-1200$.

Summary of Distortion Facts

Some facts follow from the equations for distortion products, especially as developed in Appendix K:

(1) The distortion products have frequencies of the form

$$f_{i.m.}=\pm k_1 f_1 \pm k_2 f_2, \tag{22.52}$$

where k_1 and k_2 are positive integers or zero.

(2) For distortion power n (term x^n), the sum of k values is limited:

$$k_1+k_2\leqslant n. \tag{22.53}$$

Specifically:

$$k_1+k_2=n$$

or

$$k_1+k_2=n-2$$

or

$$k_1+k_2=n-4$$

etc.

or

$$k_1+k_2=0 \text{ or } 1.$$

(3) For every difference tone with frequency $k_1 f_1 - k_2 f_2$, there is a summation tone with frequency $k_1 f_1 + k_2 f_2$ that has the same amplitude. This result comes from the memoryless nonlinearity in Appendix K. It does not apply to the dynamical distortion calculated by Helmholtz. When the nonlinearity appears in the force term of the dynamical equation, per Eq. (22.29), the difference tones come out to have amplitudes that are much larger than the summation tones. This prediction agrees with perceptual experience, which is an argument in favor of the dynamical model.

(4) For every lower difference tone there is an upper difference tone, but their amplitudes are the same only if $A_1 = A_2$. For example, there is a famous distortion product called the "lower cubic difference tone." Its frequency is $2f_1 - f_2$ ($f_2 > f_1$) and its amplitude is $A_1^2 A_2$. There is also an "upper cubic difference tone," with frequency $2f_2 - f_1$, and its amplitude is $A_2^2 A_1$. This result too is special to the memoryless model.

(5) The amplitudes of distortion products do not depend on relative phase ϕ.

(6) Symmetries 3, 4, and 5 are wrecked for any frequency where a degeneracy exists.

DISTORTION AND THE AUDIBILITY OF PHASE

As a general rule, the ear is insensitive to the relative phases among the harmonics of a complex tone. This is especially true for the low harmonics, which are relatively widely separated on a log scale and on the physiological tonotopic axis. However, the presence of nonlinear distortion and degeneracy can make relative phase audible because it can change the actual amplitudes of the harmonics. The simplest nontrivial demonstration of this fact is a signal, consisting of a fundamental plus a second harmonic, passed through a device that has a linear response and square-law distortion, as in the following example.

EXAMPLE 5: *Cancelling the fundamental*

We consider a distorter described by a power series with $c_1 = c_2 = 1$ and all other $c_n = 0$. Therefore,

$$y = x + x^2. \tag{22.54}$$

The input is a two-component signal consisting of a fundamental and its second harmonic:

$$x(t) = A_1 \cos(\omega_1 t) + A_2 \cos(2\omega_1 t + \phi). \tag{22.55}$$

For this simple system, it is only a matter of some algebra and a few trigonometric identities to calculate the amplitudes of the components in the output $y(t)$ as follows:

$$\text{DC term (0):} \quad \frac{1}{2}(A_1^2+A_2^2) \tag{22.56}$$

$$\text{fundamental } (f_1)\text{:} \quad A_1\sqrt{1+A_2^2+2A_2 \cos \phi}$$

$$\text{2nd harmonic } (2f_1)\text{:} \quad \sqrt{A_1^4/4+A_2^2+A_1^2A_2 \cos \phi}$$

$$\text{3rd harmonic } (3f_1)\text{:} \quad A_1A_2.$$

$$\text{4th harmonic } (4f_1)\text{:} \quad \frac{1}{2}A_2^2$$

Because the frequency ratio is a simple factor of 2, and both even and odd powers of n are present, there is degeneracy for both the fundamental and the second harmonic. These amplitudes then depend on the relative phase angle ϕ. If amplitude A_2 equals 1, the fundamental component in the output can be cancelled entirely when ϕ=180 degrees. Such a dramatic change is clearly audible. A demonstration of this effect appears on compact disc (Houtsma *et al.*, 1987) for f_1=440 Hz and f_2=880 Hz.

Example 5 demonstrates an effect of phase on tone color caused by an electronic distortion device. If the signals delivered to the listener are not distorted then changes in tone color caused by changes in phase can be interpreted as evidence in favor of nonlinear distortion in the auditory system. However, phase dependence that appears because auditory distortion alters the effective amplitudes may not be regarded as ''genuine'' phase sensitivity by psychoacousticians.

EXAMPLE 6: *Even-order distortion and the reproduction of music*

Audiophiles frequently remark that even-order distortion in an audio system is less objectionable than odd-order distortion (Olsher, 1991). Power amplifiers can be designed to minimize odd-order distortion while paying the price of increased even-order distortion.

The power series that describes a system with even-order distortion, of course, has a linear term x, which has odd symmetry, but this is the only term in the series with odd symmetry. All the deviations from linearity are an even function of x. Terms like x^2 and x^4 are allowed. Therefore, the input/output function looks like Fig. 22.6 panel (b). It does not look like panel (a).

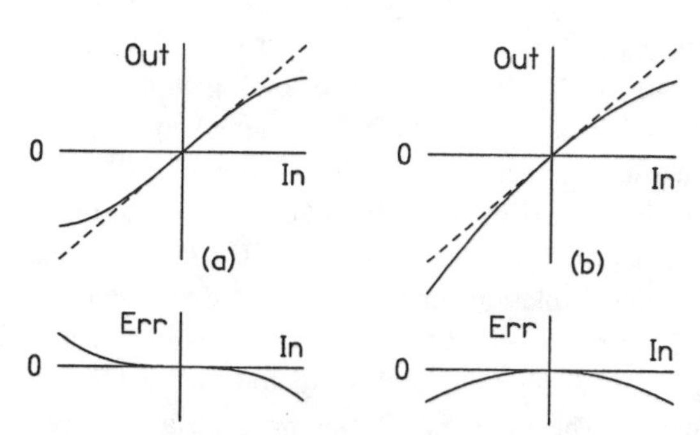

FIGURE 22.6. (a) Input/output characteristic for a system with odd-order distortion and the deviation of this characteristic from linear. (b) Input/output characteristic for a system with even-order distortion and the deviation of this characteristic from linear. Compared to a real audio system, the deviations shown here are exaggerated.

The preference for even-order distortion may be understood from some basic musical considerations. We suppose that the even and odd order distortions are dominated by their lowest order terms, quadratic distortion and cubic distortion, respectively. First, there is harmonic distortion. For each input spectral component the quadratic distortion produces a component with twice the frequency. Thus each component acquires a spurious octave companion, but this does not lead to musical dissonances. It mainly makes the music sound brighter. On the other hand, the cubic distortion multiplies frequencies by three, which means that the music is accompanied by a weak artifact that has been transposed up by the interval of a twelfth. Added twelfths are more likely to cause dissonance than are added octaves.

More important is the intermodulation distortion. If tones are played in close harmony, cubic distortion will create difference tones of the form $2f_1 - f_2$ which (for f_2 close to f_1) will not be very different from f_1 and f_2 themselves. This leads to many possibilities for dissonance. By contrast, quadratic distortion will create difference tones of the form $f_2 - f_1$, which have low frequency when f_2 is close to f_1. These difference tones may well be masked by the powerful bass notes in most music.

AUDITORY NONLINEARITY

The mechanical operation of the auditory system begins with the vibration of the eardrum and middle ear and ultimately leads to the bending of the stereocillia bundles of the haircells. All stages of this process are potential sites of mechanical nonlinearity.

Most measurements of the transfer function of the middle ear indicate linear behavior over a large intensity range. Laser-Doppler velocimetry measurements by Buunen and Vlaming (1981) show linear response of the eardrum for sound pressure levels from 40 to 110 dB SPL. A plot of stapes displacement versus sound pressure level shows linear behavior from low levels up to 70 dB SPL. For higher levels the attenuating characteristics of the acoustic reflex compress the stapes displacement function, but if the acoustic reflex is defeated the linearity of the plot extends up to 120 dB SPL (Guinan and Peake, 1967).

Plots showing a linear relationship between output amplitude and input amplitude over a wide dynamic range are impressive evidences of linearity, but they are not entirely adequate to determine that audible distortion products are not the result of middle ear mechanics. The audible distortion products are in any case low level, and the nonlinearities needed to make them would not be prominent on an input-output amplitude plot. Better experimental evidence would be obtained from measuring the amplitudes of the distortion products themselves in the output. Nevertheless, the appearance of linearity up to *high* sound pressure levels suggests that nonlinear distortion products are not generated by the middle ear at *moderate* levels. When the middle ear does distort, the distortion seems to be primarily quadratic, just as Helmholtz suggested, with the outward motion of the stapes during refraction larger than the inward motion during condensation (Moeller, 1974). However, contrary to the view of Helmholtz, this distortion is not important at normal listening levels.

The inner ear transforms mechanical vibration of the basilar membrane into neural impulses primarily by a row of about 3,500 inner haircells. The basilar membrane also has three rows with a total of about 12,000 outer haircells that serve to increase the gain of the system for low-level signals. Therefore, the system is compressive, and this compressive nonlinearity is tonotopically local. Thus, the sine tone response of most of the basilar membrane is a linear function of the acoustical signal, but near the place tuned to the signal frequency the compressive action leads to a nonlinear contribution to the displacement of the basilar membrane (Rhode, 1971). The effect may enhance the local displacement by as much as 40 dB (Johnstone *et al.*, 1986—Mössbauer; Ruggero and Rich, 1991—laser-Doppler velocimetry). Brief reviews of this contemporary cochlear physiology have been given by Dallos (1988, 1992) and by Allen and Neely (1992). An intriguing by-product of the nonlinear cochlear action is the phenomenon of cochlear emissions. When the cochlea is stimulated by an input click, it responds with a delayed (5 to 10 ms) ringing that can be recorded from the ear canal (Kemp, 1978).

Nonlinearities lead to difference tones, as would be expected from the mathematics of intermodulation distortion. Because the nonlinearity is tonotopically local, a difference tone is generated at the place of the primary tones f_1 and f_2 and propagates on the basilar membrane to the place tuned to the difference-tone frequency (Gibian and Kim, 1982). This is one reason why difference tones are perceived to be so much stronger than summation tones. Summation tones have frequencies greater than the primaries and would have to propagate in the direction opposite to the traveling wave to be detected at their best place.

Aural Harmonics

The term ''aural harmonics'' refers to harmonic distortion created by nonlinearities in the auditory system itself. There are two essential questions concerning aural harmonics: The first is whether they exist, the second is whether they should exist.

The existence of aural harmonics was suggested by the tone-on-tone masking study by Wegel and Lane (1924). This work showed that when the frequency of the probe sine tone was near an octave (or other integer multiple) of the frequency of the masking sine tone then a beating effect caused a significant reduction in the threshold level of the probe. The natural explanation for this effect (Fletcher, 1930; Stevens and Davis, 1938/1983) is that harmonics of the masker beat with the (weaker) probe. Wegel and Lane studied the effect by adjusting the level of the probe to maximize the beating sensation. According to the equation for beats [see Eq. (17.11)], the strongest beats should occur when the probe tone has an amplitude equal to the amplitude of the aural harmonic. This technique is known as the *method of best beats*, and it can be used to measure the amplitudes of distortion products (or ''subjective tones'') of all kinds.

The results of Wegel and Lane suggested a rather strong aural harmonic. For example, a 1000-Hz tone at 80 dB SPL apparently produces a second harmonic at 63 dB. The problem with this interpretation is that an aural harmonic of such intensity is not observed by other means, for example when the intense tone masks a narrow band of noise (Plomp, 1976).

An alternative explanation for the beating sensation, suggested by Plomp, is that it represents a periodic variation in the firing pattern of a set of neurons that respond to both the masker and the probe. Such neurons presumably have best frequencies that lie between the two tones. This explanation denies that there are actual beats, and implies a sensitivity to dynamical variation in the relative phase of mistuned consonances, like the rolling image on an oscilloscope screen. Evidence in favor of this explanation is that the beating sensation occurs only at low or medium frequencies where neural synchrony is preserved.

The rolling pattern cannot be used to explain the fact that masked threshold for a probe that is an exact harmonic of the masker depends on the phase of the probe (Clack *et al.*, 1972; Nelson and Bilger, 1974). Here the only alternative is to postulate a true phase sensitivity of the auditory system. The change of tone color with phase has recently been studied by Ozawa *et al.* (1993).

There would seem to be little doubt that a listener should hear aural harmonics when presented with a sine tone greater than 90 dB SPL. At this level there is distortion in the middle ear (if not in the electroacoustic stimulus transducer), and the harmonic distortion products should be heard just as though they were part of the physical stimulus. It is not clear whether listeners should hear aural harmonics for lower levels. The theory of nonlinear systems implies that harmonics are indeed generated in the cochlea. Harmonic distortion might be particularly prominent for low signal levels where the local compressive action of the organ of Corti is most important. However, to be heard at their proper place on the basilar membrane, the aural harmonics have to propagate the wrong way and it is not known how effective backward propagation can be.

Difference Tones

There are two kinds of difference tone that are important in hearing. One is the simple difference tone with frequency $f_2 - f_1$. The other has a frequency given by the formula $(n+1)f_1 - nf_2$, e.g. $2f_1 - f_2$, $3f_1 - 2f_2$, $4f_1 - 3f_2$, etc., and the most prominent is the lower cubic difference tone (LCDT), $2f_1 - f_2$. These difference tones can be heard by normal listeners given the correct ratios of f_2/f_1 and appropriate levels. Their amplitudes can be measured using the method of best beats or by using the cancellation method. In the cancellation method an extra tone, with the frequency of the difference tone, is added to the signal. The listener adjusts the amplitude and phase of this extra tone (the cancellation tone) to cause the distortion product to disappear (Zwicker, 1955).

The simple difference tone $(f_2 - f_1)$ is normally the most readily heard, probably because its frequency is far from the primaries. In several ways the simple difference tone behaves like a classic quadratic nonlinearity. It cannot be heard at all unless the primaries have adequate level (greater than 50 dB SPL). Its amplitude grows proportional to the product $A_1 A_2$. Thus, if $A_1 = A_2$ the level of $f_2 - f_1$ grows by 2 dB for every 1 dB increase in the level of the primaries. The level of $f_2 - f_1$ does not depend much on the frequency ratio f_2/f_1. All these features are what would be expected for a quadratic nonlinearity.

In contrast to the difference tone $f_2 - f_1$, the LCDT, $2f_1 - f_2$, does not behave at all like a simple cubic nonlinearity (Goldstein, 1967b). It is tuned, and to observe it well requires that the ratio of the primaries be in the range $1 < f_2/f_1 < 1.3$. With the correct frequency ratio, the LCDT can be heard for low levels of the primaries, considerably lower than 40 dB SPL; the level of the f_2 component can even be less than 10 dB. Whereas the cubic nonlinearity in Eq. (22.50) has an amplitude of $A_1^2 A_2$, the cancellation level of the LCDT does not necessarily increase when one primary amplitude is fixed and the other increases. Instead, the LCDT amplitude has a maximum when $A_1 = A_2$. When the two primaries are constrained to have the same level, then as the level of the primaries increases the level of the LCDT increases by the same amount (i.e. 1 dB/dB) or somewhat less. By contrast, the cubic law would predict 3 dB/dB. The phase of the LCDT, as measured by the cancellation tone, varies greatly as the amplitude of the primaries changes.

The LCDT appears to be a by-product of functioning outer haircells in a normal cochlea. Because the outer hair cells are vulnerable, they are likely points of failure in cases of abnormality. Therefore, testing for a normal LCDT is potentially useful as a diagnostic tool. What makes it particularly attractive is that the LCDT can be found as an objective tone in the cochlear emissions (Kemp, 1979a), which leads to an objective measure of outer haircell function (Lonsbury-Martin *et al.*, 1990; Zurek and Rabinowitz, 1995). Like the audible distortion products, the emitted distortion products can be cancelled by adjusting the amplitude and phase of an external tone. However, the amplitudes and phases that cancel the audible difference tone do not cancel the emission and vice versa (Furst *et al.*, 1988). To appear in the ear canal, the emissions must travel the wrong way on the basilar membrane. There is evidence that they are able to do this only by combining with other sources of cochlear emission located near the place corresponding to $2f_1-f_2$. Alas, it seems that there is nothing about the nonlinear auditory system that is simple.

Nonlinear Auditory Filtering

A filter, as described in Chapter 9, is a linear device. Its frequency response, defined as the ratio of output to input as a function of frequency, is independent of the signal level. Auditory filtering, however, is nonlinear, and one expects this nonlinearity to be particularly important at low levels where the outer haircells act to boost the gain of the cochlea. On the basis of physiological observations on the tuning of cells in the auditory nerve one would expect that the bandwidth of auditory filters increases with increasing level.

Despite expectation, the nonlinearity of auditory filtering has not been easy to demonstrate psychophysically. In fact, on the basis of loudness summation and lateralization experiments, Scharf and Meiselman (1977) concluded that critical bandwidths depend on intensity only at very *high* levels, above 80 dB SPL. In 1950 Hawkins and Stevens studied the masking of sine tones by broadband noise as the spectrum level of the noise varied from −10 dB to 60 dB (30−100 dB SPL). If the bandwidth of an auditory filter increases with increasing level then in order to maintain a constant signal-to-noise ratio in an auditory filter, the level of the signal should increase faster than the level of the noise. In fact, however, Hawkins and Stevens found that increasing the level of the noise by 10 dB always led to a 10-dB increase in signal threshold, just as expected for a linear filter.

Tone-on-tone masking and masking of a tone by narrowband noise (e.g. Egan and Hake, 1950) appear to show a nonlinear growth of masking with level. For masker levels above 60 dB there is a great deal more masking for signals with frequencies above the masker than for signals below the masker. This asymmetrical effect, known as the *upward spread of masking*, increases significantly with increasing level. The problem with this observation as evidence for nonlinear filtering is the problem of off-frequency listening. As the masker level changes the listener very likely uses different auditory filters, located at different patches of the basilar membrane, to detect the signal. It is not clear whether the level-dependent masking observed for signals above the masker represents an actual change of bandwidth for a filter at the signal frequency, or whether it is caused by the fact that the listener is using a filter centered at a higher frequency, hence wider bandwidth, with increasing level. As a further complication, physiological observations suggest that the locus of maximum excitation shifts basally with increasing level. Zwislocki (1991) has reported enormous shifts, equivalent to one octave for a 60-dB increase in level. Such a shift would lead to increased masking for signals above the masker. The shift would make it appear that the lower skirt of an auditory filter had a decreasing slope with increasing level, while the upper skirt might actually become steeper.

Using the notched noise paradigm described in Chapter 10, Patterson *et al.* (1982) found wider auditory filters with increasing level. Similarly Glasberg and Moore (1990) found that the lower skirt of a roex filter had a decreasing slope with increasing level, while the upper skirt became slightly steeper. As the level of the notched noise increased from 60 to 90 dB, the slope of the lower skirt decreased almost by a factor of two on a scale of dB/Hz. The calculation has not been replicated by others, and it is not clear whether it adequately discounts the maximum excitation shift.

The difficulties of proving nonlinear effects in masking are, in part, responsible for the continued importance of linear filters in auditory models. When nonlinearities are included in models, they are usually incorporated as a second stage that follows a stage of *linear* filtering, as in the revcor model (deBoer and deJongh, 1978).

DYNAMIC RANGE COMPRESSION

Dynamic range compression is an important technique in electroacoustics. A compressor acts on the amplitude of a signal so that when the amplitude of the input changes, the amplitude of the output changes in the same direction, but by a lesser amount. Figure 22.7 shows the input and the output of a dynamic range compressor. As the input amplitude changes by a factor of 10 (−20 dB) the output amplitude changes by a factor of only 3.16 (−10 dB). This compression characteristic is a ratio of two to one on a decibel scale, corresponding to the compressors made by the dBx company.

Compressors are used in broadcasting and recording of speech and music. Compression is particularly notable in music intended as background. Compression is also used in many hearing aids to boost weak signals while allowing for the recruitment that occurs for stronger signals. In these applications it is essential that the compressor not introduce nonlinear distortion. Figure 22.7 is an example; the output waveform in part (b) is a sine wave just like the input in part (a). Dynamic range compressors work by extracting a measure of the input signal strength and then scaling the output amplitude according to the desired compression law, normally with a voltage-controlled amplifier. When the signal frequency is very low, so that waveform variations are not fast compared to amplitude variations, then the dynamic range compressor introduces nonlinear signal distortion as it compresses the waveform instead of only the envelope.

The opposite of compression is dynamic range expansion. When audio is recorded by analog means, for example on film, a combination of compression on recording and compensating expansion on playback, as in the Dolby system, decreases the low-level noise of the recording medium while maintaining the dynamic range of the original material.

The cochlea operates as a dynamic range compressor at low levels in that the

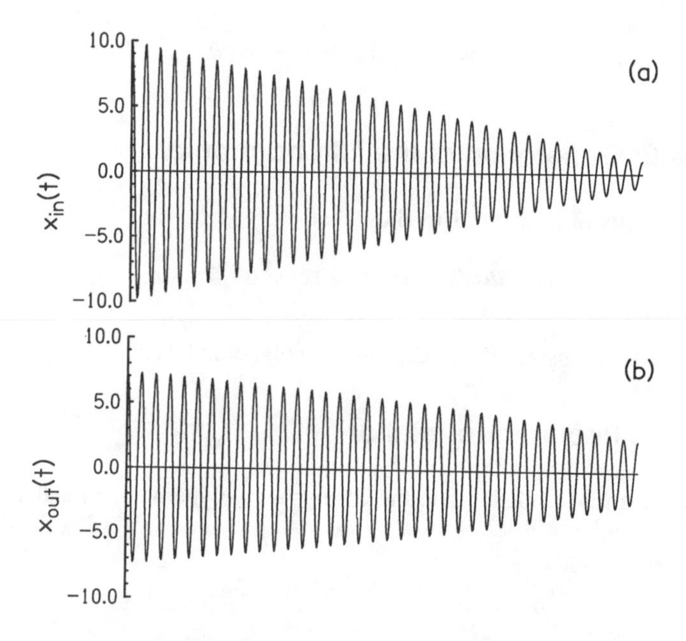

FIGURE 22.7. The operation of a dynamic range compressor that is linear on a log scale. As the input amplitude varies as in part (a), the output amplitude in part (b) varies according to the square root of the input amplitude. Thus the compression leads to an output change of 1 dB for every 2-dB change in the input.

neural spike count decreases less rapidly than the signal power. However, the cochlea does not accomplish this process without creating distortion products.

EXERCISES

HARMONIC DISTORTION EXERCISES

Exercise 1: Fourier expansion of fifth and third powers

(a) Use Eq. (22.16) for the coefficients in a Fourier series to show that

$$\cos^5 \theta = \frac{1}{16} \cos 5\theta + \frac{5}{16} \cos 3\theta + \frac{5}{8} \cos \theta.$$

Check your answer with Eq. (22.28).

(b) Similarly show that

$$\cos^3 \theta = \frac{1}{4} \cos 3\theta + \frac{3}{4} \cos \theta.$$

Exercise 2: Generation of the fifth harmonic

Use the results of Exercise 1 to show that

$$\cos 5\theta = 16 \cos^5 \theta - 20 \cos^3 \theta + 5 \cos \theta,$$

and compare your answer to the Chebyshev polynomial T_5.

Exercise 3: Recurrence relation

Use the recurrence relation and the Chebyshev polynomials T_3 and T_4 to show that

$$T_5(x) = 16x^5 - 20x^3 + 5x.$$

Exercise 4: Distortion percentage

(a) Is it ever possible for the harmonic distortion percentage to be greater than 100%?

(b) Show that the harmonic distortion of a full-wave rectifier is 100%.

(c) A device creates harmonic distortion having five harmonics of equal level. Each of the five harmonics has a level that is 30 dB below the level of the linear response. Calculate the distortion percentage.

Exercise 5: Half-wave rectifier

Show that the harmonic distortion of a half-wave rectifier is 39.9%. Note: The easy way to do this exercise is to use the Fourier series for a half-wave rectified sine. The DC component should be removed from both numerator and denominator of the fraction for distortion percentage. Then the distortion power is the total remaining power minus the power in the fundamental.

Exercise 6: General harmonic distortion formula

Use Eq. (22.16) to check the formula for harmonic generation by (a) the square-law distorter of Eq. (22.12) and (b) the cubic distorter of Eq. (22.14).

Exercise 7: Fifth-order distortion

Suppose you input an 800-Hz sine tone into a distorter that raises the signal values to the fifth power. Use Eq. (22.28) to show that you expect distortion products at 800 Hz, 2400 Hz, and 4000 Hz, with levels of 0 dB, −6 dB, and −20 dB with respect to the 800-Hz component.

Exercise 8: The triangle as an approximation to a sine

The triangle wave is not enormously different from a sine wave. Use a sum over all integers from Appendix C to show that the triangle wave differs from a sine (the fundamental of the triangle) by a total harmonic distortion that is down by 18 dB.

Figure E.8(a) shows a sine wave and the triangle approximation to it. What is wrong with the triangle, as an approximation to the sine, is that the sides are too straight and the peak is too tall. Both problems can be alleviated a little by adding a triangle with three times the frequency, as shown in part (b).

Use Fourier analysis to show that the correct amplitude for the triangle wave with triple frequency is (1/9)-th of the amplitude of the original sine. Which harmonics are cancelled? Show that the distortion is reduced from −18 dB to about −26 dB.

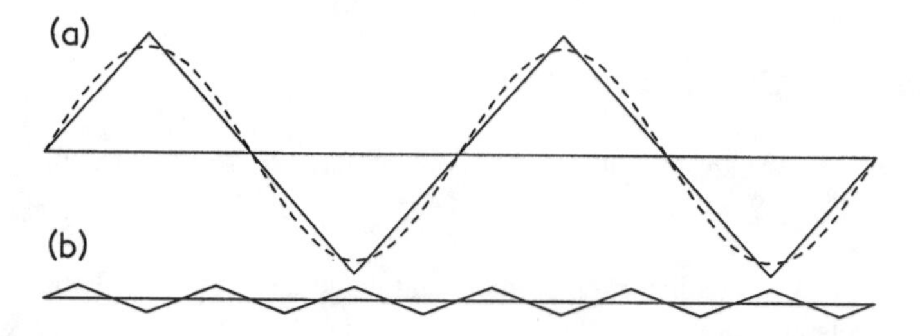

FIGURE E.8. (a) A sine wave and a triangle approximation to it. (b) A triangle that can be added to the triangle of part (a) to improve the approximation to a sine.

Exercise 9: Low-frequency audiometry

A well-known manufacturer of low-cost electronic test equipment sells a function generator that is said to be "low distortion." However, the THD is quoted to be less than one percent, which is not really low. A reasonable assumption is that this distortion is all due to spurious second and third harmonics, though it is impossible to know how much of each.

You plan to use this function generator as a sine source for testing hearing thresholds at low frequency, but you are concerned that normal-hearing listeners will detect the harmonic distortion instead of the intended frequency.

(a) Explain why a conservative evaluation of the measurement should assume that all the distortion is third harmonic.

(b) Use the threshold of hearing curve from Chapter 4 to show that you can use this function generator to measure threshold even at 50 Hz.

INTERMODULATION DISTORTION EXERCISES

Exercise 10: Distortion degeneracy

(a) You are interested in an intermodulation distortion product that arises from two primaries, f_1 and f_2, with frequency $f_{i.m.}=k_1 f_1 - k_2 f_2$. You wonder if there is some other set of integers k_1' and k_2' such that $k_1' f_1 - k_2' f_2$ is equal to the same frequency, $f_{i.m.}$, i.e. if frequency $f_{i.m.}$ is degenerate. Show that such a pair indeed does exist unless the ratio f_2/f_1 is irrational. (Recall that a rational number is defined as the ratio between two integers. For instance 193425/8739 is a rational number.)

(b) Suppose that the ratio between the two primaries is $f_2/f_1=6/5$. You are interested in the distortion product $f_{i.m.}=3f_1-2f_2$. Show that the smallest pair of integers that gives the same $f_{i.m.}$ is $9f_1-7f_2$.

Exercise 11: Cubic distortion

(a) Use trigonometric substitutions from Appendix B to verify the cubic distortion products in Eq. (22.50).

(b) Explain why cubic distortion predicts that the level of the LCDT increases with primary levels at a rate of 3 dB/dB.

Exercise 12: A single compressive power

Instead of representing auditory nonlinearity by a power series, one can represent it by a single power less than one. This exercise requires the computation of the lower cubic difference tone (LCDT) at 800 Hz, given primary tones at 1000 Hz and 1200 Hz.

Let the stimulus signal be

$$x(\xi)=A_1 \cos(2\pi \cdot 5\xi)+A_2 \cos(2\pi \cdot 6\xi),$$

where ξ is a period-scaled measure of time. Let the nonlinear transformation have the same sign as the stimulus but be proportional to the 0.6 power of the stimulus magnitude:

$$y(\xi)=\text{sign}[x(\xi)]|x(\xi)|^{0.6}.$$

(a) Use a computer to compute the amplitude of the LCDT, proportional to

$$Y_4=\sum_{\xi} y(\xi)\cos(2\pi\cdot 4\xi),$$

where ξ runs over a few hundred values between 0 and 1.

(b) Do computations to show that the LCDT has a maximum when the levels of the two primaries are the same. As Smoorenberg (1974) found, the LCDT decreases with increasing A_2 when $A_2>A_1$. Show that when $A_2=A_1$, the level of the LCDT grows by 12 dB when the amplitude of the two primaries grows by 20 dB.

(c) The computation in part (a) calculates the cosine component of the LCDT. Explain why the sine component is zero.

(d) Could the 0.6 power law account for an observed dependence of the phase of the LCDT on primary amplitudes?

Exercise 13: Two odd powers of distortion

Suppose that the cochlear response to input x is

$$y=c_1x+c_3x^3+c_5x^5.$$

Suppose that signal x is the sum of two cosines with amplitudes A_1 and A_2. Show that the amplitude of the LCDT is

$$Y_{(2f_1-f_2)}=A_1^2A_2(c_3+c_5A_2^2).$$

Exercise 14: Define compressive nonlinearity

When physiologists talk about compressive nonlinearity they mean that as the input signal is reduced, the response is reduced by a lesser amount. Doesn't that imply an expansion and not a compression?

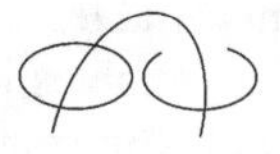

CHAPTER 23

Noise

A barbarous noise environs me.
JOHN MILTON

In everyday usage the word ''noise'' refers to any unwanted sound. In acoustics, and other communication sciences, the word usually refers to a complicated signal with a dense spectrum whose properties are defined only statistically. Therefore, an intense 1000-Hz sine tone may be objectionable, but formally anyhow, it would not be called noise. The present chapter treats some of the mathematical properties of noise and extends ideas already introduced in previous chapters, especially Chapter 3.

GAUSSIAN AND THERMAL NOISE

A noise is said to be Gaussian if repeated sampling of the noise signal leads to a Gaussian (normal) distribution. This definition can be understood from a thought experiment: We suppose that an analog-to-digital converter takes instantaneous samples of a signal and puts the sampled values into computer memory. If these samples are then plotted on a histogram showing the number of times that a value falls within each bin range, we find that the histogram is a normal distribution if the noise is Gaussian noise.

There is a tendency for noise signals to be Gaussian because noise is usually the result of many independent random events. The random rain of air molecules on a surface or the random walk of electrons in a resistor are examples. The *central limit theorem*, otherwise known as the ''law of large numbers,'' says that if a signal is created by adding together a lot of random variables, the distribution of the signal will approach a Gaussian as the number of random variables becomes large. It does not matter if the original random variables come from a distribution that is not Gaussian; in the end, adding up the variables leads to a Gaussian. The approach to a Gaussian appearance is rather rapid. For example, Appendix H gives an algorithm for generating random variables starting with a rectangular distribution from a computer random number generator. The rectangular distribution is not even remotely Gaussian, and yet when one adds twelve samples from that distribution one gets a variable whose distribution is an excellent approximation to a Gaussian.

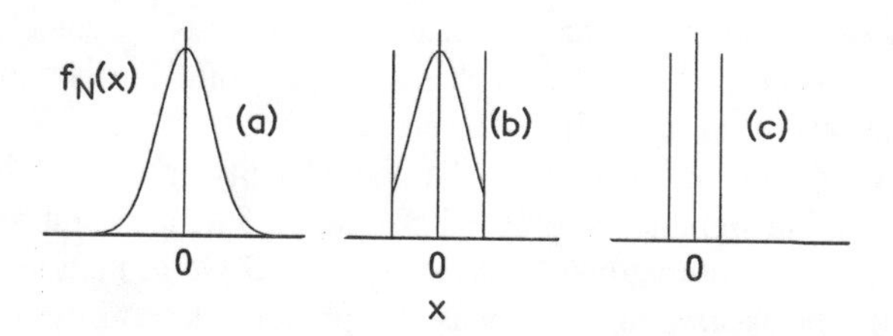

FIGURE 23.1. Waveform probability densities for noise: (a) Gaussian noise, (b) Gaussian noise that has been symmetrically peak clipped, (c) noise after passing through a comparator.

Although Gaussian noise is the norm, it is easy to generate noise with non-Gaussian distributions if the noise is in the form of an electrical voltage. One simple way is to remove the extreme values in the tails of the distribution by peak clipping the noise. Symmetrical clipping removes the positive and negative tails of the distribution, and puts their areas into delta functions at the edges, as shown in Fig. 23.1(b).

Alternatively, if noise is put into a comparator that outputs $+1$ whenever the input is positive and -1 whenever the input is negative, the output is even more clearly non-Gaussian, as shown in Fig. 23.1(c). Such a signal is known as a "random telegraph wave."

Thermal noise is the kind that is generated electrically in resistors. It is often called Johnson noise or Nyquist noise. The noise power per unit frequency is given by $4k_BT_K$ where k_B is Boltzmann's constant, and T_K is the Kelvin temperature, i.e. the temperature on an absolute scale. This noise was described in terms of a Fourier series by Einstein and Hopf (1910):

$$x(t)=\sum_{n=0}^{N} [A_n \cos(\omega_n t)+B_n \sin(\omega_n t)], \tag{23.1}$$

where n is an integer, and angular frequencies ω_n are equally spaced (see also Rice, 1954).

The coefficients A_n and B_n are random variables, and there are several ways to interpret them. The best way is in terms of an ensemble, which can be illustrated by another thought experiment. We suppose that there is a large number of independent, but stochastically identical, noise generators. This means that any average quantity (the power spectrum, the fourth moment, etc.) is the same for each generator, but the cross-correlation function between any two generators is zero. We take a long sample signal, $x(t)$, of duration T_D from each generator. Each sample signal can be represented exactly and uniquely by Eq. (23.1), where the ω_n values are harmonics of $2\pi/T_D$. We now focus on a particular component, say $n=126$, and make a histogram for A_{126} and another for B_{126}, using all the sample signals. We find that both histograms represent the same distribution, a normal distribution with mean zero and variance σ_{126}^2. A similar calculation can be performed for any other value of n with the same result. Further, if the noise is white the values of all the σ_n^2 will be the same.

An alternative approach is to suppose that there is only one noise source, and that the individual sample signals, $x(t)$, all with duration T_D, are acquired at different times. That is why it is sometimes said that A_n and B_n are functions of time (e.g. Beranek, 1988). The assumption that the two approaches lead to the same answer is the *ergodic hypothesis*. Still another approach, possible only if the noise is white, is to assume that within any given sample of noise the A and B coefficients themselves are normally distributed.

Any valid statistical approach to the distributions of coefficients A_n and B_n for white noise must preserve the relationship between the variance of the distribution, σ^2, and the average power. To find that relationship we find the total power by treating $x(t)$ as a voltage that looks into a resistance of one ohm. The power is therefore the square of $x(t)$. The energy over a span of time is the integral of the power, and the average power is obtained by dividing the energy by the duration of the time span like this:

$$\bar{P}=\frac{1}{T_D}\int_0^{T_D} dt\ x^2(t), \tag{23.2}$$

$$=\frac{1}{2}\sum_{n=0}^{N}(A_n{}^2+B_n{}^2)=N\sigma^2. \tag{23.3}$$

An alternative to Eq. (23.1) for the description of noise is the amplitude and phase form,

$$x(t)=\sum_{n=0}^{N} C_n\cos(\omega_n t+\phi_n), \tag{23.4}$$

where $C_n=\sqrt{A_n^2+B_n^2}$, and ϕ_n is the argument of the complex number A_n+iB_n. Because quantities A_n and B_n are normally distributed with zero mean and variance σ^2, the quantities C_n are distributed according to a Rayleigh distribution, with most probable value σ, as described in Appendix I. The phase angles ϕ are uniformly distributed (rectangular distribution) over the entire range from 0 to 2π.

MAKING NOISE

Psychoacoustical experiments frequently require well-controlled sources of noise. An analog thermal noise source with a very flat frequency response can be made by amplifying the voltage appearing across a Zener diode biassed near the breakdown knee. Zeners differ greatly in their noise characteristics, even if they come from the same lot, and it is necessary to select. A Zener with a breakdown voltage above 10 volts is much more likely to be successful than one with a lower breakdown voltage. A package of ten such Zeners normally contains several winners with high noise output, flat spectrum, Gaussian density, and a smooth sound. The analog noise can be filtered to obtain the desired frequency band. If broadband noise is needed, the output of a Zener generator should be lowpass filtered anyway to eliminate the power that is above the audible range of frequencies. Only then does a physical measurement of the noise reveal something useful about the effective noise power in the experiment.

Noise with more precisely controlled characteristics can be created digitally. To create an ensemble of noise waveforms with a given average spectral density, one lets σ_n^2 be the desired average power for frequency component n and then applies Eq. (I.16) to the output of a standard random number generator. This gives the amplitude for component n in a single noise waveform, according to the Rayleigh distribution. Repeating the process gives the amplitude for component n for the other waveforms in the ensemble. The ensemble average of noise signals generated in this way is a Gaussian noise, no matter how many components are in the waveform. Even if there is only a single component, the ensemble-average of waveform values has a Gaussian distribution.

At least as popular as Rayleigh-distributed noise is *fixed-amplitude random-phase* noise. In this case, the amplitude of the n-th component is not a random variable. It is determined entirely by the desired power for the component. Only the phases of the components are random variables. The advantage of this kind of noise is that every noise waveform in the ensemble has the same power spectrum. This method of generating noise is particularly attractive in a frozen noise experiment, where only a single noise waveform is used throughout, because any waveform will exactly represent the entire ensemble.

The price that one pays for fixed-amplitude random-phase noise is that coefficients A_n and B_n are no longer normally distributed, and they are no longer independent. Coefficients A_n and B_n are determined by the phase angle: $A_n = C_n \cos \phi_n$ and $B_n = C_n \sin \phi_n$, where C_n is fixed. Because ϕ_n follows a rectangular distribution, A_n (or B_n) follows the sine probability density function with square-root singularities, as described in Chapter 16. This density, with peaks at extreme values, is very different from a normal density with a peak at zero. However, the values in the waveforms themselves are normally distributed so long as the number of components is not small. Figure 23.2 shows the densities for $N=1$, 3, and 5 components in equal-amplitude random-phase noise. Obviously for $N=1$ or 3 the densities are not Gaussian, but for $N=5$ and higher the densities depart only insignificantly from Gaussian. Thus the law of large numbers comes to the rescue of fixed-amplitude random-phase noise, even if the number isn't large.

POWER SPECTRUM

If one takes a long enough sample of thermal noise one finds that all frequencies contribute to the noise signal, more or less. The Fourier spectrum appears to be a

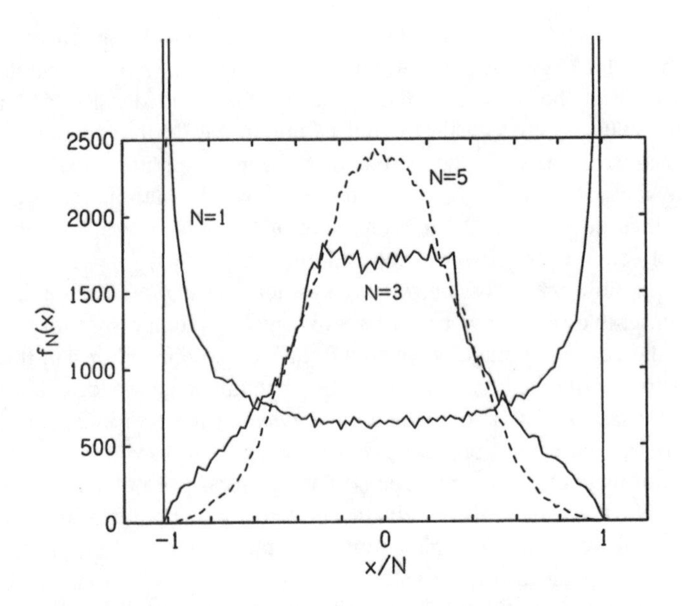

FIGURE 23.2. Histograms, with 100 bins, showing the distribution of sampled values from 100,000 waveforms. The number of sine components in each waveform is N (N=1, 3, or 5) and the amplitude of each sine component is the same (1/N). The component phases are random variables.

continuum of components. As described in Chapter 3, this continuum leads to the concept of the power spectrum. The power spectrum, or spectral density, is symbolized by N_0, and has units of watts/Hz. The total power in a band with lower frequency f_l and upper frequency f_u is given by the area under the power spectrum plot,

$$\bar{P}=\int_{f_l}^{f_u} df\, N_0(f) \tag{23.5}$$

or

$$\bar{P}=\int_{\omega_l}^{\omega_u} d\omega\, \mathcal{N}_0(\omega). \tag{23.6}$$

The power spectrum in terms of angular frequency is simply related to the power spectrum in terms of frequency by the equation $\mathcal{N}_0(\omega)=N_0(f)/2\pi$. It is common to deal with the density in Hz, as given in the first equation. Therefore, the spectrum level in dB is given by

$$L=10\log(N_0/10^{-12}), \tag{23.7}$$

where N_0 is the density of the noise in units of watts per Hz, and the reference density is taken to be 10^{-12} watts per Hz.

COLOR OF NOISE

White Noise

If the power spectrum is independent of frequency then the noise is called "white," in analogy to white light which contains all frequencies equally. In that case, the integral in Eq. (23.5) for the total power is easy to do, and the result is

$$\bar{P}=N_0(f_u-f_l)=N_0(BW), \tag{23.8}$$

where BW is the bandwidth, over which the power spectrum is constant.

Pink Noise

The color pink is a light color, closer to being white than are saturated colors, but with an obvious reddish (low-frequency) tint. Similarly, the spectrum of "pink" noise is tilted, slightly, in favor of the lower frequency end of the spectrum. The power spectrum for pink noise is inversely proportional to frequency,

$$N_0(f)=P_0/f, \tag{23.9}$$

where P_0 is a constant power (watts) and f is the frequency in Hertz. The spectrum level of pink noise decreases at a rate of -3 dB/octave. The power in a band that extends from f_l to f_u is

$$\bar{P}=\int_{f_l}^{f_u} df\, N_0(f)=P_0 \ln(f_u/f_l). \tag{23.10}$$

This result for the total power makes it evident that a power spectrum cannot be pink all the way down to a lower frequency of $f_l=0$. Nor can it be pink all the way to an upper frequency of $f_u=\infty$ because the log function is not defined for zero or infinity. Power spectra for white and pink noise bands are shown in Fig. 23.3.

UNIFORM EXCITING NOISE

The idea of uniform exciting (UE) noise originates with the concept of auditory filtering. UE noise has a power spectrum that is shaped so that every auditory filter in the auditory system receives the same amount of power. Because auditory filters at low frequency have a bandwidth that is relatively narrow (units of Hz), the

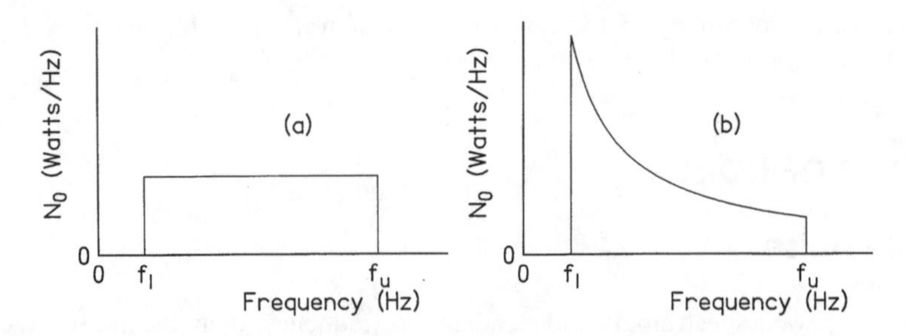

FIGURE 23.3. Power spectra (spectral densities) for (a) white noise and (b) pink noise. The figures have been drawn to have the same areas. Therefore, the noises have equal power.

spectral density of UE noise must be particularly intense at low frequency. To find a formula for the spectrum of UE noise we begin with a general transformation of densities.

Suppose that variable u represents critical band units (cbu). These might be units of Bark or Cam, but we shall use the general term cbu. We assume that there is a one-to-one mapping of u onto frequency f in Hertz by a continuous function $u(f)$, such that the inverse function, $f(u)$, also exists. For instance, if the cbu are Bark then $u(1000\text{ Hz})=8.5$ Bark, and $f(8.5\text{ Bark})=1000$ Hz.

We let $N_u(u)$ be the density in units of watts per cbu, and let $N_0(f)$ be the usual spectral density in units of watts per Hz. Our goal is to discover the relationship between $N_u(u)$ and $N_0(f)$ given the mapping $u(f)$. The key is that an element of power at a particular u (or the corresponding f) must be the same when calculated with either density, i.e.

$$N_u(u)\Delta u = N_0(f)\Delta f, \tag{23.11}$$

where $u=u(f)$.

Then, given a particular form for $N_u(u)$, the spectral density becomes

$$N_0(f) = N_u(u)\,\frac{du}{df}, \tag{23.12}$$

or

$$N_0(f) = N_u(u) \Big/ \frac{df}{du}. \tag{23.13}$$

Function df/du is the change in frequency for a unit change in cbu; in other words, it is the critical bandwidth as a function of frequency.

The transformation among densities can be illustrated with a few examples of UE noise. By the definition of UE noise, $N_u(u)$ is a constant equal to the power per

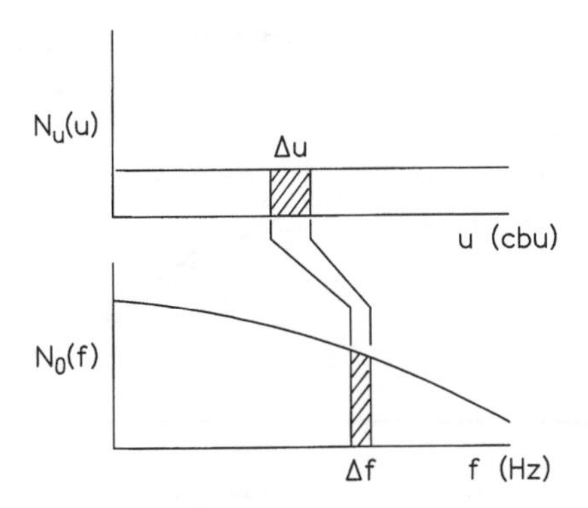

FIGURE 23.4. The top graph shows the cbu density of UE noise, which is constant. An increment Δu at cbu value u implies a value of the increment Δf at frequency f, where $f=f(u)$. The spectral density shown in the bottom graph is determined by the requirement that the two descriptions must lead to the same area under the curves.

critical band, N_u. Graphically, a transformation looks like Fig. 23.4.

Case 1: Constant-Q auditory filters

If auditory filters have constant Q then the bandwidth is given by $\Delta f=Qf\Delta u$, and $df/du=Qf$. With N_u equal to the power per critical band, the density becomes

$$N_0(f)=(N_u/Q)\,\frac{1}{f}. \tag{23.14}$$

This spectrum has the $1/f$ dependence seen in Eq. (23.9). Therefore, if auditory filters are constant-Q, UE noise is pink noise.

Case 2: Munich auditory filters

Auditory filters determined by Zwicker (Bark scale) have a bandwidth that is approximated ($\pm 10\%$) by the formula of Zwicker and Terhardt (1980),

$$\frac{df}{du}=25+75[1+1.4\times 10^{-6}f^2]^{0.69}, \tag{23.15}$$

where f is the frequency in Hz. Because $N_u(u)$ is a constant, N_u, Eq. (23.13) gives the density as

$$N_0(f)=N_u\{25+75[1+1.4\times 10^{-6}f^2]^{0.69}\}^{-1}, \tag{23.16}$$

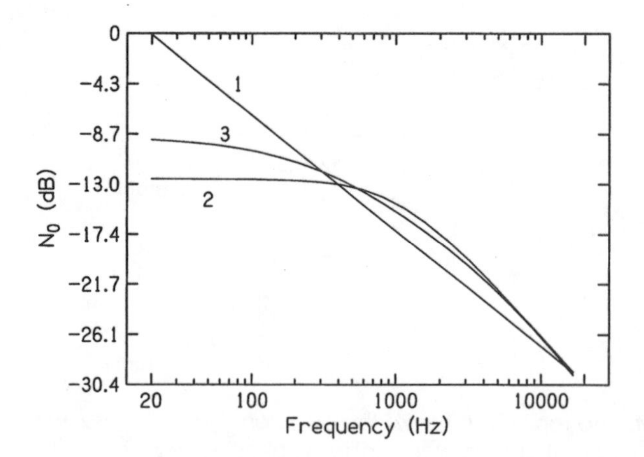

FIGURE 23.5. Spectral density for uniform exciting noise for auditory systems that are tuned according to (1) constant-Q, (2) Munich filters (Bark scale), (3) Cambridge filters (Cam scale). The densities are plotted here on a log-log scale. All densities have the same area, hence the same total power.

where constant N_u determines the total power.

Case 3: Cambridge auditory filters

Auditory filter bandwidths determined by Patterson, Moore, and colleagues (Cam bands) are given by the ERB formula from Glasberg and Moore (1990),

$$\frac{df}{du} = 24.7(0.00437f + 1). \tag{23.17}$$

Therefore, the formula for the spectral density of UE noise is

$$N_0(f) = N_u\{24.7(0.00437f + 1)\}^{-1}. \tag{23.18}$$

Spectral densities for Cases 1, 2, and 3 are shown in Fig. 23.5.

In practice, delivering uniform exciting noise to a listener requires that one further equalize the spectrum for middle ear effects, as observed in the free-field equal-loudness contours of Chapter 4, or for the headphone transfer function in headphone listening. That is because the critical band calculation for uniform exciting noise has not otherwise considered these effects.

ERNEST GLEN WEVER (1902–1991)*

Ernest Glen Wever was born on October 16, 1902 in Benton, IL. In 1927 he joined the Psychology Department at Princeton University where he remained for the rest of his scientific career. Working with Charles W. Bray in 1930, he placed an electrode near the auditory nerve of a cat and monitored potentials using an audio system. Bray heard Wever's voice transduced by the hair cells and nerve fibers of the cat's ear. In this way they discovered the cochlear microphonic and thus began the field of cochlear electrophysiology.

Wever critically evaluated contemporary theories of hearing and developed the "volley principle" in which time and place mechanisms were combined to account for frequency analysis. This work was published in his book *Theory of Hearing* in 1949. Wever founded, defined, and for several decades maintained, the modern field evolutionary biology of hearing, publishing *The Reptile Ear* in 1978 and *The Amphibian Ear* in 1985. These books are unparalleled treatises on the comparative study of the ear.

*The author thanks Professor Richard R. Fay, who was one of Wever's students and wrote this biography.

NOISE FLUCTUATIONS

The discussion of fluctuations in this section is in terms of moments of distributions, as introduced in Chapter 16. It is assumed that the noise has zero average value ($\bar{x}=0$) so that moments and central moments are the same. Most of the results below are given without proof; for further details the reader may see two papers by Hartmann and Pumplin (1988) and (1991).

Power Fluctuations

The average power in a noise waveform is $\overline{x^2}$, where the bar denotes an average over time. At any given time the power differs from the average by an amount ΔP,

$$\Delta P(t)=x^2(t)-\overline{x^2}. \tag{23.19}$$

Quantity ΔP has an average value of zero by definition, but even powers of ΔP have a finite average. The mean-square power fluctuation, as found in Chapter 16, is

$$\overline{\Delta P^2}=\overline{x^4}-\overline{x^2}^2. \tag{23.20}$$

If waveform $x(t)$ is multiplied by a factor of 2, the mean-square power fluctuation is multiplied by a factor of 16, which makes it awkward to discuss the size of fluctuations. To avoid that situation, the fluctuation can be normalized by the square of the average power, leading to the dimensionless quantity

$$\overline{\Delta P^2}/\overline{x^2}^{\,2} = \overline{x^4}/\overline{x^2}^{\,2} - 1. \tag{23.21}$$

Quantity $\overline{x^4}/\overline{x^2}^{\,2}$ is W_P, the kurtosis. Depending on the phases and amplitudes of the components, W_P may be large or small.

If the phases and amplitudes are random variables, then different sample waveforms have different values of W_P, and general remarks can only be made about ensemble averages. For a noise waveform constructed from N Fourier components having Rayleigh-distributed amplitudes and random phases, as in thermal noise, the kurtosis is

$$\langle W_P \rangle = 3 - 3/(N+1). \tag{23.22}$$

For equal-amplitude random-phase noise

$$\langle W_P \rangle = 3 - 3/(2N). \tag{23.23}$$

The noise fluctuation is always $\langle W_P \rangle - 1$, by definition.

The smallest values of W_P, approaching unity, are obtained with low-noise noise invented in 1985 by Jon Pumplin. Pumplin's technique begins with a fixed, deterministic, amplitude spectrum and finds the set of phases that minimizes the kurtosis. Although the algorithm specifically minimizes the kurtosis (fourth moment), it also leads to small values of the higher moments: the sixth, the eighth, etc. The latter is an empirical result; it is not entirely understood at this time.

In contrast to low-noise noise are waveforms generated with constant phase angles. Such waveforms differ from random-phase signals because the kurtosis grows as the number of components increases. For example, if the components have equal amplitudes and phases of zero (sum of N cosines) the kurtosis is

$$W_P = \left[\frac{4}{3} N^2 - N + \frac{7}{6}\right] \Big/ N. \tag{23.24}$$

Signals like this are impulsive. The kurtosis appears to be a useful measure of impulsive sounds as they are occur in the workplace (Erdreich, 1986).

Envelope Fluctuations

The fluctuation from Eq. (23.21), $\langle W_P \rangle - 1$, correctly calculates the mean-square power fluctuation about the average power, but this measure of fluctuation may not be perceptually meaningful. By this measure even a pure tone has fluctuations. [Equations 23.22, 23.23, and 23.24 all agree that for $N=1$ the kurtosis is $W_P = 3/2$, and the fluctuation becomes 1/2.] But a pure tone sounds entirely smooth, and one would expect a meaningful measure of the fluctuation to be zero. The mathematical problem arises from the fact that $\cos^2 \omega t$ is equal to $1/2 + (1/2)\cos(2\omega t)$. The fluctuation about the average value of 1/2 occurs at twice the frequency of the tone, which is too fast to be perceived as a loudness fluctuation. When there is more than one cosine in the waveform, a similar situation occurs because the power fluctuation contains both sum and difference frequencies. For example, for two components:

$$(\cos \omega_1 t + \cos \omega_2 t)^2 = \cos^2 \omega_1 t + \cos^2 \omega_2 t + \cos(\omega_2 - \omega_1)t + \cos(\omega_2 + \omega_1)t, \tag{23.25}$$

but the sum frequencies are too high to be perceived as fluctuations. Only the difference frequencies can lead to audible beats.

A measure of fluctuation that comes closer to perception is the fluctuation in envelope. As shown in Chapter 18, the envelope is given by

$$E(t) = \left| \sum_{n=1}^{N} C_n e^{i(\omega_n t + \phi_n)} \right|. \tag{23.26}$$

The average envelope power is $\overline{P_E} = \overline{E^2}$. The normalized fourth moment is

$$W_E = \overline{E^4} / \overline{E^2}^{\,2}, \tag{23.27}$$

and the relative fluctuation is $W_E - 1$. For a single sine, $E(t)$ and all its powers are unity so that the fluctuation is zero. For equal-amplitude random-phase noise the ensemble average value is

$$\langle W_E \rangle = 2[1 - 1/(2N)]. \tag{23.28}$$

The RMS fluctuation in envelope power resembles a standard deviation; relative to the envelope power itself, it is equal to

$$\sqrt{\langle W_E \rangle - 1} = \sqrt{1 - 1/N}. \tag{23.29}$$

As the number of components becomes infinite, the RMS fluctuation in the envelope becomes as large as the envelope power itself.

Noise Envelope Model

This section expands the treatment of envelope fluctuations to incorporate a temporal modulation transfer function (TMTF), as described in Chapter 20. The development requires only elementary statistics and leads to an analytic expression for envelope fluctuation and its dependence on spectral density, bandwidth, and integration time.

It is assumed that the perceptually relevant character of a band of noise is its envelope power. As shown in Eq. (18.21), the envelope power is given by the square of Eq. (23.26),

$$P_E(t)=E^2=\sum_{n=1}^{N} C_n^2+2\sum_{n=1}^{N}\sum_{n'=1}^{n-1} C_nC_{n'}\cos[(\omega_n-\omega_{n'})t+\phi_n-\phi_{n'}]. \tag{23.30}$$

In the double sum above, with no loss of generality, $\omega_n>\omega_{n'}$.

The fluctuations with higher frequencies are attenuated by the TMTF, which operates like a filter. Therefore, the output envelope power is

$$P_{out}(t)=\int_0^{\infty} dt' P_E(t-t')h(t'), \tag{23.31}$$

where the filter is modeled by an impulse response with time constant τ,

$$h(t)=\frac{1}{\tau}e^{-t/\tau} \quad (\text{for } t>0) \tag{23.32}$$

$$h(t)=0 \quad (\text{for } t<0).$$

Doing the integral leads to

$$P_{out}(t)=\sum_{n=1}^{N} C_n^2+2\sum_{n=1}^{N}\sum_{n'=1}^{n-1} C_nC_{n'}\frac{\cos(y_{nn'})+\tau\Delta_{nn'}\sin(y_{nn'})}{1-(\tau\Delta_{nn'})^2}, \tag{23.33}$$

where the instantaneous phase difference is given the notation

$$y_{nn'}=(\omega_n-\omega_{n'})t+\phi_n-\phi_{n'}, \tag{23.34}$$

and the difference in angular frequency is defined by the positive quantity

$$\Delta_{nn'}=\omega_n-\omega_{n'}. \tag{23.35}$$

For any ensemble, the average value of P_{out} over time is $\overline{P_E}=\Sigma C_n^2$. The fluctuation terms in the double sum of Eq. (23.33) average to zero because of the time dependence in the sine and cosine functions. Likewise, for any time, the average value of P_{out} over ensembles is also ΣC_n^2, i.e., $\langle P_{out}(t)\rangle=\overline{P_E}$. The ensemble average of the fluctuation terms in Eq. (23.33) is zero because of the phase dependence in the sine and cosine functions.

The mean-square deviation of P_{out} from its mean is given by the ensemble average $\langle[\Delta P(t)]^2\rangle$,

$$\langle[\Delta P(t)]^2\rangle=\langle P_{out}^2(t)\rangle-\overline{P_E}^2. \tag{23.36}$$

We next need to calculate $\langle P_{out}^2(t)\rangle$ by squaring Eq. (23.33) and averaging. There are two terms: The first is just $\overline{P_E}^2$, and the second is the square of the double sum. Because of the randomness of component phases, all the cross terms in the double sum average to zero. Therefore, we find that the mean-square fluctuation in envelope power is

$$\langle[\Delta P(t)]^2\rangle=\left\langle 4\sum_{n=1}^{N}\sum_{n'=1}^{n-1} C_n^2 C_{n'}^2 \frac{\cos^2(y_{nn'})+(\tau\Delta_{nn'})^2\sin^2(y_{nn'})}{[1-(\tau\Delta_{nn'})^2]^2}\right\rangle. \tag{23.37}$$

If the noise is white over the band, then the amplitudes C_n are statistically independent of component number n, and the ensemble average of C_n^2 is $\langle C_n^2\rangle = \overline{P_E}/N$. Because phases are uncorrelated with amplitudes, the ensemble averages of the squared sine and cosine functions can be done independently; both are equal to 1/2.

Therefore, Eq. (23.37) reduces to

$$\langle[\Delta P(t)]^2\rangle=2(\overline{P_E}/N)^2\left\langle \sum_{n=1}^{N}\sum_{n'=1}^{n-1} \frac{1}{1-(\tau\Delta_{nn'})^2}\right\rangle. \tag{23.38}$$

The ensemble average of the double sum depends only on the distribution of frequency differences, and we replace the average summation by an integral over a normalized frequency difference density $\rho(\Delta)$. From the finite series formulas in Appendix C, we find that there are $\Sigma_{n=1}^{N}(n-1)=N(N-1)/2$ terms in the double sum. Therefore, the mean-square fluctuation becomes

$$\langle[\Delta P(t)]^2\rangle=(\overline{P_E}/N)^2 N(N-1)\int_0^\infty d\Delta\, \frac{\rho(\Delta)}{1-(\tau\Delta)^2}. \tag{23.39}$$

The integral in Eq. (23.39) depends on the difference density. There are several possibilities. This envelope fluctuation model was originally developed to explain the results of an experiment in spectral density discrimination (Hartmann *et al.*, 1986). In that experiment, the frequency band (width W) was divided into N equal bins, and one component was placed randomly in each bin. Then the density of differences looks like the function in Fig. 23.6 that has a peak at W/N, which is the most probable frequency difference. If the spectral components are not binned and fall randomly within the band then the difference density is shown by the monotonically decreasing function in Fig. 23.6.

The case of binned components was treated in the 1986 article. In that case the number of components N enters the calculation in two ways, first as a factor in Eq. (23.39) and second as a parameter in the density ρ. This leads to some algebraic simplification. The calculation presented below is more general; it introduces parameter M such that the peak of ρ occurs at a frequency difference of W/M. If M equals N, the density corresponds to the binned-components case. If M is infinite, the density corresponds to the case of unbinned frequencies.

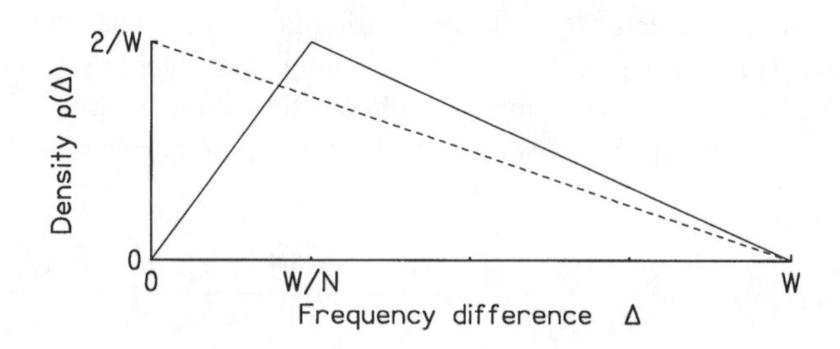

FIGURE 23.6. The solid line shows the density of frequency differences when N components in a band of width W are binned. The dashed line shows the density when there is no binning. Both densities have areas of unity.

Doing the integrals over the density with peak at W/M leads to a mean-square fluctuation of

$$\langle[\Delta P(t)]^2\rangle=\overline{P_E}^2\,\frac{(1-1/N)}{W\tau}\,\frac{2}{1-1/M}\left\{\tan^{-1}(W\tau)-\tan^{-1}\left(\frac{W\tau}{M}\right)\right.$$
$$\left.+\frac{1}{2W\tau}\left[M\ \ln\left[1+\left(\frac{W\tau}{M}\right)^2\right]-\ln[1+(W\tau)^2]\right]\right\}. \tag{23.40}$$

The mean-square fluctuation relative to the square of the envelope power is obtained by dividing,

$$\langle W_E-1\rangle=\frac{\langle[\Delta P(t)]^2\rangle}{\overline{P_E}^2}. \tag{23.41}$$

The quantity $\langle W_E-1\rangle$ has several interesting properties:

(1) Its value is never greater than 1. The RMS fluctuation in the envelope power is never greater than the envelope power itself.

(2) It depends upon the bandwidth W and the time constant τ only through their product, $W\tau$.

(3) In the limit that $W\tau$ is zero the relative mean-square fluctuation is $1-1/N$. The interpretations of the individual limits are different, however. In the limit that W equals zero the limit is simply wrong. The ensemble averaging assumptions made in the derivation are invalid for zero bandwidth. In the limit that τ is zero the limit is correct. For a given N, this limit gives the largest possible fluctuation because the TMTF filter does no smoothing in this limit. Therefore, $\langle W_E-1\rangle$ agrees with Eq. (23.29).

(4) Fluctuation $\langle W_E-1\rangle$ is zero if $N=1$. There is no envelope fluctuation for a single component.

(5) When the components are not binned (limit of infinite M) the relative mean-square fluctuation is given by

$$\langle W_E-1\rangle=\frac{2(1-1/N)}{W\tau}\left\{\tan^{-1}(W\tau)-\frac{1}{2W\tau}\ln[1+(W\tau)^2]\right\}. \quad (23.42)$$

(6) The relative mean-square fluctuation increases with increasing number of components N and with decreasing bandwidth W, but it is not a simple function of the spectral density N/W.

(7) When the product $W\tau$ becomes large there is a lot of smoothing and the fluctuation decreases as

$$\langle W_E-1\rangle\approx\pi\frac{1-1/N}{W\tau}. \quad (23.43)$$

Because W is 2π times the bandwidth in Hz, BW,

$$\langle W_E-1\rangle\approx\frac{1-1/N}{2BW\tau}. \quad (23.44)$$

This large-$W\tau$ behavior occurs for $W\tau$ in the neighborhood of 2. Therefore, if the time constant is 3 ms, the limit is in place for a bandwidth of 100 Hz or greater. If the bandwidth is less than 100 Hz, perhaps due to critical-band filtering, deviations from the large-bandwidth approximation may be seen.

(8) When the number of components becomes large, M must also become large. Therefore, the limit of Eq. (23.42) applies whether the components are binned or not.

This envelope fluctuation model proved successful in accounting for the ability of listeners to discriminate between noise bands with different spectral densities. At the limit of discrimination the listeners apparently used the size of the fluctuations to make their decisions. This model might also be used in a signal detection theory where d' is given by the ratio of the signal envelope power to the square root of the fluctuation from Eq. (23.40). Unlike energy detection theory, this model deals in power because the temporal integration is assumed to be done by the modulation transfer function.

EXERCISES

Exercise 1: Integrating noise

To find the power in a band, one integrates the spectral density.

(a) Consider a noise that is white over the audible range. Show that the power in the octave band from 100 to 200 Hz is ten times less than the power in the octave band from 1000 to 2000 Hz.

(b) Show that if the noise is pink then there is equal power in those two octave bands. [Note: That is why pink noise generators are used to set up audio equipment using octave band or one-third-octave band equalizers.]

Exercise 2: Spectral tilt of pink noise

(a) Use Eq. (23.9) for the density of pink noise to show that the spectrum level of pink noise decreases at a rate of -3 dB/octave, that is, the spectrum level at frequency $2\,f_1$ is 3 dB less than the spectrum level at a frequency of f_1. As luck would have it, a slope of -3 dB/octave is not a natural response for any simple filter. To convert white noise to pink requires a tricky filter network. Horowitz and Hill (1989) give an example on page 452.

(b) Compare the spectral tilt for pink noise with the low-frequency rise in the equal loudness contours in Fig. 4.1. At what level does pink noise best match the equal loudness contour between 100 and 400 Hz?

Exercise 3: The pink spectrum is established by a single point

Consider a noise that is pink over the audible range from 20 Hz to 20,000 Hz and is zero outside that range. Suppose that the spectrum level at 1000 Hz is 30 dB. What is the total power in the noise? [Ans: 9.2×10^{-6} watt.]

Exercise 4: Spectral tilt in hearing aids

Consider a hearing aid that passes signals in a frequency band extending from f_l to f_u. Suppose that the listener can adjust the spectral tilt of the signal by means of a filter that is described only by its power response, $(f/f_0)^m$. The constant f_0 defines the frequency where the filter gain is unity. The exponent m is positive for a high-pass filter and negative for a lowpass filter.

(a) Show that if the input is pink noise then the output power is given by

$$\overline{P}=\frac{P_0}{m}\left[\left(\frac{f_u}{f_0}\right)^m-\left(\frac{f_l}{f_0}\right)^m\right].$$

(b) Suppose that f_0 is the geometric mean of the upper and lower band frequencies, i.e. $f_0=\sqrt{(f_l f_u)}$. Show that it does not matter whether the filter is high pass or low pass, the output power is independent of the sign of m.

(c) Show also that the power increases as the absolute value of m increases. In a simulated hearing aid study Punch and Rakerd (1993) found that spectral tilt of either sign increased loudness.

More:
Additional problems on spectral density can be found in Chapter 3.

Exercise 5: Interaural coherence

To study binaural hearing an experimenter creates a left-ear stimulus, $x_L(t)$, consisting of a noise, $n(t)$, to which a sine tone is added, so that $x_L(t)=n(t)+s(t)$. The right-ear stimulus, $x_R(t)$, is the same except that the signal is added with a phase inversion so that $x_R(t)=n(t)-s(t)$. Presenting signals in this way leads to an advantage in detection called the binaural masking level difference, $N0S\pi$. Show that the normalized interaural coherence function is given by

$$a_0=(1-S/N)/(1+S/N),$$

where S is the power in the signal and N is the power in the noise.

This expression was derived by Durlach, Gabriel, Colburn, and Trahiotis (1986) and used by Koehnke, Colburn, and Durlach (1986) in a successful demonstration of the idea that binaural signal detection could be understood as a reduction in interaural coherence as the signal was added.
[Hint: Assume that the signal is not correlated with the noise. Then the integral of the product of the signal and noise is zero, and both x_L and x_R have the same power.]

Exercise 6: Difference densities

(a) Use the convolution theorem to show that if frequencies are uniformly distributed in a band then the difference density is given by the dashed line in Fig. 23.6. Then write a computer program to check that conclusion.

(b) Write a computer program to verify the solid line in Fig. 23.6 for the difference density when the components are binned.

Exercise 7: Envelope fluctuations for short time constant

Use series approximations for the inverse tangent and logarithm functions to verify that Eq. (23.40) goes to $\overline{P_E}^2(1-1/N)$ in the limit that $W\tau$ is small.

CHAPTER 24

Signal Detection Theory

Did ye not hear it?—No; 'twas but the wind.
LORD BYRON

It is almost impossible to overestimate the importance of the role that signal detection theory (TSD) has played in guiding the theory and practice of psychoacoustics in the latter half of the 20th century. Its influence is everywhere. The ideas of TSD may be broadly interpreted because the ''signal'' can be anything that one likes. It could be the presence of a tone, or it could be a difference between two tones, perhaps their frequency, or intensity, or spectral shape, or position in space. Therefore, TSD can be applied to discrimination as well as detection.

Signal detection theory asserts that decisions about perception are fundamentally probabilistic. Given the large variability that is typically observed in perceptual experiments, this approach seems inherently reasonable. The stochastic character of the decision process is modeled as noise, either as part of the input received by the decision-making system or as part of the operation of the system itself. Such a model fits well with the observed behavior of sensory neurons, exhibiting spontaneous random firings in the absence of external stimuli and a noisy response when a stimulus is present. Therefore, TSD can serve as an interface between psychophysical observations and neural models of perception.

Signal detection theory, as applied to psychophysics (Green and Swets, 1966), provides a framework that can unify the results of different experimental methods. The most important link is the parameter d' (d-prime), which is the ratio of the size of the response to the signal to the size of the standard deviation in the response to noise. To see how this works out in practice we consider a forced-choice task. The most popular is the two-alternative forced-choice task.

TWO-ALTERNATIVE FORCED-CHOICE

In a two-alternative forced-choice experiment (2AFC, or 2IFC, where I stands for ''interval'') a subject is presented with two observation intervals in succession. One of these intervals (randomly the first or the second) contains the signal; the other has no signal. Both intervals contain noise, either part of the stimulus or internal to the subject. After the two stimulus intervals the subject must say which one contained

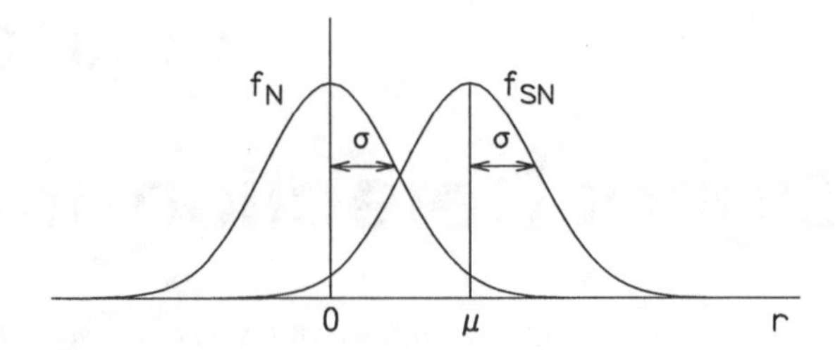

FIGURE 24.1. Probability density for excitation along an internal coordinate for two intervals, noise alone (*N*) and signal plus noise (*SN*). Both distributions have the same standard deviation σ.

the signal. If the experiment is done correctly, there is no reason to expect that one interval or the other will be favored. The experiment is symmetrical and independent of the subject's *a priori* inclination to claim that a signal is present or not present.

It is evident that scores on a 2AFC task should range from 50% correct to 100% correct. If the signal is much too weak to detect, the subject can only guess, and over the long term he or she will be correct half the time. At the other extreme, a strong signal will lead to perfect performance. The function that shows the dependence of the percentage of correct responses on the strength of the signal is the *psychometric function.*

Signal detection theory represents a subject's decision process by the construction shown in Fig. 24.1.

It is assumed that there is an internal coordinate, r, that the listener uses for a decision. Possibly this coordinate is neural firing rate, possibly it is a synchrony index, or it might be something else. What matters is that there is a distribution representing excitation on different trials, and the presence of the signal displaces the distribution along the internal coordinate by an amount μ. The direction of this displacement defines the positive direction for r. It is assumed that the only effect of the signal is to displace the distribution, and so the standard deviations for f_N and f_{SN} are the same, namely σ.

Given that the subject chooses the interval associated with the larger response, the probability of a correct response is the probability that a sample chosen from distribution f_{SN} is greater than the sample chosen from f_N. This can be rephrased in terms of a distribution of differences. If r_{SN} is a sample from f_{SN} and r_N is a sample from f_N, then a correct response occurs if $\Delta r = r_{SN} - r_N$ is greater than zero. As shown in Chapter 16, the distribution for this difference has a standard deviation that is $\sigma_D = \sigma\sqrt{2}$. The difference distribution is shown in Fig. 24.2.

The probability of a correct response, indicated by the shaded area, is the probability that Δr is greater than one. It can be calculated from the cumulative normal function F_G, where subscript "G" stands for Gaussian.

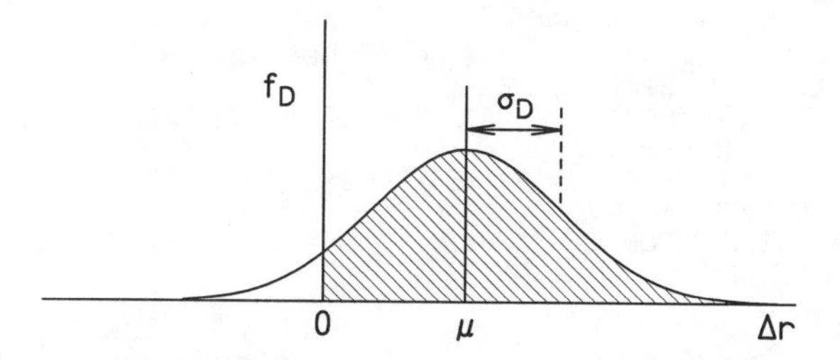

FIGURE 24.2. The difference distribution, with a standard deviation, σ_D larger than σ. The probability of a correct response is given by the shaded area.

$$P_c = F_G[\mu/(\sigma\sqrt{2})]. \tag{24.1}$$

Equation 24.1 says that the positive area of the Gaussian displaced by μ in Fig. 24.2 can be calculated as the area from minus infinity to μ of an undisplaced Gaussian.

By definition, μ/σ is d', and so

$$P_c = F_G(d'/\sqrt{2}). \tag{24.2}$$

This expression holds for d' both positive and negative. A negative value of d' occurs when the subject can detect the signal but is cOnFuSed about how to respond to it and preferentially chooses the non-signal interval. By the symmetry of the cumulative normal function,

$$P_c(d') + P_c(-d') = 1. \tag{24.3}$$

EXAMPLE 1: *Use the cumulative normal function to calculate percent correct on 2AFC as a function of d′*

Solution:

According to Eq. (24.2) we find the percent correct from the cumulative normal function,

$$P_c = CUMNOR(d'/\sqrt{2})$$

where the cumulative normal function is given by the following Fortran code:

```
      FUNCTION CUMNOR(Z)
C     CALCULATE THE CUMULATIVE NORMAL FOR Z.
C     IT IS IN UNITS OF S.D.
      DATA P,B1,B2,B3/0.33267,0.1740121,-0.0479399,0.3739278/
      T=1./(1.+P*ABS(Z))
      AREA = (B1*T+B2*T**2+B3*T**3)*EXP(-Z**2/2.)
      CUMNOR=AREA
      IF(Z.GT.0.)CUMNOR=1.-AREA
      RETURN
      END
```

The function $P_c(d')$ is shown in Fig. 24.3. Approximately half way between guessing ($P_c=0.5$) and certainty ($P_c=1$) is $P_c=0.76$, corresponding to $d'\approx 1$, shown by a dot.

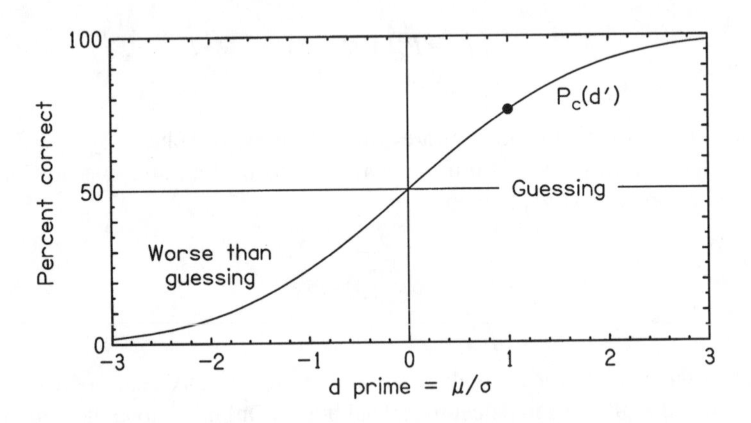

FIGURE 24.3. Percent correct on 2AFC as a function of d' calculated from the cumulative normal function.

Whereas the cumulative normal function calculates P_c given d', a more common problem is to find d' given P_c. This requires inverting Eq. (24.2) to get

$$d'(P_c)=\sqrt{2}F_G^{-1}(P_c). \tag{24.4}$$

The inverse cumulative normal function F_G^{-1} generates a *z-score*. Equation (24.4) says that d' on a 2AFC experiment is $\sqrt{2}$ times the z-score. Because F_G is monotonic, a simple method of converging bracketing can be used to compute its inverse. This technique was applied to the $(M+1)$-alternative forced-choice task to calculate Fig. 24.4.

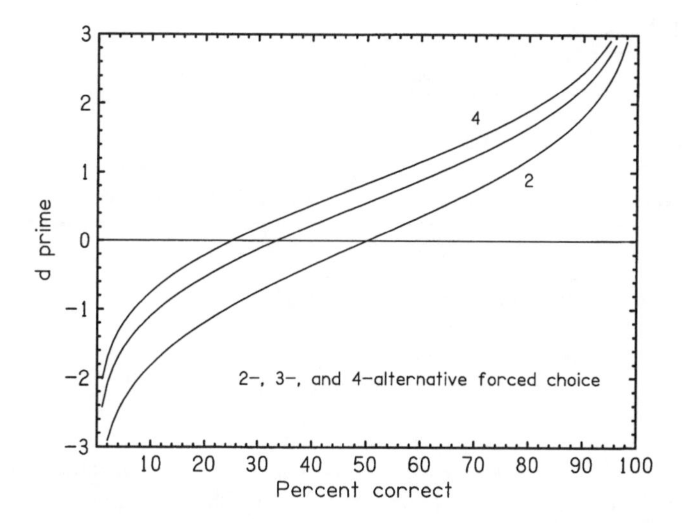

FIGURE 24.4. Sensitivity parameter d' as a function of the percentage of correct responses for (M+1)AFC where M is 1, 2, or 3.

(*M*+1)-ALTERNATIVE FORCED-CHOICE

In recent years psychoacousticians have discovered that the 2AFC task, especially as it is used in staircase methods, is less efficient than a task with three or four observation intervals (Schlauch and Rose, 1990). In the 3IFC (3AFC) task there are three intervals, one with a signal and two with noise alone. In 4AFC there are three intervals with noise alone and one with an added signal. We refer to such tasks as $(M+1)$AFC.

The probability of a correct response in $(M+1)$AFC is the probability that the excitation on the signal interval exceeds the excitation on all of the M non-signal intervals. The calculation goes like this: The probability that the excitation on the signal interval is within the ranger r to $r+dr$ is given by $f_{SN}(r)dr$. The probability that the value r exceeds the excitation on a non-signal interval is given by the cumulative normal, $F_G(r/\sigma)$, and the probability of exceeding the excitation on M non-signal intervals is this function raised to the power M. Therefore, the probability of a correct response is

$$P_C=\int_{-\infty}^{\infty} dr\ f_{SN}(r)[F_G(r/\sigma)]^M. \tag{24.5}$$

Because f_{SN} is a Gaussian displaced by the signal strength μ,

$$P_C=\frac{1}{\sigma\sqrt{2\pi}}\int_{-\infty}^{\infty} dr\ e^{[-(r-\mu)^2/2\sigma^2]}[F_G(r/\sigma)]^M. \tag{24.6}$$

TABLE 24.1. Yes-No Possibilities

Stimulus	Response	Result	Rate
Signal (SN)	Yes	Hit	P_H
Signal (SN)	No	Miss	P_M
No Signal (N)	Yes	False Alarm	P_{FA}
No Signal (N)	No	Correct Rejection	P_{CR}

Letting $z = r/\sigma$ we obtain a form in terms of d',

$$P_C = \frac{1}{\sqrt{2\pi}} \int_{-\infty}^{\infty} dz\ e^{[-(z-d')^2/2]} F_G^M(z). \tag{24.7}$$

Unlike 2AFC ($M=1$), where P_C could be calculated from a tabulated function, computing P_C for general M requires an integral. Therefore, inverting the expression to find the function $d' = d'(P_C)$ can be time consuming. Tables are given for many values of M by Hacker and Ratcliff (1979). A Fortran program that agrees with the tables is available from the author. A plot of $d'(P_C)$, calculated from that program, is given in Fig. 24.4 for 2AFC, 3AFC, and 4AFC.

YES-NO

The Yes-No experiment involves only a single observation interval. This interval may contain a signal (SN) or it may be noise alone (N). After the observation interval the subject must decide whether the signal was present (Yes) or not present (No). Unlike the 2AFC task, the Yes-No experiment involves the subject's criterion as well as the purely sensory character of the task. Therefore, if a subject gives a series of "Yes" responses it is not clear whether that is because the signal is much stronger than the noise or because the listener is eager to say, "Yes," or reluctant to say "No." The virtue of signal detection theory is that it enables the experimenter to distinguish between decisions based upon sensory characteristics and response proclivities.

The TSD analysis is as follows. Given a Yes-No observation interval, there are four possible outcomes, as given in Table 24.1.

The conditional probabilities in Table 24.1 are called rates, and they are not independent. Because the response to an SN trial must be either Yes or No, we have $P_H + P_M = 1$, and similarly $P_{FA} + P_{CR} = 1$. Therefore, two rates are enough to characterize the results of an experiment. Signal detection theorists typically choose to work with the hit rate P_H and the false alarm rate P_{FA}.

The subject's tendency to say "Yes" is modelled as a criterion. By setting a low criterion for a positive response, a subject can ensure that almost no signals are missed, and the hit rate is almost 100%. The penalty for this criterion is that the false alarm rate will also be high. Ideally, a subject should adopt a criterion that

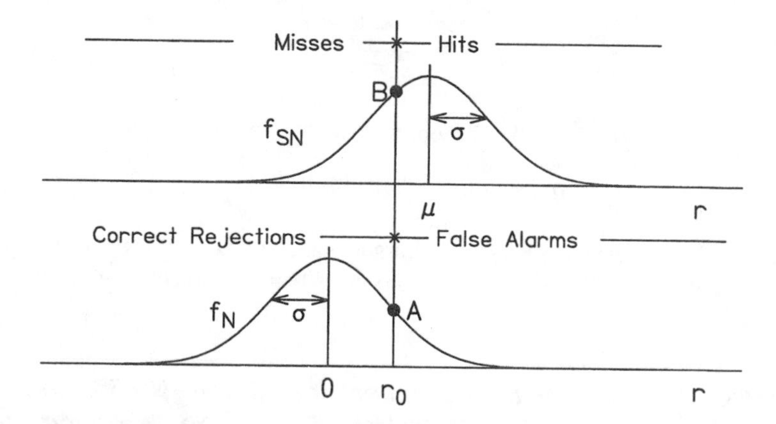

FIGURE 24.5. Distribution of excitation for noise-alone and noise-plus-signal intervals with the criterion set at r_0, establishing the boundary between saying No ($r<r_0$) and saying Yes ($r>r_0$). This boundary, together with the distributions, determines the boundaries between hits and misses and between correct rejections and false alarms.

maximizes the hit rate while minimizing the false alarm rate. The construction in Fig. 24.5 shows the Yes-No experiment with a sensible choice of criterion.

From the definition of the cumulative normal,

$$P_{FA}=F_G(-r_0/\sigma), \tag{24.8}$$

and

$$P_H=F_G[(\mu-r_0)/\sigma]. \tag{24.9}$$

Because $d'=\mu/\sigma$, we can eliminate the criterion, r_0. We take the inverses of Eqs. (24.8) and (24.9) and combine them to get μ/σ,

$$d'=F_G^{-1}(P_H)-F_G^{-1}(P_{FA}). \tag{24.10}$$

The two experimental variables P_H and P_{FA} can be used to calculate a second parameter of interest, a quantity known as the likelihood ratio, or beta (β), that describes the criterion chosen by the subject. By definition, β is the ratio of the ordinates of the f_{SN} and f_N distributions at the criterion point, i.e. it is the ratio B/A from Fig. 24.5. By this definition, β is given by

$$\beta = \frac{e^{-(\mu - r_0)^2/(2\sigma^2)}}{e^{-r_0^2/(2\sigma^2)}}. \tag{24.11}$$

From Eqs. (24.8) and (24.9) this becomes

$$\beta = e^{-1/2\{[F_G^{-1}(P_H)]^2 - [F_G^{-1}(P_{FA})]^2\}}. \tag{24.12}$$

Parameter β is 1 when $P_H + P_{FA} = 1$. Larger values of β occur when the subject readily responds Yes, smaller values of β occur when the subject is reluctant to say Yes.

EXAMPLE 2: *In a Yes-No experiment the hit rate is 85% and the false alarm rate is 45%. Calculate d′ and the likelihood ratio using Fig. 24.3.*

Solution:

From Eq. (24.10) for d' and Eq. (24.12) for β we realize that we need an inverse cumulative normal function. Figure 24.3 can give us that function except that that horizontal axis is multiplied by $\sqrt{2}$ to give d' for a different kind of experiment. We can use Fig. 24.3 if we divide by $\sqrt{2}$.

For $P_H = 0.85$ we estimate $F_N^{-1}(P_H) = 1.15/\sqrt{2} = 1.06$.

For $P_{FA} = 0.45$ we estimate $F_N^{-1}(P_{FA}) = -0.15/\sqrt{2} = -0.11$.

From Eq. (24.10) we find $d' = 1.06 - (-0.11) = 1.17$.

From Eq. (24.11) we find $\beta = \exp[-0.5(1.06)^2 + 0.05(-0.11)^2] = 0.57$.

These estimates can be compared with more exact answers $d' = 1.16$ and $\beta = 0.59$.

The Yes-No task does not have the artificial symmetry of the forced-choice tasks. It might even be said to mimic some of the anguish of decision-making in ordinary life. The surgeon, looking at an x-ray, needs to decide whether to operate or not to operate. She does not look at two x-rays and make a forced choice about which patient to open up. At least we hope not. For the most part, however, psychoacousticians are interested in the ability of the auditory system to detect signals and are not interested in the criteria that a listener may use in reporting whether detection is taking place or not. Therefore, there is little motivation to use the Yes-No method. Even though TSD makes it possible to extract a d' from Yes-No data, the Yes-No method is less used than forced-choice methods.

G. T. FECHNER (1801–1887)

Gustav Theodor Fechner is known as the father of psychophysics. He was born in Gross-Särchen, Germany, on April 19, 1801 and received a degree in medicine/biology in 1822 from the University of Leipzig. His interests then turned to physics, and he made such significant contributions to the infant science of electrical measurements that he was appointed professor of physics at Leipzig in 1834. Within a decade he had changed interests again, this time to philosophical inquiries that led him to pantheism and a metaphysics that asserted the unity between the material world and the world of the human mind. While pondering the possibility of a quantitative connection between mind and matter, in bed, on the morning of October 22, 1850, Fechner achieved the insight that sensation could be measured quantitatively. His main published work, *Elemente der Psychophysik* (1860), was intended to support his philosophy. Primarily, however, it is notable for the classical methods of psychophysics introduced there. Fechner died in Leipzig on November 18, 1887. October 22 is known as Fechner Day and is celebrated each year by psychophysicists around the world.

TRIALS AND PROCEDURES

The tasks described above, 2AFC, 3AFC, Yes-No, etc., define the nature of a single trial in an experiment. In a trial there is a sequence of observation intervals that contain the stimulus; there is a response and an outcome. The outcome contributes a datum to the results of the experiment. On a forced-choice task, the nature of a single trial determines the details of the psychometric function that we expect to get. For instance, if the trial-type is 2AFC then we expect to get 76% correct responses when the d' is 1. This figure drops to 63% if the task is 3AFC.

Trials are concatenated to make an experimental run, typically lasting about five minutes. There are different rules that can be used to construct a run. If the stimulus parameters depend upon the outcome of previous trials, the procedure may be said to be "adaptive." Otherwise, the procedure is known as a "method of constant stimuli." The procedure that has been the most popular in recent years is the up-down staircase, which is described next.

STAIRCASE METHODS

The up-down staircase method is a method of limits wherein the magnitude of the stimulus is varied automatically in a way that leads to an efficient estimation of the threshold for detection or discrimination. The staircase method works for forced-choice tasks, where the outcome of a trial has a binary character; the judgement is either correct (C) or incorrect (I).

The staircase method assumes that the subject is more likely to make correct responses when the stimulus magnitude is larger, i.e. that the psychometric function is monotonically increasing with stimulus magnitude. Therefore, the method increases the magnitude of the stimulus (staircase going up) at times when the listener is making incorrect responses and decreases the magnitude (staircase going down) when the listener is making correct responses. Stimulus magnitudes where the staircase changes direction are called "turnarounds." The result of this automatic stimulus variation is to keep the stimulus in a range where the subject is uncertain. Typical values of the staircase visited frequently during the run then give an estimate of the threshold magnitude for detection or discrimination. The most commonly used estimate is the so-called "Wetherill estimate," (Wetherill and Levitt, 1965) which is obtained by running the staircase for an even number of turnarounds (equal number of tops and bottoms) and then taking the average of all turnaround magnitudes as the threshold.

The threshold magnitude is related to the psychometric function for individual trials in a way that depends upon the staircase rules. The key to the relationship is the percentage of correct responses on individual trials that corresponds to the equilibrium point for the staircase. The staircase reaches equilibrium at a stimulus magnitude that causes the staircase to go up or down with equal probability. Equivalently, the probability that the staircase will go down is set equal to 0.5.

The simplest staircase rule is based upon a single trial: the stimulus magnitude increases after each incorrect response and decreases after each correct response. The probability that the staircase will go down is the probability of a correct response on any single trial. Therefore, the staircase settles at a magnitude where the psychometric function is 50% correct. It should be evident that this staircase is useless if the trial structure is 2AFC, where 50% on the psychometric function corresponds to guessing.

The most commonly used staircase rule is "1-up 2-down," which increases the magnitude after each incorrect response and decreases the magnitude after two successive correct responses. The probability that the staircase will go down after a trial is equal to the probability that the subject gets the trial correct and also got the previous trial correct. Because trials are assumed to be independent, the probability of both responses being correct is the square of the probability that one is correct, and the probability that the staircase will go down is P_c^2. Setting this probability equal to 1/2 means that $P_c=\sqrt{1/2}$ or 0.707. Therefore, the "1-up 2-down" staircase equilibrium occurs at a stimulus magnitude corresponding approximately to the 71% correct point on the psychometric function. Seventy-one percent correct is an attractive equilibrium point if the trial structure is 2AFC. It is approximately half way between 50% (guessing) and 100% (perfect) and corresponds to $d'=0.77$.

Another attractive staircase is "1-up 3-down," where the subject must make three consecutive correct judgements in order to cause the staircase magnitude to go down. Any incorrect judgment, at any time, causes the staircase to go up. This staircase settles at the magnitude where $P_c^3=1/2$, or $P_c \approx 79\%$. If the trial structure is 2AFC, this corresponds to $d'=1.16$, which is conveniently close to unity. Table 24.2 summarizes staircase rules of the form "1-up n-down." The table concludes with one example of a different kind of rule, where an incorrect response is ignored

TABLE 24.2. Staircase Rules—Sequences and Statistics

Rule	Sequences that cause staircase to go up	Sequences that cause staircase to go down	Probability of going down	Percent correct at the equilibrium point
1-up 1-down	I	C	P_C	P_C=0.5
1-up 2-down	I, CI	CC	P_C^2	P_C=0.7071
1-up 3-down	I, CI, CCI	CCC	P_C^3	P_C=0.7937
1-up 4-down	I, CI, CCI, CCCI	CCCC	P_C^4	P_C=0.8409
1-up 3-down with reprieve	I, CI, CCI, CCII	CCC, CCIC	$P_C^3(2-P_C)$	P_C=0.7336

if it is preceded by two correct responses. For this rule, there are two conditions that cause the staircase to go down. Therefore, the probability that the staircase goes down is the sum, $P_c^3 + P_c^3(1-P_c) = P_c^3(2-P_c)$. Other staircases are given by Levitt (1971); they are not much used in practice, however.

The increment by which the magnitude of the stimulus increases or decreases as the staircase goes up or down is usually kept constant. In a signal detection task where the varying magnitude is the level of the signal, it is common for the increment to be 1 or 2 dB. Generally, at the start of an experiment run, the stimulus magnitude is set high enough that early responses are all correct. Therefore, the first turnaround is a staircase minimum, occurring at an incorrect response. In order to reach useful values of the stimulus more quickly, the increment may be larger (perhaps twice its normal size) prior to the first turnaround. Alternatively, the rule may be changed so that each correct response reduces the stimulus magnitude prior to the first turnaround.

EXAMPLE 3: *Staircase experiment*

Figure 24.6 shows an example of a "1-up 2-down" staircase used to find the absolute threshold for a sine tone. The staircase starts at 20 dB because the experimenter is rather sure that the listener can hear this level without difficulty. To reach interesting levels quickly, each correct response results in a decrease of 2 dB until the first incorrect response. Thereafter, the "1-up 2-down" rule applies, with an increment of ±2 dB. The run proceeds until 18 turnarounds have occurred. This corresponds to 75 trials, but the number of trials might have been more or less depending upon the tightness of the staircase, determined largely by luck. Table 24.3 shows a typical analysis of the results. The first four turnarounds (at trial numbers 8, 10, 13, and 17) are discarded and the remaining 14 are averaged to find an estimate of the threshold. The table shows separate averages for staircase bottoms and tops, but the final answer is simply their average, 4.57 dB.

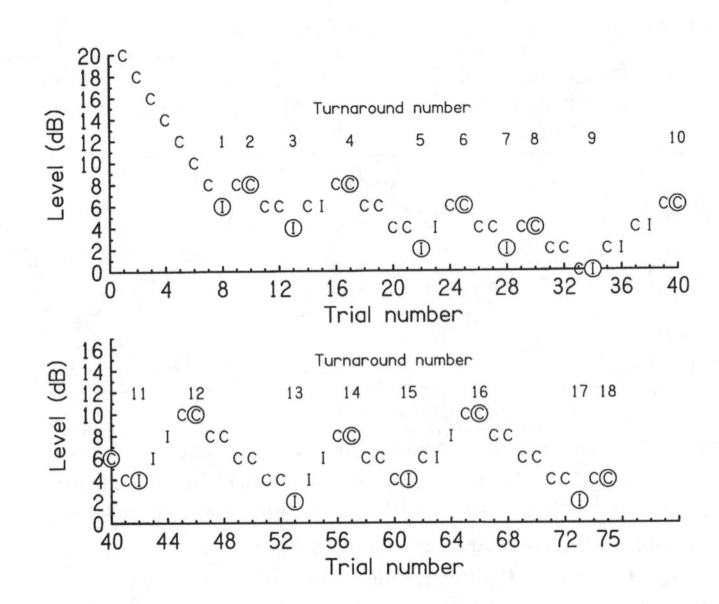

FIGURE 24.6. The operation of a staircase. The level begins high (20 dB) and decreases on each correct response until the first turnaround at the 8th trial. Thereafter, a "1-up 2-down" staircase rule applies. Trials continue until 18 turnarounds have occurred. Turnarounds are circled.

TABLE 24.3. Analysis of the staircase run

	Bottom L (dB)	Top L (dB)	
	6	8	
	4	8	
	2	6	
	2	4	
	0	6	
	4	10	
	2	8	
	4	10	
	2	4	
Average at 7,7	2.29	6.84	Overall=4.57

By knowing only the sequence of turnaround levels, as in Table 24.3, and knowing the size of the increment one can make a reasonable sketch of the entire staircase history, but one cannot reproduce it completely. Figure 24.6 illustrates the problem. From trials 42 through 46 the staircase rises from a bottom at 4 dB to a top at 10 dB. From trials 61 through 66 the staircase has the same bottom and top values. However, there is a correct response at trial 62 that cannot be predicted by knowing just the turnaround levels. By knowing both the turnaround levels and their

TABLE 24.4. Block rules

Block outcome		
First trial	Second trial	Effect
C	C	decrease level by ΔL
C	I	increase level by ΔL
I	C	increase level by ΔL
I	I	increase level by ΔL

trial numbers one has more information, but still not enough to characterize the staircase completely. For example, there is no way that one would know that a correct response occurred on trial 62 (rising sequence C,I,I) and not on 63 (rising sequence I,C,I).

Complete data from a staircase can be used to find the percentage of correct responses for each level visited by the staircase. In principle, these data give an estimate of a portion of the underlying psychometric function. The only problem is that the choice of levels is not systematic but follows the vagaries of the staircase. A study of the distribution of staircase values was given by de Boer and van Breugel (1984). A maximum-likelihood method for estimating threshold from staircase values was given in Appendix C of Versfeld's thesis (1992).

Staircase Equilibrium Points Reconsidered

The sequences in Table 24.2 describe the staircase rules, and these are adequate to design a program to run a staircase experiment. However, it is not evident that they provide a way to calculate the percentage of correct responses at the equilibrium point. The problem is that the staircase rules are not a consistent basis for calculating probability because the outcomes cannot be numbered. To explain this statement, the analogy of rolling dice is helpful.

If we roll a single die, the probability of getting a "3" is 1/6 because there are six equally likely outcomes and only one of them is a "3." If we roll two dice, the probability of getting a "3" is 1/18 because there are 36 possible outcomes and two of them give "3." Obviously the calculation of probabilities depends upon the number of dice, which determines the number of possible outcomes. To return to the staircase method, consider the 1-up 2-down staircase, where a single incorrect response is enough to send the level up, whereas it takes two correct responses to send the level down. Calculating probabilities for this staircase resembles a game in which we haphazardly roll one or two dice and do not know how to count the number of outcomes.

The solution to this problem is that the staircase is equivalent to a different kind of experiment, with different rules, for which the number of possible outcomes *is* known. Consider an experiment consisting of blocks with two trials per block. The two trials of a block are always done at the same level. The outcomes and effects of a block are given in Table 24.4.

The experiment has a countable structure with four possible outcomes for each block, but only two effects. After a block, the staircase level goes up or down to set the level for the next block. There is no outcome that leaves the level the same. These block rules replace the staircase rules.

We next observe that this experiment is not as efficient as it could be. If the first trial leads to an incorrect response, the second trial is of no consequence. The staircase level will increase by ΔL regardless of the outcome of the second trial. Therefore, in this case the second trial can be omitted. But if we omit the second trial we recover the original staircase experiment with the original staircase rules. We are back where we started.

The benefit of this exercise is that we now know how to calculate the probabilities for the staircase. Using P_c for the probability of a correct trial, the probability that the level goes down after a block is P_c^2. The probability that the level goes up after a block is the sum of the probabilities for the outcomes on the last three lines of Table 24.4, namely

$$P_{up}=P_c(1-P_c)+(1-P_c)P_c+(1-P_c)(1-P_c). \tag{24.13}$$

Summing the last two terms and repeating the first, we have

$$P_{up}=P_c(1-P_c)+(1-P_c). \tag{24.14}$$

The right-hand side is now seen to be the sum of the two probabilities given by the staircase rule: go up if the response sequence is CI or I. Because this rule has been derived from the block with a countable structure, we know that in the end the probabilities of going up and of going down must sum to one. Therefore, the two terms in Eq. (24.14) must sum to $1-P_c^2$, which they do.

The final step in calculating the percent correct at the staircase equilibrium is to define the equilibrium point by that value of P_c where the probability of going up after a block is equal to the probability of going down. Equivalently, we set one of these probabilities equal to 1/2. It is then a simple matter to solve for P_c at equilibrium.

The purpose of this rather lengthy reconsideration is to show how the probabilities for the staircases in Table 24.2, or any other staircase, can be calculated. The key is to find an equivalent block structure with rules that either send the level up or send the level down.

Average Values of Staircases

Assuming that the psychometric function is monotonic, we expect that the average value of the levels visited by the staircase will be close to the equilibrium value. For example, we expect that the average percent correct for 1-up 2-down will be close to 0.707. In fact, however, the average value of a staircase depends upon

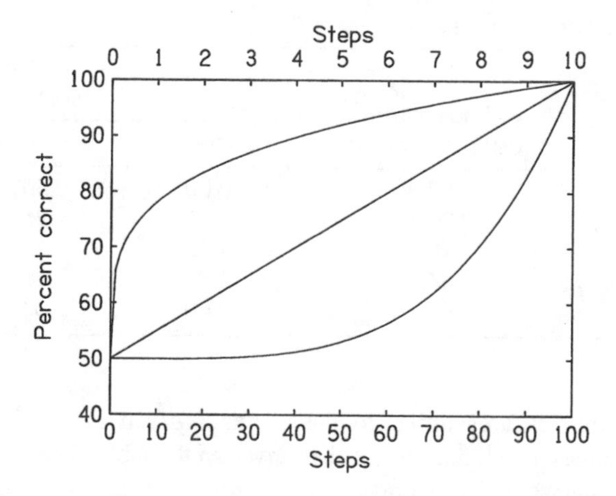

FIGURE 24.7. Three psychometric functions described by a linear law, a square law, and a square-root law.

the shape of the psychometric function and upon the step size. By contrast, the equilibrium depends only upon the psychometric function at a single point.

The dependence of the average value on psychometric function shape and step size can be seen in numerical experiments that simulate the staircase operation. Consider Fig. 24.7 for an experiment with two-alternative forced-choice trials. There are three psychometric functions, characterized by power laws with exponents 2 (square law), 1 (linear), and 1/2 (square-root law). The horizontal axis shows the stimulus magnitude in arbitrary units called "steps" from 1 to 100. Simulations of the 100-step 1-up 2-down staircase, running 10 experiments of 10,000 trials, show that for the linear law one can expect an average value between 0.706 and 0.712. For the square-law psychometric function one can expect an average value greater than 0.707, but for the square-root psychometric function the average is usually less than 0.707. These data are shown in Table 24.5, which also gives the results for power laws with exponents 4 and 1/4. The table further shows that the effect of reducing the number of staircase steps from 100 to 10 is to magnify all the trends observed for 100 steps.

TABLE 24.5. Average 2AFC staircase values for power-law psychometric functions

Power	100 Steps	10 Steps
1/4	0.701–0.709	0.667–0.677
1/2	0.701–0.709	0.686–0.694
1	0.706–0.712	0.706–0.712
2	0.710–0.718	0.737–0.743
4	0.737–0.741	0.796–0.800

TABLE 24.6. Equilibrium values of d' for different staircases

Staircase rule	Equilibrium percent correct	d' Trial task		
		2AFC	3AFC	4AFC
1-up 1-down	50.0	0.00	0.56	0.84
1-up 2-down	70.7	0.77	1.26	1.52
1-up 3-down	79.4	1.16	1.62	1.87
1-up 4-down	84.1	1.41	1.86	2.09
1-up 5-down	87.1	1.60	2.03	2.26

Wetherill estimates are similar to the averages given above. When they are different, the ranges are slightly lower than given in the table. The main message of Table 24.5 is that staircase averages do not differ much from their theoretical equilibrium points, even if the psychometric functions are badly curved.

EFFICIENCY

The staircase method is an efficient means of determining a threshold because it concentrates the trials at stimulus values near equilibrium. Little time is spent at values far from the equilibrium. By choosing the task for a trial and the rule for the staircase, the experimenter can set the equilibrium value at will. Table 24.6 shows some possible options.

The traditional practice has been to set the equilibrium point near the threshold, as it would be determined in a method of adjustment or Yes-No procedure, approximately $d'=1$. Therefore, a 2AFC task in a 1-up 2-down staircase ($d'=0.77$) has been the standard. However, it has been found that it is more efficient to set the equilibrium at a considerably higher value of d'. Greater efficiency means that for a given number of trials, the standard error of the estimate of a point on the psychometric function is smaller. For example, Kollmeier, Gilkey, and Sieben (1988) found that the 3AFC task in the 1-up 3-down staircase was optimum.

Recent methodological studies have made several points. First, for many psychoacoustical investigations the thresholds *per se* are not particularly important. These investigations vary the way in which the stimulus is presented and find the stimulus level that is required for the detection performance to meet a given standard. Whether the standard is low (e.g. $d'=1$) or high, the same information is learned about the effect of stimulus context. Because a higher standard of performance is found to lead to a smaller variability, the higher standard is preferred. Second, Green (1990) has argued that even if the goal is to find a threshold value, it is more efficient to concentrate stimuli at levels where performance is high and to estimate threshold from the psychometric function. Further, a staircase with equilibrium at larger d' is less affected by the shape of the psychometric function than the staircase based on 2AFC and a 1-up 2-down studied in Table 24.5.

At this time the future direction of psychoacoustical methods is uncertain. Many experimenters will no doubt continue to use a 2AFC task and 1-up 2-down staircase. This technique has been used so frequently that everyone understands it, and results can be compared across studies. Others will be seduced by the promise of greater efficiency and explore techniques that target a larger value of d'. There is likely to be a little more anarchy than normal for awhile.

EXERCISES

YES-NO

Exercise 1: Indecision—same as noise

The percentage of hits and false alarms in a Yes-No experiment will vary if the criterion used by the subject varies during the course of an experiment. Show that if the criterion varies with a normal probability density then the criterion variation appears merely as additional noise, i.e. simply as a larger standard deviation in the equations for P_H and P_{FA}.

Exercise 2: Likelihood ratio of one

Show that if $\beta=1$ then misses and correct rejections are related by $P_M+P_{CR}=1$.

FORCED CHOICE

Exercise 3: Psychometric function

Use Fig. 24.4 to sketch a psychometric function for 3AFC, where percent correct is on the vertical axis and stimulus level is on the horizontal axis.

Exercise 4: A perfect differential

Show that if $d'=0$ then Eq. (24.5) becomes

$$P_c=\int_{-\infty}^{\infty}(dF_G)F_G^M=\frac{1}{M+1}.$$

Explain why this is the "guessing limit" and compare with Fig. 24.4 where possible.

STAIRCASES

Exercise 5: Going up or down

Table 24.2 gives the probability that a staircase goes down. For each staircase in the table, calculate the probability of going up.

Exercise 6: 1-up 3-down

Consider the 1-up 3-down staircase. Write the outcome table, for the equivalent block experiment, with nine possible outcomes. Show that the sum of the probabilities for going up in level is $1-P_c^3$.

Exercise 7: Unhelpful suggestions

Here are two suggestions for modifying the 1-up 3-down staircase to make the requirements for going down less stringent:

(a) Go down if CCC or CCI; otherwise go up.

(b) Go down if CCC or CIC; otherwise go up.

Show that these two suggestions are equivalent to 1-up 2-down.

Appendices

A. Greek Alphabet
B. Trigonometric Functions
C. Series
   infinite—reciprocal, exp, log, trig
   binomial
   geometric—complex geometric
   special finite series
   special infinite series
   series exercises
D. Integrals—Even and Odd Functions
E. Integrals in the Complex Plane
   Impulse response and the residue
   Hilbert transform integrals
F. Hilbert Transforms
   Components of the analytic signal
   Dispersion relations
   Minimum-phase filters, gain and phase shift
G. Electrical Filters
   Filter exercises
H. The Normal Distribution
   The error function
   The cumulative normal
   The central limit theorem
I. The Rayleigh Distribution
   Generating random numbers having Gaussian and Rayleigh distributions
J. Standards
   ISO frequencies,
   Mathematical constants
   units of pressure and hearing threshold,
   physical constants
K. Calculation of Intermodulation Distortion

Appendix A Greek Alphabet

A	α	Alpha
B	β	Beta
Γ	γ	Gamma
Δ	δ	Delta
E	ϵ	Epsilon
Z	ζ	Zeta
H	η	Eta
Θ	θ	Theta
I	ι	Iota
K	κ	Kappa
Λ	λ	Lambda
M	μ	Mu
N	ν	Nu
Ξ	ξ	Xi
O	o	Omicron
Π	π	Pi
P	ρ	Rho
Σ	σ	Sigma
T	τ	Tau
Υ	υ	Upsilon
Φ	ϕ	Phi
X	χ	Chi
Ψ	ψ	Psi
Ω	ω	Omega

Appendix B
Trigonometric Functions

DEFINITIONS

The trigonometric functions are defined in terms of a right triangle, shown in Fig. B.1.

By definition, the hypotenuse (with length hyp) is the side opposite to the right angle ($\pi/2$ radians or 90 degrees) and it is the longest side. The angle of interest (θ) determines the definition of the "opposite side" (op) and the "adjacent side" (adj). The basic trigonometric functions are the ratios of the lengths of these sides. Because they are ratios, the trigonometric functions depend only upon angle θ; they do not depend upon the size of the triangle:

$$\sin\,\theta = (\text{op/hyp}) \quad \text{The sine function.} \tag{B.1}$$

$$\cos\,\theta = (\text{adj/hyp}) \quad \text{The cosine function.} \tag{B.2}$$

$$\tan\,\theta = (\text{op/adj}) \quad \text{The tangent function.} \tag{B.3}$$

Using Fig. (B.1) and the definitions of the sine, cosine, and tangent, we see that the sine and cosine functions can only have values from $+1$ to -1. The tangent function, however, can have any value between plus and minus infinity.

The definitions immediately relate the tangent function to the sine and cosine functions:

$$\tan\,\theta = \frac{\sin\,\theta}{\cos\,\theta} \quad \text{for all } \theta. \tag{B.4}$$

The Pythagorean relation for a right triangle relates the lengths of the sides:

$$\text{adj}^2 + \text{op}^2 = \text{hyp}^2. \tag{B.5a}$$

Divided by the square of the hypotenuse, this equation becomes

$$\cos^2\,\theta + \sin^2\,\theta = 1 \quad \text{for all } \theta. \tag{B.5b}$$

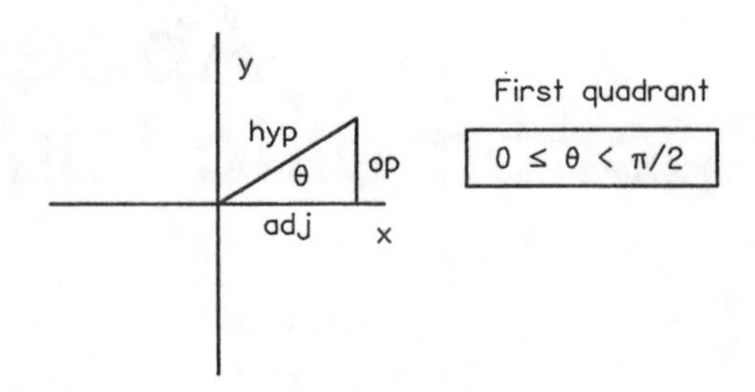

FIGURE B.1. An angle in the first quadrant where the opposite side and the adjacent side are both positive.

Figure B.1 shows angle θ in the first quadrant of the x-y plane. The adjacent and opposite sides are both positive numbers in this quadrant. Further, the hypotenuse is always a positive number. Therefore, both sine and cosine functions are positive in

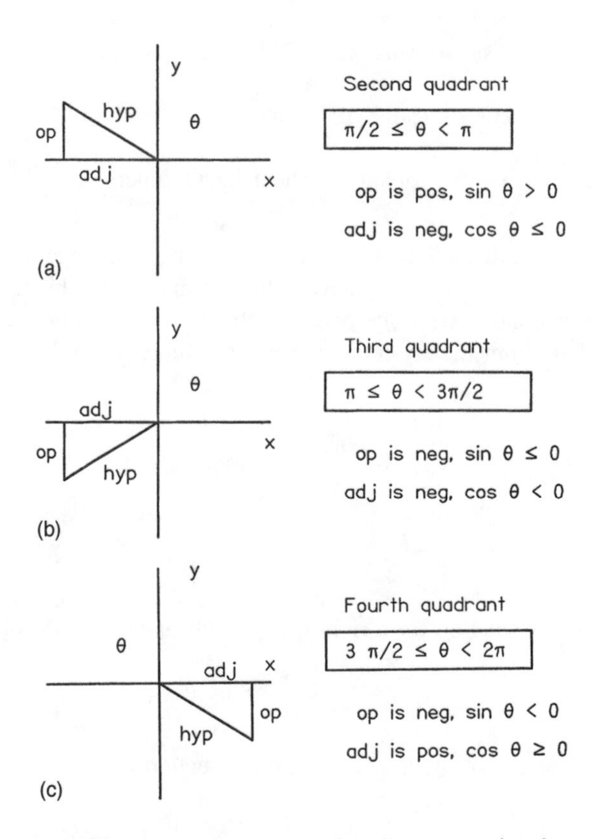

FIGURE B.2. Angles in the other three quadrants.

this quadrant. Figure B.2 shows angle θ in other quadrants and gives the signs for sine and cosine functions.

Relationships between trigonometric functions have frequent application in signal processing mathematics and acoustics. The most useful relationships are the angle sum and difference formulas:

ANGLE SUM AND DIFFERENCE FORMULAS

$$\sin(A \pm B) = \sin A \cos B \pm \sin B \cos A \tag{B.6}$$

$$\cos(A \pm B) = \cos A \cos B \mp \sin A \sin B. \tag{B.7}$$

Note that by the use of $\pm$ signs and $\mp$ signs each equation above becomes two equations. In reading an equation one either uses all the top signs or all the bottom signs. For instance, using the top signs in Eq. (B.7) we have

$$\cos(A + B) = \cos A \cos B - \sin A \sin B. \tag{B.8}$$

A third sum and difference formula is for the tangent function:

$$\tan(A \pm B) = \frac{\tan A \pm \tan B}{1 \mp \tan A \tan B}. \tag{B.9}$$

EXAMPLE: *Find the sine and cosine of 90 degrees, given that the sine and cosine of 45 degrees both equal $1/\sqrt{2}$.*

Solution: From Eq. (B.6), we have

$$\sin(45 + 45) = \sin 45 \cos 45 + \sin 45 \cos 45$$

or

$$\sin(90) = \frac{1}{\sqrt{2}} \times \frac{1}{\sqrt{2}} + \frac{1}{\sqrt{2}} \times \frac{1}{\sqrt{2}} = 1.$$

From Eq. (B.7), we have

$$\cos(45+45) = \cos 45 \cos 45 - \sin 45 \sin 45$$

or

$$\cos(90) = \frac{1}{\sqrt{2}} \times \frac{1}{\sqrt{2}} - \frac{1}{\sqrt{2}} \times \frac{1}{\sqrt{2}} = 0.$$

Double-Angle Formulas

$$\sin 2A = 2 \sin A \cos A \tag{B.10}$$

$$\cos 2A = \cos^2 A - \sin^2 A = 2 \cos^2 A - 1 = 1 - 2 \sin^2 A \tag{B.11}$$

$$\tan 2A = \frac{2 \tan A}{1 - \tan^2 A} \tag{B.12}$$

EXAMPLE: *An angle of 135 degrees is equal to 3π/4 radians. Show that the tangent of 3π/4 is −1 by first calculating the tangent of 3π/8 and then using the double angle formula.*

Solution: We use a calculator and put it in ''radian mode.'' We find $\tan(3\pi/8) = 2.4142$. Then from Eq. (B.12)

$$\tan(3\pi/4) = \frac{2 \tan(3\pi/8)}{1 - \tan^2(3\pi/8)}$$

or

$$\tan(3\pi/4) = \frac{2 \times 2.4142}{1 - 5.828} = -1.$$

Half-Angle Formulas

$$\sin \frac{A}{2} = \pm \sqrt{\frac{1 - \cos A}{2}} \tag{B.13}$$

$$\cos \frac{A}{2} = \pm \sqrt{\frac{1 + \cos A}{2}} \tag{B.14}$$

$$\tan \frac{A}{2} = \frac{\sin A}{1 + \cos A} = \frac{1 + \cos A}{\sin A} \tag{B.15}$$

Squared-Function Formulas

$$\sin^2 A = \frac{1}{2}(1 - \cos 2A) \tag{B.16}$$

$$\cos^2 A = \frac{1}{2}(1 + \cos 2A) \tag{B.17}$$

$$\tan^2 A = \frac{1 - \cos 2A}{1 + \cos 2A} \tag{B.18}$$

Function Product Formulas

$$\sin A \sin B = \frac{1}{2}\cos(A - B) - \frac{1}{2}\cos(A + B) \tag{B.19}$$

$$\cos A \cos B = \frac{1}{2}\cos(A - B) + \frac{1}{2}\cos(A + B) \tag{B.20}$$

$$\sin A \cos B = \frac{1}{2}\sin(A + B) + \frac{1}{2}\sin(A - B) \tag{B.21}$$

Function Sum and Difference Formulas

$$\sin A + \sin B = 2\sin\frac{1}{2}(A + B)\cos\frac{1}{2}(A - B) \tag{B.22}$$

$$\sin A - \sin B = 2\cos\frac{1}{2}(A + B)\sin\frac{1}{2}(A - B) \tag{B.23}$$

$$\cos A + \cos B = 2\cos\frac{1}{2}(A + B)\cos\frac{1}{2}(A - B) \tag{B.24}$$

$$\cos A - \cos B = -2\sin\frac{1}{2}(A + B)\sin\frac{1}{2}(A - B) \tag{B.25}$$

Functions of Angles Increased/Decreased by π (180 Degrees)

$$\sin(A \pm \pi) = -\sin A \tag{B.26}$$

$$\cos(A \pm \pi) = -\cos A \tag{B.27}$$

Functions of Angles Increased/Decreased by $\pi/2$ (90 Degrees)

$$\sin\left(A \pm \frac{\pi}{2}\right) = \pm \cos A \tag{B.28}$$

$$\cos\left(A \pm \frac{\pi}{2}\right) = \mp \sin A \tag{B.29}$$

$$\tan\left(A \pm \frac{\pi}{2}\right) = -1/\tan A = -\cot A \tag{B.30}$$

where the reciprocal of the tangent function is the cotangent function, cot.

There are other reciprocal functions, the secant and cosecant, respectively:

$$\sec\,\theta = 1/\cos\,\theta \tag{B.31}$$

$$\csc\,\theta = 1/\sin\,\theta. \tag{B.32}$$

Because these functions are so simply related to sine and cosine functions there is little need for them and they are not used in this book.

EULER'S FORMULA AND EQUIVALENTS

Euler's formula expresses the sine and cosine function in terms of e, the natural logarithm base. Here i represents the square root of minus 1.

$$e^{i\theta} = \cos\,\theta + i\,\sin\,\theta \tag{B.33}$$

It follows that

$$\cos\,\theta = \frac{e^{i\theta} + e^{-i\theta}}{2} \tag{B.34}$$

and

$$\sin\,\theta = \frac{e^{i\theta} - e^{-i\theta}}{2i}. \tag{B.35}$$

Appendix C
Series

INFINITE SERIES

There are two ways to think about the infinite series formulas below. First, a series formula provides an approximation. If the value of the argument of the function (here called x) is small, then a finite number of terms in the series approximates the function of x. Second, if an infinite number of terms is kept then the series is exactly equal to the function. Writing the function in terms of the series can sometimes give useful insights into the function itself.

Reciprocal

$$\frac{1}{(1-x)}=1+x+x^2+x^3+x^4+\ldots(\text{for } x^2<1) \tag{C.1}$$

Exponential and Logarithmic

$$e^x=1+x+\frac{x^2}{2!}+\frac{x^3}{3!}+\frac{x^4}{4!}+\ldots \tag{C.2}$$

$$\ln(1-x)=-x-\frac{x^2}{2}-\frac{x^3}{3}-\frac{x^4}{4}-\ldots(\text{for } x^2<1) \tag{C.3}$$

Trigonometric

$$\sin(x)=x-\frac{x^3}{3!}+\frac{x^5}{5!}-\frac{x^7}{7!}+\ldots \tag{C.4}$$

$$\cos(x)=1-\frac{x^2}{2!}+\frac{x^4}{4!}-\frac{x^6}{6!}+\ldots \tag{C.5}$$

In the above equations, $n!$ means ''n factorial,'' that is, $n!=1\cdot2\cdot3\cdot\ldots(n-1)\cdot n$, and ''ln'' stands for the natural (base e) logarithm.

Square Root

$$(1+x)^{1/2}=1+\frac{1}{2}x+\frac{1\cdot 1}{2\cdot 4}x^2+\frac{1\cdot 1\cdot 3}{2\cdot 4\cdot 6}x^3+\frac{1\cdot 1\cdot 3\cdot 5}{2\cdot 4\cdot 6\cdot 8}x^4+\ldots \quad \text{(C.6)}$$

BINOMIAL SERIES

The binomial series can be written in a form similar to an infinite series:

$$(1+x)^n=1+nx+\frac{n(n-1)}{2!}x^2+\frac{n(n-1)(n-2)}{3!}x^3+\ldots(\text{for } x^2<1). \quad \text{(C.7)}$$

However, unlike the infinite series expansions for functions in Eqs. (C.1)–(C.6), the binomial series, with integral n, has a finite number of terms. That can be shown by rewriting Eq. (C.7) including both ends:

$$(1+x)^n=1+nx+\frac{n(n-1)}{2!}x^2+\frac{n(n-1)(n-2)}{3!}x^3+\ldots+\frac{n(n-1)}{2!}x^{n-2}+nx^{n-1}+x^n \quad (\text{for all } x). \quad \text{(C.8a)}$$

The symmetry can be emphasized by replacing the number 1 by a variable y. Then

$$(y+x)^n=y^n+ny^{n-1}x+\frac{n(n-1)}{2!}y^{n-2}x^2+\frac{n(n-1)(n-2)}{3!}y^{n-3}x^3+\ldots+\frac{n(n-1)}{2!}y^2x^{n-2}+nyx^{n-1}+x^n \quad (\text{for all } x \text{ and } y). \quad \text{(C.8b)}$$

For a given power n, there are $n+1$ terms. Therefore, the series can be written as a finite sum,

$$(y+x)^n=\sum_{m=0}^{n}\frac{n!}{m!(n-m)!}y^{(n-m)}x^m. \quad \text{(C.9)}$$

The factors in the sum that multiply the powers of $y^{(n-m)}x^m$ are the binomial coefficients, $\binom{n}{m}$, defined as

$$\binom{n}{m}=\frac{n!}{m!(n-m)!}. \quad \text{(C.10)}$$

TABLE C.1. The binomial coefficients $\binom{n}{m}$ for $n \leq 5$

m=.	0	1	2	3	4	5
n=0	1					
n=1	1	1				
n=2	1	2	1			
n=3	1	3	3	1		
n=4	1	4	6	4	1	
n=5	1	5	10	10	5	1

Pascal's Triangle

To calculate the binomial coefficients, the factorials can be reduced leaving a single simple product:

$$\binom{n}{m} = \prod_{l=1}^{m} \frac{(n-l+1)}{l} \quad (m \neq 0)$$

$$\binom{n}{0} = 1. \tag{C.11}$$

This formula can readily be computed by machine.

A table of these coefficients, called *Pascal's triangle*, shows their symmetry.

GEOMETRIC SERIES

A geometric series is a sum of terms that are related by a common factor. Starting with any term in the series, the next term is obtained by multiplying by the common factor, called the "common ratio," r. For example, the series $[1+1/3+1/9+1/27+1/81]$ is a geometric series with five terms in which the common ratio is $r=1/3$. The series $[1-5+25-125]$ is a geometric series with four terms in which the common ratio is $r=-5$.

The geometric series with N terms is of the general form

$$x_N = 1 + r + r^2 + r^3 + r^4 + \ldots + r^{N-1}. \tag{C.12}$$

There is a formula for summing a geometric series of the form of Eq. (C.12) with N terms. It is

$$x_N = \frac{1-r^N}{1-r}. \tag{C.13}$$

Proof of Eq. (C.13)

The proof of the result in Eq. (C.13) is not hard. We begin with the general form from Eq. (C.12)

$$x_N = 1 + r + r^2 + r^3 + r^4 + \ldots + r^{N-1}.$$

If we multiply this by r we get

$$rx_N = r + r^2 + r^3 + r^4 + r^5 + \ldots r^N. \tag{C.14}$$

If we subtract Eq. (C.14) from Eq. (C.12), we find that most of the terms on the right-hand side cancel leaving us with

$$(1-r)x_N = 1 - r^N, \tag{C.15}$$

or $x_N = (1-r^N)/(1-r)$, which is what we set out to prove.

There are a few observations that one can make about the result in Eq. (C.13). First, it is not necessary that a geometric series start at the number 1. It could start with some general number, call it a_1. But starting the series at a_1 merely has the effect of multiplying each term in the series by a_1, and so the sum is just multiplied by a_1.

In particular, if the first term in the series is r^{n_B} and the last term is r^{n_T} then the sum becomes $x_N = (r^{n_B} - r^{n_T+1})/(1-r)$.

Second, the series converges for an infinite number of terms ($N=\infty$) if the magnitude of the common ratio is less than 1. That can be seen from Eq. (C.13) where, in the limit that N is infinite, the term r^N is zero for $|r|<1$. Therefore, for the infinite series Eq. (C.13) becomes

$$x_\infty = \frac{1}{1-r}. \tag{C.16}$$

For example, the series $x_\infty = 1 + 1/3 + 1/9 + 1/27 + 1/81 + \ldots$ converges to $1/(1-1/3)$ or 3/2.

Complex Geometric Series

What has been said about the geometric series so far has assumed that all the numbers involved are real numbers. But looking at the proof of the important result for the series sum in Eq. (C.13), one realizes that it applies also if the common ratio r is a complex number.

Quite generally, we can write a complex ratio r as the product of its absolute value and a phase factor, $r=|r|e^{i\theta}$. Then the series of N terms can be written

$$x_N=1+|r|e^{i\theta}+|r|^2e^{i2\theta}+|r|^3e^{i3\theta}+\ldots+|r|^{N-1}e^{i(N-1)\theta}, \tag{C.17}$$

and Eq. (C.13) for the sum becomes

$$x_N=\frac{1-|r|^Ne^{iN\theta}}{1-|r|e^{i\theta}}. \tag{C.18}$$

The infinite series converges if the absolute value $|r|$ is less than 1. It converges to the value

$$x_\infty=\frac{1}{1-|r|e^{i\theta}}. \tag{C.19}$$

For example, if the common ratio is $r=(1/3)e^{i\pi/4}$ then the infinite series is

$$x_\infty=1+\frac{1}{3}\,e^{i\pi/4}+\frac{1}{9}\,e^{i2\pi/4}+\frac{1}{27}\,e^{i3\pi/4}+\ldots \tag{C.20}$$

and from Eq. (C.19) this sums to $1/(1-1/3e^{i\pi/4})$ which is $[1-1/(3\sqrt{2}) - i1/(3\sqrt{2})]^{-1}$.

What is interesting about the sum with the form of Eq. (C.17) is that we can use it to represent a signal that is sum of harmonics. If we just think of θ as ωt and take the real part of x_N we have a sum of cosines. This is done in Chapter 13.

SPECIAL FINITE SERIES

$$\sum_{n=1}^{N} n=\frac{N(N+1)}{2} \tag{C.21}$$

$$\sum_{n=1}^{N} n^2=\frac{N(N+1)(2N+1)}{6} \tag{C.22}$$

$$\sum_{n=1}^{N} n^3=\left[\frac{N(N+1)}{2}\right]^2 \tag{C.23}$$

SPECIAL INFINITE SERIES

$$\sum_{n=1}^{\infty} (-1)^{n+1} \frac{1}{n} = \ln(2) \tag{C.24}$$

$$\sum_{n=1}^{\infty} (-1)^{n+1} \frac{1}{2n-1} = \pi/4 \tag{C.25}$$

Riemann Zeta Function

$$\sum_{n=1}^{\infty} \frac{1}{n^p} = \zeta(p) \tag{C.26}$$

$$\sum_{n=1}^{\infty} (-1)^{n+1} \frac{1}{n^2} = \pi^2/12 \tag{C.27}$$

Power in a Square Wave

$$\sum_{n=1}^{\infty} \frac{1}{(2n-1)^2} = \pi^2/8 \tag{C.28}$$

Catalan's Constant

$$\sum_{n=1}^{\infty} (-1)^{n+1} \frac{1}{(2n-1)^2} = 0.915965594 \tag{C.29}$$

$$\sum_{n=1}^{\infty} (-1)^{n+1} \frac{1}{(2n-1)^3} = \pi^3/32 \tag{C.30}$$

Power in a Triangle Wave

$$\sum_{n=1}^{\infty} (-1)^{n+1} \frac{1}{(2n-1)^4} = \pi^4/96 \tag{C.31}$$

SERIES EXERCISES

Exercise 1

Equations (C.1) and (C.3) for the reciprocal and logarithmic functions have been written in terms of $1-x$, which leads to a series where all the terms are positive.

Show that if the series is written for $1+x$ instead, then successive terms alternate in sign. In particular, find the series for $1/(1+x)$.

Exercise 2

It is a fact of differential calculus that

$$\frac{d}{dx}\sin(x)=\cos(x).$$

Show that Eqs. (C.4) and (C.5) are consistent with this fact.

Exercise 3

Another fact from differential calculus is that

$$\frac{d}{dx}\ln[y(x)]=\frac{1}{y(x)}\frac{dy}{dx}.$$

Show that Eqs. (C.1) and (C.3) are consistent with this fact, by letting $y=1-x$.

Exercise 4

Consider the function e^x when x is equal to 1. Calculate the percentage error made by keeping only the first four terms in Eq. (C.2).

Exercise 5

Consider the function $1/(1-x)$ when $x=0.5$. Calculate the percentage error made by keeping only the first four terms in Eq. (C.1).

Exercise 6

Given the series for $\sin x$ in Eq. (C.4), find the series for the sinc function, $\text{sinc}(x)=\sin(\pi x)/(\pi x)$.

Exercise 7

Use Eq. (C.11) to verify a few lines of Pascal's triangle.

Exercise 8

A series of real numbers in which the sign of successive terms alternates between positive and negative is called, appropriately enough, an alternating series. A geometric series is alternating if the common ratio is negative. Show that the complex equations (C.17) and (C.18) for the special case $\theta = \pi$ agree with the real equations (C.12) and (C.13) for a negative common ratio.

Exercise 9

Show that the complex series for $\theta = \pi/4$ can be written as the sum of two alternating geometric series, one real and the other imaginary such that

$$x_\infty = \frac{1}{1+|r|^2} + i\,\frac{|r|}{1+|r|^2}.$$

Appendix D Integrals—Even and Odd Functions

An even function is defined by the property that $x(-t)=x(t)$. The cosine is an example of an even function.

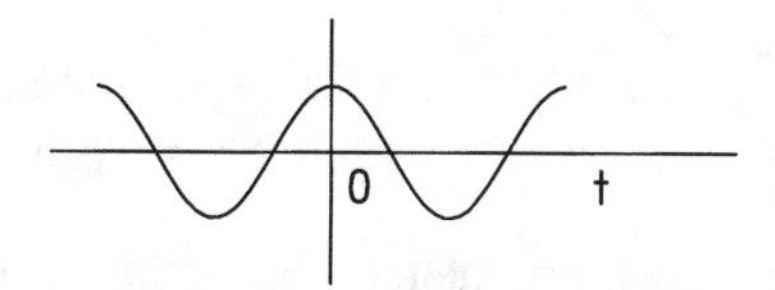

FIGURE D.1. The cosine function is even.

An odd function is defined by the property that $x(-t)=-x(t)$. The sine is an example of an odd function.

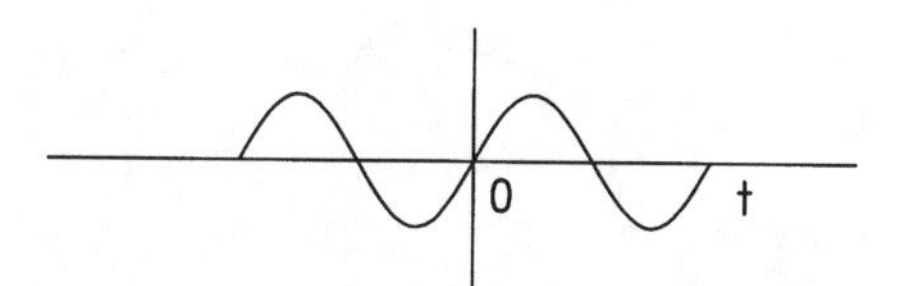

FIGURE D.2. The sine function is odd.

In the study of the Fourier transform, we make extensive use of three facts:

(1) Multiplication of odd and even functions is similar to the multiplication of negative and positive numbers:

The product of an even function and an even function is an even function.
The product of an odd function and an odd function is an even function.
The product of an even function and an odd function is an odd function.

(2) When an odd function, no matter how complicated, is integrated over symmetrical limits, the area under the curve cancels out, and the resulting integral is zero. Symmetrical limits are of the form $-T_1$ and $+T_1$. The figure below shows such cancellation, whereupon

$$\int_{-T_1}^{T_1} dt\, x(t) = 0. \tag{D.1}$$

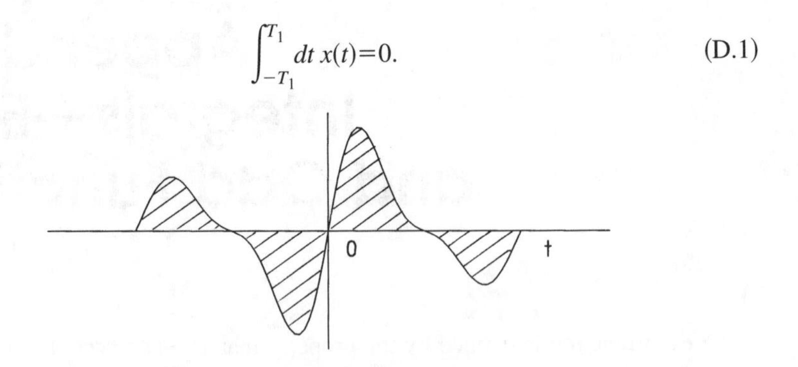

FIGURE D.3. The integral is given by the hatched area under the curve which is equally positive and negative, causing the integral to be zero.

(3) When an even function, no matter how complicated, is integrated over symmetrical limits, the area under the curve is exactly double the area on the positive half side.

$$\int_{-T_1}^{T_1} dt\, x(t) = 2\int_{0}^{T_1} dt\, x(t). \tag{D.2}$$

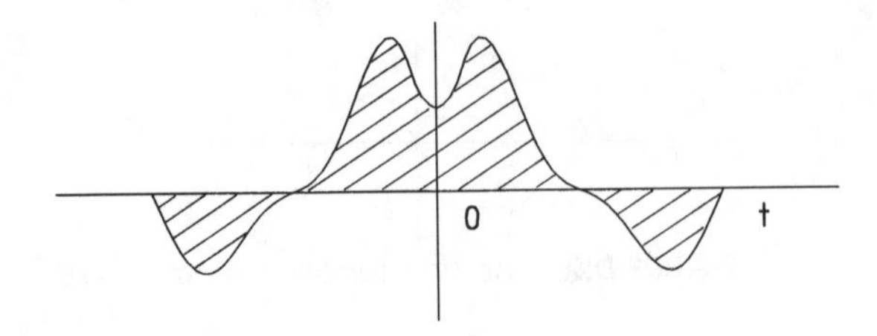

FIGURE D.4. The integral is given by the hatched area under the curve which is the same on positive and negative *t* axes.

EXAMPLE 1: ***The product of two even functions is even***

The function $x(t) = t^2$ is even. So is the cosine. Therefore,

$$\int_{-T_1}^{T_1} dt\, t^2 \cos(\omega t) = 2\int_{0}^{T_1} dt\, t^2 \cos(\omega t). \tag{D.3}$$

EXAMPLE 2: ***The product of two odd functions is even***

The function $x(t)=t^3$ is odd. So is the sine. Therefore,

$$\int_{-T_1}^{T_1} dt\, t^3 \sin(\omega t) = 2\int_0^{T_1} dt\, t^3 \sin(\omega t). \quad \text{(D.4)}$$

EXAMPLE 3: ***The product of an even function and an odd function is odd***

(3a) The function $x(t)=t^2$ is even. The sine function is odd. Therefore,

$$\int_{-T_1}^{T_1} dt\, t^2 \sin(\omega t) = 0. \quad \text{(D.5)}$$

(3b) The function $x(t)=t^3$ is odd. The cosine function is even. Therefore,

$$\int_{-T_1}^{T_1} dt\, t^3 \cos(\omega t) = 0. \quad \text{(D.6)}$$

The statements in Example 3 show that the Fourier series for an even function contains no sine terms (it is a cosine series), and the Fourier series for an odd function contains no cosine terms (it is a sine series).

Appendix E Integrals in the Complex Plane

Some signal calculations in this book require mathematics of complex variables beyond the scope of the text. Those calculations are presented here.

IMPULSE RESPONSE AND THE RESIDUE

The impulse response of a filter is the inverse Fourier transform of the transfer function, i.e.

$$h(t)=\frac{1}{2\pi}\int_{-\infty}^{\infty} d\omega\ e^{i\omega t}H(\omega). \tag{E.1}$$

The transfer function $H(\omega)$ is a fraction with a polynomial function of angular frequency ω in the numerator and another polynomial function of ω in the denominator. Zeros of these polynomials become zeros and poles of the transfer function, respectively. The values of ω for which the poles and zeros occur are complex, which enlarges somewhat the concept of what is meant by frequency.

Finding the inverse Fourier transform of the transfer function fraction is the topic of this section. The method of integration used to find this transform also regards angular frequency ω to be a complex variable.

Key to the transformation is the Cauchy residue theorem as follows: Let $f(z)$ be an analytic function (satisfying the Cauchy-Riemann conditions) within and on a simple closed contour C in the complex z plane. If z_0 is any point interior to C, then

$$\int_C dz\,\frac{f(z)}{z-z_0}=2\pi i f(z_0), \tag{E.2}$$

where $f(z_0)$ is the "residue" at the pole where $z=z_0$ (Pennisi, 1963, p. 118).

We illustrate the method with a simple application to a first-order lowpass filter. The transfer function is given by Eq. (9.22),

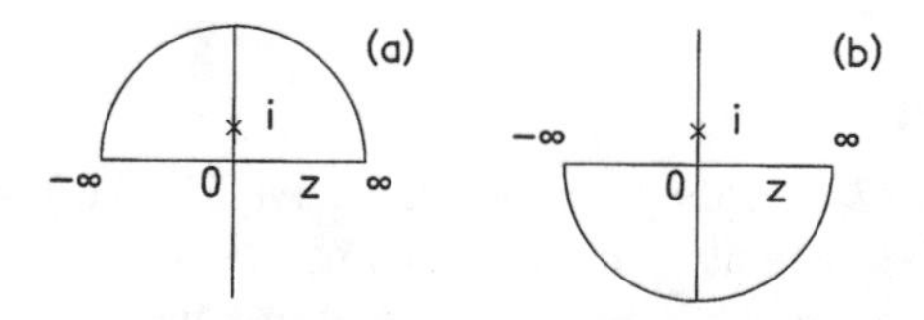

FIGURE E.1. (a) For positive time t, the contour for integration includes the entire real axis and a semicircle in the upper half plane at infinity. The transfer function has a single pole within the contour at x. (b) For negative time the contour must be closed at infinity by a semicircle in the lower half plane. There are no poles inside this contour.

$$H(\omega)=\frac{1}{1+i\omega\tau}, \tag{E.3}$$

which has a single pole and no zeros.

Therefore, the impulse response is

$$h(t)=\frac{1}{2\pi}\int_{-\infty}^{\infty}d\omega\ e^{i\omega t}\,\frac{1}{1+i\omega\tau}. \tag{E.4}$$

We put this integral in a form suitable for application of the residue theorem by the change of variables $z=\omega\tau$, so that

$$h(t)=\frac{1}{2\pi i}\,\frac{1}{\tau}\int_{-\infty}^{\infty}dz\ e^{izt/\tau}\,\frac{1}{z-i}. \tag{E.5}$$

The integrand has a pole at $z=i$, as shown in Fig. E.1. Otherwise it is an analytic function of z for all values of z, so long as $|z|$ remains finite. However, in order to do the integral on the real axis between infinite limits, we need to close the contour at infinite radius, which causes the integrand either to diverge exponentially or to be analytic, depending upon how the contour is closed.

If time t is positive then the integrand is analytic when the imaginary part of z is positive, i.e. in the upper half plane. (The integrand actually becomes zero on the contour at infinite radius, because izt/τ becomes equal to $-\infty$.) We therefore do the integral by closing the contour at infinity in the upper half plane, as shown in Fig. E.1(a). Because the integrand vanishes on the semicircle at infinity, the integral along the real axis is the residue at the pole, where $z=i$, and therefore,

$$h(t)=\frac{1}{\tau}\,e^{-t/\tau}\quad(t>0). \tag{E.6a}$$

If time t is negative then the integrand is analytic in the lower half plane. We therefore must do the integral on the real axis by closing the contour at infinity in the lower half plane, as shown in Fig. E.1(b). Because the integrand vanishes on the semicircle at infinity, and because there are no poles at all within the contour, the integral along the real axis is zero.

$$h(t)=0 \quad (t<0). \tag{E.6b}$$

This transfer function agrees with what was found in Chapter 9.

HILBERT TRANSFORM INTEGRALS

The Hilbert transform of signal $x(t)$ is

$$\mathcal{H}[x(t)]=\frac{-1}{\pi}\int_{-\infty}^{\infty} dt' \, \frac{x(t')}{t'-t}. \tag{E.7}$$

To find the Hilbert transform of a sine or a cosine or anything that has a Fourier series, we need to evaluate the integral where $x(t)=\exp(\pm i\omega t)$. We consider first the integral for $x(t)=\exp(i\omega t)$,

$$\xi_1(t)=\frac{-1}{\pi}\int_{-\infty}^{\infty} dt' \, \frac{e^{i\omega t'}}{t'-t}. \tag{E.8}$$

This integral has a pole, but unlike the integral for the lowpass filter, the pole lies on the real axis, where $t'=t$. The integral is analytic in the upper half plane, where the imaginary part of t' is positive. Therefore, we choose the contour of Fig. E.2(a).

We take the limit where the small semicircle C is infinitesimal and the large semicircle has infinite radius. Because the contour includes no poles, Cauchy's residue theorem says that the integral must be zero. That is, the sum of the integral along the real axis ($\int_{-\infty}^{\infty} dt...$), around the small semicircle, and around the large semicircle adds up to zero. But the integrand vanishes on the large semicircle, so it gives zero, and

$$\int_{-\infty}^{\infty} dt... = -\int_{C} dt... \tag{E.9}$$

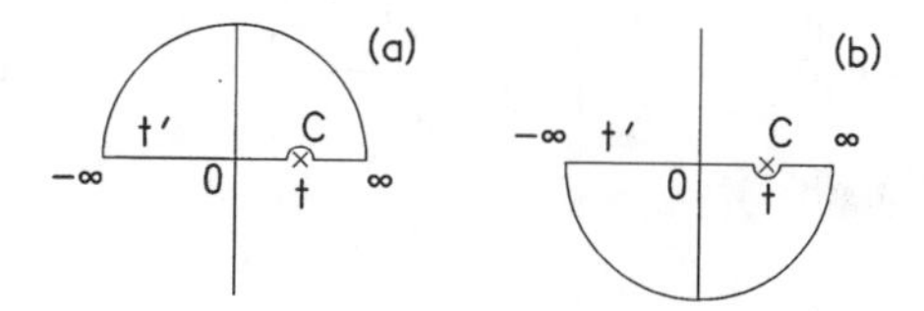

FIGURE E.2. (a) For the kernel $\exp(i\omega t')$, the contour for integration includes the entire real time axis and a semicircle in the upper half plane at infinity. The integrand has a single pole on the real axis and little indented contour C is drawn to avoid it. (b) For the kernel $\exp(-i\omega t')$, the contour must be closed at infinity by a semicircle in the lower half plane.

Because the small semicircle goes only halfway around the pole, the integral is equal to $i\pi$Residue (instead of $2\pi i$Residue as before). Because the path is backward (clockwise) the integral gets another negative sign, and therefore,

$$\xi_1(t) = \frac{-1}{\pi} \int_{-\infty}^{\infty} dt' \, \frac{e^{i\omega t'}}{t' - t} = -ie^{i\omega t}. \tag{E.10}$$

Next we consider the integral for $x(t) = \exp(-i\omega t)$,

$$\xi_2(t) = \frac{-1}{\pi} \int_{-\infty}^{\infty} dt' \, \frac{e^{-i\omega t'}}{t' - t}. \tag{E.11}$$

This integral is analytic in the lower half plane. Therefore, we choose the contour of Fig. E.2(b). The evaluation is similar to the evaluation of the contour in Fig. E.2(a), except that this time the direction around the infinitesimal semicircle C is counterclockwise. Therefore,

$$\xi_2(t) = \frac{-1}{\pi} \int_{-\infty}^{\infty} dt' \, \frac{e^{-i\omega t'}}{t' - t} = +ie^{-i\omega t}. \tag{E.12}$$

Integrals ξ_1 and ξ_2 enable us to find the Hilbert transforms of sine and cosine functions:

$$\mathcal{H}[\cos(\omega t)] = \frac{-1}{2\pi} \int_{-\infty}^{\infty} dt' \, \frac{e^{i\omega t'} + e^{-i\omega t'}}{t' - t} = \sin(\omega t), \tag{E.13}$$

and

$$\mathcal{H}[\sin(\omega t)] = \frac{-1}{2\pi i} \int_{-\infty}^{\infty} dt' \, \frac{e^{i\omega t'} - e^{-i\omega t'}}{t' - t} = -\cos(\omega t). \tag{E.14}$$

The Hilbert transform of the cosine is the sine, and the Hilbert transform of the sine is the negative of the cosine. It is left as an exercise to show that this sign reversal is just what one needs to extend the transforms so that

$$\mathcal{H}[\cos(\omega t + \phi)] = \sin(\omega t + \phi), \tag{E.15}$$

and

$$\mathcal{H}[\sin(\omega t + \phi)] = -\cos(\omega t + \phi). \tag{E.16}$$

Having found Hilbert transforms for cosine and sine, we can find the Hilbert transform of any function for which a Fourier transform exists.

Appendix F
Hilbert Transforms

The Hilbert transform is used in two ways in this book. It is used in the time domain to find the analytic signal beginning with the real signal. The analytic signal, in turn, leads to the envelope. The Hilbert transform is also used in the frequency domain to relate real and imaginary parts of the transfer function of a filter.

In this appendix we deal first with time-domain applications and then with frequency-domain applications.

COMPONENTS OF THE ANALYTIC SIGNAL

If $x(t)$ is a real signal then we define $\tilde{x}(t)$ as the analytic signal. The analytic signal is complex, and the real signal $x(t)$ is its real part. The imaginary part of the analytic signal will be defined as $x_I(t)$. Therefore, by definition, the analytic signal is

$$\tilde{x}(t) = x(t) + i x_I(t). \tag{F.1}$$

The analytic signal was introduced in Chapter 18 as a means of calculating the envelope of real signal $x(t)$. The imaginary part of the analytic signal is the Hilbert transform of $x(t)$,

$$x_I = \mathcal{H}[x(t)] = \frac{1}{\pi} \int_{-\infty}^{\infty} dt' \, \frac{x(t')}{t-t'}. \tag{F.2}$$

Therefore, the analytic signal is

$$\tilde{x}(t) = x(t) + i\mathcal{H}[x(t)]. \tag{F.3}$$

There is an inverse relation too,

$$x(t)=\frac{-1}{\pi}\int_{-\infty}^{\infty}dt'\,\frac{x_I(t')}{t-t'}. \tag{F.4}$$

Thus, the real part of the analytic signal $x(t)$ and the imaginary part $x_I(t)$ are Hilbert transforms of one another.

This transformation is reminiscent of the Fourier transform, where $X(\omega)$ is calculated from $x(t)$ by an integral equation that differs only slightly from the inverse operation that calculates $x(t)$ from $X(\omega)$. In the case of the Hilbert transform, the difference between the transform and its inverse is a mere matter of a minus sign.

The Hilbert transform differs from the Fourier transform in that the Fourier transform of $x(t)$ introduces a new dimension, namely the frequency dimension, whereas the Hilbert transform of $x(t)$ remains a function of time.

Some intuition about the Hilbert transform can be gained from writing the signal as a Fourier sum:

$$x(t)=\sum_{n=1}^{N} A_n\cos(\omega_n t)+B_n\sin(\omega_n t). \tag{F.5}$$

It is not necessary that $x(t)$ be periodic, only that it can be adequately represented by a finite number of components, with frequencies ω_n. We recall from Appendix E that the Hilbert transforms of cosine and sine functions are simple,

$$\mathcal{H}[\cos(\omega_0 t)]=\sin(\omega_0 t) \tag{F.6}$$

$$\mathcal{H}[\sin(\omega_0 t)]=-\cos(\omega_0 t). \tag{F.7}$$

Because the Hilbert transform is linear, the transform of a sum is equal to the sum of the transforms. This can be applied to the Fourier sum in Eq. (F.5) to give,

$$\mathcal{H}[x(t)]=\sum_{n=1}^{N} A_n\sin(\omega_n t)-B_n\cos(\omega_n t). \tag{F.8}$$

Symmetry

One of the implications of Eqs. (F.5)–(F.8) is that if function $x(t)$ is an even function of time (all the B_n are zero) then its Hilbert transform is an odd function of time. Likewise, if $x(t)$ is odd then its Hilbert transform is even.

Positive Frequencies

A second implication is that only positive frequencies are included in the analytic signal. It is easy to show that this is so: Forming $\widetilde{x}(t)=x(t)+i\mathcal{H}[x(t)]$, we find

$$\tilde{x}(t)=\sum_{n=1}^{N} A_n[\cos(\omega_n t)+i\ \sin(\omega_n t)]-iB_n[\cos(\omega_n t)+i\ \sin(\omega_n t)], \quad \text{(F.9)}$$

or

$$\tilde{x}(t)=\sum_{n=1}^{N} (A_n-iB_n)[\cos(\omega_n t)+i\ \sin(\omega_n t)] \quad \text{(F.10)}$$

which, by Euler's formula, is

$$\tilde{x}(t)=\sum_{n=1}^{N} (A_n-iB_n)e^{i\omega_n t}. \quad \text{(F.11)}$$

It is evident that only positive frequencies enter Eq. (F.11). Because of the Euler relation, it happened that the negative frequency components that occurred in the sine and cosine functions in Eq. (F.9) cancelled out.

Orthogonality

A third implication of Eq. (F.8) is that function $x(t)$ and its Hilbert transform are orthogonal. We prove that by showing that the integral of the product of the two functions is zero, i.e.

$$\text{I'gral}=\int_{-\infty}^{\infty} dt\ x(t)\mathcal{H}[x(t)]=0. \quad \text{(F.12)}$$

Substituting from Eqs. (F.5) and (F.8), we find

$$\text{I'gral}=\int_{-\infty}^{\infty} dt\left[\sum_{n=1}^{N} A_n\ \cos(\omega_n t)+B_n\ \sin(\omega_n t)\right]\left[\sum_{n'=1}^{N} A_{n'}\ \sin(\omega_{n'} t) - B_{n'}\ \cos(\omega_{n'} t)\right]. \quad \text{(F.13)}$$

We exchange the order of integration and summation so that the expression is a double sum (over n and n') of an integral. By the orthogonality of the trigonometric functions, we know that the integral of a sine function with a cosine function will be zero. We also know that none of these integrals will be finite unless index n is equal to index n'. This leaves us with a single sum,

$$\text{I'gral}=\sum_{n=1}^{N} \int_{-\infty}^{\infty} dt[A_n B_n\ \sin^2\ \omega_n t-A_n B_n\ \cos^2\ \omega_n t]. \quad \text{(F.14)}$$

But the integrals of $\sin^2$ and $\cos^2$ are both equal to 1/2, and the subtraction operation makes each term in the sum equal to zero. Therefore, the entire sum is zero, which shows that $x(t)$ and its Hilbert transform are orthogonal.

DISPERSION RELATIONS

A second important application of the Hilbert transform is in frequency space, where the transform relates the real and imaginary parts of a transfer function, $H(\omega)$ (see Chapter 9). The key to the relationship is that the impulse response $h(t)$ is causal.

To see the implications of causality in the frequency domain, we consider the inverse Fourier transform,

$$h(t)=\frac{1}{2\pi}\int_{-\infty}^{\infty} d\omega\ e^{i\omega t}H(\omega). \tag{F.15}$$

When t is negative, the above integral must be zero because of causality. Therefore, we concentrate entirely on the case for $t<0$.

We imagine that the integral is done as a contour integral in the complex plane. For $t<0$ the exponential factor $\exp(i\omega t)$ becomes

$$e^{-i|t|\mathrm{Re}(\omega)}e^{|t|\mathrm{Im}(\omega)}. \tag{F.16}$$

If the contour is closed in the upper-half plane then the exponential factor diverges as $\mathrm{Im}(\omega)$ goes to infinity. Therefore, we must close the contour in the lower-half plane, as shown in Fig. F.1. Here the exponential factor goes to zero at infinity.

Because this contour integral is zero, the function $H(\omega)$ must be analytic in the lower-half plane. In the end, this is what causality means in the frequency domain. The transfer function has no poles or other singularities in the lower-half frequency plane.

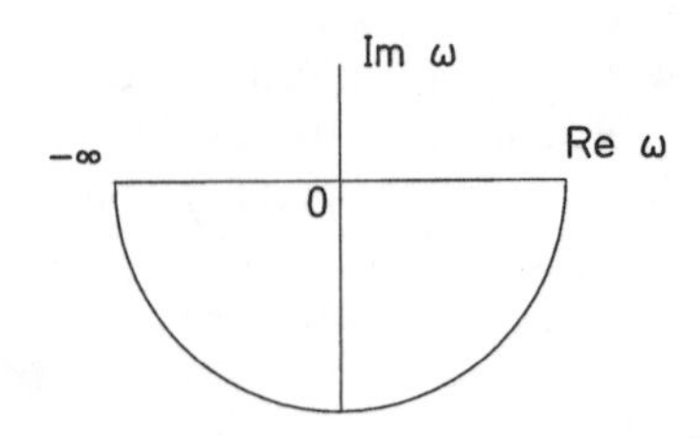

FIGURE F.1. Contour for doing the inverse Fourier integral for negative time.

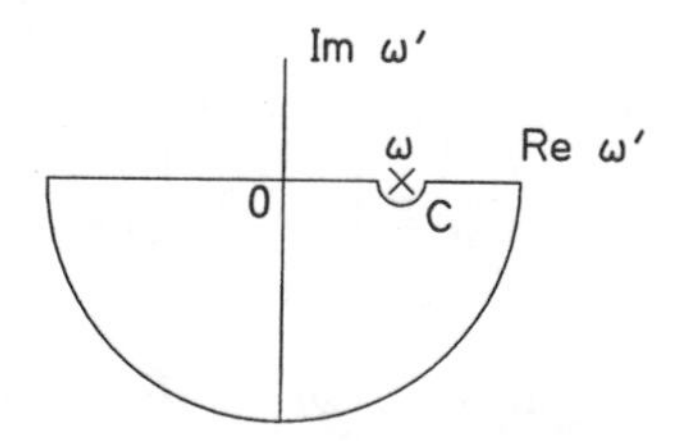

FIGURE F.2. Contour for doing the integral in Eq. (F.17). There is one simple pole, when ω' is equal to ω.

To find the connection between real and imaginary parts of $H(\omega)$ we form an integral like the Hilbert transform of the transfer function:

$$\int_{-\infty}^{\infty} d\omega' \, \frac{H(\omega')}{\omega - \omega'}. \tag{F.17}$$

The integrand has a simple pole at ω, as shown in Fig. F.2 and is otherwise analytic in the lower-half plane.

The entire integral is equal to the principal value integral plus the integral on the small contour C plus the integral on the large semicircle at infinity. Because the integrand is analytic within the contour, the integral is zero. Since the contour at infinity gives zero, we have

$$\mathcal{P}\int_{-\infty}^{\infty} d\omega' \, \frac{H(\omega')}{\omega - \omega'} = -\int_{C} d\omega' \, \frac{H(\omega')}{\omega - \omega'}, \tag{F.18}$$

i.e. the principal value integral is the negative of the integral on the tiny half-circle contour C. The latter is equal to $-i\pi$ times the residue at ω, and so

$$\mathcal{P}\int_{-\infty}^{\infty} d\omega' \, \frac{H(\omega')}{\omega - \omega'} = i\pi H(\omega). \tag{F.19}$$

Only one step remains. We write the transfer function as a sum of its real and imaginary parts,

$$H(\omega) = H_R(\omega) + iH_I(\omega), \tag{F.20}$$

where H_R and H_I are both real functions.

Equating real and imaginary parts in Eq. (F.19), we have

$$H_R(\omega) = \frac{1}{\pi} \, \mathcal{P}\int_{-\infty}^{\infty} d\omega' \, \frac{H_I(\omega')}{\omega - \omega'}, \tag{F.21a}$$

and

$$H_I(\omega) = \frac{-1}{\pi} \mathcal{P} \int_{-\infty}^{\infty} d\omega' \frac{H_R(\omega')}{\omega - \omega'}. \qquad \text{(F.21b)}$$

Thus the real and imaginary parts of $H(\omega)$ are Hilbert transforms of one another. These relationships are called "dispersion relations," or Krammers-Kronig relations.

An important qualification should be added to the dispersion relations in Eq. (F.21). It may happen that the transfer function does not go to zero as the frequency goes to infinity. This is true, for example, for a highpass filter, where the real part of the transfer function goes to a finite limit, $H(\infty)$. In this case, the assumption that the integral on the contour at infinity is zero fails. However, it is always possible to define a new function

$$H'(\omega) = H(\omega) - H(\infty). \qquad \text{(F.22)}$$

The dispersion relations may now hold for function $H'(\omega)$.

The dispersion relations in Eq. (F.21) are integral equations where the integrals cover the entire frequency axis. It is possible to use an additional symmetry of the transfer functions to reduce these integrals to the positive part of the frequency axis only. We recall that changing the sign of ω turns $H(\omega)$ into its complex conjugate. This is a result of the fact that the impulse response $h(t)$ is real. Therefore, $H_R(-\omega) = H_R(\omega)$, and $H_I(-\omega) = -H_I(\omega)$. As a result, Eqs. (F.21) can be written as

$$H_R(\omega) = \frac{2}{\pi} \mathcal{P} \int_0^{\infty} d\omega' \frac{\omega' H_I(\omega')}{\omega^2 - \omega'^2}, \qquad \text{(F.23a)}$$

and

$$H_I(\omega) = \frac{-2\omega}{\pi} \mathcal{P} \int_0^{\infty} d\omega' \frac{H_R(\omega')}{\omega^2 - \omega'^2}. \qquad \text{(F.23b)}$$

This form of the dispersion relations is particularly useful if one has to subtract off a finite limit at infinite frequency.

MINIMUM-PHASE FILTERS, GAIN AND PHASE SHIFT

As noted in Chapter 9, it is possible to find a dispersion relation that relates the gain and the phase shift of a filter, so long as the filter transfer function satisfies the minimum phase requirements. The gain of the filter in nepers, $n(\omega)$, and the phase shift in radians, $\phi(\omega)$, are the real and imaginary parts of the natural logarithm of the transfer function,

$$\ln H(\omega) = n(\omega) + i\phi(\omega). \tag{F.24}$$

If $H(\omega)$ has no zeros in the lower half plane (in addition to having no poles there) then the filter is minimum phase, and the gain and phase shift are related by Eqs. (9.89), which are

$$n(\omega) = n(0) - \frac{\omega^2}{\pi} \mathcal{P}\int_{-\infty}^{\infty} d\omega' \, \frac{\phi(\omega')}{\omega'(\omega'^2 - \omega^2)} \tag{9.89a}$$

and

$$\phi(\omega) = \frac{\omega}{\pi} \mathcal{P}\int_{-\infty}^{\infty} d\omega' \, \frac{n(\omega')}{\omega'^2 - \omega^2}. \tag{9.89b}$$

It is the purpose of this section to derive these equations.

Because the logarithm of the transfer function is expected to vanish at infinity only logarithmically, we form an integral with a squared frequency in the denominator,

$$\text{I'gral} = \int_{-\infty}^{\infty} d\omega' \, \frac{n(\omega') + i\phi(\omega')}{\omega'^2 - \omega^2}. \tag{F.25}$$

If the filter is minimum phase, the numerator of the integrand is analytic in the lower-half plane, and we close the contour there, just as in Fig. (F.2). The only difference is that the denominator now contributes *two* poles on the real axis.

The entire integral consists of a principal value integral plus contributions from the two poles, and, by the Cauchy integral theorem, this all adds up to zero. Therefore, the principal value integral itself is the negative of the contributions at the poles,

$$\mathcal{P}\int_{-\infty}^{\infty} d\omega' \, \frac{n(\omega') + i\phi(\omega')}{\omega'^2 - \omega^2} = \frac{-i\pi}{2\omega} \{[n(\omega) - n(-\omega)] + i[\phi(\omega) - \phi(-\omega)]\}. \tag{F.26}$$

This equation can be simplified by symmetry. Because $n(\omega)$ is the log of the absolute value of H it is even and vanishes from the right-hand side. Because $\phi(\omega)$ is the argument of the ratio of the imaginary part of H (odd) to the real part of H (even), ϕ is odd and vanishes from the left-hand side. Further, the two ϕ terms on the right can be combined into one. What remains is precisely Eq. (9.89b).

To find the other dispersion relation we form the integral

$$\text{I'gral} = \int_{-\infty}^{\infty} d\omega' \, \frac{n(\omega') + i\phi(\omega')}{\omega'(\omega'^2 - \omega^2)}. \tag{F.27}$$

Again we close the contour in the lower half plane and set the entire integral equal to zero. This time we have three poles on the real axis, including one at the origin. The poles can be put into separate terms with a partial fraction expansion, so that the principal value integral is given by

$$\mathcal{P}\int_{-\infty}^{\infty} d\omega' \, \frac{n(\omega') + i\phi(\omega')}{\omega'(\omega'^2 - \omega^2)} = \frac{-i\pi}{2\omega}\left[-2\frac{n(0) + i\phi(0)}{\omega} + \frac{n(\omega) + i\phi(\omega)}{\omega} - \frac{n(-\omega) + i\phi(-\omega)}{-\omega}\right]. \tag{F.28}$$

By symmetry $n(\omega')$ vanishes from the integral on the left-hand side, and $\phi(\omega)$ vanishes from the right-hand side. Further, the two terms in $n(\omega)$ on the right are equal, and therefore Eq. (9.89a) follows.

If a filter is not minimum phase, due to simple zeros in the lower half plane, the dispersion relation in Eqs. (9.89) may still be applied if one deals with the additional phase shift contributed by the zeros as described in Chapter 9.

Appendix G Electrical Filters

The study of filtering requires concrete examples. These can be realized in either mechanical, acoustical, or electrical form, but the electrical form is easiest, and all of the discussion below will be about electrical circuits.

Electrical Elements

Electrical filters employ three kinds of passive linear elements: resistors, capacitors, and inductors. These three elements can be described in a variety of different ways, but when the context is filtering, our first interest is in determining a transfer function. The transfer function describes the response of a system to a sine wave of given frequency, and therefore, our representation of the resistors, capacitors, and inductors will be one which describes their response to a sine wave as a function of frequency. Such a description is given by the *impedance*, the ratio of the voltage across the element to the current through it, as shown in Fig. G.1. The relationship between voltage (V), current (I), and impedance (Z) is *Ohm's law*,

$$V = IZ. \tag{G.1}$$

The impedance of a resistor (R) is independent of frequency. The impedance of a capacitor (C) decreases with increasing frequency, and the impedance of an inductor (L) increases with increasing frequency. Specifically, for R, C, and L, the impedances are given in Fig. G.2.

If R is resistance in ohms, C is capacitance in farads, and L is inductance in henrys, while ω is the angular frequency in radians per second, then all the impedances in the box above are in units of ohms.

Resistive impedance is real, but the impedance of a capacitor or of an inductor is purely imaginary. An imaginary impedance is called a "reactance." The concept of an imaginary impedance requires some explanation. For instance, it doesn't exactly correspond to the procedure suggested in Fig. G.1, which only measures a magnitude. The interpretation of the real and imaginary character of these impedances is that if an AC voltage is impressed across a resistor then the current will track the voltage in time and be in phase with the voltage. In a reactive element the current through the element is out of phase with the voltage across it. Thus, the magnitude

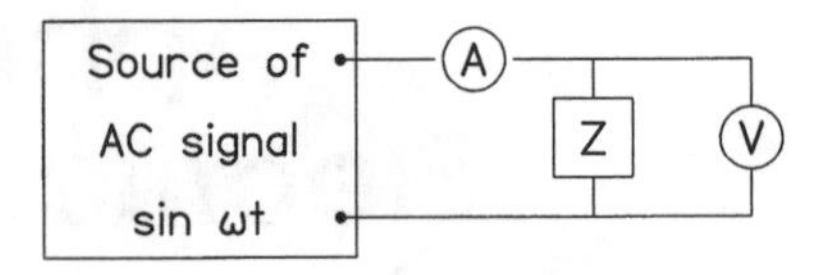

FIGURE G.1. Method of measuring the impedance of a passive linear element at angular frequency ω. The AC voltmeter reads the RMS voltage (V) **across** the element, and the AC ammeter (A) reads the RMS current (I) **through** the element. The ratio V/I is equal to the impedance $|Z(\omega)|$.

of the impedance determines the ratio of RMS voltage to RMS current (as in Fig. G.1), whereas the phase of the impedance determines the phase difference between the voltage and current.

In polar form, the impedance of a capacitor is $Z_C(\omega) = 1/(\omega C)e^{-i\pi/2}$. From Ohm's law, $V = IZ_C$, we see that in order to find the phase of the voltage from the phase of the current we need to subtract an angle of $\pi/2$. Therefore, in a capacitor the current leads the voltage by a phase shift of $\pi/2$. This means that any waveform feature, a positive-going zero crossing for example, occurs in a plot of the current vs. time before it occurs in a plot of the voltage vs. time.

In polar form, the impedance of an inductor is $Z_L(\omega) = \omega L e^{i\pi/2}$. From Ohm's law, $V = IZ_L$, we see that in an inductor the current *lags* the voltage by a phase shift of $\pi/2$.

An important physical result of the phase relationship between voltage and current is that resistors dissipate electrical energy (turn it into heat) whereas capacitors and inductors do not. They simply store energy for awhile. It is not hard to see why this is so, using the concepts of average power from Chapter 3. The power in any element is the product of the voltage and the current. If the voltage across a resistor is a sine function of time then the current in the resistor will also be a sine function. The product of the two is the square of a sine function, and this has an average value of 1/2. It is this average value of power that measures the rate at which energy is used. But when a sine voltage is across a capacitor or an inductor, the current is a cosine function. The average value of the product of a sine function and a cosine function is zero because of the orthogonality of these two trigono-

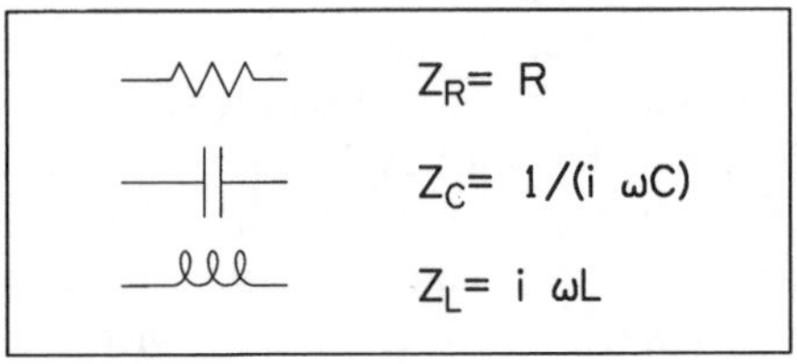

$Z_R = R$

$Z_C = 1/(i\ \omega C)$

$Z_L = i\ \omega L$

FIGURE G.2. From top to bottom, impedances of Resistor, Capacitor, and Inductor.

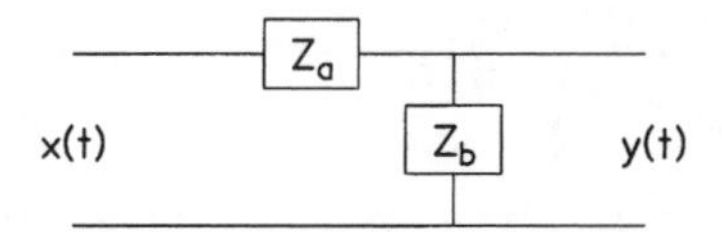

FIGURE G.3. The voltage divider puts a signal across two passive elements, with impedances Z_a and Z_b, and extracts the signal across one of them, Z_b.

metric functions. Therefore, the average power dissipated by one of these reactive elements is zero.

Filter Circuits

A filter has an input $x(t)$ and an output $y(t)$. The Fourier transforms of x and y are related by the equation that defines a filter transfer function H,

$$Y(\omega)=H(\omega)X(\omega). \tag{G.2}$$

To construct a filter one combines at least two passive elements into a circuit. Two is the minimum number; there is no maximum number, and filters of enormous complexity are often constructed. The simplest filter is the voltage divider shown in Fig. G.3.

The transfer function is given by

$$H(\omega)=\frac{Z_b(\omega)}{Z_a(\omega)+Z_b(\omega)}. \tag{G.3}$$

First-Order Passive Filter

A first-order passive filter can be either lowpass or highpass. The circuit for a lowpass filter is given in Fig. G.4.

Using the impedances from Fig. G.2, we find that $Z_b=1/(i\omega C)$ and $Z_a=R$. Therefore, from Eq. (G.3) we find a transfer function

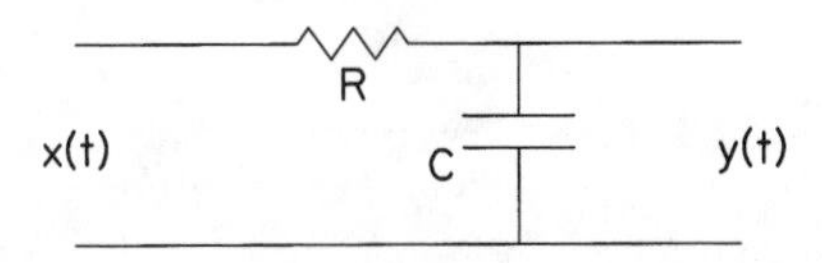

FIGURE G.4. The first-order lowpass filter is made with a resistor and capacitor.

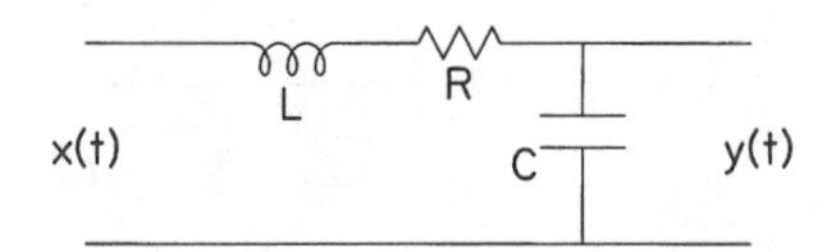

FIGURE G.5. A two-pole lowpass filter.

$$H(\omega)=\frac{1/(i\omega C)}{R+1/(i\omega C)} \tag{G.4}$$

or

$$H(\omega)=\frac{1}{1+i\omega\tau}, \tag{G.5}$$

where $\tau=RC$ is called the ''time constant'' and is measured in units of seconds. This transfer function was the basis for the discussion of the one-pole lowpass filter in Chapter 9.

Second-Order Passive Filter

A two-pole lowpass filter can be made by adding an inductor, as in Fig. G.5.

Because the inductor and the resistor are in series, their impedances add algebraically to form a new value of Z_a, and the transfer function becomes

$$H_{LP}(\omega)=\frac{1/(i\omega C)}{1/(i\omega C)+i\omega L+R}. \tag{G.6}$$

Multiplying numerator and denominator by $i\omega/L$, we find

$$H_{LP}(\omega)=\frac{1/LC}{1/LC-\omega^2+iR\omega/L}, \tag{G.7}$$

which is

$$H_{LP}(\omega)=\frac{\omega_0^2}{\omega_0^2-\omega^2+i\omega\omega_0/Q}, \tag{G.8}$$

where $\omega_0^2=1/LC$, and $Q=\omega_0 L/R$.

FILTER EXERCISES

Exercise 1: One-Pole Highpass Filter

The first-order highpass filter is shown in Fig. G.E1. Show that the transfer function is

$$H(\omega)=\frac{i\omega\tau}{1+i\omega\tau},$$

where $\tau=RC$.

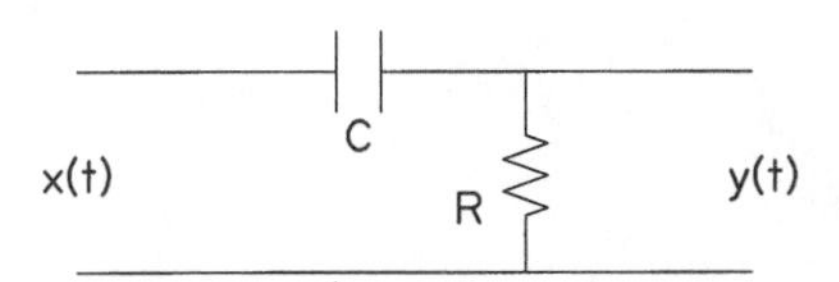

FIGURE G.E1. The one-pole highpass filter is made by exchanging the components of the one-pole lowpass filter.

Exercise 2: Two-Pole Filters

By permuting the passive components of the two-pole lowpass filter in Fig. G.5, one can make highpass and bandpass filters, as in Fig. G.E2(a) and Fig. G.E2(b), respectively.

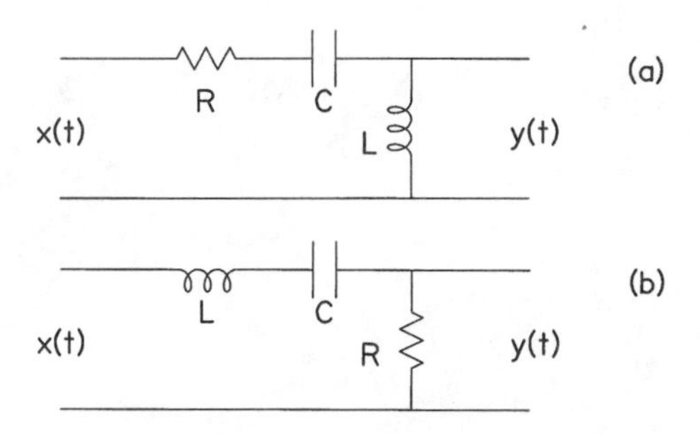

FIGURE G.E2. Two-pole filters: (a) The highpass. (b) The bandpass.

Show that the transfer functions are: (a) for the highpass:

$$H_{HP}(\omega) = \frac{-\omega^2}{\omega_0^2 - \omega^2 + i\omega\omega_0/Q},$$

and (b) for the bandpass:

$$H_{BP}(\omega) = \frac{i\omega\omega_0/Q}{\omega_0^2 - \omega^2 + i\omega\omega_0/Q}.$$

Appendix H
Normal Distribution

The normal (or Gaussian) density is given by the expression

$$f_N(x) = \frac{1}{\sigma\sqrt{2\pi}} e^{-x^2/(2\sigma^2)}, \tag{H.1}$$

where σ is a measure of the width of the distribution known as the standard deviation. Function $f_N(x)$ is the classic bell-shaped curve. Because it is defined as a density, the value of $f_N(x)dx$ gives the probability that a normally distributed variable x lies in the range between x and $x+dx$, as shown in Fig. H.1.
Normalization: Because it is certain that the value of x lies somewhere between $-\infty$ and $+\infty$ the function $f_N(x)$ must be normalized

$$\int_{-\infty}^{\infty} dx\ f_N(x) = 1. \tag{H.2}$$

Mean: The average (mean) value of the distribution described by $f_N(x)$ above is zero because $f_N(x)$ is symmetrical and centered on zero. To obtain a distribution with non-zero mean x_0, one shifts the function $f_N(x)$,

$$f_N(x-x_0) = \frac{1}{\sigma\sqrt{2\pi}} e^{-(x-x_0)^2/(2\sigma^2)}. \tag{H.3}$$

This shift does not affect the normalization or the width of the Gaussian function.
It is easy to see that the average value of x, distributed according to $f_N(x-x_0)$, is the first moment of f_N:

$$\frac{\int_{-\infty}^{\infty} dx\ f_N(x-x_0)x}{\int_{-\infty}^{\infty} dx\ f_N(x)}. \tag{H.4}$$

The denominator is unity because of the normalization. The numerator is simply the number x_0. Therefore, the first moment is x_0.
Variance: The variance of $f_N(x)$ is given by the second moment of $f_N(x)$,

$$\text{Variance} = \int_{-\infty}^{\infty} dx\ f_N(x-x_0)(x-x_0)^2 = \sigma^2. \tag{H.5}$$

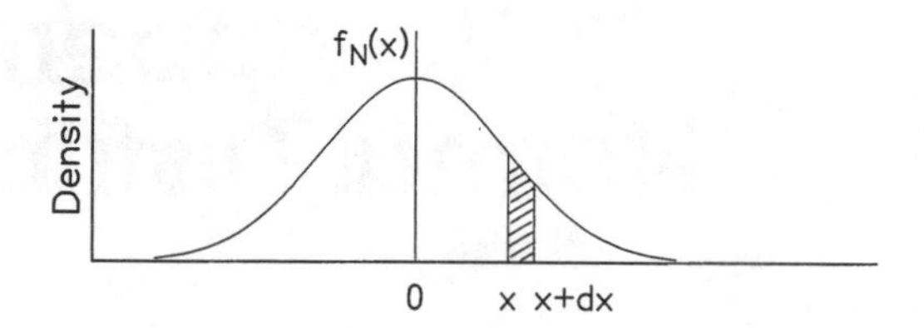

FIGURE H.1. The hatched area under the Gaussian density gives the probability that a normally distributed variable x lies in the range between x and $x+dx$.

Thus parameter σ^2, as it appears in the definition of $f_N(x)$, is seen to be the variance. It is independent of the value of x_0. The standard deviation is the square root of the variance, namely σ.

The full width at half height of the Gaussian is related to the standard deviation by a constant factor,

$$FWHH = 2\sigma\sqrt{2\ln(2)}, \tag{H.6}$$

or

$$FWHH = 2.35482\sigma.$$

ERROR FUNCTION

The probability that variable x lies in a range of $\pm X$ around the mean is $P(X)$, given by the area shown in Fig. H.2.

This area can be calculated from an integral of the Gaussian called the error function, defined as

$$\mathrm{erf}(q) = \frac{2}{\sqrt{\pi}} \int_0^q d\zeta\, e^{-\zeta^2}. \tag{H.7}$$

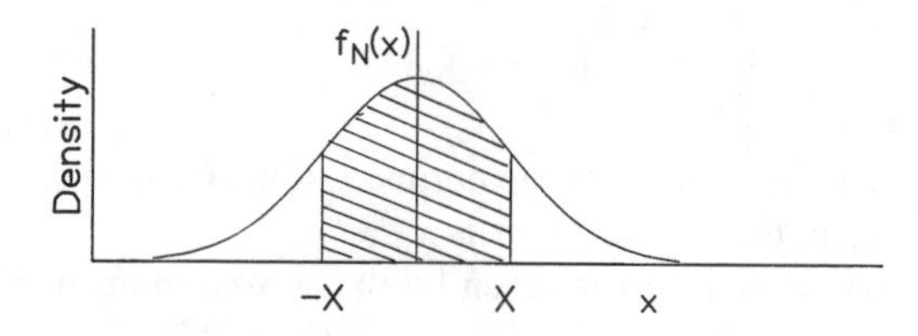

FIGURE H.2. The hatched area is $P(X)$, the probability that x lies in a range between $-X$ and $+X$ around the mean. It is given by $\mathrm{erf}(X/(\sigma\sqrt{2}))$. The values of $\pm X$ chosen for this figure happen to be $\pm\sigma$, plus and minus one standard deviation.

TABLE H.1. Selected values of the area about the mean $P(X)$ as calculated from the error function

If $X=0$ then $P(X)=\mathrm{erf}(0)=0$

If $X=\sigma/2$ then $P(X)=\mathrm{erf}\left(\frac{1}{2\sqrt{2}}\right)=0.36$

If $X=0.67451\sigma$ then $P(X)=\mathrm{erf}(0.47695)=0.5$

If $X=\sigma$ then $P(X)=\mathrm{erf}\left(\frac{1}{\sqrt{2}}\right)=0.68$

If $X=2\sigma$ then $P(X)=\mathrm{erf}(\sqrt{2})=0.95$

If $X=\infty$ then $P(X)=\mathrm{erf}(\infty)=1.$

Error Function.

By substituting $x/(\sigma\sqrt{2})$ for ζ in Eq. (H.7) we recover the integral of the Gaussian in Eq. (H.1). Therefore, the probability for the range $\pm X$ is given by

$$P(X)=\mathrm{erf}\left(\frac{X}{\sigma\sqrt{2}}\right). \qquad \text{(H.8)}$$

Table H.1 gives the probability for several values of X.

Table H.1 says, for example, that if data points are normally distributed then 68% of the data lie within a range that is plus or minus one standard deviation from the mean. As another practical example, we expect that in a pitch-matching experiment, where the listener adjusts the frequency of a matching tone to match the pitch of a target, half the matches will lie in a range of $\pm 2/3\sigma$ around the average frequency ($2/3 \approx 0.67451$).

Computing the Error Function

The error function can be computed throughout its range with an accuracy of better than 25 parts per million from a formula given by Abramowitz and Stegun (1964):

$$\mathrm{erf}(q)=1-(a_1 t+a_2 t^2+a_3 t^3)e^{-q^2}, \qquad \text{(H9)}$$

where $t=1/(1+pq)$, $p=0.47047$, $a_1=0.3480242$, $a_2=-0.0958798$, and $a_3=0.7478556$.

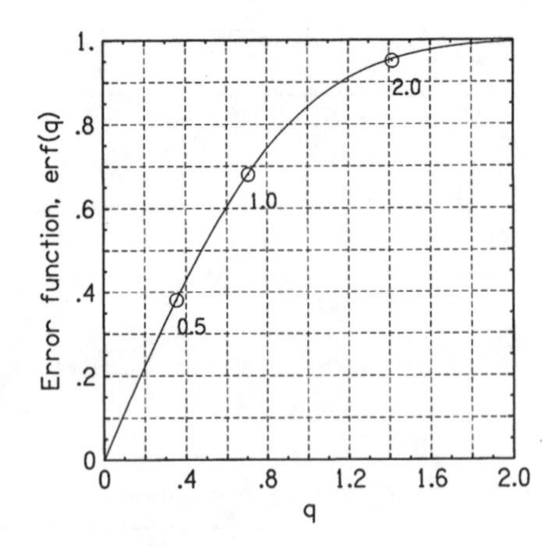

FIGURE H.3. The error function erf(q) gives the area under a Gaussian curve from $-X$ to $+X$ about the mean value. The value of q is $X/(\sigma\sqrt{2})$. Three special points are marked with the half-width X in units of standard deviation, σ, X/σ=0.5, 1.0, and 2.0.

A few values for testing purposes are erf (0.5)=0.5204998778, erf(1) =0.8427007929, and erf(2)=0.9953222650. A plot of this function is given in Fig. H.3.

CUMULATIVE NORMAL

The idea of the cumulative normal can be understood from a thought experiment where we measure random noise. Every ten microseconds we take a sample of the noise using an electronic device. The device includes an adjustable ceiling, with height x. If the sample is below the ceiling height x the device registers TRUE. If the sample is above the ceiling the device registers FALSE.

It is evident that we can control the percentages of TRUE and FALSE readings by adjusting the ceiling. For a very high ceiling almost 100% of the readings will be TRUE. The percentage of sample values that are below the ceiling is the cumulative distribution, F_C. As a function of ceiling height x, function $F_C(x)$ is a monotonically increasing function.

If the noise that we are measuring is Gaussian, then the probability that a sample has value between x and $x+dx$ follows the Normal distribution $f_N(x)$. In that case, the cumulative distribution is $F_G(x)$, the cumulative *normal* distribution. The idea is given in Fig. H.4.

Figure H.4 immediately shows us how to calculate $F_G(x)$ from the error function. It is easiest to work with the function $1-\text{erf}$ and to express the argument in terms of a z-score, x/σ. Starting with the numerical parameters for the error function, we calculate the Area function, which gives the area under the normal distri-

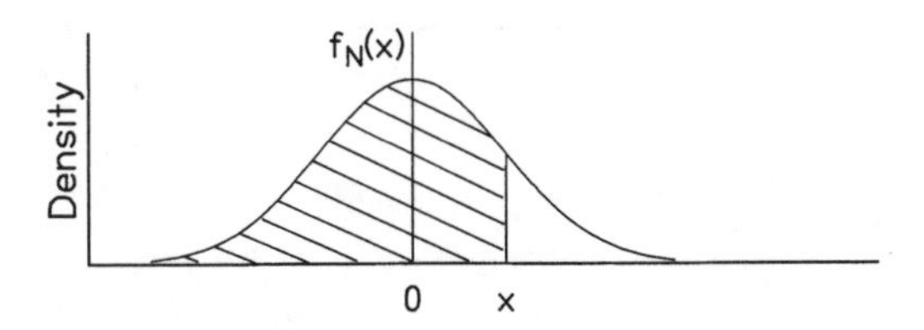

FIGURE H.4. The hatched area gives the probability that a sample of Gaussian noise is less than the ceiling value x. This area is given by the cumulative normal function $F_G(x)$.

bution starting at the extreme left. Then the cumulative normal distribution is given by

$$F_G(x) = \text{Area}(x/\sigma) \quad x \leqslant 0 \tag{H.10a}$$

$$F_G(x) = 1 - \text{Area}(x/\sigma) \quad x > 0 \tag{H.10b}$$

where

$$\text{Area}(z) = 1 - (b_1 t + b_2 t^2 + b_3 t^3) e^{-z^2/2}, \tag{H.11}$$

where $t = 1/(1 + \rho z)$, $\rho = 0.33267$, $b_1 = 0.1740121$, $b_2 = -0.0479399$, and $b_3 = 0.3739278$.

Selected values of $F_G(x)$ appear in Table H.2.

Figure H.5 shows the cumulative normal function. The value of x in units of the standard deviation is called the "z-score." The horizontal axis of Fig. H.5 goes from a z-score of -2 to $+2$.

TABLE H.2. Selected values of the cumulative normal function $F_G(x)$ as calculated from the error function

If $x=-2\sigma$ then $F_G(x)=0.025$
If $x=-\sigma$ then $F_G(x)=0.16$
If $x=-\sigma/2$ then $F_G(x)=0.31$
If $x=0$ then $F_G(x)=0.5$
If $x=\sigma/2$ then $F_G(x)=0.69$
If $x=\sigma$ then $F_G(x)=0.84$
If $x=2\sigma$ then $F_G(x)=0.975$

Cumulative Normal.

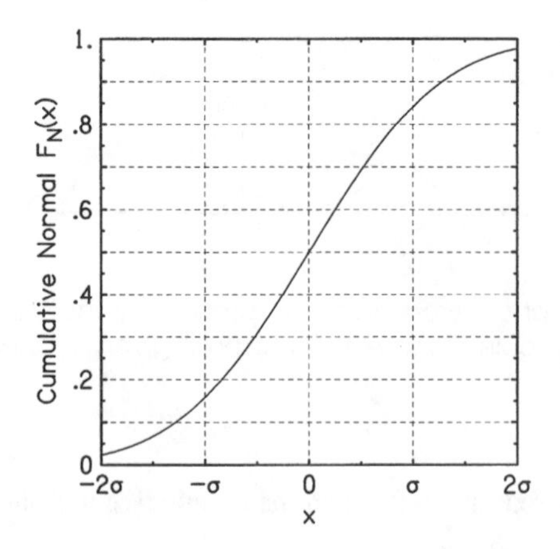

FIGURE H.5. The cumulative normal $F_G(x)$ with x expressed in units of σ.

CENTRAL LIMIT THEOREM

The *central limit theorem*, otherwise known as the "law of large numbers," says that if one adds together N random variables, then the sum tends toward a normal distribution when N becomes large. What is remarkable about this theorem is that it does not matter what the distribution of the original random variables is.

The central limit theorem also says that the average value of a large number, N, of samples from *any* distribution is a value that is approximately normally distributed. This statement of the theorem follows immediately from the first because the average is equal to the sum of the random variables divided by N.

We do not prove the central limit theorem in this appendix, but we can illustrate its operation. We begin with a rectangular distribution, as implemented in computer programming languages: All numbers between 0 and 1 are equally probable, and no numbers outside that range are possible. Therefore, the probability density is

$$R(x)=1 \quad 0\leqslant x<1$$

$$R(x)=0 \quad \text{otherwise.} \tag{H.12}$$

The density is normalized because the area under $R(x)$ is unity. The average value is 1/2 and the variance is 1/12. Fig. H.6 shows the density $R(x)$ along with a Gaussian density with mean of 1/2 and variance 1/12. The two densities are not similar, but we are not done yet.

Suppose we generate a random variable by taking two samples from the rectangular distribution R and adding them together. We call the sum x. To find the distribution of variable x we do the convolution integral of two rectangular distri-

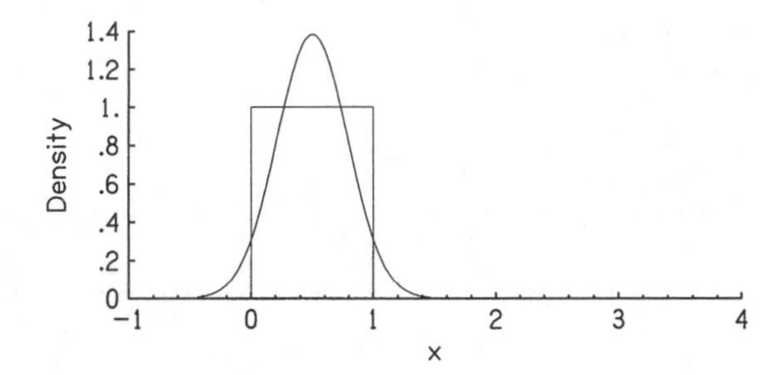

FIGURE H.6. A rectangular density and a Gaussian with the same mean and variance. Both are normalized to unit area.

butions, as described in Chapter 16. From this we find that x is distributed with a triangular density $T(x)$, given by

$$T(x)=x \quad 0 \leqslant x < 1$$

$$T(x)=2-x \quad 1 \leqslant x < 2$$

$$T(x)=0 \quad \text{otherwise.} \tag{H.13}$$

The minimum value is 0, and the maximum is 2. The density is normalized because the area under $T(x)$ is unity. The average value is 1, and the variance is 2/12. Figure H.7 shows the density $T(x)$ along with a Gaussian density that also has a mean of 1 and variance 2/12. The two densities are not very similar, but we are not done yet.

Suppose we generate a random variable by taking three samples from the rectangular distribution R and adding them together. We call the sum x. To find the distribution of variable x we do the convolution integral of three rectangular distributions, or, what is the same thing, the convolution of the rectangular density R with the triangular density T. The integrals are a little tricky, but eventually we find

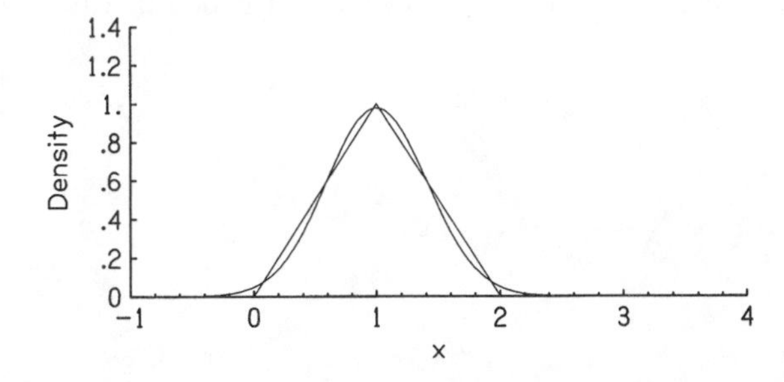

FIGURE H.7. A triangular density and a Gaussian with the same mean and variance and area.

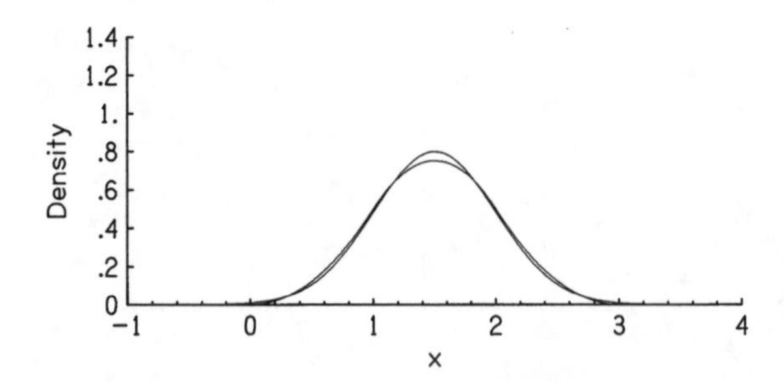

FIGURE H.8. A parabolic density and a Gaussian with the same mean and variance. The Gaussian is slightly taller.

that x is distributed with a parabolic density $\Pi(x)$, given by

$$\Pi(x)=x^2/2 \quad 0 \leqslant x < 1$$

$$\Pi(x)=\frac{3}{4}-\left(x-\frac{3}{2}\right)^2 \quad 1 \leqslant x < 2$$

$$\Pi(x)=(x-3)^2/2 \quad 2 \leqslant x < 3$$

$$\Pi(x)=0 \quad \text{otherwise.} \tag{H.14}$$

The minimum value is 0, and the maximum is 3. The density is normalized because the area under $\Pi(x)$ is unity. The average value is 3/2 and the variance is 3/12. Figure H.8 shows the density $\Pi(x)$ along with a Gaussian density with the same mean and variance. The two densities are so similar that a casual observer could not say which was the Gaussian and which was not. Only the fact that the parabolic distribution is strictly limited to the range 0–3 gives it away. If we were to continue this progression and add four or more random variables from a rectangular distribution then the sum would have a distribution that is even closer to a Gaussian.

Appendix I
Rayleigh Distribution

If a and b are two quantities that are both normally distributed with zero mean and variance σ^2, then the quantity

$$y=\sqrt{a^2+b^2} \tag{I.1}$$

is distributed according to the Rayleigh distribution. Another name for the Rayleigh distribution is Chi-squared with two degrees of freedom. The density is

$$f_R(y)=\frac{y}{\sigma^2}\, e^{-y^2/(2\sigma^2)}, \tag{I.2}$$

where σ is the standard deviation of the distribution of a and of the distribution of b. The density is normalized so that

$$\int_0^\infty dy\; f_R(y)=1. \tag{I.3}$$

The importance of the Rayleigh distribution to psychoacoustics is that it is the distribution for the amplitudes of the components of random noise. Random noise, $x(t)$, with duration T_D can be presented by a Fourier sum

$$x(t)=\sum_n a_n \cos(2\pi nt/T_D)+b_n \sin(2\pi nt/T_D). \tag{I.4}$$

A series of papers by Einstein and von Laue (1910–1915), cited by Rice (1954), concluded that the sets $\{a_n\}$ and $\{b_n\}$ can be regarded as independent, normally distributed variables. From Chapter 5 on Fourier series, we know that the amplitude of the n-th component is

$$X_n=\sqrt{a_n^2+b_n^2}. \tag{I.5}$$

Therefore, amplitude X_n has a Rayleigh distribution.

Working with the Rayleigh distribution is pleasant because, unlike the normal distribution, important integrals can be done analytically. The average value is given by

$$\mu=\int_0^\infty dy\ y\ f_R(y)=\sigma\sqrt{\frac{\pi}{2}}=1.2533\sigma. \tag{I.6}$$

The second moment is given by

$$\mu_2=\int_0^\infty dy\ y^2\ f_R(y)=2\sigma^2. \tag{I.7}$$

This is a sensible result because it is the average value of a^2+b^2, and by hypothesis both a and b distributions have variance σ^2.

Therefore, the second moment about the mean is

$$\mu_2'=\int_0^\infty dy(y-\mu)^2 f_R(y)=\sigma^2(2-\pi/2)=0.42920\sigma^2. \tag{I.8}$$

The fourth moment is given by

$$\mu_4=\int_0^\infty dy\ y^4\ f_R(y)=8\sigma^4. \tag{I.9}$$

The peak of the density f_R occurs where the first derivative of f_R=0, and this occurs for $y=\sigma$. Therefore, σ is the most probable value of y.

The cumulative Rayleigh distribution is given by

$$F_R(y)=\int_0^y dy' f_R(y'). \tag{I.10}$$

It too is an easy integral to do,

$$F_R(y)=1-e^{-y^2/(2\sigma^2)}. \tag{I.11}$$

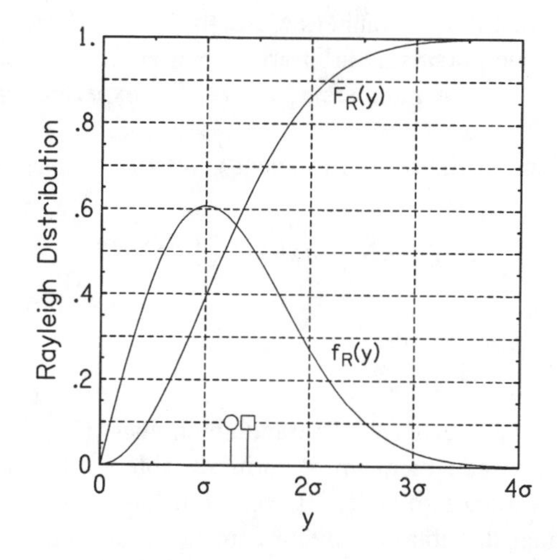

FIGURE I.1. The Rayleigh density $f_R(y)$ and the cumulative Rayleigh distribution $F_R(y)$. The most probable value occurs at $y=\sigma$. Other landmarks of interest are the mean value at $y=\sigma\sqrt{(\pi/2)}$ (circle symbol) and the RMS value at $y=\sigma\sqrt{2}$ (square symbol).

GENERATING RANDOM NUMBERS HAVING GAUSSIAN AND RAYLEIGH DISTRIBUTIONS

One sometimes needs to compute a set of random variables that follow a particular distribution. The basis for computing random numbers is usually the random number generator of a high-level computer language such as Fortran. Such generators return a number x in the range $0 \leq x < 1$, and the distribution is rectangular. In a rectangular distribution all allowed values are equally probable.

Normal Distribution

Because of the central limit theorem one can easily generate a set of normally distributed random numbers, y, starting with samples x_i, from a computer's random number generator, shifting the samples, and then adding them. For one such output value, y,

$$y = \sum_{i=1}^{N} \left(x_i - \frac{1}{2} \right). \tag{I.12}$$

Because there are only N numbers in the sum, no value of y can be greater than

$N/2$. Therefore, the distribution cannot be exactly Gaussian. However, with a large enough value of N one obtains an arbitrarily good approximation to a Gaussian.

Because the mean of the numbers x_i is 1/2, the expected value of y is zero, $E(y)=0$.

The expected value of the second moment gives the variance

$$E(y^2)=N\int_0^1 dx\left(x-\frac{1}{2}\right)^2=N/12. \tag{I.13}$$

Therefore, if we take twelve ($N=12$) random numbers ($0\leqslant x_i<1$) from a random number generator and add them up to form variable y, the values of y will be distributed with a variance of unity. If we subtract six from the sum we obtain values with a normal distribution having zero mean and unit variance. Although $N=12$ is convenient, a larger number would give a better distribution.

Rayleigh Distribution

It is possible to generate a set of random variables y that have a Rayleigh distribution starting with a set of variables x that have a rectangular distribution, like the variables from a random number generator. This section derives the mapping from x to y. We already know something about this mapping: We know that as the values of x go from zero to 1, the values of y go from zero to infinity. The key to the details of the mapping lie in the cumulative distributions.

We define $f_R(y)$ as the Rayleigh density for variables y and $R(x)$ as the rectangular density for variables x. Then the mapping of x values into y values is such that the cumulative rectangular distribution up to some arbitrary value x is equal to the cumulative Rayleigh distribution up to the value y that is the mapping $y=y(x)$. In symbols, we write this statement as

$$\int_0^x dx' R(x')=\int_0^y dy' f_R(y'). \tag{I.14}$$

The function $R(x)$, however, is simply the number 1. And the cumulative Rayleigh distribution on the right-hand side is given by F_R. Therefore, from Eq. (I.11) we find

$$x=F_R(y)=1-e^{-y^2/(2\sigma^2)}, \tag{I.15}$$

or, by inverting,

$$y = \sigma\sqrt{-2\ln(1-x)}. \tag{I.16}$$

The solution to the mapping is thus to take values of x from the random number generator and plug them into Eq. (I.16) to get values of y that are Rayleigh distributed. This simple transformation is made possible by the fact that the cumulative Rayleigh distribution can be expressed as a closed form expression, $x = F_R(y)$, and that this expression is invertible, $y = F_R^{-1}(x)$.

Appendix J
Standards

ISO FREQUENCIES

The International Standards Organization has established standard center frequencies for one-third octave bands. In units of Hz they are:

	100	**1000**	10000
12.5	**125**	1250	12500
16	160	1600	**16000**
20	200	**2000**	20000
25	**250**	2500	25000
31.5	315	3150	31500
40	400	**4000**	40000
50	**500**	5000	
63	630	6300	
80	800	**8000**	

For reference, notice that dividing an octave into three equal parts gives frequency ratios that are $2^{1/3}$, $2^{2/3}$, and $2^{3/3}$, respectively 1.259, 1.587, and 2 (compare 1.25, 1.6, and 2).

The frequencies listed in bold are the preferred one-octave band centers.

MATHEMATICAL CONSTANTS

π=3.1415 92653 58979 32384 62643
Natural logarithm base, e=2.7182 81828 45904 52353 60287

UNITS OF PRESSURE AND HEARING THRESHOLD

The fundamental MKS unit of pressure is the Pascal:

$$1 \text{ Pascal} = 1 \text{ Pa} = 1 \text{ Newton/meter}^2.$$

Other units of pressure are the standard cgs unit of dynes/cm^2 and the bar:

$$1 \ \text{dyne/cm}^2 = 0.1 \ \text{Pa},$$

$$1 \ \text{bar} = 10^5 \ \text{Pa}.$$

The bar is very close to atmospheric pressure at sea level, which is

$$1 \ \text{atm} = 760 \ \text{mmHg} = 1.01325 \times 10^5 \ \text{Pa}.$$

The nominal threshold of hearing, approximating the average threshold for normal listeners for a 1000-Hz sine tone, corresponds to an RMS pressure of

$$p_0 = 0.0002 \ \text{dyne/cm}^2 = 2 \times 10^{-5} \ \text{Pa} = 20 \ \mu\text{Pa} = 2 \times 10^{-10} \ \text{bar}.$$

This RMS pressure is the reference for the decibel scale known as Sound Pressure Level or SPL. The last equation above says that the threshold of hearing corresponds to a pressure variation that is about ten orders of magnitude (ten thousand million times) less than normal atmospheric pressure.

A sound pressure level of 120 dB SPL is often called the threshold of feeling, where listeners experience a tickling sensation in the ear. Sound at this level may be painful, and an extended exposure is almost certainly damaging. A level ratio of 120 dB corresponds to six orders of magnitude (120/20=6) so that the RMS pressure for feeling is

$$p_f = 200 \ \text{dyne/cm}^2 = 20 \ \text{Pa} = 2 \times 10^{-4} \ \text{bar}.$$

PHYSICAL CONSTANTS

Absolute zero of temperature: −273.16 degrees Centigrade=0 Kelvin
Boltzmann's constant: $k_B = 1.38066$ Joules/Kelvin
Gas constant: $R = 8.31441$ Joules/Kelvin
Speed of sound in air at 0 degrees Centigrade: $c = 331.30$ meters/s
For temperatures other than zero degrees Centigrade,

$$c = 331.30\sqrt{1 + T/273.16},$$

where T is the Centigrade temperature.

The characteristic impedance of air at 0 degrees Centigrade and 760 mm Hg is $\rho_0 c = 428$ rayls. For other conditions, the characteristic impedance is given by

$$\rho_0 c = 428.0\sqrt{\frac{273}{T}} \times \frac{P}{760},$$

where T is the absolute temperature in Kelvin, and P is the barometric pressure in mm Hg. At room temperature (20 degrees C=293 degrees K) and 760 mm Hg, $\rho_0 c = 413.3$ rayls. The unit of rayls used here is an MKS unit. One occasionally finds impedance expressed in CGS (centimeter-gram-second) units, which are ten times larger. The corresponding impedance is then 41.33 CGS rayls.

Appendix K Calculation of Intermodulation Distortion

Intermodulation distortion occurs when a signal having more than one spectral component is put into a nonlinear device. This appendix presents the mathematics of intermodulation distortion for the case of two input components only. The two input components are called *primaries*. They have frequencies f_1 and f_2, and by definition $f_2 > f_1$. The input is therefore,

$$x(t) = \cos(2\pi f_1 t) + r \cos(2\pi f_2 t + \phi), \tag{K.1}$$

where r and ϕ are the relative amplitude and phase.

We consider the memoryless distorter, where the output is related to the input by the power series in Eq. (22.1), namely

$$y(t) = c_1 x(t) + c_2 x^2(t) + c_3 x^3(t) + \ldots \tag{K.2}$$

The general term in that series is $c_n x^n$, which we can write as

$$c_n x^n = c_n \left(\frac{1}{2}\right)^n [(e^{i\theta_1} + e^{-i\theta_1}) + r(e^{i\theta_2} + e^{-i\theta_2})]^n, \tag{K.3}$$

where we have adopted the shortened notation:

$$\theta_1 = 2\pi f_1 t, \tag{K.4}$$

and

$$\theta_2 = 2\pi f_2 t + \phi. \tag{K.5}$$

The distortion term given by x^n can be divided into four kinds of terms:

$$x^n = DC + H1 + H2 + IM. \tag{K.6}$$

The terms DC are zero-frequency terms. Terms $H1$ and $H2$ are harmonics of f_1 and f_2 respectively. Terms IM are cross terms involving sums and differences of f_1 and f_2. We consider these terms one by one. Each of them involves a product of three binomial coefficients (see Chapter 22).

DC Terms

The DC terms are given by

$$DC = \left(\frac{1}{2}\right)^n \sum_{m=0}^{n} r^m \binom{n}{m} \binom{n-m}{(n-m)/2} \binom{m}{m/2}. \tag{K.7}$$

Binomial coefficients are zero unless both indices are positive integers or zero. Therefore, the last binomial coefficient is zero unless $m/2$ is an integer. Therefore, only even values of m remain in the sum. As a result, the second binomial coefficient in the sum insists that $n-m$ be even. Therefore, a DC term exists only for even powers n. This is exactly what we would expect based on our experience with harmonic distortion in Chapter 22. However, it is important to realize that the DC term is not the same as what one would get by considering the two components separately. If only the f_1 component exists then $r=0$ and only the $m=0$ term contributes to the sum above. If only the f_2 component exists then $r=\infty$, and only the $m=n$ term contributes to the sum. In both cases the equation for DC reduces to

$$DC = \left(\frac{1}{2}\right)^n \binom{n}{n/2}, \tag{K.8}$$

The terms in the sum with $m \neq 0$ and $m \neq n$ correspond to cross terms which make a contribution to the DC value because both f_1 and f_2 components exist.

*H*1 Terms

The $H1$ terms are given by

$$H1 = \left(\frac{1}{2}\right)^{n-1} \sum_{l_1=1}^{\infty} \sum_{m=0}^{n-l_1} r^m \binom{n}{m} \binom{n-m}{(n-m+l_1)/2} \binom{m}{m/2} \cos l_1 \theta_1. \tag{K.9}$$

The summation variable l_1 is a harmonic number for frequency f_1; the harmonic distortion product occurs at frequency $l_1 f_1$. The last binomial coefficient insists that m be even. The second binomial coefficient then insists that l_1 be even or odd as n is even or odd. Therefore, as expected, even harmonics appear when n is even and odd harmonics appear when n is odd. The sum on m prevents the harmonic number from being greater than n, the order of the distortion.

The $m=0$ term in the sum corresponds to the harmonic distortion arising from the f_1 term alone. All the other terms in the sum on m are contributions to harmonics of f_1 that arise because the f_2 component is also present.

H2 Terms

The $H2$ terms are given by

$$H2=\left(\frac{1}{2}\right)^{n-1}\sum_{l_2=1}^{\infty}\sum_{m=0}^{n-l_2} r^{m-n}\binom{n}{m}\binom{n-m}{(n-m+l_2)/2}\binom{m}{m/2}\cos l_2\theta_2. \quad \text{(K.10)}$$

The summation variable l_2 is a harmonic number for frequency f_2; the harmonic distortion product occurs at frequency $l_2 f_2$. Apart from the power of r and the additional phase in θ_2, the $H2$ terms are the same as the $H1$ terms. Although θ_2 depends on phase ϕ, there is no dependence on ϕ in the amplitudes of the various harmonics of f_2.

IM Terms

The IM terms are given by

$$IM=\left(\frac{1}{2}\right)^{n-1}\sum_{k_1=1}^{\infty}\sum_{k_2=1}^{\infty}\sum_{m=k_2}^{n-k_1} r^{m}\binom{n}{m}\binom{n-m}{(n-m+k_1)/2}\binom{m}{(m+k_2)/2} \quad \text{(K.11)}$$

$$\times[\cos(k_1\theta_1+k_2\theta_2)+\cos(k_1\theta_1-k_2\theta_2)]. \quad \text{(K.12)}$$

Although it appears that the two indices k_1 and k_2 range over all values, the binomial coefficients enforce the requirements that k_1+k_2 is less than or equal to n and that the sum k_1+k_2 is even or odd as n is even or odd. The disposition of the cosines shows that for every difference tone there is a summation tone with the same amplitude, regardless of relative amplitude r or relative phase ϕ.

Degeneracy

The above equations for DC, $H1$, $H2$, and IM can be used individually to find particular distortion products. For instance, $H1$ contains all the harmonics of f_1, at frequencies $f=l_1 f_1$, and IM contains all the sum and difference tones, at frequencies $f=\pm k_1 f_1 \pm k_2 f_2$. The amplitudes of the harmonics and of the sum and difference tones do not depend on the phase ϕ.

There is an important exception to the statements in the immediately preceding paragraph. The exception arises in cases of degeneracy. Degeneracy occurs when two or more combinations of l_1, l_2, k_1, and k_2 lead to the same frequency. Whether degeneracy occurs or not depends on the frequencies of the primaries, in particular on the ratio of the primary frequencies.

The way in which degeneracy affects a calculation of amplitudes of distortion products can best be illustrated with several concrete examples. We suppose that the two primary components have frequencies of $f_1=1000$ Hz and $f_2=1200$ Hz.

EXAMPLE 1: *Harmonic degeneracy*

Given $f_1=1000$ Hz and $f_2=1200$ Hz, the equation for $H1$ can be used to find the amplitude of the harmonics of 1000 Hz, and the equation for $H2$ can be used to find the amplitude of the harmonics of 1200 Hz, all independent of the relative phase. However this technique fails for harmonic distortion at 6000 Hz because there is a degeneracy, $6\times1000=5\times1200$. In general, degeneracy occurs when

$$1000\, l_1=1200\, l_2. \tag{K.13}$$

The index 6 is even and index 5 is odd, and so this degeneracy is never seen for a given power of n. But if the distortion is represented by a power series with both even and odd powers of n then this degeneracy might well be seen. Terms $H1$ and $H2$ have to be combined, with the added complication that different powers of n are weighted with different values of c_n.

At 12000 Hz there is a degeneracy for $l_1=12$ and $l_2=10$. Both indices are even and this degeneracy can occur for a single power of n, introducing a phase dependence into the amplitude for this frequency.

EXAMPLE 2: *Degeneracy among IM terms*

Degeneracy among the IM terms occurs when there are two sets of indices (k_1,k_2) and (k_1',k_2') such that

$$\pm1200k_2\pm1000k_1=\pm1200k_2'\pm1000k_1'. \tag{K.14}$$

An example occurs for n even and $n\geqslant10$ where there is the possibility

$$1000(10)-1200(2)=1200(8)-1000(2). \tag{K.15}$$

This degeneracy means that more than one term in the equation for IM contributes to the amplitude at $f=7600$ Hz, which introduces a phase dependence into the amplitude.

Again, if the power series contains both odd and even powers n, there is a possibility of degeneracy at lower values of k_1 and k_2:

$$1000(5)-1200(1)=1200(4)-1000(1). \tag{K.16}$$

Coefficients (5,1) come from n as low as 6 and coefficients (4,1) come from n as low as 5.

EXAMPLE 3: *Degeneracy between harmonics and IM terms*

The final possibility for degeneracy is across harmonic terms ($H1$ or $H2$) and IM terms. For instance,

$$1200(8)=1000(12)-1200(2). \tag{K.17}$$

This degeneracy, at 9600 Hz, between the 8th harmonic of f_2 and a difference tone $12f_1-2f_2$ can occur for n even and $n \geqslant 12$.

The cases of degeneracy shown above involve rather high values of n and rather high frequencies because the ratio of the primaries is 6/5. It should be evident that if the primaries are in a ratio of smaller integers then degeneracy will occur for lower values of n and frequency.

References

Abramowitz, M. and Stegun, I. (1964) *Handbook of Mathematical Functions*, National Bureau of Standards Applied Math Series no. 55., USGPO, Washington, DC.

Allen, J. B. (1995) ''Introduction'' in Fletcher, H. *Speech and Hearing in Communication*, ASA edition, Acoustical Society of America, New York.

Allen, J. B. and Neely, S. T. (1992) ''Micromechanical models of the cochlea,'' Physics Today, July, 40–47.

Allman, W. F. (1989) ''A matter of imperfect pitch,'' U.S. News and World Report, June 26, **106**, 55–56.

American National Standards (1960) ''Acoustical Terminology S1,'' Acoustical Soc. of Am. New York.

American National Standards (1994) ''Acoustical Terminology S1,'' Acoustical Soc. of Am. New York.

Anantharaman, J. N., Krishnamurthy, A. K., and Feth, L. L. (1993) ''IWAIF as a model for frequency discrimination,'' J. Acoust. Soc. Am. **94**, 723–729.

Assmann, P. F. and Summerfield, Q. (1990) ''Modeling the perception of concurrent vowels: Vowels with different fundamental frequencies,'' J. Acoust. Soc. Am. **88**, 680–697.

Attneave, F. and Olson, R. K. (1971) ''Pitch as a medium: A new approach to psychophysical scaling,'' Am. J. Psych. **84**, 147–165.

Aures, W. (1985) ''Ein Berechnungsverfahren der Rauhigkeit,'' Acustica **58**, 268–281.

Baggeroer, A. and Munk, W. (1992) ''The Heard Island feasibility test,'' Physics Today **45**, no. 9, 22–30.

Balzano, G. J. (1980) ''The group-theoretic description of twelvefold and microtonal pitch systems,'' Computer Music Jnl. **4**, 66–84.

Beauchamp, J. W. (1975) ''Analysis and synthesis of cornet tones using nonlinear interharmonic relationships,'' J. Audio Engr. Soc. **23**, 778–795.

Beauchamp, J. W. (1979) ''Brass-tone synthesis by spectrum evolution matching with nonlinear functions,'' Computer Music Jnl. **3**, 35–43.

Beauchamp, J. W. (1982) ''Synthesis by spectral amplitude and 'brightness' matching of analyzed musical instrument tones,'' J. Audio Engr. Soc. **30**, 396–406.

Beck, J. and Shaw, W. A. (1961) ''The scaling of pitch by magnitude estimation,'' Am. J. Psych. **74**, 242–251.

Beck, J. and Shaw, W. A. (1963) ''Single estimates of pitch magnitude,'' J. Acoust. Soc. Am. **35**, 1722–1724.

Békésy, G. von (1947) ''A new audiometer,'' Acta Otolaryngol. **35**, 411–422.

Békésy, G. von (1960) *Experiments in Hearing*, tr. E. G. Wever, McGraw Hill, reprinted by the Acoustical Society of America.

Benade, A. H. (1976) *Fundamentals of Musical Acoustics*, Oxford, New York.

Bendat, J. S. and Piersol, A. G. (1971) *Random Data Analysis and Measurement Procedures*, Wiley, New York.

Beranek, L. L. (1988) *Acoustical Measurements*, American Institute of Physics, New York.

Bernstein, L. R. and Trahiotis, C. (1996) ''Binaural beats at high frequencies: Listeners use of envelope-based interaural temporal and intensive disparities,'' J. Acoust. Soc. Am. **99**, 1670–1679.

Bialek, W. S. and Wit, H. P. (1984) ''Quantum limits to oscillator stability: Theory and experiments on otoacoustic emissions from the human ear,'' Phys. Letts. **104A**, 1973–1978.

Bilsen, F. A. (1977) ''Pitch of noise signals: Evidence for a central spectrum,'' J. Acoust. Soc. Am. **61**, 150–161.

Bilsen, F. A. (1980) ''Atonal periodicity sensation for comb filtered noise signals,'' *Psychophysical Physiological and Behavioral Studies in Hearing*, (eds. G. van den Brink and F. A. Bilsen) Delft University Press, Delft, Netherlands, (pp. 379–383).

Bilsen, F. A. and Goldstein, J. L. (1974) ''Pitch of dichotically delayed noise and its possible spectral basis,'' J. Acoust. Soc. Am. **55**, 292–296.

Bilsen, F. A. and Ritsma, R. J. (1969/70) ''Repetition pitch and its implication for hearing theory,'' Acustica **22**, 63–73.

Blauert, J. (1972) "On the lag of lateralization caused by interaural time and intensity differences," Audiology, **11**, 265–270.

Boer, E., de (1956) "On the Residue in Hearing," Doctoral thesis, University of Amsterdam, The Netherlands.

Boer, E., de and Jongh, H. R., de (1978) "On cochlear encoding: Potentialities and limitations of the reverse-correlation technique," J. Acoust. Soc. Am. **63**, 115–135.

Boer, E., de and van Breugel H. (1984) "Distributions of judgements in adaptive testing," Biol. Cybern. **50**, 343–355.

Bordone-Sacerdote, C. and Sacerdote, G. G. (1971) "The structure of choral sing," Proc Seventh International Congress on Acoustics, **3**, 245–248, Akademiai Kiado, Budapest.

Bregman, A. S. (1981) "Asking the 'what for' question in auditory perception," in *Perceptual Organization*, ed. M. Kubovy, J. R. Pomerantz, and L. Erlbaum, Hillsdale, N. J.

Brink, G. van den (1970) "Experiments on binaural diplacusis and tone perception," in *Frequency Analysis and Periodicity Detection in Hearing*, ed. R. Plomp, G. F. Smoorenburg, and A. W. Sijthoff, Leiden, The Netherlands.

Brink, G. van den (1975a) "The relation between binaural diplacusis for pure tones and for complex sounds under normal conditions and with induced monaural pitch shift," Acustica **32**, 159–165.

Brink, G. van den (1975b) "Monaural frequency-pitch relations as the origin of binaural diplacusis for pure tones and residue sounds," Acustica **32**, 166–173.

Brink, G. van den (1980) "Cochlear mechanics as the possible cause of binaural diplacusis," in *Psychophysical, Physiological and Behavioral Studies in Hearing*, ed. G. van den Brink and F. A. Bilsen, Delft University Press, Delft, The Netherlands.

Brokx, J. P. L. and Nooteboom, S. G. (1982) "Intonation and the perceptual segregation of simultaneous voices," J. Phonetics **10**, 23–36.

Brown, J. C. and Puckette, M. S. (1993) "A high-resolution fundamental frequency determination based on phase changes of the Fourier transform," J. Acoust. Soc. Am. **94**, 662–667.

Brown, J. C. and Vaughn, K. V. (1995) "Pitch center of frequency-modulated musical sounds," Private communication.

Burns, E. M. (1981) "Circularity in relative pitch judgments for inharmonic complex tones: The Shepard demonstration revisited, again," Percept. and Psychophys. **30**, 467–472.

Burns, E. M. (1982) "Pure-tone pitch anomalies. I. Pitch-intensity effects and diplacusis in normal ears," J. Acoust. Soc. Am. **72**, 1394–1402.

Burns, E. M. and Feth, L. M. (1983) "High frequency pitch perception," J. Acoust. Soc. Am. (abst.) **73**, S44.

Burns, E. M. and Viemeister, N. F. (1976) "Nonspectral pitch," J. Acoust. Soc. Am. **60**, 863–869.

Burns, E. M. and Viemeister, N. F. (1981) "Played again SAM: Further observations on the pitch of amplitude modulated noise," J. Acoust. Soc. Am. **70**, 1655–1660.

Burns, E. M. and Ward, W. D. (1982) "Intervals, scales and tuning," in *The Psychology of Music*, ed. D. Deutsch, Academic Press, New York, (pp. 241–269).

Burns, E. M., Hoberg, K. E., Schlauch, R. S., and Bargones, J. Y. (1986) "Frequency microstructure of pitch-level functions," J. Acoust. Soc. Am. (abst.) **80**, S93.

Buunen, T. J. F. and Vlaming, M. S. M. G. (1981) "Laser-Doppler velocity meter applied to tympanic membrane vibrations in cat," J. Acoust. Soc. Am. **69**, 744–750.

Buus, S., Florentine, M., and Zwicker, T. (1995) "Psychometric functions for level discrimination in cochlearly impaired and normal listeners with equivalent-threshold masking," J. Acoust. Soc. Am. **98**, 853–861.

Carlyon, R. P., Buus, S., and Florentine, M. (1990) "Temporal integration of trains of tone pulses by normal and by cochlear impaired listeners," J. Acoust. Soc. Am. **87**, 260–268.

Carney, L. H. and Yin, T. C. T. (1988) "Temporal coding of resonances by low-frequency auditory nerve fibers: single-fiber responses and a population model," J. Neurophys. **60**, 1653–1677.

Chowning, J. M. (1973) "The synthesis of complex audio spectra by means of frequency modulation," J. Audio Engr. Soc. **21**, 526–534.

Christman, R. J. and Williams, W. E. (1963) "Influence of the time interval on experimentally-induced shifts of pitch," J. Acoust. Soc. Am. **35**, 1030–1033.

Clack, T. D., Erdreich, J., and Knighton, R. W. (1972) "Aural harmonics: The monaural phase effects at 1500 Hz, 2000 Hz and 2500 Hz observed in tone-on-tone masking when $f_1 = 1000$ Hz," J. Acoust. Soc. Am. **52**, 536–541.

Clifton, R. (1987) "Breakdown of echo suppression in the precedence effect," J. Acoust. Soc. Am. **82**, 1834–1835.

Cohen, E. (1984a) "Some effects of inharmonic partials on interval perception," Music Percept. **1**, 323–349.
Cohen, M. F. and Schubert, E. D. (1987) "Influence of place synchrony on the detection of sinusoids," J. Acoust. Soc. Am. **81**, 452–458.
Cohen, M. H. (1984b) *Quantifying Music—The Science of Music at the First Stage of the Scientific Revolution 1580–1650*, D. Reidel Publishing Co., Dordrecht, Holland, volume 23 of the University of Western Ontario Series in Philosophy of Science.
Colburn, H. S. (1973) "Theory of binaural interaction based on auditory-nerve data I. General strategy and preliminary results on interaural discrimination," J. Acoust. Soc. Am. **54**, 1458–1470.
Colburn, H. S. (1977) "Theory of binaural interaction based on auditory-nerve data II. Detection of tones in noise," J. Acoust. Soc. Am. **61**, 525–533.
Corey, V. B. (1950) "Change in pitch of sustained musical note by clenching jaws of observer," J. Acoust. Soc. Am. (letter) **22**, 640.
Cramer, E. M. and Huggins, W. H. (1958) "Creation of pitch through binaural interaction," J. Acoust. Soc. Am. **30**, 413–417.
Dai, H. (1993) "On the pitch of two–tone complexes," J. Acoust. Soc. Am. **94**, 730–734.
Dallos, P. (1988) "The cochlea and the auditory nerve," in *Auditory Function*, ed. G. M. Edelman, W. E. Gall, and W. M. Cowen, Wiley, New York.
Dallos, P. (1992) "The active cochlea," J. Neurosci. **12**, 4575–4585.
Darwin, C. J. (1984) "Perceiving vowels in the presence of another sound: Constraints on formant perception," J. Acoust. Soc. Am. **76**, 1636–1647.
De Togne, G. A. (1992) "Adding sound to the Mirage volcano," Sound and Video Contractor **10**, No. 11, Nov., p. 33.
Demany, L. and Semal, C. (1986) "On the detection of amplitude modulation and frequency modulation at low modulation frequencies," Acustica **61**, 243–255.
Demany, L. and Semal, C. (1989) "Detection thresholds for sinusoidal frequency modulation," J. Acoust. Soc. Am. **85**, 1295–1301.
Deng, L., Geisler, C. D., and Greenberg, S. (1987) "Responses of auditory-nerve fibers to multiple-tone complexes," J. Acoust. Soc. Am. **82**, 1989–2000.
Deutsch, D. (1978) "Octave generalization and melody identification," Percept. and Psychophys. **23**, 91–92.
Deutsch, D. (1986) "A musical paradox," Music Percept. **3**, 275–280.
Deutsch, D. (1990) "The tritone paradox: An influence of language on music perception," Music Percept. **8**, 335–347.
Dirac, P. A. M. (1958) *Quantum Mechanics*, Oxford, fourth ed.
Doughty, J. M. and Garner, W. R. (1947) "Pitch characteristics of short tones I—Two kinds of pitch threshold," J. Exp. Psych. **37**, 351–365.
Doughty, J. M. and Garner, W. R. (1948) "Pitch characteristics of short tones II—Pitch as a function of tonal duration," J. Exp. Psych. **38**, 478–494.
Duifhuis, H., (1970a) "Audibility of high harmonics in a periodic pulse," J. Acoust. Soc. Am. **48**, 888–893.
Duifhuis, H., (1970b) "Audibility of high harmonics in a periodic pulse II, Time effect," J. Acoust. Soc. Am. **49**, 1155–1162.
Duifhuis, H., Willems, L. F., and Sluyter, R. J. (1982) "Measurement of pitch in speech: An implementation of Goldstein's theory of pitch perception," J. Acoust. Soc. Am. **71**, 1568–1580.
Durlach, N. I. and Colburn, H. S. (1978) "Binaural Phenomena" in *Handbook of Perception*, vol. 4, ed. E. Carterette, Academic, New York, (pp. 365–466).
Durlach, N. I., Gabriel, K. J., Colburn, H. S., and Trahiotis, C. (1986) Interaural correlational discrimination I. Bandwidth and level dependence," J. Acoust. Soc. Am. **79**, 1548–1557.
Dye, R. H., Niemiec, A. J., and Stellmack, M. A. (1994) "Discrimination of interaural envelope delays: The effect of randomizing component starting phase," J. Acoust. Soc. Am. **95**, 463–470.
Eadie, W. T., Dryard, D., James, F. E., Roos, M., and Sadoulet, B. (1971) *Statistical Methods in Experimental Physics*, North-Holland, Amsterdam.
Ebata, M., Tsumura, N., and Okda, J. (1983) "Pitch shift of a tone burst on the presence of a preceding tone," Proc. 11th ICA, Revue d'Acoustique **3**, 111–114.
Eddins, D. and Green, D. M. (1995) "Temporal integration and temporal resolution," in *Hearing—Handbook of Perception and Cognition* second edition, ed. B. C. J. Moore, Academic Press, San Diego (pp. 207–239).
Edwards, B. W. and Viemeister, N. F. (1994) "Modulation detection and discrimination with three-

component signals,'' J. Acoust. Soc. Am. **95**, 2202–2212.
Egan, J. P. and Hake, H. W. (1950) ''On the masking pattern of a simple auditory stimulus,'' J. Acoust. Soc. Am. **22**, 622–630.
Egan, J. P. and Meyer, D. R. (1950) ''Changes in pitch of tones of low frequency as a function of the pattern of excitation produced by the band of noise,'' J. Acoust. Soc. Am. **22**, 827–833.
Einstein, A. and Hopf, L. (1910), ''A principle of the calculus of probabilities and its application to radiation theory,'' Ann. d. Physik **33**, 1096–1115.
Elliott, D. N., Sheposh, J., and Frazier, J. (1964) ''The effect of monaural fatigue on pitch matching and discrimination,'' J. Acoust. Soc. Am. **36**, 752–756.
Ellis, A. J. (1885) ''Some effects of inharmonic partials on interval perception, The History of Musical Pitch in Europe,'' Addition H to his translation of Helmholtz, H. *On the Sensations of Tone*, Dover, ed. 1954.
Erdreich, J. (1986) ''A distribution based definition of impulsive noise,'' J. Acoust. Soc. Am. **79**, 990–998.
Evans, J. E. (1968) ''Characterization of synchronization to low frequency tones by fibres in the cat's auditory nerve,'' (unpublished).
Fastl, H. (1978) ''Frequency discrimination for pulsed vs modulated tones,'' J. Acoust. Soc. Am. **63**, 275–277.
Faulkner, A. (1985) ''Pitch discrimination of harmonic complex signals: Residue pitch or multiple component discriminations?'' J. Acoust. Soc. Am. **78**, 1993–2004.
Feth, L. L. (1974) ''Frequency discrimination of complex periodic tones,'' Percept. and Psychophys., **15**, 375–378.
Feth, L. L. and O'Malley, H. (1977) ''Two-tone auditory spectral resolution,'' J. Acoust. Soc. Am. **62**, 940–947.
Feth, L. L., O'Malley, H., and Ramsey, J. W. (1982) ''Pitch of unresolved two-tone complex tones,'' J. Acoust. Soc. Am. **72**, 1403–1412.
Flanagan, J. L. (1972) *Speech analysis, synthesis and perception*, Springer, Heidelberg (second ed.).
Flanagan, J. L. and Guttman, N. (1960) ''On the pitch of periodic pulses,'' J. Acoust. Soc. Am. **32**, 1308–1319.
Fletcher, H. (1930) ''A space-time pattern theory of hearing,'' J. Acoust. Soc. Am. **1**, 313–343.
Fletcher, H. (1934) ''Loudness, pitch and timbre of musical tones and their relations to the intensity, the frequency and the overtone structure,'' J. Acoust. Soc. Am. **6**, 59–69.
Fletcher, H. (1938) ''Loudness, masking and their relation to the hearing process and the problem of noise measurement,'' J. Acoust. Soc. Am. **9**, 275–293.
Fletcher, H. (1940) ''Auditory Patterns,'' Rev. Mod. Phys. **12**, 47–65.
Fletcher, H. (1953) *Speech and Hearing in Communication*, Van Nostrand, New York, reprinted (1995) by the Acoustical Society of America, New York, ed. J. B. Allen.
Fletcher, H. (1964) ''Normal vibration frequencies of a stiff piano string,'' J. Acoust. Soc. Am. **36**, 203–209.
Fletcher, H. and Steinberg, J. C. (1924) ''The dependence of the loudness of a complex sound upon the energy in various frequency regions of the sound,'' Phys. Rev. **24**, 306–317.
Florentine, M., Buus, S., and Mason, C. R. (1987) ''Level discrimination as a function of level and frequency between 0.25 and 16 kHz,'' J. Acoust. Soc. Am. **81**, 1528–1541.
Florentine, M., Fastl, H., and Buus, S. (1988) ''Temporal integration in normal hearing, cochlear impairment, and impairment simulated by masking,'' J. Acoust. Soc. Am. **84**, 195–203.
Forrest, T. and Green, D. M. (1987) ''Detection of partially-filled gaps in noise and the temporal modulation transfer function,'' J. Acoust. Soc. Am. **82**, 1933–1943.
Freyman, R. L., Clifton, R. K., and Litovsky, R. Y. (1991) ''Dynamic processes in the precedence effect,'' J. Acoust. Soc. Am. **90**, 874–884.
Furst, M., Rabinowitz, W. M., and Zurek, P. M. (1988) ''Ear canal acoustic distortion at $2f_1-f_2$ from human ears: Relation to other emissions and perceived combination tones,'' J. Acoust. Soc. Am. **84**, 215–221.
Furst, M., Reshef (Haran), I., and Attias, J. (1992) ''Manifestations of intense noise stimulation on spontaneous otoacoustic emission and threshold microstructure: Experiment and model,'' J. Acoust. Soc. Am. **91**, 1003–1014.
Gabor, D. (1946) ''Theory of communications,'' J. Inst. Elec. Engrs. (London) **93**, Part III, 429.
Gerken, G. M., Bhat, V. K. H., and Hutchison-Clutter, M. (1990) ''Auditory temporal integration and the power function model,'' J. Acoust. Soc. Am. **88**, 767–778.
Gibian, G. L. and Kim, D. O. (1982) ''Cochlear microphonic evidence for mechanical propagation of

distortion products (f_1-f_2) and ($2f_1-f_2$),'' Hearing Res. **6**, 35–59.
Glasberg, B. R. and Moore, B. C. J. (1990) ''Derivation of auditory filter shapes from notched noise data,'' Hearing Res. **47**, 103–138.
Gold, T. and Pumphrey, R. J. (1948) ''Hearing I. The cochlea as a frequency analyzer,'' Proc. Roy. Soc. (London) **135**, 462–491.
Goldstein, J. L. (1967a) ''Auditory spectral filtering and monaural phase perception,'' J. Acoust. Soc. Am. **41**, 458–479.
Goldstein, J. L. (1967b) ''Auditory nonlinearity,'' J. Acoust. Soc. Am. **41**, 676–689.
Goldstein, J. L. (1973) ''An optimal processor theory for the central formation of the pitch of complex tones,'' J. Acoust. Soc. Am. **54**, 1496–1516.
Goldstein, J. L. and Srulovicz, P. (1977) ''Auditory-nerve spike intervals as an adequate basis for aural spectrum analysis,'' in *Psychophysics and Physiology of Hearing*, ed. E. F. Evans and J. P. Wilson, Academic, New York (pp. 337–345).
Grantham, D. W. and Wightman, F. L. (1978) ''Detectability of varying interaural temporal differences,'' J. Acoust. Soc. Am. **63**, 511–523.
Grantham, D. W. and Wightman, F. L. (1979) ''Detectability of a pulsed tone in the presence of a masker with time-varying interaural correlation,'' J. Acoust. Soc. Am. **65**, 1509–1517.
Green, D. M. (1965) ''Masking with two tones,'' J. Acoust. Soc. Am. **37**, 802–813.
Green, D. M. (1968) ''Sine and cosine masking,'' J. Acoust. Soc. Am. **44**, 168–175.
Green, D. M. (1976) *An Introduction to Hearing*, Lawrence Erlbaum, Hilsdale, NJ.
Green, D. M. (1990) ''Stimulus selection in adaptive psychophysical procedures,'' J. Acoust. Soc. Am. **87**, 2662–2674.
Green, D. M. and Swets, J. A. (1966) *Signal Detection Theory and Psychophysics*, J. Wiley, New York, reprinted (1988) by Peninsula Publishing, Los Altos, CA.
Green, D. M., Mason, C. R., and Kidd, G. (1984) ''Profile analysis: Critical bands and duration,'' J. Acoust. Soc. Am. **75**, 1163–1167.
Green, D. M., Wier, C. C., and Wightman, F. L. (1975) ''Gold and Pumphrey revisited, again,'' J. Acoust. Soc. Am. **57**, 935–938.
Greenwood, D. D. (1961) ''Critical bandwidth and the frequency coordinates of the basilar membrane,'' J. Acoust. Soc. Am. **33**, 1344–1356.
Greenwood, D. D. (1990) ''A cochlear frequency-position function for several species—29 years later,'' J. Acoust. Soc. Am. **87**, 2592–2605.
Greenwood, D. D. (1991) ''Critical bandwidth and consonance in relation to cochlear frequency-position coordinates,'' Hearing Res. **54**, 164–208.
Greenwood, D. D. (1995) Personal communication.
Grey, J. M. (1977) ''Multidimensional scaling of musical timbre,'' J. Acoust. Soc. Am. **61**, 1270–1277.
Guinan, J. and Peake, W. T. (1967) ''Middle ear characteristics of anesthetized cats,'' J. Acoust. Soc. Am. **41**, 1237–1261.
Guinness Book of World Records (1990): Vampire bat 120–210 kHz, Common dolphin 280 kHz, Bantam, New York, (p. 50).
Gulick, W. L., Gescheider, G. A., and Frisina, R. D. (1989) *Hearing*, Oxford University Press, New York.
Guttman, N. and Julesz, B. (1963) ''Lower limits of auditory periodicity analysis,'' J. Acoust. Soc. Am. **35**, 610.
Hacker, M. J. and Ratcliff, R. (1979) ''A revised table of d' for M-alternative forced choice,'' Percept. and Psychophys. **26**, 168–170.
Hafter, E. R., Dye, R. H., Wenzel, E. M., and Knecht, K. (1990) ''The combination of interaural time and intensity in the lateralization of high-frequency complex signals,'' J. Acoust. Soc. Am. **87**, 1702–1708.
Hall, D. (1991) *Musical Acoustics*, second ed. Brooks Cole, Pacific Grove, CA.
Hall, J. L. (1974) ''Two-tone distortion products in a nonlinear model of the basilar membrane,'' J. Acoust. Soc. Am. **56**, 1818–1828.
Hall, J. L., Mathews, M. V., and Roberts, L. A. (1983) ''Phantom beats in mistuned chords,'' J. Acoust. Soc. Am. abst. **74**, S22
Hall, J. W. and Peters, R. W. (1981) ''Pitch for nonsimultaneous successive harmonics in quiet and noise,'' J. Acoust Soc. Am. **69**, 509–513.
Hall, J. W., Haggard, M. P., and Fernandes, M. A. (1984) ''Detection in noise by spectro-temporal pattern analysis,'' J. Acoust. Soc. Am. **76**, 50–56.

Hanna, T. E. and Robinson, D. E. (1985) "Phase effects for a sine wave masked by reproducible noise," J. Acoust. Soc. Am. **77**, 1129–1140.
Harris, D. J. (1960) "Scaling of pitch intervals," J. Acoust. Soc. Am. **32**, 1575–1581.
Hartmann, W. M. (1978a) "Flanging and phasers," J. Audio Engr. Soc. **26**, 439–443.
Hartmann, W. M. (1978b) "The effect of amplitude envelope on the pitch of sine wave tones," J. Acoust. Soc. Am. **63**, 1105–1113.
Hartmann, W. M. (1979) "Some psychoacoustical experiments with all-pass networks," Am. J. Phys. **47**, 29–34.
Hartmann, W. M. (1981) "A model for the perception of reiterated noise at low repetition frequency," J. Acoust. Soc. Am. (abst.) **69**, S65, and ΦΨ Report 36.
Hartmann, W. M. (1982) "On the detection of a tone masked by two tones," J. Acoust. Soc. Am. **71**, 127–132.
Hartmann, W. M. (1985) "The frequency-domain grating," J. Acoust. Soc. Am. **78**, 1421–1425.
Hartmann, W. M. (1988) "Pitch perception and the segregation and integration of auditory entities," in *Auditory Function*, ed. G. M. Edelman, W. E. Gall, and W. M. Cowan (Wiley, New York) (pp. 623–645).
Hartmann, W. M. (1993a) "Auditory demonstrations on compact disk for large N," J. Acoust. Soc. Am. **93**, 1–16.
Hartmann, W. M. (1993b) "On the origin of the enlarged melodic octave," J. Acoust. Soc. Am. **93**, 3400–3409.
Hartmann, W. M. (1996) "Listening in a room and the precedence effect," in *Binaural and Spatial Hearing*, ed. R. H. Gilkey and T. B. Anderson, Lawrence Erlbaum, Hilsdale, NJ.
Hartmann, W. M. and Blumenstock, B. J. (1976) "Time dependence of pitch perception: Pitch step experiment," J. Acoust. Soc. Am. (abst.) **60**, S40.
Hartmann, W. M. and Hnath, G. M. (1982) "Detection of mixed modulation," Acustica **50**, 297–312.
Hartmann, W. M. and Johnson, D. (1991) "Stream segregation and peripheral channeling," Music Percept. **9**, 155–184.
Hartmann, W. M. and Kanistanaux, D. C. (1979) "The effect of a prior tone on the pitch of a short tone," *Proc. Res. Symposium on the Psychology and Acoustics of Music*, ed. W. V. May, pp. 199–215, also J. Acoust. Soc. Am. (abst.) **65**, S37.
Hartmann, W. M. and Klein, M. A. (1980) "Theory of frequency modulation detection for low modulation frequencies" J. Acoust. Soc. Am. **67**, 935–946.
Hartmann, W. M. and Klein, M. A. (1981) "The effect of uncertainty on the detection of FM for low modulation frequencies," Percept. and Psychophys. **30**, 417–424.
Hartmann, W. M. and Lin, J-Y. (1994) "Experiments with the tritone paradox," ΦΨ Report number 85.
Hartmann, W. M. and Pumplin, J. (1988) "Noise power fluctuations and the masking of sine signals," J. Acoust. Soc. Am. **83**, 2277–2289.
Hartmann, W. M. and Pumplin, J. (1991) "Periodic signals with minimal power fluctuations," J. Acoust. Soc. Am. **90**, 1986–1999.
Hartmann, W. M. and Sartor, D. (1991) "Turning on a tone," J. Acoust. Soc. Am. **90**, 866–873.
Hartmann, W. M., McAdams, S., and Smith, B. K. (1990) "Matching the pitch of a mistuned harmonic in an otherwise periodic complex tone," J. Acoust. Soc. Am. **88**, 1712–1724.
Hartmann, W. M., McAdams, S., Gerszo, A., and Boulez, P. (1986) "Discrimination of spectral density," J. Acoust. Soc. Am. **79**, 1915–1925.
Hartmann, W. M., Rakerd, B., and Packard, T. N. (1985) "On measuring the frequency difference limen for short tones," Percept. and Psychophys. **38**, 199–207.
Hauser, M. W. (1991) "Principles of oversampling A/D conversion," J. Audio Engr. Soc. **39**, 3–26.
Hawkins, J. E. and Stevens, S. S. (1950) "The masking of pure tones and speech by white noise," J. Acoust. Soc. Am. **22**, 6–13.
Heinbach, W. (1988) "Aurally adequate signal representation. The part-tone time pattern," Acustica **67**, 113–121.
Heisenberg, W. (1927) Zeits. f. Physik **43**, 172.
Hellman, R. P. and Zwislocki, J. J. (1963) "Monaural loudness function at 1000 cps and interaural summation," J. Acoust. Soc. Am. **35**, 856–865.
Hellman, W. S. and Hellman, R. P. (1990) "Intensity discrimination as the driving force for loudness. Application to pure tones in quiet," J. Acoust. Soc. Am. **87**, 1255–1265.
Helmholtz, H. von (1877) *On the sensations of tone*, tr. Ellis, A. J., 1885, reprinted (1954) by Dover, New York.
Hellström, P.-A. (1993) "The relationship between sound transfer functions from free sound field to the

eardrum and temporary threshold shift,'' J. Acoust. Soc. Am. **94**, 1301–1306.
Henning, G. B. (1966) ''Frequency discrimination of random-amplitude tones,'' J. Acoust. Soc. Am. **39**, 336–339.
Henning, G. B. (1970) ''Effects of duration on frequency and amplitude discrimination,'' in *Frequency Analysis and Periodicity Detection in Hearing*, ed. R. Plomp and G. Smoorenburg, (pp. 350–361), A. W. Sijthoff, Leiden.
Henning, G. B. (1974) ''Detectability of interaural delay in high-frequency complex waveforms,'' J. Acoust. Soc. Am. **55**, 84–90.
Henning, G. B. and Grosberg, S. L. (1968) ''Effect of harmonic components on frequency discrimination,'' J. Acoust. Soc. Am. **44**, 1386–1389.
Hiesey, R. W. and Schubert, E. D. (1972) ''Cochlear resonance and phase-reversed signals,'' J. Acoust. Soc. Am. **51**, 518–519.
Holdsworth, J., Nimmo-Smith, I., Patterson, R., and Rice, P. (1988) Appendix C in Patterson *et al.* (1991).
Honrubia, V. and Ward, R. (1968) ''Longitudinal distribution of the cochlear microphonics inside the cochlear duct (Guinea pig),'' J. Acoust. Soc. Am. **44**, 951–958.
Horowitz, P. and Hill W. (1989) *The Art of Electronics*, Cambridge University Press, Cambridge, UK, second edition.
Houtgast, T. (1972) ''Psychophysical evidence for lateral inhibition in hearing,'' J. Acoust. Soc. Am. **51**, 1885–1894.
Houtgast, T. (1974) ''Lateral suppression in hearing,'' Thesis, University of Amsterdam.
Houtsma, A. J. M. (1981) ''Noise-induced shifts in the pitch of pure and complex tones,'' J. Acoust. Soc. Am. **70**, 1661–1668.
Houtsma, A. J. M. and Goldstein, J. L. (1972) ''The central origin of the pitch of complex tones: Evidence from musical interval recognition,'' J. Acoust Soc. Am. **51**, 520–529.
Houtsma, A. J. M., Durlach, N. I., and Braida, L. D. (1980) ''Intensity perception XI. Experimental results on the relation of intensity resolution to loudness matching,'' J. Acoust. Soc. Am. **68**, 807–813.
Houtsma, A. J. M., Rossing, T. D., and Wagenaars, W. M. (1987) *Auditory Demonstrations*, compact disc, Acoustical Society of America, Philips Eindhoven, The Netherlands.
Hsueh, K. D. and Hamernik, R. P. (1990) ''A generalized approach to random noise synthesis: Theory and computer simulation,'' J. Acoust. Soc. Am. **87**, 1207–1217.
Huggins, W. H. (1952) ''A phase principle for complex-frequency analysis and its implications in auditory theory,'' J. Acoust. Soc. Am. **24**, 582–589.
Hykes, D. and Chemirani, D. *Windhorse Riders*, New Albion Records Inc., 584 Castro #515, San Francisco, CA, 94114, compact disc number NA024.
Jaroszewski, A. and Rakowski, A. (1976) ''Pitch shifts in post-stimulatory masking,'' Acustica **34**, 220–223.
Javel, E. (1980) ''Coding of AM tones in the chinchilla auditory nerve: Implications for the pitch of complex tones,'' J. Acoust. Soc. Am. **68**, 133–146.
Jeffress, L. A. (1948) ''A place theory of sound localization,'' J. Comp. Physiol. Psychol. **41**, 35–39.
Jeffress, L. A. (1968) ''Beating sinusoids and pitch changes,'' J. Acoust. Soc. Am. **43**, 1464.
Jesteadt, W. and Sims, S. L. (1975) ''Decision processes in frequency discrimination,'' J. Acoust. Soc. Am. **57**, 1161–1168.
Johannesma, P. I. M. (1972) ''The pre-response stimulus ensemble of neurons in the cochlear nucleus,'' *Hearing Theory 1972*, Institute for Perception Research, Eindhoven, Holland, (pp. 58–69).
Johnson, D. H. (1974) *The response of single auditory-nerve fibers in the cat to single tones; synchrony and average rate*, thesis MIT.
Johnson, D. H. (1980) ''The relationship between spike rate and synchrony in responses of auditory-nerve fibers to single tones,'' J. Acoust. Soc. Am. **68**, 1115–1122.
Johnstone, B. M., Patuzzi, R., and Yates, G. K. (1986) ''Basilar membrane measurements and the traveling wave,'' Hearing Res. **22**, 147–153.
Jones, K., Tubis, A., and Burns, E. M. (1983) ''Temporal-neural-response correlates of pitch-intensity effects and diplacusis,'' J. Acoust. Soc. Am. (abst.) **74**, S8.
Jones, K., Tubis, A., and Burns, E. M. (1985) ''On the extraction of the signal excitation function from a non-Poisson cochlear neural spike chain,'' J. Acoust. Soc. Am. **78**, 90–94.
Jorgensen, O. (1977) *Tuning the Historical Temperaments by Ear*, Northern Michigan University Press, Marquette, MI.
Julesz, B. (1971) *Foundations of Cyclopean Perception*, University of Chicago Press, Chicago.

Kaernbach, C. (1992) ''On the consistency of tapping to repeated noise,'' J. Acoust. Soc. Am. **92**, 788–793.

Kaernbach, C. (1993) ''Temporal and spectral basis of the features perceived in repeated noise,'' J. Acoust. Soc. Am. **94**, 91–97.

Kay, R. H. and Matthews, D. R. (1972) ''On the existence in the human auditory pathway of channels selectively tuned to the modulation present in FM tones,'' J. Physiol. **255**, 657–677.

Kemp, D. T. (1978) ''Stimulated acoustic emissions from within the human auditory system,'' J. Acoust. Soc. Am. **64**, 1386–1391.

Kemp, D. T. (1979a) ''Evidence of mechanical nonlinearity and frequency selective wave amplification in the cochlea,'' Arch. Otorhinolaryngol, **224**, 37–45.

Kemp, D. T. (1979b) ''The evoked cochlear mechanical response and the auditory microstructure—Evidence for a new element in cochlear mechanics,'' in *Models of the Auditory System and Related Signal Processing Techniques,* eds. M. Hoke and E. deBoer, Scand. Audiol. Suppl. **9**, 35–47.

Kiang, N. Y-S., Watanabe, T., Thomas, E. C., and Clark, L. F. (1965) *Discharge Patterns of Single Fibers in the Cat's Auditory Nerve*, MIT Press, Cambridge, MA.

Kidd, G., Mason, C. R., Brantley, M. A., and Owen, G. A. (1989) ''Roving-level tone-in-noise detection,'' J. Acoust. Soc. Am. **86**, 1310–1317.

King, B. H. (1994) ''Accuphase DP-90 cd transport and DC-91 D/A converter,'' Audio, **78**, number 4, April, 44–54.

Klasco, M. (1993) ''Compression Drivers,'' Sound and Communications, **39**, No. 12, (pp. 32).

Klein, M. A. (1981) ''Two Studies of Pitch Perception,'' Ph.D. Thesis, Michigan State University, unpublished.

Klein, M. A. and Hartmann, W. M. (1979) ''Perception of frequency modulation at low modulation frequencies—the effect of phase angle in a complex modulation,'' J. Acoust. Soc. Am. (abst.) **65**, S37.

Klein, M. A. and Hartmann, W. M. (1981) ''Binaural edge pitch,'' J. Acoust. Soc. Am. **70**, 51–61.

Koehnke, J., Colburn, H. S., and Durlach, N. I. (1986) ''Performance in several binaural-interaction experiments,'' J. Acoust. Soc. Am. **79**, 1558–1562.

Kohlrausch, A. and Houtsma, A. J. M. (1992) ''Pitch related to spectral edges of broadband signals,'' Phil. Trans. Roy. Soc. **336**, 375–382. [Also J. Acoust. Soc. Am. (abst.) **93**, 2409 (1993).]

Kollmeier, B., Gilkey, R. H., and Sieben, U. K. (1988) ''Adaptive staircase techniques in psychoacoustics: A comparison of human data and a mathematical model,'' J. Acoust. Soc. Am. **83**, 1852–1862.

Lamoré, P. J. J., Muijser, H., and Keemink, C. J. (1986) ''Envelope detection of amplitude-modulated high-frequency sinusoidal signals by skin mechanoreceptors,'' J. Acoust. Soc. Am. **79**, 1082–1085.

Larkin, W. D. (1978) ''Pitch shifts following tone adaptation,'' Acustica **41**, 110–116.

Lee, J. and Green, D. M. (1994) ''Detection of a mistuned component in a harmonic complex,'' J. Acoust. Soc. Am. **96**, 716–725.

Leshowitz, B. and Wightman, F. L. (1971) ''On-frequency masking with continuous sinusoids,'' J. Acoust. Soc. Am. **49**, 1180–1190.

Levitt, H. (1971) ''Transformed up-down methods in psychoacoustics,'' J. Acoust. Soc. Am. **49**, 467–477.

Liang, C-A. and Chistovich, L. A. (1961) ''Frequency difference limens as a function of tonal duration,'' Sov. Phys. Acoust. (tr.) **6**, 75–80.

Liberman, M. C. and Oliver, M. (1984) ''Morphometry of intracellularly-labeled neurons of the auditory nerve: correlations with functional properties,'' J. Comparative Neurology **223**, 163–176.

Licklider, J. C. R. (1951) ''The duplex theory of pitch perception,'' Experientia **7**, 128–133.

Licklider, J. C. R. (1959) ''Three auditory theories'' in *Psychology a Study of a Science*, ed. S. Koch, McGraw Hill, NY, pp. 41–144.

Lin, J-Y. (1996) ''Psychoacoustical theory and experiments on human auditory organization of complex sounds and the critical bandwidth,'' Thesis, Michigan State University (unpublished).

Lin, J-Y. and Hartmann, W. M. (1995a) ''Roughness and the critical bandwidth at low frequency,'' J. Acoust. Soc. Am (abst.) **97**, 3274.

Lin, J-Y. and Hartmann, W. M. (1995b) ''Duifhuis pitch and other anomaly pitches,'' J. Acoust. Soc. Am. (abst.) **98**, 2926.

Lindemann, W. (1986) ''Extension of a binaural cross-correlation model by contralateral inhibition I. Simulation of lateralization for stationary signals,'' J. Acoust. Soc. Am. **80**, 1608–1622.

Long, G. R., Tubis, A., and Jones, K. L. (1991) ''Modeling synchronization and suppression of spontaneous otoacoustic emissions using Van der Pol oscillations: Effects of aspirin administration,'' J. Acoust. Soc. Am. **89**, 1201–1212.

Longuet-Higgins, H. C. (1978) ''The perception of music,'' Interdisciplinary Science Reviews **3**, 148–156.
Lonsbury-Martin, B. L., Harris, F. P., Stagner, B. B. Hawkins, M. D., and Martin, G. K. (1990) ''Distortion product emissions in humans. I Basic properties in normal hearing subjects,'' Ann. Otol. Rhinol. Laryngol. (Suppl. 147), 3–14.
Maiwald, D. (1967a) ''Ein Funktionsschema des Gehörs zur Beschreibung der Erkennbarkeit kleiner Frequenz und Amplitudenänderungen,'' Acustica **18**, 81–92.
Maiwald, D. (1967b) ''Die Berechnung von Modulationsschwellen mit Hilfe eines Funktionsschemas,'' Acustica **18**, 193–207.
Marks, L. (1974) *The New Psychophysics*, Academic Press, New York.
Martens, J.-P. (1981) ''Audibility of harmonics in a periodic complex,'' J. Acoust. Soc. Am. **70**, 234–237.
Martin, D. W., Meyer, A., Duncan, R. K., and Broxon, E. C. (1956a) ''An experimental 9000-watt airborne sound system,'' IRE Transactions on Audio Nov.–Dec. (pp. 146–155).
Martin, D. W., Murphy, R. L., and Meyer, A. (1956b) ''Articulation reduction by combined distortions of speech waves,'' J. Acoust. Soc. Am. **28**, 597–601.
Mathews, M. V. and Kohut, J. (1973) ''Electronic simulation of violin resonances,'' J. Acoust. Soc. Am. **53**, 1620–1626.
Mathews, M. V. and Pierce, J. R. (1980) ''Harmony and nonharmonic partials,'' J. Acoust. Soc. Am. **68**, 1252–1257.
McAdams, S. (1984a) ''Spectral fusion, spectral parsing and the formation of auditory images,'' Thesis, Stanford University.
McAdams, S. (1984b) ''The auditory image: A metaphor for musical and psychological research on auditory organization,'' in *Cognitive Processes in the Perception of Art*, North Holland, Amsterdam.
McFadden, D. (1975) ''Duration-intensity reciprocity for equal loudness,'' J. Acoust. Soc. Am. **57**, 702–704.
McFadden, D. and Pasanen E. G. (1976) ''Lateralization at high frequencies based on interaural time differences,'' J. Acoust. Soc. Am. **59**, 634–639.
McGill, W. J. and Goldberg, J. P. (1968) ''Pure tone intensity discrimination and energy detection,'' J. Acoust. Soc. Am. **44**, 576–581.
McIntyre, M. E., Schumacher, R. T., and Woodhouse, J. (1981) ''Aperiodicity in bowed-string motion,'' Acustica **49**, 13–32.
Meddis, R. (1988) ''Simulation of auditory neural transduction: Further studies,'' J. Acoust. Soc. Am. **83**, 1056–1063.
Meddis, R. and Hewitt, M. J. (1991a) ''Virtual pitch and phase sensitivity of a computer model of the auditory periphery I,'' J. Acoust. Soc. Am. **89**, 2866–2882.
Meddis, R. and Hewitt, M. J. (1991b) ''Virtual pitch and phase sensitivity of a computer model of the auditory periphery II,'' J. Acoust. Soc. Am. **89**, 2883–2894.
Middlebrooks, J. C. and Green, D. M. (1990) ''Directional dependence of interaural envelope delays,'' J. Acoust. Soc. Am. **87**, 2149–2162.
Moeller, A. R. (1974) ''Function of the middle ear,'' in *Handbook of Sensory Physiology*, Vol. 5, Springer-Verlag, Berlin.
Moore, B. C. J. (1973) ''Frequency difference limens for short duration tones,'' J. Acoust. Soc. Am. **54**, 610–619.
Moore, B. C. J. (1982) *An Introduction to the Psychology of Hearing*, second edition. Academic, London. (Third edition, 1989).
Moore, B. C. J. (1995) ''Frequency analysis and masking'' in *Hearing*, ed. B. C. J. Moore, Handbook of Perception and Cognition, 2nd ed., Academic Press, San Diego.
Moore, B. C. J. and Glasberg, B. R. (1990) ''Frequency discrimination of complex tones with overlapping and non-overlapping harmonics,'' J. Acoust. Soc. Am. **87**, 2163–2177.
Moore, B. C. J. and Sek, A. (1992) ''Detection of combined frequency and amplitude modulation,'' J. Acoust. Soc. Am. **92**, 3119–3131.
Moore, B. C. J. and Sek, A. (1995) ''Effects of carrier frequency, modulation rate, and modulation waveform on the detection of modulation and the discrimination of modulation type (AM vs FM),'' J. Acoust. Soc. Am. **97**, 2468–2478.
Moore, B. C. J., Glasberg, B. R., and Peters, R. W. (1985a) ''Relative dominance of individual partials in determining the pitch of complex tones,'' J. Acoust. Soc. Am. **77**, 1853–1859.
Moore, B. C. J., Glasberg, B. R., and Shailer, M. J. (1984) ''Frequency and intensity difference limens for harmonics within complex tones,'' Acoust. Soc. Am. **75**, 550–561.

Moore, B. C. J., Peters, R. W., and Glasberg, B. R. (1985c) ''Thresholds for the detection of inharmonicity in complex tones,'' J. Acoust. Soc. Am. **77**, 1861–1867.

Moore, B. C. J., Peters, R. W., and Glasberg, B. R. (1986) ''Thresholds for hearing mistuned partials as separate tones in harmonic complexes,'' J. Acoust. Soc. Am. **80**, 479–483.

Moore, B. C. J., Peters, R. W., and Glasberg, B. R. (1990) ''Auditory filter shapes at low center frequencies,'' J. Acoust. Soc. Am. **88**, 132–140.

Morgan, C. T., Garner, W. R., and Galambos, R. (1951) ''Pitch and intensity,'' J. Acoust. Soc. Am. **23**, 658–663.

Munson, W. A. (1947) ''The growth of auditory sensation,'' J. Acoust. Soc. Am. **19**, 584–591.

Naeve, S. L., Margolis, R. H., Levine, S. C., and Fournier, E. M. (1992) ''Effect of ear-canal air pressure on evoked otoacoustic emissions,'' J. Acoust. Soc. Am. **91**, 2091–2095.

Nedzelnitsky, V. (1980) ''Sound pressure in the basal turn of the cat cochlea,'' J. Acoust. Soc. Am. **68**, 1676–1689.

Nelson, D. A. and Bilger, R. C. (1974) ''Pure-tone octave masking in normal hearing listeners,'' J. Speech and Hearing Res. **17**, 223–251.

Ohgushi, K. (1978) ''On the role of spatial and temporal cues in the perception of the pitch of complex sounds,'' J. Acoust. Soc. Am. **64**, 764–771.

Ohgushi, K. (1983) ''The origin of tonality and a possible explanation of the octave enlargement phenomenon,'' J. Acoust. Soc. Am. **73**, 1694–1700.

Olsher, D. (1991) ''VAC PA90 monoblock power amplifier,'' Stereophile, **14**, 129–139 (Nov.) *ibid*, Sept. 1992, p. 43.

Ozawa, K., Suzuki, Y., and Sone, T. (1993) ''Monaural phase effects on timbre of two-tone signals,'' J. Acoust. Soc. Am. **93**, 1007–1011.

Papoulis, A. (1962) *The Fourier Integral and its Applications*, McGraw-Hill, New York.

Patterson, R., and Nimmo-Smith, I. (1980) ''Off-frequency listening and auditory-filter asymmetry,'' J. Acoust. Soc. Am. **67**, 229–245.

Patterson, R., Holdsworth, J., Nimmo-Smith, I., and Rice, P. (1991) ''The auditory filter bank,'' MRC-APU Report 2341, Cambridge, England.

Patterson, R. D. (1976) ''Auditory filter shapes derived with noise,'' J. Acoust. Soc. Am. **74**, 640–654.

Patterson, R. D. (1987) ''A pulse ribbon model of monaural phase perception,'' J. Acoust. Soc. Am. **82**, 1560–1586.

Patterson, R. D. (1993) ''What determines the sound of a sinusoid?,'' J. Acoust. Soc. Am. (abst.) **93**, 2293.

Patterson, R. D., Nimmo-Smith, I., Weber, D. L., and Milroy, R. (1982) ''The deterioration of hearing with age: Frequency selectivity, the critical ratio, the audiogram and speech threshold,'' J. Acoust. Soc. Am. **72**, 1788–1803.

Patterson, R. D., Robinson, K., Holdsworth, J., McKeown, D., Zhang, C., and Allerhand, M. (1992) ''Complex sounds and auditory images,'' in *Auditory Physiology and Perception*, (ed) Y. Cazals, L. Demany, and K. Horner, Pergamon, Oxford, (pp. 429–446).

Penner, M. J. (1978) ''A power law transformation resulting in a class of short-term integrators that produce time-intensity trades for noise bursts,'' J. Acoust. Soc. Am. **63**, 195–201.

Perrott, D. R. and Musicant, A. D. (1977) ''Rotating tones and binaural beats,'' J. Acoust. Soc. Am. **61**, 1288–1292.

Pennisi, L. L. (1963) *Elements of Complex Variables*, Holt, Rinehart, Winston, New York.

Phipps, A. R. and Henning, G. B. (1976) ''Effect of signal phase on the detectability of a tone masked by two tones,'' JASA **59**, 442–447.

Plomp, R. (1964) ''The ear as a frequency analyzer,'' J. Acoust. Soc. Am. **36**, 1628–1636.

Plomp, R. (1967) ''Beats of mistuned consonances,'' J. Acoust. Soc. Am. **42**, 462–474.

Plomp, R. (1976) *Aspects of Tone Sensation—A Psychophysical Study*, Academic, London.

Plomp, R. and Bouman, M. A. (1959) ''Relation between hearing threshold and duration for tone pulses,'' J. Acoust. Soc. Am. **31**, 749–758.

Plomp, R. and Levelt, W. J. M. (1965) ''Tonal consonance and critical bandwidth,'' J. Acoust. Soc. Am. **38**, 548–560.

Plomp, R. and Mimpen, A. M. (1968) ''The ear as a frequency analyzer II,'' J. Acoust. Soc. Am. **43**, 764–767.

Plomp, R. and Steeneken, H. J. M. (1969) ''Effect of phase on the timbre of complex tones,'' J. Acoust. Soc. Am. **46**, 409–421.

Pohlmann, K. C. (1985) *Principles of Digital Audio*, H. Sams, Indianapolis, IN.

Press, W. H., Flannery, B. P., Teukolsky, S. A., and Vetterling, W. T. (1986) *Numerical Recipes*, Cambridge U. Press, Cambridge, UK.
Price, R. (1958) ''A useful theorem for nonlinear devices having Gaussian inputs,'' IRE Trans. on Information Theory IT-4, 69–72.
Punch, J. and Rakerd, B. (1993) ''Loudness matching of signals spectrally shaped by a simulated hearing aid,'' J. Speech and Hearing Res., **36**, 357–364.
Pumplin, J. (1985) ''Low-noise noise,'' J. Acoust. Soc. Am. **78**, 100–104.
Rajcan, E. (1972) ''The pitch of short sinusoidal pulses as a function of initial phase,'' Acustica **26**, 147–152.
Rajcan, E. (1974) ''The role of the initial phase angle by pitch perception of short tonal pulses as a function of sound pressure level,'' Acustica **30**, 270–274.
Rakowski, A. and Hirsh, I. J. (1980) ''Post-stimulatory pitch shifts for pure tones,'' J. Acoust. Soc. Am. **68**, 467–474.
Rakowski, A. and Jaroszewski, A. (1974) ''On some secondary masking effect (Pitch shifts in post-stimulatory masking,)'' Acustica **31**, 325–329.
Rasch, R. A. (1978) ''The perception of simultaneous notes as in polyphonic music,'' Acustica **40**, 21–33.
Rasch, R. A. (1979) ''Synchronization in performed ensemble music,'' Acustica **43**, 121–131.
Rees, A. and Moller, A. R. (1983) ''Responses of neurons in the inferior colliculus of the rat to AM and FM tones,'' Hearing Res. **10**, 301–330.
Repp, B. H. (1994) ''The tritone paradox and the pitch range of the speaking voice: a dubious connection,'' Music Percept. **12**, 227–255.
Rhode, W. S. (1971) ''Observations of the vibration of the basilar membrane in squirrel monkeys using the Mössbauer technique,'' J. Acoust. Soc. Am. **49**, 1218–1231.
Rhode, W. S. (1995) ''Interspike intervals as a correlate of periodicity pitch in cat cochlear nucleus,'' J. Acoust. Soc. Am. **97**, 2414–2429.
Rice, S. O. (1954) ''Mathematical analysis of random noise,'' in *Stochastic Processes*, ed. N. Wax, Dover, New York.
Richards, V. M. (1988) ''Components of monaural envelope correlation perception,'' Hearing Res. **35**, 47–58.
Richards, V. M. (1990) ''The role of single-channel cues in synchrony perception: The summed waveform,'' J. Acoust. Soc. Am. **88**, 786–795.
Richards, V. M. (1992) ''The detectability of a tone added to narrow bands of equal-energy noise,'' J. Acoust. Soc. Am. **91**, 3424–3435.
Richardson, L. F. and Ross, J. S. (1930) ''Loudness and telephone current,'' J. General Psychol. **3**, 288.
Risset, J.-C. and Mathews, M. V. (1969) ''Analysis of musical instrument tones,'' Physics Today, **22**, 23–30.
Ritsma, R. J. (1967) ''Frequencies dominant in the perception of pitch of complex sounds,'' J. Acoust. Soc. Am. **42**, 191–198.
Romani, G. L., Williamson, S. J., and Kaufman, L. (1982) ''Tonotopic organization of the human auditory cortex,'' Science **216**, 1339–1340.
Ronken, D. A. (1971) ''Some effects of bandwidth-duration constraints on frequency discrimination,'' J. Acoust. Soc. Am. **49**, 1232–1242.
Rose, J. E., Hind, J. E., Anderson, D. J., and Brugge, J. F. (1971) ''Some effects of stimulus intensity on response of auditory nerve fibers in the squirrel monkey,'' J. Neurophysiol. **34**, 685–699.
Rose, J. E., Brugge, J. F., Anderson, D. J., and Hind, J. E. (1967) ''Phase-locked response to low-frequency tones in single auditory nerve fibers of the squirrel monkey,'' J. Neurophys. **30**, 769–793.
Rossing, T. D. (1990) *The Science of Sound*, second ed., Addison Wesley, Reading, MA.
Rossing, T. D. and Houtsma, A. J. M. (1986) ''Effects of signal envelope on the pitch of short sinusoidal tones,'' J. Acoust. Soc. Am. **79**, 1926–1933.
Ruggero, M. A. and Rich, N. (1991) ''Application of a commercially-manufactured Doppler-shift laser velocimeter to the measurement of basilar membrane motion,'' Hearing Res. **51**, 215–230.
Scharf, B. (1970) ''Critical bands,'' in *Foundations of Modern Auditory Theory*, vol. 1, ed. J. Tobias, Academic, New York.
Scharf, B. (1983) ''Loudness adaptation,'' in *Hearing Research and Theory*, vol. 2, ed. J. V. Tobias and E. D. Schubert, Academic, New York.
Scharf, B. and Meiselman, C. H. (1977) ''Critical bandwidth at high intensities,'' in *Psychophysics and Physiology of Hearing*, ed. E. F. Evans and J. P. Wilson, Academic, London.

Scharf, B., Florentine, M., and Meiselman, C. H. (1976) ''Critical band in auditory lateralization,'' Sensory Processes, **1**, 109–126.
Scheffers, M. T. M. (1983) *Sifting vowels: Auditory pitch analysis and sound segregation*, Doctoral dissertation, Groningen University, Groningen, The Netherlands.
Schlauch, R. S. and Rose, R. M. (1990) ''Two-, three-, and four-interval forced-choice staircase procedures: Estimator bias and efficiency,'' J. Acoust. Soc. Am. **88**, 732–740.
Schloth, E. and Zwicker, E. (1983) ''Mechanical and acoustical influences on spontaneous otoacoustic emissions,'' Hearing Res. **11**, 285–293.
Schneider, B. (1981) ''Determining individual loudness scales from binary comparisons of loudness intervals,'' J. Acoust. Soc. Am. **64**, 1208–1209.
Schouten, J. F. (1940) *Five articles on the perception of sound (1938–1940)*, Institute for Perception, Eindhoven.
Schouten, J. F. (1970) ''The residue revisited,'' *Frequency Analysis and Periodicity Detection in Hearing*, ed. R. Plomp and G. Smoorenburg, A. W. Sijthoff, Leiden (pp. 41–58).
Schroeder, M. R. (1970) ''Synthesis of low-peak-factor signals and binary sequences with low autocorrelation,'' IEEE Trans. on Information Theory, IT-16 (pp. 85–89).
Schroeder, M. R. (1979) ''Integrated-impulse method measuring sound decay without using impulses,'' J. Acoust. Soc. Am. **62**, 497–500.
Schubert, E. D. (1950) ''The effect of thermal masking noise on the pitch of a pure tone,'' J. Acoust. Soc. Am. **22**, 497–499.
Seashore, C. E. (1938) *Psychology of Music*, McGraw-Hill, New York, reprinted (1967) by Dover, New York.
Sek, A. and Moore, B. C. J. (1994) ''The critical modulation frequency and its relationship to auditory filtering at low frequencies,'' J. Acoust. Soc. Am. **95**, 2606–2615.
Sellick, P. and Russell, I. (1980) ''The response of inner hair cells to basilar membrane velocity during low-frequency auditory stimulation in the guinea pig,'' Hearing Res. **2**, 227–236.
Semal, C. and Demany, L. (1990) ''The upper limit of musical pitch,'' Music Percept. **8**, 165–176.
Shaw, E. A. G. and Teranishi, R. (1968) ''Sound pressure generated in an external ear replica and real human ears by a nearby point source,'' J. Acoust. Soc. Am. **44**, 240–249.
Shepard, R. N. (1964) ''Circularity in judgments of relative pitch,'' J. Acoust. Soc. Am. **36**, 2346–2353.
Shepard, R. N. (1982) ''Structural representations of musical pitch,'' in *The Psychology of Music*, ed. D. Deutsch, Academic, New York (pp. 344–390).
Shofner, W. P. (1991) ''Temporal representation of rippled noise in the anteroventral cochlear nucleus of the chinchilla,'' J. Acoust. Soc. Am. **90**, 2450–2466.
Shower, E. G. and Biddulph, R. (1931) ''Differential pitch sensitivity of the ear,'' J. Acoust. Soc. Am. **3**, 275–287.
Siebert, W. M. (1970) ''Frequency discrimination in the auditory system—Place or periodicity mechanisms?'' Proc. IEEE **58**, 723–730.
Smoorenburg, G. F. (1974) ''On the mechanisms of combination tone generation and lateral inhibition in hearing,'' in *Facts and Models in Hearing*, ed. E. Zwicker and E. Terhardt, Springer-Verlag, Berlin.
Srulovicz, P. and Goldstein, J. L. (1983) ''A central spectrum model: A synthesis of auditory-nerve timing and place cues in monaural communication of frequency spectrum,'' J. Acoust. Soc. Am. **73**, 1266–1276.
Stephenson, M. R., Billings, B. L., and Jutila, G. A. (1990) ''Voyager flight crew hearing threshold levels resulting from 5- and 9-day continuous in-flight noise exposure,'' J. Acoust. Soc. Am. **87**, 901–904.
Stern, R. M. and Colburn, H. S. (1978) ''Theory of binaural interaction based on auditory-nerve data IV. A model for subjective lateral position,'' J. Acoust. Soc. Am. **64**, 127–140.
Stern, R. M. and Shear, G. D. (1992) ''Lateralization and detection of low-frequency binaural stimuli: Effects of distribution of internal delay,'' (unpublished).
Stern, R. M. and Trahiotis, C. (1995) ''Models of binaural interaction,'' in *Hearing—Handbook of Perception and Cognition*, second edition, ed. B. C. J. Moore, Academic Press, San Diego. p. 378.
Stevens, S. S. (1935) ''The relation of pitch to intensity,'' J. Acoust. Soc. Am **6**, 150–154.
Stevens, S. S. (1955) ''The measurement of loudness,'' J. Acoust. Soc. Am. **27**, 815–829.
Stevens, S. S. (1961) ''Procedure for calculating loudness: Mark VI,'' J. Acoust. Soc. Am. **33**, 1577–1585.
Stevens, S. S. (1966) ''On the operation known as judgment,'' Amer. Sci. **54**, 385–401.
Stevens, S. S. (1971) ''Issues in psychophysical measurements,'' Psych. Rev. **78**, 426–450.

Stevens, S. S. (1972) ''Perceived level of noise by Mark VII and decibels'' J. Acoust. Soc. Am. **51**, 575–601.
Stevens, S. S. and Davis H. (1938) *Hearing: its Psychology and Physiology*, John Wiley and Sons, New York. Reprinted (1983) by the Acoustical Society of America.
Stevens, S. S. and Greenbaum, H. B. (1966) ''Regression effect in psychophysical judgement,'' Percept. and Psychophys. **1**, 439–446.
Stevens, S. S. and Volkman, J. (1940) ''The Relation of Pitch to Frequency,'' Am. J. Psych. **53**, 329–353.
Stevens, S. S., Volkman, J., and Newman, E. B. (1937) ''A scale for the measurement of the psychological magnitude of pitch,'' J. Acoust. Soc. Am. **8**, 185–190.
Stewart, G. W. (1931) ''Problems suggested by an uncertainty principle in acoustics,'' J. Acoust. Soc. Am. **2**, 325–329.
Strong, W. J. and Plitnik, G. R. (1992) *Music, Speech, Audio*, Soundprint, Provo, UT.
Strube, H. W. (1986) ''The shape of the nonlinearity generating the combination tone $2f_1-f_2$,'' J. Acoust. Soc. Am. **79**, 1511–1518.
Tansley, B. W. and Suffield, J. B. (1983) ''Time course of adaptation and recovery of channels selectively sensitive to frequency and amplitude modulation,'' J. Acoust. Soc. Am. **74**, 765–775.
Teich, M. C. and Khanna, S. M. (1985) ''Pulse number distribution for the neural spike train in the cat's auditory nerve,'' J. Acoust. Soc. Am. **77**, 1110–1128.
Teich, M. C., Khanna, S. M., and Guiney, P. C. (1993) ''Spectral characteristics and synchrony in primary auditory-nerve fibers in response to pure-tone acoustic stimuli,'' J. Stat. Phys. **70**, 257–279.
Terhardt, E. (1971) ''Pitch shifts of harmonics, an explanation of the octave enlargement phenomenon,'' *Proc. 7th ICA*, Budapest, **3**, 621–624.
Terhardt, E. (1974a) ''On the perception of periodic sounds (Roughness),'' Acustica, **30**, 202–213.
Terhardt, E. (1974b) ''Pitch, consonance, and harmony.'' J. Acoust. Soc. Am. **55**, 1061–1069.
Terhardt, E. and Fastl, H. (1971) ''Zum Einfluss von Störtönen und Störgeräuschen auf die Tonhöhe von Sinustönen,'' Acustica **25**, 53–61.
Terhardt, E., Stoll, G., and Seewann, M. (1982a) ''Pitch of complex tonal signals according to virtual pitch theory: Tests, examples and predictions,'' J. Acoust. Soc. Am. **71**, 671–678.
Terhardt, E., Stoll, G., and Seewann, M. (1982b) ''Algorithm for extraction of pitch and pitch salience from complex tonal signals,'' J. Acoust. Soc. Am. **71**, 679–688.
Thomas, I. B. (1975) ''Microstructure of the pure-tone threshold,'' J. Acoust. Soc. Am. (abst.) **57**, S26–27.
Thurlow, W. R. (1957) ''An auditory figure-ground effect,'' Am. J. Psych. **70**, 653–654.
Thurlow, W. R. and Elfner, L. F. (1959) ''Continuity effects with alternately sounding tones,'' J. Acoust. Soc. Am. **31**, 1337–1339.
Traunmüller, H. (1990) ''Analytical expressions for the tonotopic sensory scale,'' J. Acoust. Soc. Am. **88**, 97–100.
van de Par, S. and Kohlrausch, A. (1995) ''Analytical expressions for the envelope correlation of certain narrow-band stimuli,'' J. Acoust. Soc. Am. **98**, 3157–3169.
van der Heijden, M. and Kohlrausch, A. (1995) ''The role of envelope fluctuations in spectral masking,'' J. Acoust. Soc. Am. **97**, 1800–1807.
Verschuure, J. and van Meeteren, A. A. (1975) ''The effect of intensity on pitch,'' Acustica **32**, 33–44.
Versfeld, N. J. (1992) ''On the auditory discrimination of spectral shape,'' Thesis, Technical University of Eindhoven (unpublished).
Versfeld, N. J. and Houtsma, A. J. M. (1992) ''Spectral shape discrimination of two-tone complexes,'' Advances in Bioscience, **83**, 363–371.
Versfeld, N. J. and Houtsma, A. J. M. (1995) ''Discrimination of changes in the spectral shape of two-tone complexes,'' J. Acoust. Soc. Am. **98**, 807–816.
Viemeister, N. F. (1979) ''Temporal modulation transfer functions based upon modulation thresholds,'' J. Acoust. Soc. Am. **66**, 1364–1380.
Viemeister, N. F. (1988) ''Intensity coding and the dynamic range problem,'' Hearing Res. **34**, 267–274.
Viemeister, N. F. and Bacon, S. P. (1988) ''Intensity discrimination, increment detection and magnitude estimation for 1-kHz tones,'' J. Acoust. Soc. Am. **84**, 172–178.
Viemeister, N. F. and Wakefield, G. H. (1991) ''Temporal integration and multiple looks,'' J. Acoust. Soc. Am. **90**, 858–864.
Vogten, L. L. M. (1978) ''Simultaneous pure-tone masking: The dependence of masking asymmetries on intensity,'' J. Acoust. Soc. Am. **63**, 1509–1519.
Wada, H., Metoki, T., and Kobayashi, T. (1992) ''Analysis of dynamic behavior of human middle-ear

using a finite-element method,'' J. Acoust. Soc. Am. **92**, 3157–3168.
Walliser, K. (1969) ''Zusammenhange zwischen dem Schallreiz und der Periodentonhohe,'' Acustica **21**, 319–329.
Ward, W. D. (1955) ''Tonal monaural diplacusis,'' J. Acoust. Soc. Am. **27**, 365–372.
Ward, W. D. (1973) ''Adaptation and fatigue,'' in *Modern Developments in Audiology*, ed. J. Jerger, second edition, Academic, New York.
Warren, R. M. (1977) ''Subjective loudness and its physical correlate,'' Acustica **37**, 334–346.
Warren, R. M. (1984) ''Perceptual restoration of obliterated sounds,'' Psychological Bulletin, **96**, 371–383.
Warren, R. M. and Bashford, J. A. (1981) ''Perception of acoustic iterance: Pitch and infrapitch,'' Percept. and Psychophys. **29**, 395–402.
Warren, R. M. and Bashford, J. A. Jr. (1988) ''Broadband repetition pitch: Spectral dominance or pitch averaging,'' J. Acoust. Soc. Am. **68**, 1301–1305.
Warren, R. M. and Wrightson, J. M. (1981) ''Stimuli producing conflicting temporal and spectral cues to frequency,'' J. Acoust. Soc. Am. **70**, 1020–1024.
Warren, R. M., Obusek, C. J., and Ackroff, J. M. (1972) ''Auditory induction: Perceptual synthesis of absent sounds,'' Science, **176**, 1149–1151.
Watkinson, J. (1994) *The Art of Digital Audio*, second ed., Focal Press, Oxford, UK.
Watson, C. S. and Gengel, R. W. (1969) ''Signal duration and signal frequency in relation to auditory sensitivity,'' J. Acoust. Soc. Am. **46**, 989–997.
Webster, J. C. and Muerdter, D. R. (1965) ''Pitch shifts due to low-pass and high-pass noise bands,'' J. Acoust. Soc. Am. **37**, 382–383.
Webster, J. C. and Schubert, E. D. (1954) ''Pitch shifts accompanying certain auditory threshold shifts,'' J. Acoust. Soc. Am. **26**, 754–758.
Wegel, R. L. and Lane, C. E. (1924) ''The auditory masking of one pure tone by another and its probable relation to the dynamics of the inner ear,'' Phys. Rev. **23**, 266–285.
Wetherill, G. B. and Levitt, H. (1965) ''Sequential estimation of points on a psychometric function,'' Brit. J. Math. Stat. Psychol. **53**, 1–10.
Whitfield, I. C. and Evans, E. F. (1965) ''Responses of auditory cortical neurons to stimuli of changing frequency,'' J. Neurophys. **28**, 655–672.
Wier, C. C., Jesteadt, W., and Green, D. M. (1977a) ''A comparison of method-of-adjustment and forced-choice procedures in frequency discrimination,'' Percept. and Psychophys. **19**, 75–79.
Wier, C. C., Jesteadt, W., and Green, D. M. (1977b) ''Frequency discrimination as a function of frequency and sensation level.'' J. Acoust. Soc. Am. **61**, 178–184.
Wightman, F. L. (1973) ''The pattern transformation model of pitch,'' J. Acoust. Soc. Am. **54**, 548–557.
Wightman, F. L. and Green D. M. (1974) ''The perception of pitch,'' Am. Sci. **62**, 208–215.
Wilson, A. S., Hall, J. W., and Grose, J. H. (1990) ''Detection of FM in the presence of another FM tone,'' J. Acoust. Soc. Am. **88**, 1333–1338.
Yin, T. C. T., Chan, J. C. K., and Carney, L. H. (1987) ''Effects of interaural time delays of noise stimuli on low-frequency cells in the cat's inferior colliculus. III. Evidence for cross-correlation,'' J. Neurophys. **58**, 562–583.
Yin, T. C. T., and Chan, J. C. K. (1990) ''Interaural time sensitivity in medial superior olive of cat,'' J. Neurophysiol. **64**, 465–488.
Yost, W. A. (1982) ''The dominance region and ripple noise pitch: A test of the peripheral weighting model,'' J. Acoust. Soc. Am. **72**, 416–425.
Yost, W. A. (1994) *Fundamentals of Hearing*, Academic, San Diego, third edition.
Yost, W. A. and Hill, R. (1978) ''Strength of the pitches associated with ripple (sic) noise'' J. Acoust. Soc. Am. **64**, 485–492.
Yost, W. A. and Sheft, S. (1990) ''A comparison among three measures of cross-spectral processing of AM with tonal signals'' J. Acoust. Soc. Am. **87**, 897–900.
Yost, W. A., Hill, R., and Perez-Falcon, T. (1978) ''Pitch and pitch discrimination of broadband signals with rippled power spectra,'' J. Acoust. Soc. Am. **63**, 1166–1173.
Yost, W. A., Sheft, S., and Opie, J. (1989) ''Modulation interference in detection and discrimination of AM,'' J. Acoust. Soc. Am. **86**, 2138–2147.
Yost, W. A., Sheft, S., Shofner, B., and Patterson, R. (1995) ''A temporal account of complex pitch'' in Proc. ATR Workshop on a Biological Framework for Speech Perception and Production,'' ed. H. Kawahara, ATR, Kyoto, Japan.
Zurek, P. M. (1981) ''Spontaneous narrow-band acoustic signals emitted by human ears,'' J. Acoust. Soc. Am. **69**, 514–523.

Zurek, P. M. (1985) ''Acoustic emissions from the ear: A summary of results from humans and animals,'' J. Acoust. Soc. Am. **78**, 340–344. [AKA ''Basic Research in Hearing,'' Proceedings of the 1983 CHABA Symposium.]

Zurek, P. M. and Rabinowitz, W. M. (1995) ''System for testing adequacy of human hearing,'' US Patent 5,413,114, reviewed in J. Acoust. Soc. Am. **98**, 1835–1836.

Zwicker, E. (1952) ''Die Grenzen der Horbarkeit der Amplitudenmodulation und der Frequenz-modulation eines Tones,'' Acustica **2**, 125–133.

Zwicker, E. (1955) ''Der ungewöhnliche Amplitudengang der nichtlinearen Verzerrungen des Ohres,'' Acustica, **5**, 67–74.

Zwicker, E. (1956) ''Die elementaren Grundlagen zur Bestimmung der Informationskapazitat des Gehörs,'' Acustica **6**, 365–381.

Zwicker, E. (1961) ''Subdivision of the audible frequency range into critical bands (Frequenz-gruppen)'' J. Acoust. Soc. Am. **33**, 248.

Zwicker, E. (1970) ''Masking and psychological excitation as consequences of the ear's frequency analysis,'' in *Frequency Analysis and Periodicity Detection in Hearing*, ed. R. Plomp and G. F. Smoorenburg, A. W. Sijthoff, Leiden, The Netherlands.

Zwicker, E. and Schloth, E. (1984) ''Interrelation of different oto-acoustic emissions,'' J. Acoust. Soc. Am. **75**, 1148–1154.

Zwicker, E. and Fastl, H. (1990) *Psychoacoustics—Facts and Models*, Springer-Verlag, Berlin.

Zwicker, E. and Terhardt, E. (1980) ''Analytical expressions for critical-band rate and critical bandwidth as a function of frequency,'' J. Acoust. Soc. Am. **68**, 1523–1525.

Zwicker, E., Flottorp, G., and Stevens, S. S. (1957) ''Critical bandwidth in loudness summation,'' J. Acoust. Soc. Am. **28**, 548–557.

Zwislocki, J. J. (1962) ''Analysis of middle ear function. Part I: Input impedance,'' J. Acoust. Soc. Am. **34**, 1514–1523.

Zwislocki, J. J. (1965) ''Analysis of some auditory characteristics,'' in *Handbook of Math. Psych.* vol III, ed. R. D. Luce, B. R. Bush, and E. Galanter, J. Wiley, New York.

Zwislocki, J. J. (1991) ''What is the cochlear place code for pitch?'' Acta Otolaryngol. (Stockholm) **111**, 256–262.

Index

*Brackets denote pages where biographies may be found.

D

E

G

H

K

L

M

N

O

P

Q

RITTER LIBRARY
BALDWIN-WALLACE COLLEGE